D1543202

ARISTOTLE

XII

LCL 323

ARISTOTLE

PARTS OF ANIMALS

WITH AN ENGLISH TRANSLATION BY

A. L. PECK

MOVEMENT OF ANIMALS
PROGRESSION OF ANIMALS

WITH AN ENGLISH TRANSLATION BY

E. S. FORSTER

HARVARD UNIVERSITY PRESS

CAMBRIDGE, MASSACHUSETTS

LONDON, ENGLAND

First published 1937
Revised and reprinted 1945, 1955
Revised and reprinted 1961
Reprinted 1968, 1983, 1993, 1998, 2006

LOEB CLASSICAL LIBRARY® is a registered trademark
of the President and Fellows of Harvard College

ISBN 0-674-99357-8

Printed on acid-free paper and bound by
Edwards Brothers, Ann Arbor, Michigan

CONTENTS

PARTS OF ANIMALS

FOREWORD

ARISTOTLE refers to the *De partibus animalium* as an inquiry into the causes that in each case have determined the composition of animals. He does not, however, employ the category of causation in the manner normally adopted by men of science, since in this book causes are always considered in relation to ends or purposes, and design is regarded as having had a far larger share in the origin and development of living structures than that allotted to necessity.

In the *Historia animalium* the parts themselves are described, for although this work is to some extent physiological, its main object was to deal with the anatomy of the organism. The *De partibus animalium*, on the other hand, is almost exclusively physiological and teleological, and treats of the functions of the parts. But Aristotle's position was that of a teleologist only in a limited degree, for he appears to have taken that view of life which Bergson calls the doctrine of internal finality (that is to say, that each individual, or at any rate each species, is made for itself, that all its parts conspire for the greatest good of the whole, and are intelligently organized in view of that end but without regard for other organisms or kinds of organisms). Since every organ or part of the body was held to have its peculiar function, the existence of vestigial or rudimentary organs was

4

unrecognized. This was the doctrine of internal finality which was generally accepted until Darwin elaborated his theory of Natural Selection. The wider doctrine of external finality, according to which living beings are ordered in regard to one another, never gained acceptance among scientific philosophers, and the only indication that Aristotle ever adopted it is furnished by a passage in which he suggests that the mouth in Selachians is placed on the under surface so as to allow their prey to escape while the fish are turning on their backs before taking their food ; but even this he qualified by the suggestion that the arrangement served a useful end for the fishes in question by preventing them from indulging in the harmful habit of gluttony.

The *De partibus animalium* opens with an introduction devoted to general considerations. This is followed by a discussion of the three degrees of composition, the first degree being composition of physical substances, the second degree, of homogeneous parts or tissues, and the third, of heterogeneous parts or organs. The tissues referred to are blood, fat, marrow, brain, flesh, and bone. After describing these, the organs are dealt with, and a consideration of their respective functions, first in sanguineous animals (*i.e.* in Vertebrates), and secondly in bloodless animals (*i.e.* Invertebrates), occupies the remainder of the book. The account given of the physiology of the blood is especially interesting, and it is noteworthy that Aristotle understood something of the nature of the process of absorption whereby the food becomes converted into nutriment which is carried by the blood to all parts of the body. He supposed, however, that the matter derived from the gut passed first to the heart in the form of vapour or serum, and that it was there converted into true blood by a process of concoction. Aristotle knew nothing of the real nature of respiration, and he regarded the lungs as serving to temper the bodily heat by means of the inspired air. He was also entirely ignorant of the fact that the blood passes back to the heart and lungs after supplying the tissues and organs with nourishment. On the other hand, he fully appreciated the existence of excretory organs, the function of which was to remove from the body such substances as could not be utilized. In this category are included fluids such as bile, urine, and sweat. In the section on the gall-bladder, as in so many other passages in his works on natural history, it is truly remarkable how correct Aristotle is in his statements. He points out that the gall-bladder is not found either in the horse and ass or in the deer and roe, but is generally present in the sheep and goat. In the light of the knowledge that he possessed, therefore, Aristotle could scarcely have adopted a theory about this organ which has found expression in certain modern writings. According to this theory the gall-bladder is present in the sheep and ox because, these being ruminating animals, bile is only required at certain particular times when food passes into the intestine, whereas in the horse, which does not chew the cud, but yet is constantly eating, food is continually passing into the intestine and consequently a perpetual flow of bile is desirable. Since the gall-bladder is present in the non-ruminating pig but absent in the ruminating deer and roe, it is obvious that this theory cannot be consistently applied.

FOREWORD

It is interesting to speculate about the school of research workers who must have contributed in providing material for this and the other works on natural science ascribed to Aristotle—who they were, the circumstances under which they lived, and what manner of facilities were available for their investigations—for it would seem certain that no man single-handed could possibly have acquired such a vast body of knowledge, hardly any of which could have been derived from earlier observers. Yet the work in its completed whole seems to show the mark of one master hand, and its uniform character and the clear line of teleological reasoning that runs through it have been well brought out in Dr. Peck's translation. But putting aside its philosophical implications, the book consists of an attempt at a scientific record of all the apparently known facts relating to animal function. These are considered comparatively and as far as possible are brought into relation with one another. And thus, as the earliest text-book on animal physiology in the world's history, this treatise will ever make its appeal, not only to the classical philosopher, but to all who are interested in the origin and growth of biological science.

F. H. A. M.

INTRODUCTION

Title. THE traditional title of this treatise is not a very informative one. The subject of the work is, however, stated quite clearly by Aristotle at the beginning of the second Book in these words : " I have already described with considerable detail in my *Researches upon Animals* what and how many are the parts of which animals are composed. We must now leave on one side what was said there, as our present task is to consider what are the causes through which each animal is as I there described it " (646 a 7 foll.). The title ought therefore to be " *Of the Causes of the Parts of Animals*," and this is the title actually applied to it by Aristotle himself (at *De gen. an.* 782 a 21).[a] Even so, the word " parts " is misleading : it includes not only what we call parts, such as limbs and organs, but also constituents such as blood and marrow.[b] Perhaps, therefore, no harm is done by leaving the accepted (and convenient) Latin title untranslated.

Zoological works. The *De partibus*, as well as the other treatises contained in this volume, forms a portion of Aristotle's zoological works. The foundation of these is the *Historia animalium*, or *Researches about Animals*, in nine books (the tenth is generally held to be

---

[a] For the meaning of Cause see note below, p. 24.
[b] See note on " part " below, p. 28.

spurious), in which observations are recorded, and consequent upon this are the treatises in which Aristotle puts forward theories founded upon these observations.

An animal is, according to Aristotle, a " concrete entity " made up of " matter " and " form." Hence, in the *De partibus* Aristotle treats of the causes on account of which the bodies—the " matter "—of animals are shaped and constructed as they are, in general ; in the *De incessu* he deals specially with the parts that subserve locomotion. In the *De anima* he proceeds to consider Soul—the " form " of an animal. In the remaining treatises, of which *De motu*, included in this volume, is one, he deals with what he calls the functions " common to body and Soul," among which he includes sensation, memory, appetite, pleasure, pain, waking, sleeping, respiration, and so forth (see *De sensu* 436 a). The complete scheme is set out below :

I. Record of observations.
　　Historia animalium.　10 (9) books.

II. Theory based upon observations.

(a)
De partibus animalium　4 books
De incessu animalium　1 book
} treating of the way in which the " matter " of animals is arranged to subserve their various purposes.

(b)　*De anima*　　3 books { treating of the " form " of animals — the Soul.

9

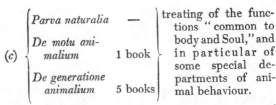

(c)
Parva naturalia	—	treating of the functions "common to body and Soul," and in particular of some special departments of animal behaviour.
De motu animalium	1 book	
De generatione animalium	5 books	

The section (b) is necessary to the completeness of the scheme, but as it has given rise to a whole department of study, it is usually treated apart from the rest. Thus the main bulk of the zoological and biological works may be taken to consist of the three great treatises, *Historia animalium*, *De partibus animalium*, and *De generatione animalium*. It was these which, through translations made from the Arabic, were restored to the West by those who revived scientific studies at the beginning of the thirteenth century.

Date of composition.

The late D'Arcy W. Thompson, in the prefatory note to his translation of *H.A.*,[a] wrote : " I think it can be shown that Aristotle's natural history studies were carried on, or mainly carried on, in his middle age, between his two periods of residence at Athens," *i.e.* in the Troad, in Lesbos and in Macedonia, between the years 347 and 335 : and this view has recently received convincing support from Mr. H. D. P. Lee,[b] who bases his argument upon an examination of the place-names in *H.A.* This is opposed to the view which has been current for some years past,[c] that the zoological works belong to a late period in Aristotle's life, and has important consequences for the reconstruction of Aristotle's philosophical develop-

[a] *The Works of Aristotle translated*, vol. iv., Oxford, 1910.
[b] *C.Q.* xlii. (1948), 61 ff.
[c] See W. D. Ross, *Aristotle*, and W. W. Jaeger, *Aristotle*.

ment, which cannot be dealt with here. It may, however, be remarked that, as Thompson said, it would follow that we might legitimately proceed to interpret Aristotle's more strictly philosophical work in the light of his work in natural history. But apart from these considerations, the great importance of the zoological works is that they represent the first attempt in Europe to observe and describe in a scientific way the individual living object.

Throughout the *De partibus* Aristotle endeavours to provide a Final Cause [a] to explain the facts which he records—some purpose which they are supposed to answer; and Causes of this sort are by far the most common in his treatise. His outlook is therefore justly described as " teleological "; but it is important not to read too much into this description. Aristotle is never tired of telling us that Nature makes nothing and does nothing " without a purpose "; but if we ask what that purpose is we may find that the answer is not quite what we had expected. Plato's notion of the " form " tended to divert his attention from individuals through a hierarchy of successive " forms "; but for Aristotle " form " is not independent of matter : form must be embodied in some matter, that is, in individuals. Thus we find all through that Aristotle cannot long keep his eyes from the individual wherein the form is actually embodied, because it, after all, is the End, the crowning achievement of the efforts of the four Causes. This outlook controls the arrangement of Aristotle's treatise. Since all processes of production are determined by the nature of the product which is to result from them, it is the fully developed product which we must first make it our business to observe,

Teleology.

[a] The four Causes are dealt with in a separate note, p. 24.

and when we have discovered what are its actual characteristics we may then go on to work out its Causes and to examine the processes by which it was produced.

Synopsis
and
Summary.
I give a brief synopsis and a contents-summary of the *De partibus* :

BRIEF SYNOPSIS OF *DE PARTIBUS*

Introduction : Methods.

Composition of Substances : Three modes :
    (1) The primary substances.
    (2) The " uniform " parts.
    (3) The " non-uniform " parts.
Consideration of (1) Hot, cold, solid, fluid.
      (2) Uniform parts : (*a*) fluid, (*b*) solid.
      (3) Non-uniform parts, as follows :—
External parts of animals.
Internal parts of blooded animals.
Internal parts of bloodless animals.
External parts of bloodless animals.
External parts of blooded animals (resumed).
   (*a*) Vivipara. (*b*) Ovipara.

SUMMARY

Book I.

639 a 15 ch. 1 *Introduction. On the Method of Natural Science.*

     Two questions propounded :

     (1) Are we to begin with the ultimate species and describe its characteristics, or with those that are common to many species ?

639 b 8

(2) (Put in three ways) :

    (a) Are we to take first the phenomena, and then proceed to their Causes ?

    (b) Which is the primary Cause, the Final or the Efficient (Motive) ? (Answered immediately : The Final ; with a reference also to the influence of Necessity.)

    (c) Are we to discuss first the processes by which the animal is formed, or the characteristics of it in its completed state ?

Answer to question (2).

We must begin with the phenomena, then go on to the Causes, and the formative processes—or, in other words, the Final Cause concerns us first and foremost. This differs from the practice of the early philosophers, who concerned themselves with the Material Cause, though sometimes also with the Efficient (Motive) Cause. We must begin at the End, not at the beginning.

640 b 17

Thus we must consider not merely the *primary* substances, but the " *uniform* " *parts*, which are made out of them, and also the " *non-uniform* " *parts*. In doing this, we shall be paying attention to the Formal Cause, which is more important than the Material Cause : the animal as a finished whole is more significant than the substances out of which it was made.

640 b 30

But mere form or shape is not enough : " shaped matter " is not an animal. " Form " in its full and true sense involves " Soul " : " Soul " somehow is the animal's Efficient and Final Cause. Actually, it is not Soul in its entirety, but

13

some " portion " of Soul which fulfils this
office.

641 b 10   Thus the universe and the living objects
in it are the products of something
analogous to human art : they are controlled by a *Final Cause.*

642 a 1   But *Necessity* also has its place in the
universe—

      not (1) " absolute " necessity
      nor (2) " coercive " necessity
      but (3) " conditional " necessity.
These two Causes, the *Final Cause* and
Necessity, set the stage for our piece.

---

642 b 5   ch. 2   Criticisms of dichotomy as a method of
classification of animals.
644 a 11   ch. 4   The correct method of classification is by
groups, such as Birds and Fishes.

---

644 a 23   *Answer* to question (1).
We must deal with *groups*, not *species* (*e.g.*
Bird, not Crane), and where a species does
not belong to a larger group, we must deal
with *species*, not *individuals* (*e.g.* Man,
not Socrates).

---

644 b 21   ch. 5   An Exhortation to the study of animals.

---

645 b 1   Final summary of the Method, combining
answers to both the original questions :
(1) First we discuss the attributes common
    to a group ;
(2) Then we give the explanation of them.

Book II.
646 a 8   ch. 1   Purpose and outline of the Treatise : Our
subject is the CAUSES of the parts of
animals.

697 a 15   (*c*) Intermediate Creatures:
                            Cetacea.
                            Seals and Bats.
                            Ostrich.

---

697 b 27   Conclusion.

Method of classi-fication.
   A glance at the summary will show clearly the order of subjects which Aristotle lays down in the first book to be followed in a treatise such as the one in which he is engaged.

First, (A) to describe the parts of animals as they are observed to be ; and

then, (B) to give an account of their causes, and their formative processes.[a]

Under (A) the order of preference is to be : first, the parts (1) common to all animals ; (2) where necessary, those common to a group of animals only ; and lastly, (3) in exceptional instances, those peculiar to a single species.

Also, it will be seen how Aristotle works out this scheme in the three books which follow. Before considering that, however, we should notice that Aristotle has a great deal to say about the correct classification of animals—or rather, against the incorrect classification of them. Chiefly, he inveighs against the method of dichotomy ; and his chief objection to it is a simple and effective one—that it does not work. It forces us to assign to each species one distinguishing mark, and one only (642 b 21— 643 a 24). And it cuts off kindred species from each other on the strength of some quite subordinate

   [a] *De partibus* is concerned chiefly with the causes and less with the processes.

characteristic (642 b 10 foll.). The right method, says Aristotle, is to follow popular usage and divide the animals up into well-defined groups such as Birds and Fishes.[a] And this leads him to distinguish two stages of difference :

(a) Cases in which the parts differ " by excess or defect "—as in different species of the same genus or group.

(b) Cases in which the resemblance is merely one of analogy—as in different genera.

Examples of (a) : differences of colour and shape ; many or few ; large or small ; smooth or rough ; e.g. soft and firm flesh, long and short bill, many or few feathers.

(b) bone and fish-spine ; nail and hoof ; hand and claw ; scale and feather.

(Reff. for the above, De part. an. 644 a 11–b 15 ; Hist. an. 486 a 15–b 21. See also Gen. An. (Loeb), Introd.)

The doctrine of differences of " excess and defect," "The more or, as Aristotle also calls them, of " the more and and less." less," may usefully be compared with that which underlies the modern theory of Transformations, and the comparison of related forms. Indeed, Professor D'Arcy Thompson asserts that " it is precisely . . . this Aristotelian ' excess and defect ' in the case of form which our co-ordinate method is especially adapted to analyse, and to reveal and demonstrate as the main cause of what (again in the Aristotelian sense) we term ' specific ' differences " (Growth and

[a] And of course, into Blooded and Bloodless, though there are, as Aristotle points out, no popular names for these groups.

Form, p. 726). The co-ordinates to which he refers are those of the Cartesian method, on which is based the theory of Transformations. By means of them it is possible to exhibit, say, the cannon-bones of the ox, the sheep, and the giraffe as strictly proportionate and successive deformations of one and the same form. These deformations can be either simple elongations, as in the instance just cited, or they may occur according to an oblique or a radial system of co-ordinates, etc.[a] In this way, differences of " excess and defect " are reduced to the terminology of mathematics ; and it is especially interesting to notice this, as the phrase " excess and defect " itself had, in the Greek of Aristotle's time, a mathematical connexion. With it may be compared the well-known Platonic phrase, " the great and small." But this is not the place to enlarge upon such topics.[b]

Classification of parts.

To return to Aristotle's classification. We find that he implements his preliminary outline in the following way :

I. First, he treats of the parts which are found in many different groups of animals, and also those which are to be considered counterparts of each other in different groups. This corresponds to A (1) above.

II. As he proceeds with this, he comes to the Viscera, which occur only in blooded animals.[c] This provides a convenient point for embarking upon his second main division—corresponding

[a] For details see D'Arcy Thompson, *op. cit.* ch. xvii.
[b] The reader is referred to A. E. Taylor, " Forms and Numbers," in *Mind*, xxxv. 419 foll. ; xxxvi. 12 foll. ; D'Arcy Thompson, " Excess and Defect," in *Mind*, xxxviii. 43 foll.
[c] By " viscera " Ar. means the blood-like ones only.

to A (2) above—the parts common to a group of animals, and we have first :
The Internal Parts of Blooded Animals.

III. This is followed by—
The Internal Parts of Bloodless Animals. Then,

IV. *The External Parts of Bloodless Animals.* Then,

V. *The External Parts of Blooded Animals,*
which includes—

   (*a*) Vivipara.
   (*b*) Ovipara.
      (i) Serpents and Quadrupeds.
      (ii) Birds.
      (iii) Fishes.
   (*c*) Intermediate Creatures.

References to exceptional instances, as to Man, corresponding to the division A (3) above, are of course to be found throughout the work.

Aristotle thus works out the main lines of his classification. And in each instance, where possible, he endeavours to assign the Cause, to name the purpose, which is responsible for the parts as he describes them. This corresponds to (B) above.

And here Aristotle is forced to admit an apparent Necessity. addition to his scheme of Causes. The purpose, the good End, the final Cause, cannot always get a free hand. There is another Cause, Necessity. Aristotle takes great care to explain what is the nature of this Necessity (642 a 2 foll.). It is what he calls Necessity " *ex hypothesi,*" or " conditional " Necessity, the sort of Necessity which is implied by any final Cause being what it is. If a piece of wood is to be split by an axe, the axe must *ex hypothesi* be hard and sharp, and that necessitates the use of bronze or

21

iron in the making of it. The same sort of Necessity applies in the works of Nature, for the living body itself is an instrument. It is thus the final Cause which necessitates the various stages of the process of formation and the use of such and such material.

Another kind of Necessity, however, makes its appearance in Natural objects, and that is " simple " Necessity. The mere presence of certain things in a living organism entails *of necessity* the presence of others (see 645 b 32, 677 a 17, b 22). Some results follow inevitably from the very nature of the material used. This " simple " Necessity can therefore be regarded as a reassertion of themselves by the motive and material Causes[a] as against the final Cause. Sometimes, however, even in circumstances where " simple " Necessity operates, Nature is able to use the resulting products to subserve a final Cause (663 b 22, 32, 677 a 15 ; see also the note on Residues, p. 32). *Cf. Gen. An.* (Loeb), Introd. §§ 6-9.

Scheme of animals. The following table will show at a glance the scheme of Animals as treated of by Aristotle in the *De partibus* :

A. Blooded Animals	B. Bloodless Animals
Man	Insects
Viviparous quadrupeds	Testacea
Oviparous quadrupeds and footless animals (reptiles and amphibians)	Crustacea
	Cephalopods
Birds	
Fishes	

---

[a] See *De gen. an.* 778 b 1.

PARTS OF ANIMALS

Intermediate between the above classes	*Intermediate*
between land and water animals 　Cetacea 　Seals between quadrupeds and birds 　Bats 　Ostrich	between animals and plants 　Ascidians 　Sponges 　Holothuria 　Acalephae

Note on the Four Classes of Bloodless Animals.—
These, in order of increasing softness, as noted
above, are the following (I give the Greek term, its
literal translation, and the term which I have used
to translate it in this volume) :

τὰ ἔντομα	insected animals	Insects
τὰ ὀστρακόδερμα	shell-skinned animals	Testacea
τὰ μαλακόστρακα	soft-shelled animals	Crustacea
τὰ μαλάκια	softies	Cephalopods

In using " Testacea " to translate τὰ ὀστρακόδερμα
(" the animals with earthenware skins "), I use
it in the old-fashioned sense, so as to include a
number of shelled invertebrates, comprising Gastero-
pods, Lamellibranchs, and some Echinoderms. It
does not refer to the Testacea of modern zoologists,
by whom the term is applied to the Foraminifera
which are shelled Protozoa. The word " Ostraco-
derms " (a transliteration of Aristotle's word) is now
given by zoologists to a group of primitive fossil
fishes.

ARISTOTLE

TERMINOLOGY

Technical terms. The following notes on some of the more difficult and important of the technical terms used by Aristotle in the *De partibus* will, I hope, help to explain my translation and also to give some indication of the background of Aristotle's thought. (A fuller account will be found in *De Gen. An.*, Loeb edn.)

Αἰτία, " cause."

I retain the traditional translation " cause," although perhaps in some contexts " reason " may be a closer rendering, but a variation in the English term might well produce more confusion than clarity. To know, says Aristotle, is to know by means of Causes (see *Anal. post.* 94 a 20). A thing is explained when you know its Causes. And a Cause is that which is responsible, in any of four senses, for a thing's existence. The four Causes, of which two are mentioned very near the beginning of the first book (639 b 11), are :

(1) The Final Cause, the End or Object towards which a formative process advances, and *for the sake of which* it advances—the *logos*, the rational purpose.

(2) The Motive (or Efficient) Cause, the agent which is responsible for having set the process in motion ; it is that *by which* the thing is made.

(3) The Formal Cause, or Form, which is responsible for the *character* of the course which the process follows (this also is described as the *logos*, expressing *what* the thing is).

(4) The Material Cause, or Matter, *out of which* the thing is made.

24

It will be seen that the first three Causes tend naturally to coalesce under the aegis of the Formal Cause, in opposition to the fourth, the Material Cause, a contrast which is clearly put by Adam of St. Victor in one of his hymns :

> *effectiva vel formalis*
> *causa Deus, et finalis,*
> *sed numquam materia.*

Hence, of course, comes the regular contrast of " form " and " matter," in which, oddly enough, in modern usage the two terms have almost exchanged meanings. " Mere form," " empty form," in contrast with " the real matter," are phrases which indicate a point of view very different from that of Aristotle. An equally drastic reversal of meaning has overtaken the term " substance," as controversies on " transubstantiation," and the existence of the word " unsubstantial " prove. " Cause " has certainly been more fortunate ; but its meaning has been narrowed down, so that " cause " now usually suggests the " efficient " cause only. At the same time, we allow ourselves a wider variety of " efficient " causes than Aristotle, and are more ready to admit actions and events or even series of actions and events. We have, in fact, applied Aristotle's precise terminology to the wider uses of everyday non-technical purposes. For Aristotle, the doctrine of the Four Causes provides an exhaustive and precise classification of the things which can be responsible for another thing's existence, and by the naming of them the thing can be completely accounted for.

As an illustration the following will serve. Suppose the object to be explained is an oak. The chronological order of the Causes is different from their logical one.

(i.) The Motive Cause: the parent oak which produced the acorn.

(ii.) The Material Cause: the acorn and its nourishment.

(iii.) The Formal Cause. The acorn as it grew into a tree followed a process of development which had the definite character proper to oaks.

(iv.) The Final Cause: the end towards which the process advanced, the perfected oak-tree.

Λόγος.

There are several places in the *De partibus* where, rather than represent λόγος by an inadequate or misleading word, I have transliterated it by *logos*. This serves the very useful purpose of reminding the reader that here is a term of very varied meanings, a term which brings into mind a number of correlated conceptions, of which one or another may be uppermost in a particular case. It is an assistance if we bear in mind that underlying the verb λέγειν, as it is most frequently used, is the conception of rational utterance or expression, and the same is to be found with λόγος, the noun derived from the same root. Λόγος can signify, simply, *something spoken or uttered*; or, with more prominence given to the rationality of the utterance, it can signify *a rational explanation, expressive of a thing's nature*, of the *plan* of it; and from this come the further meanings of *principle*, or *law*, and also of *definition*, or *formula*, as expressing the structure or character of the object defined. (Note here the application of the term *logos* to the Final and Formal Causes, recorded in the foregoing note.) Another common meaning is seen especially in the use of the dative λόγῳ (*cf.* the verb λογίζομαι and its noun)—*by reasoning, in thought,* as opposed to fact or action. (See 640 a 32, Art is the λόγος τοῦ ἔργου ὁ ἄνευ τῆς ὕλης; at 646 b 2 we read of the λόγος of a process of formation such as building, and the λόγος of the house which is built; at 678 a 35 of the λόγος which defines the essence of something, and at 695 b 19 of " the λόγος of the essence." At 639 b 15 the " Cause for the sake of which "—the Final Cause—is described as being a λόγος.)

Γένεσις, " formation," or " process of formation."
Γίγνεσθαι, " to be formed," " to go through a process of formation."

These are the translations which I normally use, as more appropriate in a biological treatise than " coming into being," and the like.

The process of formation is of course closely connected in Aristotle's thought with the doctrine of the Four Causes.

Γένεσις is a process which, at any rate in biology, results in the production of an actual object, a living creature.

Γένεσις is also contrasted with οὐσία and φύσις[a]: the order of things, we are told, in the process of formation is the reverse of the order in reality. For example, the bricks and mortar exist for the sake of the house

---

[a] Care should be taken not to regard φύσις as meaning " the process of φύεσθαι."

which is to be built out of them, but they and not it come first in the order of time and fact. Aristotle sums this up by saying that what comes last in the process comes first in " nature " (646 a 25).

Μόριον, " part."

The term which occurs in the title of the treatise and is traditionally rendered " part " includes more than is normally included in the English " part of the body." For instance, this would not normally be applied to blood, but the term μόριον is applied by Aristotle to all the constituent substances of the body as well as to the limbs and organs. For him, blood is one of the ζῴων μόρια (648 a 2 ; see also 664 a 9, 690 a 8). A striking instance of the use of μόριον in this sense is the phrase τὰ ὁμοιομερῆ μόρια, which are the subject of the next following note.

Τὰ ὁμοιομερῆ μόρια, " the uniform parts."
Τὰ ἀνομοιομερῆ μόρια, " the non-uniform parts."

Aristotle's application of the term μόριον to both these classes emphasizes the inclusiveness of its meaning. As examples of the " uniform " parts he mentions (647 b 10) blood, serum, lard, suet, marrow, semen, bile, milk, flesh—these are soft and fluid[a] ones ; also bone, fish-spine, sinew, blood-vessel— these are hard and solid ones. Of " non-uniform " parts he gives as examples (640 b 20) face, hand, foot. The relation of the " uniform " parts to the " non-uniform " he describes as follows (647 b 22 foll.) :

[a] For the meaning of " fluid " and " solid " see below, p. 32.

(a) some of the uniform are the material out of which the non-uniform are made (*i.e.* each instrumental part is made out of bones, sinews, flesh, etc.) ;

(b) some act as the nutriment of (a) ;

(c) some are the residue of (b)—faeces, urine.

It is not possible to equate the two classes with the later division into tissues and organs, since blood, for instance, though " uniform," is not a tissue ; the term " organs," however, corresponds closely with Aristotle's own description — τὰ ὀργανικὰ μέρη (647 b 23), " instrumental parts."

The practical difference between the two classes is that each of the uniform parts has its own definite character as a *substance* (in the modern sense), while each of the non-uniform parts has its own definite character as a *conformation* or organ. The heart is the only part which belongs to both classes (647 a 25 foll.) : it consists of one uniform part only, namely, flesh ; but it also has essentially a definite configuration, and thus it is a non-uniform part.

Three stages or " degrees of composition," so far as biology is concerned, are enumerated by Aristotle (at 646 a 13 foll.). What Aristotle seems to mean, though he has not expressed himself quite clearly, is that there are three stages *involved in* the composition of compound bodies, namely,

(1) the δυνάμεις (see following note) ;

(2) the uniform parts ;

(3) the non-uniform parts ;

and finally, of course, out of the non-uniform parts

(4) the animal itself is composed.

ARISTOTLE

We have thus:

(1) the simplest sorts of matter;

(2) the simplest organic substances compounded out of the foregoing (having no definite size, shape, or structure);

(3) the instrumental parts of the body constructed out of the foregoing (having definite size, shape, and structure); and

(4) the organism as a whole, assembled out of the foregoing.

Note.—For a description of the way in which the term τὰ ὁμοιομερῆ has caused confusion in the accounts of Anaxagoras's theories see *Class. Qu.*, 1931, xxv. 34 following.

Δύναμις.

This is one of the most difficult terms to render in English.

The specialized meaning of δυνάμει, " potentially," as opposed to ἐνεργείᾳ, " actually," is so well known that there is no need to enlarge upon it here. Nor need I discuss the mathematical meaning of δύναμις. Other meanings need some comment.

(1) Δύναμις was the old technical term for what were later to be called στοιχεῖα (elements). It appears in the writings of the Hippocratic corpus and in Plato's *Timaeus*. The best example of its use in *De partibus* is at the beginning of Book II. (646 a 15). The list of δυνάμεις included the substances known as τὸ ὑγρόν, τὸ ξηρόν, τὸ θερμόν, τὸ ψυχρόν, τὸ πικρόν, τὸ γλυκύ, τὸ δριμύ, etc., etc. Only the first four of these were regarded by Aristotle as

the material of compound bodies : all the " other differences," he says, are consequent upon these.

The original meaning underlying this usage of the term seems to have been " strong substance of a particular character." This would be very appropriate to τὸ δριμύ, τὸ πικρόν, etc. (see Περὶ ἀρχαίης ἰητρικῆς). There is no notion here of the substance *having* power in the sense of power to affect an external body in a particular way. (This meaning developed later.) If any effect did result, it would be described simply as the presence of the strong substance, and the remedy for it was to " concoct " the strong substance or otherwise to bring it into a harmless condition by " blending " it with other substances.

(2) As each of the substances known as δυνάμεις has its own peculiar character, sharply marked off from the others, the meaning of " peculiar and distinctive character " was naturally associated with the term. This seems to be its meaning in 655 b 12 : ἐξ ἀνάγκης δὲ ταῦτα πάντα γεώδη καὶ στερεὰν ἔχει τὴν φύσιν· ὅπλου γὰρ αὕτη δύναμις. Indeed, in this meaning, δύναμις seems to be a slightly more emphatic version of φύσις, with which it is often used in conjunction (in Hippocrates, for instance), or in a parallel way as in the passage just cited. Compare also 651 b 21, where the marrow is asserted to be αἵματός τις φύσις, not, as some suppose, τῆς γονῆς σπερματικὴ δύναμις. Other instances of this use of δύναμις will be found in *De partibus*.

(3) From this usage it is not far to the idiomatic, pleonastic usage, *e.g.* :

678 a 13 ἡ τῶν ἐντέρων δύναμις almost = τὰ ἔντερα.
682 b 15 ἡ τῶν πτερῶν δύναμις.

ARISTOTLE

657 a 4   ἡ τῶν μυκτήρων δύναμις διφυής.

This is paralleled by a similar usage of φύσις:

663 a 34 ἡ τῶν κεράτων φύσις.
676 b 11 ἡ τῶν ἐντέρων φύσις.

(Other references for δύναμις: 640 a 24, 646 a 14,
b 17, 650 a 5, 651 b 21, 652 b 8, 12, 653 a 2, 655 b 12,
658 b 34. See further *Gen. An.*, Loeb edn., Introd.
§§ 23 ff.).

Τὸ ὑγρὸν καὶ τὸ ξηρόν, " fluid substance and solid
substance," " the fluid and the solid."

These are two of the δυνάμεις.

Following Ogle, I use these renderings as being
more in conformity with the definitions given by
Aristotle than " the moist and the dry," which have
often been used. Actually neither pair of English
words quite expresses the Greek. Aristotle's definition of them (at *De gen. et corr.* 329 b 30) is this :

" ὑγρόν is that which is not limited by any limit of
its own but can be readily limited, ξηρόν is that
which is readily limited by a limit of its own but can
with difficulty be limited "—*i.e.* of course by a limit
imposed from without.
He discusses the various senses in which these
terms are used at 649 b 9 following.

Περίττωμα, " residue."

This term I have translated throughout " residue,"

as being more literal and at the same time less misleading than " excrement." " Surplus " would have
been even better if the word had been a little more
manageable.

" Residue " is so called because it is that which is
left over when the living organism, by acting upon
the nutriment which it has taken, has provided itself
with a sufficient supply for its upkeep. Some of the
surplus will be useless material contained in the food
from the outset, or else has been produced during
the process of reducing the food into a condition
suitable for its purposes in the body. The useless
residues include the excrements. In order to appreci-
ate the status of the useful residues the outlines of
the processes through which the food passes must be
kept clearly in mind. Briefly, then, the food is
masticated in the mouth, then passed on to the
stomach and then the heart, where it is concocted [a]
by means of heat—in other words, it is turned into
blood, which is the " ultimate nourishment " ; and
this, when distributed into the blood-vessels, supplies
the body with nutrition. Generally, however, more
blood is produced than is necessary for the actual
upkeep of the body, and this surplus undergoes a
further stage of concoction, and is used by Nature
in various ways. Marrow is a residue ; so are semen,
catamenia, milk. Sometimes, when nutrition is
specially abundant, the surplus blood is concocted
into fat (lard and suet). And some of the blood,
reaching the extremities of the vessels in which it
travels, makes its way out in the form of nails, claws,
or hair. The Aristotelian doctrine of residues came
down to Shakespeare, as is shown by the passage

[a] See page 34.

in *Hamlet* (III. iv.) where the Queen says to Hamlet:

> Your bedded haire, like life in excrements,
> Start up, and stand an end.

This theory, as applied to hair, is expounded by Aristotle at 658 b 14 following, and modern biochemists have reason for believing that some pigmentation in animals, such as the black melanin of mammalian hair, or the yellow xanthopterine of the butterfly's wing, is physiologically a form of excretion.

" *Concoct*," " *concoction*."

These terms, which have already appeared in these notes, are used to translate πέσσειν, πέψις. The Greek words are the same as those employed to denote the process of ripening or maturing of fruit, corn, and the like by means of heat—also that of baking and cooking.

Terms sometimes associated with these are μεταβολή and μεταβάλλειν. For example, at 650 a 5 we read that πέψις and μεταβολή take place διὰ τῆς τοῦ θερμοῦ δυνάμεως; and at 651 b 26, as the creatures grow and get " matured," the parts μεταβάλλει their colour, and so do the viscera.

Ψυχή, " Soul."

The English word " Soul," as will be seen, over-emphasizes, when compared with ψυχή, certain aspects of the Greek term, but it is by far the most convenient rendering, and I have used it in preference to " life " or " vital principle."

It will be useful to have an outline of Aristotle's general doctrine about Soul.

The different " parts " or " faculties " of Soul can

be arranged in a series in a definite order, so that the possession of any one of them implies the possession of all those which precede it in the list :

(1)	nutritive Soul	in all plants
(2)	sentient Soul	in all animals
(3)	appetitive Soul	in some animals
(4)	locomotive Soul	
(5)	rational Soul	in man only

At 641 a 23 Aristotle speaks of " parts " of the Soul, and though he often uses this phrase, the description he prefers is " faculties." In the passage which follows (641 a 33 foll.) all except appetitive Soul are mentioned. Sentient Soul is mentioned again at 650 b 24, 667 b 23, 672 b 16.

Aristotle raises the question whether it is the business of Natural science to deal with Soul in its entirety, and concludes that it is not necessary, since man is the only animal in which rational Soul is found. Thus it is only some part or parts of Soul, and not Soul in its entirety, which constitute animal nature.

In the passage 641 a 14 following, Aristotle takes for granted his doctrine about Soul, which is as follows (De anima, Book II.). Animate bodies, bodies " with Soul in them " (ἔμψυχα), are " concrete substances " made up of matter and form. In this partnership, of course, the body is the matter and the Soul is the form. Thus Soul may be described as the " form " or " realization " (ἐντελέχεια, " actuality ") of the animal (cf. De part., loc. cit.).

This statement, however, is elsewhere made more precise. It is possible to distinguish two " realizations " of an animal ; for an animal " has Soul in it "

35

even when it is asleep, but its full activity is not evident until it is awake and about its business. We must call Soul, then, the " first realization " of the animal, its waking life its " second realization." This distinction does not concern us in the *De partibus*. But an expansion of the definition is not irrelevant. Aristotle states that the Soul is the first realization of a body furnished with organs. The priority of Soul over body is emphasized in the passage just referred to (640 b 23—641 a 32), and in another interesting passage (687 a 8 foll.) Aristotle maintains that man has hands because he is the most intelligent animal, and not, as some have said, the most intelligent animal because he has hands.

With this is connected the question whether the Soul is independent of the body ; though it is not raised in *De partibus*. As we have seen already, a ζῷον is a single concrete entity made up of Soul and body, *i.e.* a certain form implanted in certain matter. The matter can exist, for it did exist, apart from the form ; and as the form that is implanted in all the individuals of a species is one and the same form, clearly it can exist apart from any one individual's matter—though of course its existence is not independent of all the individuals' matter. Furthermore, the form—the Soul—requires matter of a particular kind : not any sort of matter will do. From these considerations two conclusions seem to follow : (1) that transmigration is impossible : a human Soul cannot function in a hyena's body, any more than the carpenter's art can be executed by means of musical instruments ; (2) the Soul cannot function without a body at all ; cannot, we may say, exist (414 a 19).

So far, so good. But Aristotle is not satisfied. He feels the Soul is more than that. He finds a loophole. There may be some " part " of Soul (the rational part) which is not the " realization " of any body. The Soul, besides being the form, the formal Cause, of the body, is also its final Cause, and not only that, but the motive Cause too of all the changes originated in the body (*De anima* 415 b 7-28), for, as we saw (p. 25), the three non-material Causes tend to coalesce into one. This independent " part " of Soul " comes into the body from without " (see *De gen. an.* 736 b 25 foll.) and continues to exist after the death of the body (see *De anima* 413 a 6, b 24 foll., 430 a 22, etc.). All this, however, raises problems not touched upon in *De partibus*; indeed Aristotle himself offers no solution of them.

Ψυχή, κρᾶσις, ἀπόκρισις, σύντηξις.

I have indicated above, in the note on δύναμις, some of the older (Hippocratic) medical terminology of which traces are to be found in the *De partibus*. There is no room for an adequate discussion of such terms and theories, and the following bare references must suffice.

In the Hippocratic treatise Περὶ διαίτης the theory is put forward that the human organism, body and Soul alike, is composed of fire and water (which really consist of " the hot," " the solid," " the cold," and " the fluid ")—the function of fire being to cause motion, of water to provide nourishment. In ch. 35 we have a list of the different varieties of Blend (κρῆσις, σύγκρησις) of fire and water which may be found in the Soul in different individuals, and upon the Blend its health and sensitivity [a] depend.

With these statements may be compared the following passages in *De partibus* :

652 b 8   Some, says Aristotle, maintain that the Soul *is* fire ; but it is better to say that it subsists in some such material. " The hot " is indeed the most serviceable material for the functions which the Soul has to perform, and these include nourishing and causing motion.

647 b 30 foll.   Here is a reference to the different varieties of blood, and Aristotle tells us which sort of blood is αἰσθητικώτερον and which animals are on that account φρονιμώτερα (*cf.* 650 b 24 and 686 b 28). The phrase αἵματος κρᾶσις is actually used at 686 a 9. (*Cf.* also 650 b 29, the κρᾶσις in the heart ; 652 b 35, the parts in the head are colder than the σύμμετρος κρᾶσις ; 669 a 11, the κρᾶσις of the body ; 673 b 26, its εὐκρασία.)

The term σύντηξις, which occurs frequently in the Περὶ διαίτης, is found only once in the *De partibus* at 677 a 14—bile is said to be a residue or σύντηξις. Properly speaking, σύντηξις is the term applicable to the " colliquescence " or decay of the parts of the body themselves. (*Cf.* σύντηγμα at *De gen. an.* 724 b 26 foll. ; also σύντηξις, 456 b 34 ; *cf.* also Platt's note at the end of his translation of *De gen. an.*, on 724 b 27.) The effect of the colliquescence is to produce an

---

[a] The adjective used is φρόνιμος.

unhealthy ἀπόκρισις (abscession)—a very common term in Περὶ διαίτης (see chh. 58 foll. throughout). It occurs twice in *De partibus*. In both places it is used of a περίττωμα. At 690 a 9 the surplus earthy matter ἀπόκρισιν λαμβάνει, and forms a continuous nail or hoof. At 681 b 35 Aristotle speaks of the place where the σπερματική or the περιττωματικὴ ἀπόκρισις is effected ; and here ἀπόκρισις seems to mean simply " act of excretion." The meaning of the term seems both here and in Hippocrates to be specially associated with περιττώματα, either useful ones, or useless and even harmful ones. A great deal of Περὶ διαίτης is taken up with suggestions for getting rid of harmful ἀποκρίσεις.

The meaning of ἀπόκρισις is therefore wider than " excretion " or " secretion," as used in their present usual sense, though these are included among its meanings.

Τὸ μᾶλλον καὶ ἧττον, " the more and less," see above, p. 19, and *Gen. An.* (Loeb), Introd. §§ 70 ff.

TRANSLATIONS OF ARISTOTLE'S ZOOLOGY

The history of the translation of Aristotle's works begins with the Nestorian Christians of Asia Minor, who were familiar with the Greek language as their service-books were written in it, and before the coming of the Arabs they had translated some of the works of Aristotle and Galen into Syriac. Before

Transla-tions of Aristotle's zoological works.

435, Ibas, who in that year was made Bishop of
Edessa, had translated into Syriac the commentaries
of Theodore on the works of Aristotle. Jacob, one
of Ibas's successors at Edessa (*d.* 708), translated the
Categories into Syriac, but a much earlier version had
been made by Sergios of Resh 'Ainâ (*d.* 536), who
had studied Greek at Alexandria. In 765 the Nestorian physician Georgios was summoned to Bagdad
by the Caliph, and translated numerous Greek words
into Arabic for him. By the beginning of the ninth
century, translation was in full swing at Bagdad,
under the Caliphate of al-Mamun (813–833), son of
Harun-al-Rashid. The first leader of this school
of translators was the physician Ibn al-Batriq, who
translated the *Historia animalium*, the *De partibus
animalium*, and the *De generatione animalium* into
Arabic.

But it was through southern Italy, Sicily and Spain
that the transmission of Aristotle's works from the
Arabic into Latin was effected. Messina had been
recovered from the Saracens by 1060, and the whole
of Sicily was freed by 1091. Under the Norman
kings, Greeks, Saracens and Latins lived together
in one community, and the court was the meeting-
ground for eminent persons of all nations and
languages. The reconquest of Spain had begun in
the eighth century, so that here also an opportunity
offered for making the works of Greek science
available in Latin. Archbishop Raymond of Toledo
(1126–1151) and Bishop Michael of Tarazona (1119–
1151) were the patrons of the translators, who made
Toledo the centre of their activity. One of these
was Michael Scot.

There is in existence an Arabic translation of

the zoological works, of which there is a MS. in the British Museum.[a] It is probable that this is the translation made by Ibn al-Batriq, and that this Arabic version is the original from which Michael Scot made his Latin translation at Toledo.[b] Michael was, among his other accomplishments, astrologer to Frederick II., King of Sicily, at his court at Palermo, and before 1217 he had reached Toledo and was at work there on his translations from the Arabic. His *De animalibus* (a translation of the zoological works in nineteen books) is one of his earliest works, and two MSS. of it [c] contain a note which gives a later limit of 1220 for the work. Other evidence [d] establishes that it was certainly finished before 1217, and it may even be placed in the first decade of the century. It is probable that Michael had as collaborator one Andrew, canon of Palencia, formerly a Jew. One of the earliest to make use of Michael's translations was Robert Grosseteste,[e] Bishop of Lincoln (d. 1253), one of the leading Aristotelian scholars of the time, who quotes from Michael's version of

[a] B.M. Add. 7511 (13th–14th century). This is the MS. referred to by Steinschneider, *Die arabischen Übersetzungen* p. 64, as B.M. 437. I have seen this MS.

[b] Judging from the passages which Dr. R. Levy kindly read for me in the Arabic MS., the Latin version is a close translation from it. Also, the contents-preface which is found prefixed to Michael Scot's translation corresponds exactly with the preface which precedes the Arabic version in this MS. (see the B.M. catalogue, *Catalogus codicum manuscriptorum orientalium*, p. 215).

[c] One of them is MS. Caius 109, in the library of Gonville and Caius College, Cambridge. It is of the thirteenth century.

[d] See S. D. Wingate, *The Medieval Latin Versions*, p. 75.

[e] Born at Stradbroke, Suffolk. A Franciscan.

41

De generatione.[a] The *De animalibus* also formed the basis of a commentary in twenty-six books by Albertus Magnus.[b] This was probably written soon after the middle of the thirteenth century. Except for the portions which appear in Albertus's commentary, and the earlier part of the first chapter,[c] Michael's version has never been printed *in extenso.* Michael died in or before 1235, and is reputed to have been buried, as he was born, in the lowlands of Scotland.

About the same time, at the request of a pupil of Albertus, St. Thomas Aquinas (1227–1274), who required more accurate versions for his commentaries on the works of Aristotle, new translations, direct from the Greek, were being undertaken by William of Moerbeke.[d] William was born about 1215. He became a Dominican, was confessor to Popes Clement IV. and Gregory X., and was Archbishop of Corinth. He acted as Greek secretary at the Council of Lyons in 1274. He died in 1286. The earliest dated translation made by him is one of the *De partibus animalium.* The date 1260 occurs in a MS. of it at Florence (Faesulani 168), which also contains *Hist. an., De progressu an.,* and *De gen. an.* This translation was made at Thebes.

Among later Latin translators of the zoological

---

  [a] According to Roger Bacon, Michael appeared at Oxford in 1230, bringing with him the works of Aristotle in natural history and mathematics.
  [b] *Ed. princeps,* Rome, 1478; latest ed., H. Stadler, 1916–1921.
  [c] 639 a 1—640 a 20, printed by G. Furlani in *Rivist‧ degli Studi Orientali,* ix. (1922), pp. 246-249.
  [d] A small town south of Ghent on the borders of Flanders and Brabant.

works the names of two Greeks must be mentioned.
George of Trebizond (Trapezuntius), who was born
in Crete in 1395, visited Italy between 1430 and
1438, and was secretary to the humanist Pope
Nicholas V., an ardent Aristotelian. George's work,
however, was hurried and not over-exact, and he,
together with his predecessors, was superseded by
his contemporary Theodore of Gaza, who was born
in Thessalonica about 1400, and was professor of
Greek at Ferrara in 1447. In 1450 Theodore was
invited by the Pope to go to Rome to make Latin
versions of Aristotle and other Greek authors. His
translation of the zoological works,[a] dedicated to the
Pope, Sixtus IV., soon became the standard version,
and it is printed in the Berlin edition of Aristotle.

Translations of the *De gen.* were made by Augus-
tinus Niphus, of the University of Padua (1473–1546),
and of the *De gen.* and *De incessu* by Peter Alcyonius
(Venice, 1487–1527). The *De gen.* was also translated
by Andronicus Callixtus of Byzantium (*d.* 1478).
With the later Latin versions we need not here
concern ourselves, but something must be said of
the scientific workers who were inspired by Aristotle,
and of the translations into modern languages.

The Renaissance biologists show unmistakably the
difference in quality which there is between Aristotle's
physics and his biology. Hieronimo Fabrizio of
Acquapendente (1537–1619) knew and admired
Aristotle's work on embryology, and what is more,
himself carried out further important observations
on the same subject. His brilliant successor, William
Harvey (1578–1657), was a student of Aristotle, and

Aristotle's successors.

---

[a] In eighteen books, excluding the spurious tenth book of
the *Historia animalium.*

43

much of his inspiration came from that source. William Harvey was the first to make any substantial advance in embryology since Aristotle himself. But this is more appropriate to the *De generatione* than to the *De partibus*. In other departments of study, however, during the seventeenth century, the authority of Aristotle and the scholastic doctrine with which he was identified were being combated in the name of freedom, and thus it came about that the zoological works also, which had been brought to light by the dark ages, were allowed to pass back into oblivion by the age of enlightenment. They were not rediscovered until the end of the eighteenth century by Cuvier (1769–1832) and Saint-Hilaire (1805–1895) in the nineteenth.

MODERN EDITIONS

1. The Berlin edition of Aristotle, by Immanuel Bekker. Vol. i. (pp. 639-697) includes *P.A.* Berlin, 1831.
1A. The Oxford edition (a reprint of the preceding). Vol. v. includes *P.A.* Oxford, 1837.
2. One-volume edition of Aristotle's works, by C. H. Weise (pre-Bekker text). Leipzig, 1843.
3. The Leipzig edition. Vol. v. contains *P.A.*, edited and translated into German by A. von Frantzius. Leipzig, 1853.
4. The Didot edition. Vol. iii. includes *P.A.* Edited by Bussemaker. Paris, 1854.
5. The Teubner edition. Edited by Bernhardt Langkavel. Leipzig, 1868.
6. The Budé edition. Edited by Pierre Louis. With a French translation and notes. Paris, 1956.

TRANSLATIONS WITHOUT TEXT

7. Thomas Taylor. English translation of Aristotle in ten

volumes. Vol. vi. includes *P.A.* (pp. 3-163). London, 1810.

8. F. N. Titze. German translation of Book I. In his *Aristoteles über die wissenschaftliche Behandlungsart der Naturkunde*. Prague, 1819.

9. Anton Karsch. German translation. Stuttgart, 1855 (second ed., Berlin, 1911).

10. William Ogle. English translation, with notes. London, 1882.

11. J. Barthélemy-Saint-Hilaire. French translation, with notes. Paris, 1885.

11. William Ogle. English translation, with notes (a revision of No. 10). Oxford 1911.

12. Francisco Gallach Palés. Aristóteles : Obras completas. Vol. X contains *De partibus* and *De incessu animalium*. Spanish translation, without notes. Vol. lxii. of *Nueva Biblioteca Filosófica*. Madrid, 1932.

Langkavel reproduces almost verbatim the Berlin text, together with Bekker's *apparatus*, to which a great deal of other matter has been added, including some of Bekker's MS. notes in his copy of Erasmus's edition, and some corrected reports of the readings of the MS. E, which Langkavel himself inspected. Also, there are some emendations proposed by Bonitz.

Any English translator must stand very much indebted to the work of William Ogle, whose translation, originally published in 1882, was revised by its author and republished in the Oxford series of translations of Aristotle in 1911. It is not possible to overrate the care and exactness with which this piece of work was executed. I should like here to acknowledge my own indebtedness to it, and I have had its accuracy as a model before me. With regard to style, it will be seen that I have aimed at producing something rather different from Ogle's version.

The Text

The mss. The manuscript authorities cited by Bekker for the *De partibus* will be found on p. 50.

The dates of some of the mss. as given by different scholars vary considerably : for details I refer the reader to the various catalogues, and also to L. Dittmeyer's edition of *Hist. an.* (Leipzig, 1907) and W. W. Jaeger's edition of *De an. motu*, etc. (Leipzig, 1913).

Restoration of the text. I have relied upon the *apparatus* of Bekker and Langkavel for the readings of the Greek mss., except for those of Z, the oldest parts of which I have collated from photostats *a* ; and at several places I have inspected the ms. itself. In some places (*e.g.* 663 b 17, 685 a 2, 16) I found the reading had been defectively reported. It is clear that a more reliable collation of the chief mss. of *De partibus* is clearly needed. From a different source I have attempted to restore intelligibility to several corrupt passages with the aid of the Arabic version and the Latin version of Michael Scot, which represent an earlier stage of the Aristotelian text than our Greek mss. Among the passages dealt with in this way are the passage at 654 b 14 following, which has been dislocated by glosses and phrases imported from elsewhere, and the remarkable passage about the structure of the Cephalopods at 684 b 22 following, where considerable havoc has been done to the text by references to a diagram which were inserted at some period between the date of the ms. from which the Arabic version was made and that of the archetype of all our present Greek mss. I have been able to restore this passage, though not always the actual Greek words, by reference to the Arabic version and Michael Scot's Latin

a See additional note on p. 434.

translation made from it. Dr. Reuben Levy has most kindly read this passage for me in the 13th-14th century Arabic MS. in the British Museum, Add. 7511.

For these two passages, and for a good many other suspected places, I have consulted all the known MSS. of Michael Scot's version which are to be found in this country. They are (excluding MSS. which contain merely abridgements or extracts) :

Cambridge, Gonville and Caius College 109
     ,,   University Library Ii. 3. 16
     ,,   ,,   ,,   Dd. 4. 30
Oxford, Merton College 278
     ,,   Balliol College 252
London, British Museum Royal 12. C. XV
     ,,   ,,   ,,   Harl. 4970 [a]

All these are of the thirteenth or fourteenth century.

I have inspected at test places the following three MSS. of William of Moerbeke's version :

Oxford, Merton College 270
     ,,   ,,   271
     ,,   Balliol College 250

William's translation was made from a MS. or MSS. which had already been infected by the corruptions found in the Greek MSS. which exist to-day.

I should like here to express my thanks to the Librarians who so kindly made arrangements for me to inspect the MSS. under their care.

Where I have accepted the reading of the Berlin edition, I have not given any record of the MS. variants. These are to be found in the *apparatus criticus* of that edition and of Langkavel's edition.

Scope of apparatus criticus.

---

[a] So far as I know, this MS. has not been mentioned in any of the published lists of MSS. of Michael Scot's *De animalibus.*

I have endeavoured, except in the passage 691 b 28 to 695 a 22 in the fourth Book, to record all places where I have departed from the text of the Berlin edition, and I have given the source of the reading which I have adopted. Where Bekker himself introduced a reading different from that of the mss., this is attributed to him by name.

Punctuation. I have not recorded all of the many passages in which I have corrected the punctuation. The text has been reparagraphed throughout.

REFERENCE

Short bibliography. The following list includes authorities for statements made in the Introduction, and books which the student of the Aristotelian zoological works and their history will find useful :

C. H. Haskins, *Studies in the History of Medieval Science*, ed. 2, Cambridge, Mass., 1927.

W. Jaeger, *Aristotle* (English tr. by R. Robinson), Oxford, 1934.

L. Leclerc, *Histoire de la médecine arabe*, Paris, 1876.

T. E. Lones, *Aristotle's Researches in Natural Science*, London, 1912.

W. D. Ross, *Aristotle*, London, 1930.

J. E. Sandys, *A History of Classical Scholarship*, Cambridge, 1908–1921.

C. Singer, *Studies in the History and Method of Science*, Oxford, 1921.

C. Singer, *Greek Biology and Greek Medicine*, Oxford, 1922.

M. Steinschneider, *Die arabischen Übersetzungen aus dem Griechischen* (Beiheft XII. zum Centralblatt für Bibliothekswesen), Leipzig, 1893.

M. Steinschneider, *Die europäischen Übersetzungen aus dem Arabischen*, in *Sitzungsberichte d. kais. Akad. der Wiss.*, cxlix., Vienna, 1905.

D'Arcy W. Thompson, *Growth and Form*, Cambridge, 1917 (new ed., 1942).

PARTS OF ANIMALS

D'Arcy W. Thompson, Essay on " Natural Science " in *The Legacy of Greece*, Oxford, 1924.

S. D. Wingate, *The Medieval Latin Versions of the Aristotelian Scientific Corpus*, London, 1931.

F. Wüstenfeld, *Die Übersetzungen arabischer Werke in das Lateinische*, in *Abhandlungen der k. Gesell. d. Wiss. zu Göttingen*, xxii., 1877.

ACKNOWLEDGEMENTS

It is a great pleasure to acknowledge here the help which I have received from many friends at Cambridge, not only by way of reading typescript and proof and by discussion, but also by the interest which they have shown in the work and by their continuous encouragement. The following have read the translation either in whole or in part : Prof. F. M. Cornford, Professor of Ancient Philosophy ; Dr. F. H. A. Marshall, Reader in Agricultural Physiology (who has also kindly written the Foreword to this volume), and Dr. Joseph Needham, Reader in Biochemistry. I am under a particular obligation to my colleague Mr. H. Rackham, who has read the whole translation both in typescript and in proof. I am indebted to Dr. Sydney Smith and a number of other friends for their kindness in discussing various points and for reading certain passages. Dr. Reuben Levy, Professor of Persian, has kindly read for me some passages in the Arabic translation of the zoological works. To all of these gentlemen, without whose aid the work could not have been carried through, I record my sincerest thanks.

The present (third) edition has again been revised.

A. L. P.

July 11th 1952

E   Parisinus regius 1853 (see p. 434)
Y   Vaticanus graecus 261
Z   Oxoniensis Coll. Corp. Chr. W.A. 2. 7 (see p. 434)
U   Vaticanus 260
P   Vaticanus graecus 1339
S   Laurentianus Mediceus 81. 1
Q   Marcianus 200
b   Parisinus 1859
m   Parisinus 1921
Σ   Michael Scot's Latin version, from my own transcription.
vulg.   The usual reading, as in the Berlin edition.
Langkavel   Emendations proposed by Langkavel in his edition.
Ogle   Emendations proposed by William Ogle in footnotes to his translation.
Platt   Emendations proposed by Arthur Platt, either (*a*) in " Notes on Aristotle," in *Journal of Philology*, 1913, xxxii. 292 following, or (*b*) recorded by Ogle in footnotes to his translation.
Cornford ⎫ Suggestions in private communications
Rackham ⎬ to me from Professor Cornford and Mr.
          ⎭ Rackham.
Th(urot)   Ch. Thurot, in *Rev. Arch.*, 1867.[a]
Peck   Emendations proposed by myself.

[a] Of over 100 textual points, many being of minor importance, raised by Th., about a third had been dealt with in my first edition (before Th.'s work came to my notice), some of them more fully, by other scholars or myself. Some of Th.'s other suggestions have been adopted in this edition.

The maister Cooke was called *Concoction*.

Spenser, *Faerie Queen*

ΑΡΙΣΤΟΤΕΛΟΥΣ
ΠΕΡΙ ΖΩΙΩΝ ΜΟΡΙΩΝ

Α

639 a Περὶ πᾶσαν θεωρίαν τε καὶ μέθοδον, ὁμοίως
ταπεινοτέραν τε καὶ τιμιωτέραν, δύο φαίνονται
τρόποι τῆς ἕξεως εἶναι, ὧν τὴν μὲν ἐπιστήμην
τοῦ πράγματος καλῶς ἔχει προσαγορεύειν, τὴν δ'
5 οἷον παιδείαν τινά. πεπαιδευμένου γάρ ἐστι κατὰ
τρόπον τὸ δύνασθαι κρῖναι εὐστόχως τί καλῶς ἢ μὴ
καλῶς ἀποδίδωσιν ὁ λέγων. τοιοῦτον γὰρ δή τινα
καὶ τὸν ὅλως πεπαιδευμένον οἰόμεθ' εἶναι, καὶ τὸ
πεπαιδεῦσθαι τὸ δύνασθαι ποιεῖν τὸ εἰρημένον.
10 πλὴν τοῦτον μὲν περὶ πάντων ὡς εἰπεῖν κριτικόν
τινα νομίζομεν εἶναι ἕνα τὸν ἀριθμὸν ὄντα, τὸν δὲ
περί τινος φύσεως ἀφωρισμένης· εἴη γὰρ ἄν τις
ἕτερος τὸν αὐτὸν τρόπον τῷ εἰρημένῳ διακείμενος
περὶ μόριον. ὥστε δῆλον ὅτι καὶ τῆς περὶ φύσιν
ἱστορίας δεῖ τινὰς ὑπάρχειν ὅρους τοιούτους πρὸς
οὓς ἀναφέρων ἀποδέξεται τὸν τρόπον τῶν δεικνυ-
52

ARISTOTLE

PARTS OF ANIMALS

BOOK I

THERE are, as it seems, two ways in which a person may be competent in respect of any study or investigation, whether it be a noble one or a humble : he may have either what can rightly be called a scientific knowledge of the subject; or he may have what is roughly described as an educated person's competence, and therefore be able to judge correctly which parts of an exposition are satisfactory and which are not.   That, in fact, is the sort of person we take the " man of general education " to be ; his " education " consists in the ability to do this.   In this case, however, we expect to find in the one individual the ability to judge of almost all subjects, whereas in the other case the ability is confined to some special science; for of course it is possible to possess this ability for a limited field only.   Hence it is clear that in the investigation of Nature, or Natural science, as in every other, there must first of all be certain defined rules by which the acceptability of the method of exposition may be tested, apart from whether the statements made

639 a

15 μένων, χωρὶς τοῦ πῶς ἔχει τἀληθές, εἴτε οὕτως
εἴτε ἄλλως. λέγω δ' οἷον πότερον δεῖ λαμβάνοντας
μίαν ἑκάστην οὐσίαν περὶ ταύτης διορίζειν καθ'
αὑτήν, οἷον περὶ ἀνθρώπου φύσεως ἢ λέοντος ἢ
βοὸς ἢ καί τινος ἄλλου καθ' ἕκαστον προχειριζο-
μένους, ἢ τὰ κοινῇ συμβεβηκότα πᾶσι κατά τι
κοινὸν ὑποθεμένους—πολλὰ γὰρ ὑπάρχει ταὐτὰ
20 πολλοῖς γένεσιν ἑτέροις οὖσιν ἀλλήλων, οἷον ὕπνος,
ἀναπνοή, αὔξησις, φθίσις, θάνατος, καὶ πρὸς τούτοις ὅσα τοιαῦτα τῶν λειπομένων παθῶν τε καὶ
διαθέσεων· ἄδηλον γὰρ καὶ ἀδιόριστόν ἐστι λέγειν
νῦν περὶ τούτων· φανερὸν δ' ὅτι καὶ κατὰ μέρος
μὲν λέγοντες περὶ πολλῶν ἐροῦμεν πολλάκις ταὐτά·
25 καὶ γὰρ ἵπποις καὶ κυσὶ καὶ ἀνθρώποις ὑπάρχει
τῶν εἰρημένων ἕκαστον, ὥστε ἐὰν καθ' ἕκαστον τὰ
συμβεβηκότα¹ λέγῃ τις, πολλάκις ἀναγκασθήσεται
περὶ τῶν αὐτῶν λέγειν, ὅσα ταὐτὰ μὲν ὑπάρχει τοῖς
εἴδει διαφέρουσι τῶν ζῴων, αὐτὰ δὲ μηδεμίαν ἔχει
30 διαφοράν. ἕτερα δ' ἴσως ἐστὶν οἷς συμβαίνει τὴν
639 b μὲν κατηγορίαν ἔχειν τὴν αὐτὴν διαφέρειν δὲ τῇ
κατ' εἶδος διαφορᾷ, οἷον ἡ τῶν ζῴων πορεία· οὐ
γὰρ φαίνεται μία τῷ εἴδει· διαφέρει γὰρ πτῆσις καὶ
νεῦσις καὶ βάδισις καὶ ἕρψις.

Διὸ δεῖ μὴ διαλεληθέναι πῶς ἐπισκεπτέον, λέγω
5 δὲ πότερον κοινῇ κατὰ γένος πρῶτον, εἶθ' ὕστερον

---

¹ τὰ συμβεβηκότα Ogle: τῶν συμβεβηκότων vulg.

54

represent the truth or do not. I mean, for instance, should we take each single species severally by turn (such as Man, or Lion, or Ox, or whatever it may be), and define what we have to say about it, in and by itself ; or should we first establish as our basis the attributes that are common to all of them because of some common character which they possess ?—there being many attributes which are identical though they occur in many groups which differ among themselves, *e.g.* sleep, respiration, growth, decay, death, together with those other remaining affections and conditions which are of a similar kind. I raise this, for at present discussion of these matters is an obscure business, lacking any definite scheme. However, thus much is plain, that even if we discuss them species by species, we shall be giving the same descriptions many times over for many different animals, since every one of the attributes I mentioned occurs in horses and dogs and human beings alike. Thus, if our description proceeds by taking the attributes for every species, we shall be obliged to describe the same ones many times over, namely, those which although they occur in different species of animals are themselves identical and present no difference whatever. Very likely, too, there are other attributes, which, though they come under the same general head, exhibit specific differences ;—for example, the locomotion of animals : of which there are plainly more species than one—*e.g.* flight, swimming, walking, creeping.

Therefore we must make up our minds about the method of our investigation and decide whether we will consider first what the whole group has in

περὶ τῶν ἰδίων θεωρητέον, ἢ καθ' ἕκαστον εὐθύς.
νῦν γὰρ οὐ διώρισται περὶ αὐτοῦ, οὐδέ γε τὸ νῦν
ῥηθησόμενον, οἷον πότερον καθάπερ οἱ μαθηματικοὶ
τὰ περὶ τὴν ἀστρολογίαν δεικνύουσιν, οὕτω δεῖ καὶ
τὸν φυσικὸν τὰ φαινόμενα πρῶτον τὰ περὶ τὰ ζῷα
10 θεωρήσαντα καὶ τὰ μέρη τὰ περὶ ἕκαστον, ἔπειθ'
οὕτω λέγειν τὸ διὰ τί καὶ τὰς αἰτίας, ἢ ἄλλως πως.
πρὸς δὲ τούτοις, ἐπεὶ πλείους ὁρῶμεν αἰτίας περὶ
τὴν γένεσιν τὴν φυσικήν, οἷον τήν θ' οὗ ἕνεκα καὶ
τὴν ὅθεν ἡ ἀρχὴ τῆς κινήσεως, διοριστέον καὶ
περὶ τούτων, ποία πρώτη καὶ δευτέρα πέφυκεν.
15 φαίνεται δὲ πρώτη ἣν λέγομεν ἕνεκά τινος· λόγος
γὰρ οὗτος, ἀρχὴ δ' ὁ λόγος ὁμοίως ἔν τε τοῖς
κατὰ τέχνην καὶ ἐν τοῖς φύσει συνεστηκόσιν. ἢ
γὰρ τῇ διανοίᾳ ἢ τῇ αἰσθήσει ὁρισάμενος ὁ μὲν
ἰατρὸς τὴν ὑγίειαν ὁ δ' οἰκοδόμος τὴν οἰκίαν,
ἀποδιδόασι τοὺς λόγους καὶ τὰς αἰτίας οὗ ποιοῦσιν
ἑκάστου, καὶ διότι ποιητέον οὕτως. μᾶλλον δ'
20 ἐστὶ τὸ οὗ ἕνεκα καὶ τὸ καλὸν ἐν τοῖς τῆς φύσεως
ἔργοις ἢ ἐν τοῖς τῆς τέχνης. τὸ δ' ἐξ ἀνάγκης
οὐ πᾶσιν ὑπάρχει τοῖς κατὰ φύσιν ὁμοίως, εἰς

---

[a] This point is resumed and decided below, 644 a 23 ff.,
645 b 2 ff.
[b] " Causes." See Introduction, pp. 24 ff.
[c] " Formation." See Introduction, pp. 27 f.
[d] *i.e.* the " final " cause.
[e] *i.e.* the " motive " or " efficient " cause.
[f] See Introduction, pp. 26 f.   [g] *Cf.* 645 a 24.

common, and afterwards the specific peculiarities ; or begin straightway with the particular species.[a] Hitherto this has not been definitely settled. And there is a further point which has not yet been decided : should the student of Nature follow the same sort of procedure as the mathematician follows in his astronomical expositions — that is to say, should he consider first of all the phenomena which occur in animals, and the parts of each of them, and having done that go on to state the reasons and the causes ; or should he follow some other procedure ? Furthermore, we see that there are more causes[b] than one concerned in the formation[c] of natural things : there is the Cause *for the sake of which* the thing is formed,[d] and the Cause to which *the begin-ning of the motion* is due.[e] Therefore another point for us to decide is which of these two Causes stands first and which comes second. Clearly the first is that which we call the " Final " Cause—that for the sake of which the thing is formed—since that is the *logos*[f] of the thing—its rational ground, and the *logos* is always the beginning for products of Nature as well as for those of Art. The physician or the builder sets before himself something quite definite — the one, health, apprehensible by the mind, the other, a house, apprehensible by the senses ; and once he has got this, each of them can tell you the causes and the rational grounds for everything he does, and why it must be done as he does it. Yet the Final Cause (purpose) and the Good (Beautiful) [g] is more fully present in the works of Nature than in the works of Art. And moreover the factor of Necessity is not present in all the works of Nature in a similar sense. Almost all

639 b

ὃ πειρῶνται πάντες σχεδὸν τοὺς λόγους ἀνάγειν,
οὐ διελόμενοι ποσαχῶς λέγεται τὸ ἀναγκαῖον.
ὑπάρχει δὲ τὸ μὲν ἁπλῶς τοῖς αἰδίοις, τὸ δ᾽ ἐξ
25 ὑποθέσεως καὶ τοῖς ἐν γενέσει πᾶσιν ὥσπερ ἐν
τοῖς τεχναστοῖς, οἷον οἰκίᾳ καὶ τῶν ἄλλων ὁτῳοῦν
τῶν τοιούτων. ἀνάγκη δὲ τοιάνδε τὴν ὕλην ὑπ-
άρξαι εἰ ἔσται οἰκία ἢ ἄλλο τι τέλος· καὶ γενέσθαι
τε καὶ κινηθῆναι δεῖ τόδε πρῶτον, εἶτα τόδε, καὶ
τοῦτον δὴ τὸν τρόπον ἐφεξῆς μέχρι τοῦ τέλους καὶ
30 οὗ ἕνεκα γίνεται ἕκαστον καὶ ἔστιν. ὡσαύτως δὲ
640 a καὶ ἐν τοῖς φύσει γινομένοις. ἀλλ᾽ ὁ τρόπος τῆς
ἀποδείξεως καὶ τῆς ἀνάγκης ἕτερος ἐπί τε τῆς
φυσικῆς καὶ τῶν θεωρητικῶν ἐπιστημῶν. (εἴρηται
δ᾽ ἐν ἑτέροις περὶ τούτων.) ἡ γὰρ ἀρχὴ τοῖς μὲν τὸ
ὄν, τοῖς δὲ τὸ ἐσόμενον· ἐπεὶ γὰρ τοιόνδ᾽ ἐστὶν ἡ
5 ὑγίεια ἢ ὁ ἄνθρωπος, ἀνάγκη τόδ᾽ εἶναι ἢ γενέσθαι,
ἀλλ᾽ οὐκ ἐπεὶ τόδ᾽ ἐστὶν ἢ γέγονεν, ἐκεῖνο ἐξ

---

a " Absolute," *i.e.* simple or unconditional necessity,
belongs to the " eternal things," such as the heavenly bodies
or the eternal truths of mathematics. For further details
see *De gen. et corr.* 337 b 14 ff.

b At *Met.* 1025 b ff. Aristotle makes a threefold classification of the sciences into (*a*) theoretical (contemplative),
(*b*) practical, (*c*) productive. The result of (*a*) is knowledge
only, of (*b*) knowledge and action, of (*c*) knowledge, action,
and some article or product. The three " theoretical "
sciences are theology (*i.e.* metaphysics), mathematics, and
physics (natural science). In the present passage, however,
Aristotle contrasts natural science with the " theoretical "
sciences. This is because he is considering Nature as a
craftsman whose craft or science belongs to the third class—
the " productive " sciences. Our study of Nature's science

philosophers endeavour to carry back their explanations to Necessity; but they omit to distinguish the various meanings of Necessity. There is "absolute" Necessity,[a] which belongs to the eternal things; and there is "conditional" Necessity, which has to do with everything that is formed by the processes of Nature, as well as with the products of Art, such as houses and so forth. If a house, or any other End, is to be realized, it is necessary that such and such material shall be available; one thing must first be formed, and set in motion, and then another thing; and so on continually in the same manner up to the End, which is the Final Cause, for the sake of which every one of those things is formed and for which it exists. The things which are formed in Nature are in like case. Howbeit, the method of reasoning in Natural science and also the mode of Necessity itself is not the same as in the Theoretical sciences. (I have spoken of this matter in another treatise.[b]) They differ in the following way.[c] In the Theoretical sciences, we begin with what already *is*; but in Natural science with what *is going to be*: thus, we say, *Because* that which is going to be—health, perhaps, or man—has a certain character, *therefore* of necessity some particular thing, P, must be, or must be formed; not, *Because* P is now, or has been formed, *therefore* the other thing (health, or man) of necessity is now

---

may be a "theoretical" science, but Nature's science itself is "productive."

[c] The reasoning process in a "theoretical" science, *e.g.* mathematics, begins, say, with A, and then deduces from it the consequences B, C, D. In a "productive" science, *e.g.* building, it begins with the house which is to be built, D, and works backwards through the preliminary stages which must be realized in order to produce the house, C, B, A. *Cf.* below, 640 a 16 ff.

640 a

ἀνάγκης ἐστὶν ἢ ἔσται. οὐδ' ἔστιν εἰς ἀΐδιον συν-
αρτῆσαι τῆς τοιαύτης ἀποδείξεως τὴν ἀνάγκην,
ὥστ' εἰπεῖν, ἐπεὶ τόδ' ἐστίν, ὅτι τόδ' ἐστίν. διώρισται δὲ καὶ περὶ τούτων ἐν ἑτέροις, καὶ ποίοις
ὑπάρχει καὶ ποῖα ἀντιστρέφει καὶ διὰ τίν' αἰτίαν.

10 Δεῖ δὲ μὴ λεληθέναι καὶ πότερον προσήκει λέγειν,
ὥσπερ οἱ πρότερον ἐποιοῦντο τὴν θεωρίαν, πῶς
ἕκαστον γίνεσθαι πέφυκε μᾶλλον ἢ πῶς ἔστιν.
οὐ γάρ τι μικρὸν διαφέρει τοῦτο ἐκείνου. ἔοικε
δ' ἐντεῦθεν ἀρκτέον εἶναι (καθάπερ καὶ πρότερον
εἴπομεν, ὅτι πρῶτον τὰ φαινόμενα ληπτέον περὶ
15 ἕκαστον γένος, εἶθ' οὕτω τὰς αἰτίας τούτων
λεκτέον) καὶ περὶ γενέσεως· μᾶλλον γὰρ τάδε
συμβαίνει καὶ περὶ τὴν οἰκοδόμησιν ἐπεὶ τοιόνδ'
ἐστὶ τὸ εἶδος τῆς οἰκίας, ἢ τοιόνδ' ἐστὶν ἡ οἰκία ὅτι
γίνεται οὕτως. ἡ γὰρ γένεσις ἕνεκα τῆς οὐσίας
ἐστίν, ἀλλ' οὐχ ἡ οὐσία ἕνεκα τῆς γενέσεως. διόπερ
20 Ἐμπεδοκλῆς οὐκ ὀρθῶς εἴρηκε λέγων ὑπάρχειν
πολλὰ τοῖς ζῴοις διὰ τὸ συμβῆναι οὕτως ἐν τῇ
γενέσει, οἷον καὶ τὴν ῥάχιν τοιαύτην ἔχειν ὅτι
στραφέντος καταχθῆναι συνέβη, ἀγνοῶν πρῶτον μὲν
ὅτι δεῖ τὸ σπέρμα τὸ συνιστὰν[1] ὑπάρχειν τοιαύτην

---

[1] συνιστὰν Platt: συστὰν vulg.

---

[a] Though of course this Necessity has its place in natural
science (see 642 a 31 ff.). It is, however, not the only sort
of Necessity in Natural science, and not the paramount one.

[b] See *De gen. et corr.* 337 b 25 ff. An example of a non-
convertible proposition is : Foundations are necessary for a
house to be built. You cannot say, " If foundations are laid
a house must of necessity be built," because it is not " absolutely " and always necessary that a house should be built.

[c] *Cf.* Plato, *Philebus* 54 A–C.

or will be in the future.[a] Nor, in a process of
reasoning of this kind, is it possible to trace
back the links of Necessity to eternity, so as to say,
Because A is, therefore *Z* is. I have, however, discussed these matters in another work,[b] and I there
stated where either kind of Necessity applies, which
propositions involving Necessity are convertible, and
the reasons why.

We must also decide whether we are to discuss
the processes by which each animal comes to be
formed—which is what the earlier philosophers
studied—or rather the animal as it actually is.
Obviously there is a considerable difference between
the two methods. I said earlier that we ought first
to take the phenomena that are observed in each
group, and then go on to state their causes. This
applies just as much to the subject of the process of
formation : here too we ought surely to begin with
things as they are actually observed to be when
completed. Even in building the fact is that the
particular stages of the process come about because
the Form of the house is such and such, rather than
that the house is such and such because the process
of its formation follows a particular course : the
process is for the sake of the actual thing, the thing
is not for the sake of the process.[c] So Empedocles
was wrong when he said that many of the characteristics which animals have are due to some accident
in the process of their formation, as when he
accounts for the vertebrae of the backbone by saying [d] " the fetus gets twisted and so the backbone
is broken into pieces " : he was unaware (*a*) that
the seed which gives rise to the animal must to

---

[d] Emped. frag. 97 (Diels, *Fragmente*[5], 31 B 97).

640 a

ἔχον δύναμιν, εἶθ' ὅτι τὸ ποιῆσαν πρότερον ὑπῆρχεν
25 οὐ μόνον τῷ λόγῳ ἀλλὰ καὶ τῷ χρόνῳ· γεννᾷ γὰρ ὁ
ἄνθρωπος ἄνθρωπον, ὥστε διὰ τὸ ἐκεῖνον τοιόνδ'
εἶναι ἡ γένεσις τοιάδε συμβαίνει τῳδί. [ὁμοίως
δὲ καὶ ἐπὶ τῶν αὐτομάτως δοκούντων γίνεσθαι
καθάπερ καὶ ἐπὶ τῶν τεχναστῶν· ἔνια γὰρ¹ καὶ ἀπὸ
ταὐτομάτου γίνεται ταὐτὰ τοῖς ἀπὸ τέχνης, οἷον
30 ὑγίεια. ὧν² μὲν οὖν προϋπάρχει τὸ ποιητικὸν
[ὅμοιον],³ οἷον ἡ⁴ ἀνδριαντοποιητική, οὐ [γὰρ]⁵ γίνεται αὐτόματον. ἡ δὲ τέχνη λόγος τοῦ ἔργου ὁ
ἄνευ τῆς ὕλης ἐστίν. καὶ τοῖς ἀπὸ τύχης ὁμοίως·
ὡς γὰρ ἡ τέχνη ἔχει, οὕτω γίνεται.]⁶ διὸ μάλιστα
μὲν λεκτέον ὡς ἐπειδὴ τοῦτ' ἦν τὸ ἀνθρώπῳ εἶναι,
35 διὰ τοῦτο ταῦτ' ἔχει· οὐ γὰρ ἐνδέχεται εἶναι ἄνευ
τῶν μορίων τούτων. εἰ δὲ μή, ὅ τι ἐγγύτατα
τούτου, καὶ ἢ ὅτι ὅλως ἀδύνατον ἄλλως,⁷ ἢ καλῶς
640 b γε οὕτως. ταῦτα δ' ἔπεται· ἐπεὶ δ' ἔστι τοιοῦτον,
τὴν γένεσιν ὡδὶ καὶ τοιαύτην συμβαίνειν ἀναγ-
καῖον· διὸ γίνεται πρῶτον τῶν μορίων τόδε, εἶτα
τόδε. καὶ τοῦτον δὴ τὸν τρόπον ὁμοίως ἐπὶ πάντων τῶν φύσει συνισταμένων.

5 Οἱ μὲν οὖν ἀρχαῖοι καὶ πρῶτοι φιλοσοφήσαντες

---

¹ ἔνια γὰρ om. Z¹.
² ὧν Z : τῶν vulg.
³ om. Z¹.
⁴ ἡ Z : om. vulg.
⁵ om. Z.
⁶ ὁμοίως (l. 27) . . . γίνεται, ex Met. 1032-1034 exorta.
olim ut vid. in marg. 640 b 4 adscripta ; inepta seclusi.
⁷ ὅτι ὅλως Z¹ : ὅλως ὅτι ἀ. ἀ. vulg.

ᵃ *i.e.* the same character as the animal which it is to produce. For *dynamis* see Introduction, pp. 30 ff.
ᵇ No doubt a marginal note appended to 640 b 4.

begin with have the appropriate specific character ᵃ ; and (*b*) that the producing agent was pre-existent : it was chronologically earlier as well as logically earlier : in other words, men are begotten by men, and therefore the process of the child's formation is what it is because its parent was a man. [Similarly too with those that appear to be formed spontaneously, just as with those produced by the arts ; for some that are formed spontaneously are identical with those produced by art, *e.g.* health. As for those things whose producing agent is pre-existent, *e.g.* the art of statuary, no spontaneous formation occurs. Art is the *logos* of the article without the matter. And similarly with the products of chance : they are formed by the same process that art would employ.] ᵇ So the best way of putting the matter would be to say that *because* the essence of man is what it is, *therefore* a man has such and such parts, since there cannot be a man without them. If we may not say this, then the nearest to it must do, viz. that there cannot be a man at all otherwise than with them, or, that it is well that a man should have them. And upon this these considerations follow : *Because* man is such and such, *therefore* the process of his formation must of necessity be such and such and take place in such a manner ; which is why first this part is formed, then that. And thus similarly with all the things that are constructed by Nature.

Now those who were the first to study Nature in

περὶ φύσεως περὶ τῆς ὑλικῆς ἀρχῆς καὶ τῆς τοιαύτης αἰτίας ἐσκόπουν, τίς καὶ ποία τις, καὶ πῶς
ἐκ ταύτης γίνεται τὸ ὅλον, καὶ τίνος κινοῦντος, οἷον
νείκους ἢ φιλίας ἢ νοῦ ἢ τοῦ αὐτομάτου, τῆς δ'
10 ὑποκειμένης ὕλης τοιάνδε τινὰ φύσιν ἐχούσης ἐξ
ἀνάγκης, οἷον τοῦ μὲν πυρὸς θερμήν, τῆς δὲ γῆς
ψυχράν, καὶ τοῦ μὲν κούφην, τῆς δὲ βαρεῖαν. οὕτως
γὰρ καὶ τὸν κόσμον γεννῶσιν. ὁμοίως δὲ καὶ περὶ
τὴν τῶν ζῴων καὶ τῶν φυτῶν γένεσιν λέγουσιν,
οἷον[1] ἐν τῷ σώματι ῥέοντος μὲν τοῦ ὕδατος κοιλίαν
γενέσθαι καὶ πᾶσαν ὑποδοχὴν τῆς τε τροφῆς καὶ τοῦ
15 περιττώματος, τοῦ δὲ πνεύματος διαπορευθέντος
τοὺς μυκτῆρας ἀναρραγῆναι. ὁ δ' ἀὴρ καὶ τὸ ὕδωρ
ὕλη τῶν σωμάτων ἐστίν· ἐκ τῶν τοιούτων γὰρ
σωμάτων συνιστᾶσι τὴν φύσιν πάντες. εἰ δ' ἔστιν
ὁ ἄνθρωπος καὶ τὰ ζῷα φύσει καὶ τὰ μόρια αὐτῶν,
λεκτέον ἂν περὶ σαρκὸς εἴη καὶ ὀστοῦ καὶ αἵματος
20 καὶ τῶν ὁμοιομερῶν ἁπάντων, ὁμοίως δὲ καὶ τῶν
ἀνομοιομερῶν, οἷον προσώπου, χειρός, ποδός, ᾗ
τε τοιοῦτον ἕκαστόν ἐστιν αὐτῶν καὶ κατὰ ποίαν
δύναμιν. οὐ γὰρ ἱκανὸν τὸ ἐκ τίνων ἐστίν, οἷον
πυρὸς ἢ γῆς, ὥσπερ κἂν εἰ περὶ κλίνης ἐλέγομεν ἢ
τινος ἄλλου τῶν τοιούτων, ἐπειρώμεθα μᾶλλον ἂν
25 διορίζειν τὸ εἶδος αὐτῆς ἢ τὴν ὕλην, οἷον τὸν χαλκὸν

---

1 ὅτι post οἷον vulg.: del. Ogle.

---

a As Empedocles and Anaxagoras, whose attempts to
discover the "material" and the "efficient" causes are
mentioned a few lines below. See also *Met.* 983 b 6 ff.
b "Material" cause: see Introduction, pp. 24 ff.
c "Residue": lit. "surplus"; see Introduction, pp. 32 ff.
d *Cf.* Hippocrates, Περὶ διαίτης, i. 9.
e "Parts": see Introduction, pp. 28 ff.

the early days [a] spent their time in trying to discover what the material principle or the material Cause [b] was, and what it was like; they tried to find out how the Universe is formed out of it; what set the process going (Strife, it might be, or Friendship, Mind, or Spontaneity); assuming throughout that the underlying material had, by necessity, some definite nature : *e.g.* that the nature of Fire was hot, and light; of Earth, cold, and heavy. At any rate, that is how they actually explain the formation of the world-order. In a like manner they describe the formation of animals and plants, saying (*e.g.*) that the stomach and every kind of receptacle for food and for residue [c] is formed by the water flowing in the body, and the nostril openings are forcibly made by the passage of the breath.[d] Air and water, of course, according to them, are the material of which the body is made : they all say that Nature is composed of substances of this sort. Yet if man and the animals and their parts [e] are products of Nature, then account must be taken of flesh, bone, blood, in fact of all the " uniform parts," [f] and indeed of the " non-uniform parts " too, viz. face, hand, foot ; and it must be explained how it comes to pass that each of these is characterized as it is, and by what force this is effected. It is not enough to state simply the substances out of which they are made, as " Out of fire," or " Out of earth." If we were describing a bed or any other like article, we should endeavour to describe the form of it rather than the matter (bronze, or wood)—or, at

---

[f] " Uniform " and " non-uniform " : see Introduction, pp. 28 ff. The distinction between " uniform " and " non-uniform " parts is, historically, the predecessor of the distinction between " tissues " and " organs."

640 b

ἢ τὸ ξύλον, εἰ δὲ μή, τήν γε τοῦ συνόλου· κλίνη γὰρ
τόδε ἐν τῷδε ἢ τόδε τοιόνδε, ὥστε κἂν περὶ τοῦ
σχήματος εἴη λεκτέον, καὶ ποῖον τὴν ἰδέαν· ἡ γὰρ
κατὰ τὴν μορφὴν φύσις κυριωτέρα τῆς ὑλικῆς
φύσεως.

30   Εἰ μὲν οὖν τῷ σχήματι καὶ τῷ χρώματι ἕκαστόν
ἐστι τῶν τε ζῴων καὶ τῶν μορίων, ὀρθῶς ἂν
Δημόκριτος λέγοι· φαίνεται γὰρ οὕτως ὑπολαβεῖν.
φησὶ γοῦν παντὶ δῆλον εἶναι οἷόν τι τὴν μορφήν
ἐστιν ὁ ἄνθρωπος, ὡς ὄντος αὐτοῦ τῷ τε σχήματι
καὶ τῷ χρώματι γνωρίμου. καίτοι καὶ ὁ τεθνεὼς
35 ἔχει τὴν αὐτὴν τοῦ σχήματος μορφήν, ἀλλ᾽ ὅμως
οὐκ ἔστιν ἄνθρωπος. ἔτι δ᾽ ἀδύνατον εἶναι χεῖρα
ὁπωσοῦν διακειμένην, οἷον χαλκῆν ἢ ξυλίνην, πλὴν
641 a ὁμωνύμως, ὥσπερ τὸν γεγραμμένον ἰατρόν. οὐ γὰρ
δυνήσεται ποιεῖν τὸ ἑαυτῆς ἔργον, ὥσπερ οὐδ᾽ αὐλοὶ
λίθινοι τὸ ἑαυτῶν ἔργον, οὐδ᾽ ὁ γεγραμμένος ἰατρός.
ὁμοίως δὲ τούτοις οὐδὲ τῶν τοῦ τεθνηκότος μο-
5 ρίων οὐδὲν ἔτι τῶν τοιούτων ἐστί, λέγω δ᾽ οἷον
ὀφθαλμός, χείρ. λίαν οὖν ἁπλῶς εἴρηται, καὶ τὸν
αὐτὸν τρόπον ὥσπερ ἂν εἰ τέκτων λέγοι περὶ χειρὸς
ξυλίνης. οὕτως γὰρ καὶ οἱ φυσιολόγοι τὰς γενέσεις
καὶ τὰς αἰτίας τοῦ σχήματος λέγουσιν. ὑπὸ τίνων
γὰρ ἐδημιουργήθησαν δυνάμεων; ἀλλ᾽ ἴσως ὁ μὲν
10 τέκτων ἐρεῖ πέλεκυν ἢ τρύπανον, ὁ δ᾽ ἀέρα καὶ γῆν,

---

a See Diels, *Fragmente* [b], 68 B 165.
b *i.e.* the early writers on " Nature."

any rate, the matter, if described, would be described as belonging to the concrete whole. For example, " a bed " is a certain form in certain matter, or, alternatively, certain matter that has a certain form ; so we should have to include its shape and the manner of its form in our description of it—because the " formal " nature is of more fundamental importance than the " material " nature.

If, then, each animal and each of its parts is what it is in virtue of its shape and its colour, what Democritus says will be correct, since that was apparently his view, if one understands him aright when he says that it is evident to everyone what " man " is like as touching his shape, for it is by his shape and his colour that a man may be told.[a] Now a corpse has the same shape and fashion as a living body ; and yet it is not a man. Again, a hand constituted in any and every manner, e.g., a bronze or wooden one, is not a hand except in name ; and the same applies to a physician depicted on canvas, or a flute carved in stone. None of these can perform the functions appropriate to the things that bear those names. Likewise, the eye or the hand (or any other part) of a corpse is not really an eye or a hand. Democritus's statement, therefore, needs to be qualified, or a carpenter might as well claim that a hand made of wood really was a hand. The physiologers,[b] however, when they describe the formation and the causes of the shape of animal bodies, talk in this selfsame vein. Suppose we ask the carver " By what agency was this hand fashioned ? " Perhaps his answer will be " By my axe " or " By my auger," just as if we ask the physiologer " By what agency was this body fashioned ? " he will say " By air " and

641 a

πλὴν βέλτιον ὁ τέκτων· οὐ γὰρ ἱκανὸν ἔσται αὐτῷ
τὸ τοσοῦτον εἰπεῖν, ὅτι ἐμπεσόντος τοῦ ὀργάνου
τὸ μὲν κοῖλον ἐγένετο τὸ δὲ ἐπίπεδον, ἀλλὰ διότι
τὴν πληγὴν ἐποιήσατο τοιαύτην, καὶ τίνος ἕνεκα,
ἐρεῖ τὴν αἰτίαν, ὅπως τοιόνδε ἢ τοιόνδε ποτὲ τὴν
μορφὴν γένηται.

15 Δῆλον τοίνυν ὅτι οὐκ ὀρθῶς λέγουσι, καὶ ὅτι
λεκτέον ὡς τοιοῦτον τὸ ζῷον, καὶ περὶ ἐκείνου καὶ
τί καὶ ποῖόν τι καὶ τῶν μορίων ἑκάστου,[1] ὥσπερ
καὶ περὶ τοῦ εἴδους τῆς κλίνης.

Εἰ δὴ τοῦτό ἐστι ψυχὴ ἢ ψυχῆς μέρος ἢ μὴ ἄνευ
ψυχῆς (ἀπελθούσης γοῦν οὐκέτι ζῷόν ἐστιν, οὐδὲ
20 τῶν μορίων οὐδὲν τὸ αὐτὸ λείπεται, πλὴν τῷ
σχήματι μόνον, καθάπερ τὰ μυθευόμενα λιθοῦσθαι),
εἰ δὴ ταῦτα οὕτως, τοῦ φυσικοῦ περὶ ψυχῆς ἂν εἴη
λέγειν καὶ εἰδέναι, καὶ εἰ μὴ πάσης, κατ᾽ αὐτὸ
τοῦτο καθ᾽ ὃ τοιοῦτο τὸ ζῷον, καὶ τί ἐστιν ἡ ψυχή,
ἢ αὐτὸ τοῦτο τὸ μόριον, καὶ περὶ τῶν συμβεβη-
25 κότων κατὰ τὴν τοιαύτην αὐτῆς οὐσίαν, ἄλλως τε
καὶ τῆς φύσεως διχῶς λεγομένης καὶ οὔσης, τῆς
μὲν ὡς ὕλης, τῆς δ᾽ ὡς οὐσίας· καὶ ἔστιν αὕτη καὶ
ὡς ἡ κινοῦσα καὶ ὡς τὸ τέλος· τοιοῦτον δὲ τοῦ ζῴου

---

[1] ἑκάστου Peck : ἕκαστον vulg.

---

[a] Or, " reason " ; see Introduction, p. 24.
[b] See above, 640 b 26.
[c] " Soul " : see Introduction, pp. 34 ff.
[d] Or " motive."

" By earth." But of the two the craftsman will give a better answer, because he will not feel it is sufficient to say merely that a cavity was created here, or a level surface there, by a blow from his tool. He will state the cause[a] on account of which, and the purpose for the sake of which, he made the strokes he did ; and that will be, in order that the wood might finally be formed into this or that shape.

It must now be evident that the statements of the physiologers are unsatisfactory. We have to state how the animal is characterized, *i.e.*, what is the essence and character of the animal itself, as well as describing each of its parts ; just as with the bed we have to state its Form.[b]

Now it may be that the Form of any living creature is Soul,[c] or some part of Soul, or something that involves Soul. At any rate, when its Soul is gone, it is no longer a living creature, and none of its parts remains the same, except only in shape, just like the animals in the story that were turned into stone. If, then, this is really so, it is the business of the student of Natural science to inform himself concerning Soul, and to treat of it in his exposition ; not, perhaps, in its entirety, but of that special part of it which causes the living creature to be such as it is. He must say what Soul, or that special part of Soul, is ; and when he has said what its essence is, he must treat of the attributes which are attached to an essence of that character. This is especially necessary, because the term " nature " is used— rightly—in two senses : (*a*) meaning " matter," and (*b*) meaning " essence " (the latter including both the " Efficient "[d] Cause and the " End "). It is, of course, in this latter sense that the entire Soul or

641 a

ἤτοι πᾶσα ἡ ψυχὴ ἢ μέρος τι αὐτῆς. ὥστε καὶ
οὕτως ἂν λεκτέον εἴη τῷ περὶ φύσεως θεωρητικῷ
30 περὶ ψυχῆς μᾶλλον ἢ περὶ τῆς ὕλης, ὅσῳ μᾶλλον ἡ
ὕλη δι᾽ ἐκείνην φύσις ἐστὶν ἢ ἀνάπαλιν· καὶ γὰρ
κλίνη καὶ τρίπους τὸ ξύλον ἐστίν, ὅτι δυνάμει ταῦτά
ἐστιν.

Ἀπορήσειε δ᾽ ἄν τις εἰς τὸ νῦν λεχθὲν ἐπιβλέψας,
πότερον περὶ πάσης ψυχῆς τῆς φυσικῆς ἐστὶ τὸ
35 εἰπεῖν ἢ περί τινος.[1] εἰ γὰρ περὶ πάσης, οὐδεμία
λείπεται παρὰ τὴν φυσικὴν ἐπιστήμην φιλοσοφία.
641 b ὁ γὰρ νοῦς τῶν νοητῶν, ὥστε περὶ πάντων ἡ
φυσικὴ γνῶσις ἂν εἴη· τῆς γὰρ αὐτῆς περὶ νοῦ καὶ
τοῦ νοητοῦ θεωρῆσαι, εἴπερ πρὸς ἄλληλα, καὶ ἡ
αὐτὴ θεωρία τῶν πρὸς ἄλληλα πάντων, καθάπερ
καὶ περὶ αἰσθήσεως καὶ τῶν αἰσθητῶν. ἢ οὐκ ἔστι
5 πᾶσα ἡ ψυχὴ κινήσεως ἀρχή, οὐδὲ τὰ μόρια ἅπαντα,
ἀλλ᾽ αὐξήσεως μὲν ὅπερ καὶ ἐν τοῖς φυτοῖς, ἀλλοιώσεως δὲ τὸ αἰσθητικόν, φορᾶς δ᾽ ἕτερόν τι καὶ
οὐ τὸ νοητικόν· ὑπάρχει γὰρ ἡ φορὰ καὶ ἐν ἑτέροις
τῶν ζῴων, διάνοια δ᾽ οὐδενί. δῆλον οὖν ὡς οὐ

---

[1] τινος ⟨μορίου⟩ Rackham.

---

[a] i.e. qualitative change, which is the " motion " proper
to this part of the Soul.

some part of it is the " nature " of a living creature.
Hence on this score especially it should be the duty
of the student of Natural science to deal with Soul
in preference to matter, inasmuch as it is the Soul
that enables the matter to " be the nature " of an
animal (that is, *potentially*, in the same way as a piece
of wood " is " a bed or a stool) rather than the matter
which enables the Soul to do so.

In view of what we have just said, one may well ask
whether it is the business of Natural science to treat
of Soul in its entirety or of some part of it only ;
since if it must treat of Soul in its entirety (*i.e.*
including intellect) there will be no room left
for any other study beside Natural science—it will
include even the objects that the intellect appre-
hends. For consider : wherever there is a pair
of interrelated things, such as sensation and the
objects of sensation, it is the business of one
science, and one only, to study them both. Now
intellect and the objects of the intellect are
such a pair ; hence, the same science will study
both of them, which means that there will be
nothing whatever left outside the purview of
Natural science. All the same, it may be that
it is neither Soul in its entirety that is the
source of motion, nor yet all its parts taken
together ; it may be that one part of Soul, (*a*), viz.
that which plants have, is the source of growth ;
another part, (*b*), the " sensory " part, is the source
of change [a] ; and yet another part, (*c*), the source
of locomotion. That even this last cannot be the
intellectual part is proved, because animals other
than man have the power of locomotion, although
none of them has intellect. I take it, then, as evident

περὶ πάσης ψυχῆς λεκτέον· οὐδὲ γὰρ πᾶσα ψυχὴ
10 φύσις, ἀλλά τι μόριον αὐτῆς ἓν ἢ καὶ πλείω.

Ἔτι δὲ τῶν ἐξ ἀφαιρέσεως οὐδενὸς οἷόν τ᾽ εἶναι
τὴν φυσικὴν θεωρητικήν, ἐπειδὴ ἡ φύσις ἕνεκά του
ποιεῖ πάντα· φαίνεται γάρ, ὥσπερ ἐν τοῖς τεχναστοῖς ἐστιν ἡ τέχνη, οὕτως ἐν αὐτοῖς τοῖς πράγ-
μασιν ἄλλη τις ἀρχὴ καὶ αἰτία τοιαύτη, ἣν ἔχομεν
15 καθάπερ τὸ θερμὸν καὶ τὸ ψυχρὸν ἐκ τοῦ παντός.
διὸ μᾶλλον εἰκὸς τὸν οὐρανὸν γεγενῆσθαι ὑπὸ
τοιαύτης αἰτίας, εἰ γέγονε, καὶ εἶναι διὰ τοιαύτην
αἰτίαν μᾶλλον ἢ τὰ ζῷα τὰ θνητά· τὸ γοῦν τεταγ-
μένον καὶ τὸ ὡρισμένον πολὺ μᾶλλον φαίνεται ἐν
20 τοῖς οὐρανίοις ἢ περὶ ἡμᾶς, τὸ δ᾽ ἄλλοτ᾽ ἄλλως καὶ
ὡς ἔτυχε περὶ τὰ θνητὰ μᾶλλον. οἱ δὲ τῶν μὲν
ζῴων ἕκαστον φύσει φασὶν εἶναι καὶ γενέσθαι, τὸν
δ᾽ οὐρανὸν ἀπὸ τύχης καὶ τοῦ αὐτομάτου τοιοῦτον
συστῆναι, ἐν ᾧ ἀπὸ τύχης καὶ ἀταξίας οὐδ᾽ ὁτιοῦν
φαίνεται. πανταχοῦ δὲ λέγομεν τόδε τοῦδ᾽ ἕνεκα,
25 ὅπου ἂν φαίνηται τέλος τι πρὸς ὃ ἡ κίνησις περαίνει
μηδενὸς ἐμποδίζοντος. ὥστε εἶναι φανερὸν ὅτι ἔστι
τι τοιοῦτον, ὃ δὴ καὶ καλοῦμεν φύσιν· οὐ γὰρ δὴ
ὅ τι ἔτυχεν ἐξ ἑκάστου γίνεται σπέρματος, ἀλλὰ
τόδε ἐκ τοῦδε, οὐδὲ σπέρμα τὸ τυχὸν ἐκ τοῦ τυ-

---

[a] With this passage *cf.* Plato, *Philebus* 29-30.
[b] *Cf.* Samuel Butler, *Life and Habit*, p. 134, " A hen is only an egg's way of making another egg."

that we need not concern ourselves with Soul in its entirety ; because it is not Soul in its entirety that is an animal's " nature," but some part or parts of it.

Further, no abstraction can be studied by Natural science, because whatever Nature makes she makes to serve some purpose ; for it is evident that, even as art is present in the objects produced by art, so in things themselves there is some principle or cause of a like sort, which came to us from the universe around us, just as our material constituents (the hot, the cold, etc.) did.[a] Wherefore there is better reason for holding that the Heaven was brought into being by some such cause—if we may assume that it came into being at all—and that through that cause it continues to be, than for holding the same about the mortal things it contains—the animals ; at any rate, there is much clearer evidence of definite ordering in the heavenly bodies than there is in us ; for what is mortal bears the marks of change and chance. Nevertheless, there are those who affirm that, while every living creature has been brought into being by Nature and remains in being thereby, the heaven in all its glory was constructed by mere chance and came to be spontaneously, although there is no evidence of chance or disorder in it. And whenever there is evidently an End towards which a motion goes forward unless something stands in its way, then we always assert that the motion has the End for its purpose. From this it is evident that something of the kind really exists—that, in fact, which we call " Nature," because in fact we do not find any chance creature being formed from a particular seed, but A comes from a, and B from b ; nor does any chance seed come from any chance individual.[b] Therefore

641 b

χόντος σώματος. ἀρχὴ ἄρα καὶ ποιητικὸν τοῦ ἐξ
30 αὐτοῦ τὸ ⟨ἐξ οὗ τὸ⟩[1] σπέρμα. φύσει γὰρ ταῦτα·
φύεται γοῦν ἐκ τούτου. ἀλλὰ μὴν ἔτι τούτου
πρότερον τὸ οὗ τὸ σπέρμα· γένεσις μὲν γὰρ τὸ
σπέρμα, οὐσία δὲ τὸ τέλος. ἀμφοῖν δ' ἔτι πρότερον, ἀφ' οὗ ἐστὶ τὸ σπέρμα. ἔστι γὰρ τὸ
σπέρμα διχῶς, ἐξ οὗ τε καὶ οὗ· καὶ γὰρ ἀφ' οὗ
35 ἀπῆλθε, τούτου σπέρμα, οἷον ἵππου, καὶ τούτου
ὃ ἔσται ἐξ αὐτοῦ, οἷον ὀρέως, τρόπον δ' οὐ τὸν
αὐτόν, ἀλλ' ἑκατέρου τὸν εἰρημένον. ἔτι δὲ δυνάμει
642 a τὸ σπέρμα· δύναμις δ' ὡς ἔχει πρὸς ἐντελέχειαν
ἴσμεν.

Εἰσὶν ἄρα δύ' αἰτίαι αὗται, τό θ' οὗ ἕνεκα καὶ
τὸ ἐξ ἀνάγκης· πολλὰ γὰρ γίνεται, ὅτι ἀνάγκη.
ἴσως δ' ἄν τις ἀπορήσειε ποίαν λέγουσιν ἀνάγκην
5 οἱ λέγοντες ἐξ ἀνάγκης· τῶν μὲν γὰρ δύο τρόπων
οὐδέτερον οἷόν θ' ὑπάρχειν τῶν διωρισμένων ἐν τοῖς
κατὰ φιλοσοφίαν. ἔστι δ' ἔν γε τοῖς ἔχουσι γένεσιν
ἡ τρίτη· λέγομεν γὰρ τὴν τροφὴν ἀναγκαῖόν τι κατ'
οὐδέτερον τούτων τῶν τρόπων, ἀλλ' ὅτι οὐχ οἷόν τ'
ἄνευ ταύτης εἶναι. τοῦτο δ' ἐστὶν ὥσπερ ἐξ ὑπο-
10 θέσεως· ὥσπερ γὰρ ἐπεὶ δεῖ σχίζειν τῷ πελέκει,
ἀνάγκη σκληρὸν εἶναι, εἰ δὲ σκληρόν, χαλκοῦν ἢ

---

[1] ⟨ἐξ οὗ τὸ⟩ supplevi, Σ secutus.

---

[a] There is a reference here, which is not apparent in the
English version, to the etymological connexion between φύσις
(nature) and φύεσθαι (to grow). *Cf. Met.* 1014 b 16 ff.
[b] Viz. actuality is prior to potentiality.
[c] These treatises are referred to again in the *Politics*
(1282 b 19) and in the *Eudemian Ethics* (1217 b 23). The
two modes of necessity seem to be (1) " absolute " necessity
(mentioned here), and (2) " coercive " necessity (see *Met.*

the individual from which the seed comes is the source and the efficient agent of that which comes out of the seed. The reason is, that these things are so arranged by Nature; at any rate, the offspring *grows*[a] out of the seed. Nevertheless, logically prior to the seed stands that of which it is the seed, because the End is an actual thing, and the seed is but a formative process. But further, prior to both of them stands the creature out of which the seed comes. (Note that a seed is the seed " of " something in two senses—two quite distinct senses: it is the seed " of " that out of which it came—*e.g.* a horse—as well as " of " that which will arise out of itself—*e.g.* a mule). Again, the seed is something *by potentiality*, and we know what is the relation of potentiality to actuality.[b]

We have, then, these two causes before us, to wit, the " Final " cause, and also Necessity, for many things come into being owing to Necessity. Perhaps one might ask which " Necessity " is meant when it is specified as a cause, since here it can be neither of the two modes which are defined in the treatises written in the philosophical manner.[c] There is, however, a third mode of Necessity: it is seen in the things that pass through a process of formation; as when we say that nourishment is necessary, we mean " necessary " in neither of the former two modes, but we mean that without nourishment no animal can be. This is, practically, " conditional " Necessity. Take an illustration: A hatchet, in order to split wood, must, of necessity, be hard; if so, then it must, of necessity, be made of

1015 a 20 ff.). The third he has referred to already at 639 b 25, viz. " conditional " necessity. See pp. 21 f.

642 a

σιδηροῦν, οὕτω καὶ ἐπεὶ τὸ σῶμα ὄργανον (ἕνεκά
τινος γὰρ ἕκαστον τῶν μορίων, ὁμοίως δὲ καὶ τὸ
ὅλον), ἀνάγκη ἄρα τοιονδὶ εἶναι καὶ ἐκ τοιωνδί, εἰ
ἐκεῖνο ἔσται.

15 ῞Οτι μὲν οὖν δύο τρόποι τῆς αἰτίας, καὶ δεῖ
λέγοντας τυγχάνειν μάλιστα μὲν ἀμφοῖν, εἰ δὲ μή,
πειρᾶσθαί γε ποιεῖν τοῦτο, δῆλον,[1] καὶ ὅτι πάντες οἱ
τοῦτο μὴ λέγοντες οὐδὲν ὡς εἰπεῖν περὶ φύσεως
λέγουσιν· ἀρχὴ γὰρ ἡ φύσις μᾶλλον τῆς ὕλης.
(ἐνιαχοῦ δέ που αὐτῇ καὶ Ἐμπεδοκλῆς περιπίπτει,
ἀγόμενος ὑπ’ αὐτῆς τῆς ἀληθείας, καὶ τὴν οὐσίαν καὶ
20 τὴν φύσιν ἀναγκάζεται φάναι τὸν λόγον εἶναι, οἷον
ὀστοῦν ἀποδιδοὺς τί ἐστιν· οὔτε γὰρ ἕν τι τῶν
στοιχείων λέγει αὐτὸ οὔτε δύο ἢ τρία οὔτε πάντα,
ἀλλὰ λόγον τῆς μίξεως αὐτῶν. δῆλον τοίνυν ὅτι
καὶ ἡ σὰρξ τὸν αὐτὸν τρόπον ἐστί, καὶ τῶν ἄλλων
τῶν τοιούτων μορίων ἕκαστον. αἴτιον δὲ τοῦ μὴ
25 ἐλθεῖν τοὺς προγενεστέρους ἐπὶ τὸν τρόπον τοῦτον,
ὅτι τὸ τί ἦν εἶναι καὶ τὸ ὁρίσασθαι τὴν οὐσίαν οὐκ
ἦν, ἀλλ’ ἥψατο μὲν Δημόκριτος πρῶτος, ὡς οὐκ
ἀναγκαίου δὲ τῇ φυσικῇ θεωρίᾳ, ἀλλ’ ἐκφερόμενος
ὑπ’ αὐτοῦ τοῦ πράγματος· ἐπὶ Σωκράτους δὲ τοῦτο
μὲν ηὐξήθη, τὸ δὲ ζητεῖν τὰ περὶ φύσεως ἔληξε,

---

[1] sic Ogle: εἰ δὲ μή, δῆλόν γε πειρᾶσθαι ποιεῖν vulg.

---

[a] See Diels, *Fragmente*[5], 31 A 78.

[b] "Element": this term is normally used to denote the
four substances, earth, water, air, fire.

[c] This is probably a reference to Democritus's opposition
to the theories of Protagoras, who held that "what *appears*

76

bronze or of iron. Now the body, like the hatchet, is an instrument; as well the whole body as each of its parts has a purpose, for the sake of which it is; the body must therefore, of necessity, be such and such, and made of such and such materials, if that purpose is to be realized.

It is, therefore, evident that of Causation there are two modes; and that in our treatise both of them must be described, or at least an attempt must be made to describe them; and that those who fail herein tell us practically nothing of any value about " Nature," for a thing's " nature " is much more a first principle (or " Cause ") than it is matter. (Indeed, in some places even Empedocles, being led and guided by Truth herself, stumbles upon this, and is forced to assert that it is the *logos* which is a thing's essence or nature.[a] For instance, when he is explaining what Bone is, he says not that it is any one of the Elements,[b] or any two, or three, or even all of them, but that it is " the *logos* of the mixture " of the Elements. And it is clear that he would explain in the same way what Flesh and each of such parts is. Now the reason why earlier thinkers did not arrive at this method of procedure was that in their time there was no notion of " essence " and no way of defining " being." The first to touch upon it was Democritus; and he did so, not because he thought it necessary for the study of Nature, but because he was carried away by the subject in hand and could not avoid it.[c] In Socrates' time an advance was made so far as the method was concerned; but at that time philosophers gave up the study of Nature

to be to you, *is* for you." Protagoras had emphasized the validity of sense-data; Democritus denied it.

642 a

30 πρὸς δὲ τὴν χρήσιμον ἀρετὴν καὶ τὴν πολιτικὴν
ἀπέκλιναν οἱ φιλοσοφοῦντες.)

Δεικτέον δ' οὕτως, οἷον ὅτι ἔστι μὲν ἡ ἀναπνοὴ
τουδὶ χάριν, τοῦτο δὲ γίνεται διὰ τάδε ἐξ ἀνάγκης.
ἡ δ' ἀνάγκη ὁτὲ μὲν σημαίνει ὅτι εἰ ἐκεῖνο ἔσται
τὸ οὗ ἕνεκα, ταῦτα ἀνάγκη ἐστὶν ⟨οὕτως⟩[1] ἔχειν,
35 ὁτὲ δ' ὅτι ἔστιν οὕτως ἔχοντα καὶ πεφυκότα· τὸ
θερμὸν γὰρ ἀναγκαῖον ἐξιέναι καὶ πάλιν εἰσιέναι
ἀντικροῦον, τὸν δ' ἀέρα εἰσρεῖν· τοῦτο δ' ἤδη
642 b ἀναγκαῖόν ἐστιν, τοῦ ἐντὸς δὲ θερμοῦ ἀντικόπτοντος
ἐν τῇ ψύξει τοῦ θύραθεν ἀέρος ἡ εἴσοδος[2] καὶ ἡ
ἔξοδος. ὁ μὲν οὖν τρόπος οὗτος ὁ τῆς μεθόδου,
καὶ περὶ ὧν δεῖ λαβεῖν τὰς αἰτίας, ταῦτα καὶ
τοιαῦτά ἐστιν.

5 II. Λαμβάνουσι δ' ἔνιοι τὸ καθ' ἕκαστον, δι-
αιρούμενοι τὸ γένος εἰς δύο διαφοράς. τοῦτο δ' ἐστὶ
τῇ μὲν οὐ ῥᾴδιον, τῇ δὲ ἀδύνατον. ἐνίων γὰρ ἔσται

---

[1] οὕτως supplevi.
[2] ἡ εἴσοδος om. pr. E.

---

[a] " Goodness," or " virtue," is one of the chief topics
discussed by Socrates in the Platonic dialogues. *Cf.*
Aristotle, *Met.* 987 b 1, " Socrates busied himself about moral
matters, but did not concern himself at all with Nature as
a whole."

[b] I have not attempted, except by one insertion, to straighten out the text of this confused account, which looks
like a displaced note intended for the paragraph above
(ending " realized," p. 77). If it is to remain in the text, it
would follow at that place (after 642 a 13) least awkwardly.
For a more lucid account of the process of Respiration see
De resp. 480 a 16–b 5.

[c] This is usually held to include Plato, on the ground that

78

and turned to the practical subject of " goodness," [a] and to political science.)

[b] Here is an example of the method of exposition. We point out that although Respiration takes place for such and such a *purpose*, any one stage of the process follows upon the others *by necessity*. Necessity means sometimes (a) that if this or that is to be the final Cause and purpose, then such and such things must be so ; but sometimes it means (b) that things are as they are owing to their very nature, as the following shows : It is necessary that the hot substance should go out and come in again as it offers resistance, and that the air should flow in—that is obviously necessary. And the hot substance within, as the cooling is produced, offers resistance, and this brings about the entrance of the air from without and also its exit. This example shows how the method works and also illustrates the sort of things whose causes we have to discover.

II. Now some writers [c] endeavour to arrive at the ultimate and particular species by the process of dividing the group (genus) into two *differentiae*.[d] This is a method which is in some respects difficult and in other respects impossible. For example :

the method of dichotomy is used in the *Sophist* and *Politicus*. But the method can hardly be said to be seriously applied to the classification of animals in the *Politicus*, and in the *Sophist* it is introduced partly in a humorous way, partly to lead up to the explanation of τὸ μὴ ὄν (not-being). Either Aristotle has mistaken the purpose of the method (as he has at *An. Pr.* 46 a 31 ff.) or (much more probably) he is referring to some other writer's detailed application of it. See *e.g.* Stenzel in *Pauly-Wissowa, s.v.* Speusippus.

[d] Each stage of the division gives two *differentiae*, which are treated as " genera " for the next stage of the division, and so on.

642 b

διαφορὰ μία μόνη, τὰ δ' ἄλλα περίεργα, οἷον ὑπόπουν, δίπουν, σχιζόπουν[1]· αὕτη γὰρ μόνη κυρία.
10 εἰ δὲ μή, ταὐτὸν πολλάκις ἀναγκαῖον λέγειν. ἔτι
δὲ προσήκει μὴ διασπᾶν ἕκαστον γένος, οἷον τοὺς
ὄρνιθας τοὺς μὲν ἐν τῆδε τοὺς δ' ἐν ἄλλῃ διαιρέσει,
καθάπερ ἔχουσιν αἱ γεγραμμέναι διαιρέσεις· ἐκεῖ
γὰρ τοὺς μὲν μετὰ τῶν ἐνύδρων συμβαίνει διῃρῆσθαι, τοὺς δ' ἐν ἄλλῳ γένει. (ταύτῃ μὲν οὖν τῇ
15 ὁμοιότητι ὄρνις ὄνομα κεῖται, ἑτέρᾳ δ' ἰχθύς· ἄλλαι
δ' εἰσὶν ἀνώνυμοι, οἷον τὸ ἔναιμον καὶ τὸ ἄναιμον·
ἐφ' ἑκατέρῳ γὰρ τούτων οὐ κεῖται ἓν ὄνομα.) εἴπερ
οὖν μηδὲν τῶν ὁμογενῶν διασπαστέον, ἡ εἰς δύο
διαίρεσις μάταιος ἂν εἴη· οὕτω γὰρ διαιροῦντας
ἀναγκαῖον χωρίζειν καὶ διασπᾶν· τῶν πολυπόδων
20 γάρ ἐστι τὰ μὲν ἐν τοῖς πεζοῖς τὰ δ' ἐν τοῖς
ἐνύδροις.

III. Ἔτι στερήσει μὲν ἀναγκαῖον διαιρεῖν καὶ
διαιροῦσιν οἱ διχοτομοῦντες. οὐκ ἔστι δὲ διαφορὰ

---

[1] ἄπουν post σχιζόπουν vulg., del. Ogle; fortasse [ἄπτερον]
scribendum (cf. An. Post. 92 a 1, Met. 1037 b 34).

---

[a] Other groups will get broken up under several lines of
division, as Aristotle goes on to say, and he repeats this at
643 b 14, where he adds that " contrary " groups will get
lumped together under a single line (and " contrariety is
maximum ' difference,' " see Met. 1055 a 5 ff., cf. 1018 a 30).
[b] Aristotle holds that one is not enough; see 643 b 9 ff.
and 29 ff.

80

(a) Some [a] groups will get only one *differentia*,[b] the rest of the terms being superfluous extras,[c] as in the example : footed, two-footed, cloven-footed [d]—since this last one is the only independently valid *differentia*. Otherwise the same thing [e] must of necessity be repeated many times over.

(b) Again, it is a mistake to break up a group, as for instance the group Birds, by putting some birds in one division and some in another, as has been done in the divisions made by certain writers : in these some birds are put in with the water-creatures, and others in another class. (These two groups, each possessing its own set of characteristics, happen to have regular names—Birds, Fishes—but there are other groups which have not, *e.g.* the " blooded " and " bloodless " groups : there is no one regular name for either of these.) If, then, it is a mistake to break up any group of kindred creatures, the method of division into two will be pointless, because those who so divide are compelled to separate them and break them up, some of the many-footed animals being among the land-animals and others among the water-animals.

III. (c) Again, this method of twofold division makes it necessary to introduce privative terms, and those who adopt it actually do this. But a privation, as

---

[e] *i.e.* all terms except the final one can be dispensed with, because none of them constitutes an independent (κυρία) *differentia* ; one line of division yields one valid *differentia* and no more (*cf.* 644 a 2-10).

[d] *Cf.* 644 a 5 and *Met.* 1038 a 32.

[e] In this case, " -footed " (*cf. Met.* 1038 a 19 ff.). But Aristotle does not explain how δίπουν is " superfluous."

642 b

στερήσεως ἢ στέρησις· ἀδύνατον γὰρ εἴδη εἶναι τοῦ
μὴ ὄντος, οἷον τῆς ἀποδίας ἢ τοῦ ἀπτέρου ὥσπερ
πτερώσεως καὶ ποδῶν· δεῖ δὲ τῆς καθόλου δια-
25 φορᾶς εἴδη εἶναι· εἰ γὰρ μὴ ἔσται, διὰ τί ἂν εἴη
τῶν καθόλου καὶ οὐ τῶν καθ' ἕκαστον; τῶν δὲ
διαφορῶν αἱ μὲν καθόλου εἰσὶ καὶ ἔχουσιν εἴδη,
οἷον πτερότης· τὸ μὲν γὰρ ἄσχιστον τὸ δ' ἐσχι-
σμένον ἐστὶ πτερόν. καὶ ποδότης ὡσαύτως ἡ μὲν
πολυσχιδής, ἡ δὲ δισχιδής, οἷον τὰ δίχαλα, ἡ δ'
30 ἀσχιδὴς καὶ ἀδιαίρετος, οἷον τὰ μώνυχα. χαλεπὸν
μὲν οὖν διαλαβεῖν καὶ εἰς τοιαύτας διαφορὰς ὧν
ἔστιν εἴδη, ὥσθ' ὁτιοῦν ζῷον ἐν ταύταις ὑπάρχειν
καὶ μὴ ἐν πλείοσι ταὐτόν (οἷον πτερωτὸν καὶ
ἄπτερον· ἔστι γὰρ ἄμφω ταὐτόν, οἷον μύρμηξ καὶ
35 λαμπυρὶς καὶ ἕτερά τινα), πάντων δὲ χαλεπώτατον
ἢ ἀδύνατον εἰς τὰς ἀντικειμένας.[1] ἀναγκαῖον γὰρ
τῶν καθ' ἕκαστον ὑπάρχειν τινὶ τῶν διαφορῶν
643 a ἑκάστην, ὥστε καὶ τὴν ἀντικειμένην. εἰ δὲ μὴ
ἐνδέχεται τοῖς εἴδει διαφέρουσιν ὑπάρχειν εἶδός τι
τῆς οὐσίας ἄτομον καὶ ἕν, ἀλλ' ἀεὶ διαφορὰν ἕξει
[2](οἷον ὄρνις ἀνθρώπου—ἡ διποδία γὰρ ἄλλη καὶ
διάφορος· κἂν εἰ ἔναιμα, τὸ αἷμα διάφορον, ἢ οὐδὲν
5 τῆς οὐσίας τὸ αἷμα θετέον)—εἰ δ' οὕτως ἐστίν, ἡ

---

[1] τὰς ἀντικειμένας Peck: τὰ ἀντικείμενα Titze: τὰ ἄναιμα
vulg.: τὰ ἐναντία Ogle: τὰ ἄτομα Prantl.
[2] ll. 3-6 interpunctionem correxi.

---

[a] I have not attempted to keep a consistent translation for
πτερόν, as Aristotle applies this term to "feathers" and to
"wings" (of insects).

privation, can admit no differentiation; there cannot be species of what is not there at all, *e.g.* of " footless " or " featherless," *a* as there can be of " footed " and " feathered "; and a generic *differentia* must contain species, else it is specific not generic. However, some of the *differentiae* are truly generic and contain species, for instance " feathered " (some feathers are barbed, some unbarbed); and likewise " footed " (some feet are " many-cloven," some " twy-cloven," as in the animals with bifid hoofs, and some " uncloven " or " undivided," as in the animals with solid hoofs). Now it is difficult enough to arrange the various animals under such lines of differentiation as these, which after all do contain species, in such a way that every animal is included in them, but not the same animal in more than one of them (*e.g.* when an animal is both winged and wingless, as ants, glow-worms, and some other creatures are); but it is excessively difficult and in fact impossible to arrange them under the opposite lines of differentiation. Every *differentia* must, of course, belong to some species; and this statement will apply to the negative *differentiae* as well as to the positive. Now it is impossible for any essential characteristic to belong to animals that are specifically different and at the same time to be itself one and indivisible *b*: it will always admit of differentiation. (For example, Man and Bird are both two-footed, but this essential characteristic is not the same in both: it is differentiated.*c* And if they are both " blooded," the blood must be different, or else it cannot be reckoned as part of their essence.) If that is so, then, the one

---

b As the privative characteristic would have to be.
c See below, 693 b 2 ff.

643 a

μία διαφορὰ δυσὶν ὑπάρξει.[1] εἰ δὲ τοῦτο, δῆλον
ὅτι ἀδύνατον στέρησιν εἶναι διαφοράν.

Ἔσονται δ' αἱ διαφοραὶ ἴσαι τοῖς ἀτόμοις ζῴοις,
εἴπερ ἄτομά τε ταῦτα καὶ αἱ διαφοραὶ ἄτομοι,
κοινὴ δὲ μή ἐστιν. (εἰ δ' ἐνδέχεται ὑπάρχειν[2] καὶ
10 κοινήν, ἄτομον δέ, δῆλον ὅτι κατά γε τὴν κοινὴν ἐν
τῷ αὐτῷ ἐστιν ἕτερα ὄντα τῷ εἴδει ζῷα. ὥστ'
ἀναγκαῖον, εἰ ἴδιοι αἱ διαφοραὶ εἰς ἃς ἅπαντα
ἐμπίπτει τὰ ἄτομα, μηδεμίαν αὐτῶν εἶναι κοινήν·
εἰ δὲ μή, ἕτερα ὄντα εἰς τὴν αὐτὴν βαδιεῖται.) δεῖ
15 δ' οὔτε τὸ αὐτὸ καὶ ἄτομον εἰς ἑτέραν καὶ ἑτέραν
ἰέναι διαφορὰν τῶν διῃρημένων, οὔτ' εἰς τὴν αὐτὴν
ἕτερα, καὶ ἅπαντα εἰς ταύτας. φανερὸν τοίνυν ὅτι
οὐκ ἔστι λαβεῖν τὰ ἄτομα εἴδη ὡς διαιροῦνται οἱ εἰς
δύο διαιροῦντες τὰ ζῷα ἢ καὶ ἄλλο ὁτιοῦν γένος.
καὶ γὰρ κατ' ἐκείνους ἀναγκαῖον ἴσας τὰς ἐσχάτας
20 εἶναι διαφορὰς τοῖς ζῴοις πᾶσι τοῖς ἀτόμοις τῷ
εἴδει. ὄντος γὰρ τοῦδέ τινος γένους, οὗ διαφοραὶ
πρῶται τὰ ⟨λευκὰ καὶ τὰ μὴ⟩[3] λευκά, τούτων δ'
ἑκατέρου ἄλλαι, καὶ οὕτως εἰς τὸ πρόσω ἕως τῶν
ἀτόμων, αἱ τελευταῖαι τέτταρες ἔσονται ἢ ἄλλο τι

---

[1] ll. 3-6 interpunctionem correxi.
[2] μὴ ὑπάρχειν vulg.: corr. Titze.
[3] supplevit Cornford.

---

[a] Because it cannot fulfil the condition of admitting
differentiation. At whatever stage of the division it comes
(unless at the very end), the privative term will cover at least
two species, and therefore at the next stage the dichotomists
will have to divide it—illegitimately, as Aristotle maintains.

differentia will belong to two species. And if so, it is clear that a privative cannot be a valid *differentia.*[a]

(*d*) Now assuming that each species is indivisible : if each *differentia* also is indivisible, and none is common to more species than one, then the number of *differentiae* will be equal to the number of species. (Supposing it were possible to have a *differentia* which though indivisible was common ; clearly, in that case, animals which differed in species would be in the same division in virtue of that common *differentia.* Therefore, if the *differentiae* under which the indivisible and ultimate species fall are to be proper and private to each one, it is necessary that no *differentia* be common ; otherwise, species which are actually different will come under one and the selfsame *differentia.*) And we may not place one and the same indivisible species under two or three of the lines of differentiation given by the divisions ; nor may we include different species under one and the same line of differentiation. Yet each species must be placed under the lines of differentiation available. It is evident from this that it is impossible to arrive at the indivisible species either of animals or of any other group by the method of twofold division as these people practise it, for even on their showing the number of ultimate *differentiae* must of necessity be equal to the total number of indivisible species of animals. Thus, suppose we have some particular group of creatures whose prime *differentiae* are " pale " and " not pale " ; by that method these two will each give two other *differentiae,* and so forth, until in the end the indivisible *differentiae* are reached : these last ones will be either four in

643 a

πλῆθος τῶν ἀφ' ἑνὸς διπλασιαζομένων· τοσαῦτα δὲ καὶ τὰ εἴδη.

("Εστι δ' ἡ διαφορὰ ἐν τῇ ὕλῃ τὸ εἶδος.[1] οὔτε
25 γὰρ ἄνευ ὕλης οὐδὲν ζῴου μόριον, οὔτε μόνη ἡ ὕλη· οὐ γὰρ πάντως ἔχον σῶμα ἔσται ζῷον, οὐδὲ τῶν μορίων οὐδέν, ὥσπερ πολλάκις εἴρηται.)

"Ετι διαιρεῖν χρὴ τοῖς ἐν τῇ οὐσίᾳ καὶ μὴ τοῖς συμβεβηκόσι καθ' αὑτό, οἷον εἴ τις τὰ σχήματα διαιροίη, ὅτι τὰ μὲν δυσὶν ὀρθαῖς ἴσας ἔχει τὰς
30 γωνίας, τὰ δὲ πλείοσιν· συμβεβηκὸς γάρ τι τῷ τριγώνῳ τὸ δυσὶν ὀρθαῖς ἴσας ἔχειν τὰς γωνίας.

"Ετι τοῖς ἀντικειμένοις διαιρεῖν ⟨δεῖ⟩,[2] διάφορα γὰρ ἀλλήλοις τἀντικείμενα, οἷον λευκότης καὶ μελανία καὶ εὐθύτης καὶ καμπυλότης. ἐὰν οὖν θάτερα διάφορα ᾖ, τῷ ἀντικειμένῳ διαιρετέον, καὶ μὴ τὸ
35 μὲν νεύσει τὸ δὲ χρώματι. πρὸς δὲ τούτοις, τά γ' ἔμψυχα τοῖς κοινοῖς ἔργοις τοῦ σώματος καὶ τῆς
643 b ψυχῆς, οἷον καὶ ἐν ταῖς ῥηθείσαις νῦν πορευτικὰ καὶ πτηνά—ἔστι γάρ τινα γένη οἷς ἄμφω ὑπάρχει καὶ ἔστι πτηνὰ καὶ ἄπτερα, καθάπερ τὸ τῶν μυρμήκων

---

[1] sic Y : τὸ εἶδος ἐν τῇ ὕλῃ vulg.
[2] ⟨δεῖ⟩ supplevi.

---

[a] His point is that it is nonsensical to suppose that this numerical correspondence is bound to occur.
[b] As at 641 a 18 ff.
[c] See *Met.* 1025 a 30.
[d] These are enumerated in *De sensu*, 436 a 7 ff., and Aristotle seems here to be thinking of them as grouped together under the several faculties—nutritive, appetitive, sensory,

number, or some higher value of 2^n ; and there will be an identical number of species.[a]

(The species is the *differentia* in the Matter. There is no animal part which exists without matter ; nor on the other hand is there any which is matter only, for body in any and every condition cannot make an animal or any part of an animal, as I have often pointed out.[b])

(e) Again, the division ought to be made according to points that belong to the Essence of a thing and not according to its essential (inseparable) attributes. For instance, in making divisions of geometrical figures, it would be wrong to divide them into those whose angles are together equal to two right angles and those whose angles are together greater than two right angles ; because it is only an attribute of the triangle that its angles are together equal to two right angles.[c]

(f) Again, division should be by " opposites," opposites being mutually " different," *e.g.* pale and dark, straight and curved. Therefore, provided the two terms are truly " different," division should be by means of opposites, and should not characterize one side by ability to swim and the other side by some colour. And besides this, division of living creatures, at any rate, by the functions which are common functions of body and soul,[d] such as we actually find done in the divisions mentioned above, where animals are divided into " walkers " and " fliers "—for there are some groups, such as that of the Ants, which have both attributes, being both

locomotive, and thought (see *De an.* 414 a 28 ff.). His point is that the correct way to divide and classify animals is rather by bodily characteristics, which is what he himself does.

643 b

γένος—καὶ τῷ ἀγρίῳ καὶ[1] ἡμέρῳ ⟨οὐ δεῖ⟩[2] διαιρεῖ-
σθαι· ὡσαύτως γὰρ ἂν δόξειε ταὐτὰ εἴδη διαιρεῖν·
5 πάντα γάρ, ὡς εἰπεῖν, ὅσα ἥμερα καὶ ἄγρια τυγχάνει ὄντα, οἷον ἄνθρωποι, ἵπποι, βόες, κύνες ἐν τῇ
Ἰνδικῇ, ὕες, αἶγες, πρόβατα· ὧν ἕκαστον, εἰ μὲν
ὁμώνυμον, οὐ διῄρηται χωρίς, εἰ δὲ ταῦτα ἓν εἴδει,
οὐχ οἷόν τ᾽ εἶναι διαφορὰν τὸ ἄγριον καὶ τὸ ἥμερον.
    Ὅλως δ᾽ ὁποιανοῦν διαφορᾷ[3] μιᾷ διαιροῦντι τοῦτο
10 συμβαίνειν ἀναγκαῖον. ἀλλὰ δεῖ πειρᾶσθαι λαμβάνειν κατὰ γένη τὰ ζῷα, ὡς ὑφήγηνθ᾽[a] οἱ πολλοὶ
διορίσαντες ὄρνιθος γένος καὶ ἰχθύος. τούτων δ᾽
ἕκαστον πολλαῖς ὥρισται διαφοραῖς, οὐ κατὰ τὴν
διχοτομίαν. οὕτω μὲν γὰρ ἤτοι τὸ παράπαν οὐκ
ἔστι λαβεῖν (τὸ αὐτὸ γὰρ εἰς πλείους ἐμπίπτει
15 διαιρέσεις καὶ τὰ ἐναντία εἰς τὴν αὐτήν), ἢ μία
μόνον διαφορὰ ἔσται, καὶ αὕτη ἤτοι ἁπλῆ ἢ ἐκ
συμπλοκῆς τὸ τελευταῖον ἔσται εἶδος. ἐὰν δὲ μὴ
διαφορᾶς λαμβάνῃ τις διαφοράν,[4] ἀναγκαῖον, ὥσπερ
συνδέσμῳ τὸν λόγον ἕνα ποιοῦντας, οὕτω καὶ τὴν
διαίρεσιν συνεχῆ ποιεῖν. λέγω δ᾽ οἷον συμβαίνει
20 τοῖς διαιρουμένοις τὸ μὲν ἄπτερον τὸ δὲ πτερωτόν,
πτερωτοῦ δὲ τὸ μὲν ἥμερον τὸ δ᾽ ἄγριον, ἢ τὸ μὲν

---

[1] καὶ ΕΥ : καὶ τῷ vulg.
[2] supplevi.
[3] ὁποιανοῦν διαφορὰν alii : ὁποιαοῦν Υ : διαφορὰ vel διαφορᾷ
ESY.
[4] διαφορᾷ λ. ES : διαφορὰν λ. τῆς διαφορᾶς Ρ : διαφορᾶς λ.
διαφορὰν Υ : τις Peck : τὴν vulg.

---

[a] Cf. Plato, Politicus, 264 a 1.
[b] On this see Platt, C.Q., 1909, iii. 241.
[c] For διαφορά in the sense of "bifurcation" cf. Met.
1048 b 4, where he speaks of the two "parts" of a διαφορά.
[d] i.e. with the preceding terms. See below, 644 a 5.

88

" winged " and " wingless "—and by " wild " and
" tame," [a] is not permissible, for this similarly would
appear to divide up species that are the same, since
practically all the tame animals are also found as
wild ones : *e.g.* Man, the horse, the ox, the dog (in
India [b]), swine, the goat, the sheep ; and if, in each
of these groups, the wild and the tame bear the same
name, as they do, there is no division between them,
while if each group is specifically a unit, then it
follows that " wild " and " tame " cannot make a
valid differentiation.[c]

And generally, the same thing inevitably happens
whatever one single line of differentiation is taken for
the division. The proper course is to endeavour to
take the animals according to their groups, following the lead of the bulk of mankind, who have
marked off the group of Birds and the group of Fishes.
Each of these groups is marked off by *many differentiae*,
not by means of dichotomy. By dichotomy (*a*) either
these groups cannot be arrived at at all (because the
same group falls under several divisions and contrary
groups under the same division) or else there will be
one *differentia* only, and this either singly or in
combination [d] will constitute the ultimate species.[e]
But (*b*) if they do not take the *differentia* of the *differentia*, they are forced to follow the example of
those people who try to give unity to their prose by a
free use of conjunctions : there is as little continuity about their division. Here is an example
to show what happens. Suppose they make the
division into " wingless " and " winged," and then
divide " winged " into " tame " and " wild " or into

---

[a] And this will never *completely* represent any actual
group or species. See below, 644 a 6 ff.

λευκὸν τὸ δὲ μέλαν· οὐ γὰρ διαφορὰ τοῦ πτερωτοῦ
τὸ ἥμερον οὐδὲ τὸ λευκόν, ἀλλ' ἑτέρας ἀρχὴ δια-
φορᾶς· ἐκεῖ δὲ κατὰ συμβεβηκός. διὸ πολλαῖς τὸ
ἓν εὐθέως διαιρετέον, ὥσπερ λέγομεν. καὶ γὰρ
25 οὕτως μὲν αἱ στερήσεις ποιήσουσι διαφοράν, ἐν δὲ
τῇ διχοτομίᾳ οὐ ποιήσουσιν.

Ὅτι δ' οὐκ ἐνδέχεται τῶν καθ' ἕκαστον εἰδῶν
λαμβάνειν οὐδὲν διαιροῦσι δίχα τὸ γένος, ὥσπερ
τινὲς ᾠήθησαν, καὶ ἐκ τῶνδε φανερόν.

Ἀδύνατον γὰρ μίαν ὑπάρχειν διαφορὰν τῶν[1]
30 καθ' ἕκαστον διαιρετῶν, ἐάν θ' ἁπλᾶ λαμβάνῃ τις
ἐάν τε συμπεπλεγμένα· [λέγω δὲ ἁπλᾶ μέν, ἐὰν μὴ
ἔχῃ διαφοράν, οἷον τὴν σχιζοποδίαν, συμπεπλεγ-
μένα δέ, ἐὰν ἔχῃ, οἷον τὸ πολυσχιδὲς πρὸς τὸ[2]
σχιζόπουν·][3] τοῦτο γὰρ ἡ συνέχεια βούλεται τῶν
ἀπὸ τοῦ γένους κατὰ τὴν διαίρεσιν διαφορῶν ὡς ἓν
35 τι τὸ πᾶν ὄν, ἀλλὰ παρὰ τὴν λέξιν συμβαίνει δοκεῖν
τὴν τελευταίαν μόνην εἶναι διαφοράν [οἷον τὸ πολυ-
644 a σχιδὲς ἢ τὸ δίπουν, τὸ δ' ὑπόπουν καὶ πολύπουν
περίεργα].[4] ὅτι δ' ἀδύνατον πλείους εἶναι τοιαύτας,
δῆλον· ἀεὶ γὰρ βαδίζων ἐπὶ τὴν ἐσχάτην διαφορὰν
ἀφικνεῖται [ἀλλ' οὐκ ἐπὶ τὴν τελευταίαν καὶ τὸ
εἶδος].[5] αὕτη δ' ἐστὶν ἢ τὸ σχιζόπουν μόνον, ἢ
5 πᾶσα ἡ σύμπλεξις, ἐὰν διαιρῆται ἄνθρωπος,[6] οἷον
εἴ τις συνθείη ὑπόπουν, δίπουν, σχιζόπουν. εἰ δ'
ἦν ὁ ἄνθρωπος σχιζόπουν μόνον, οὕτως ἐγίγνετ' ἂν
αὕτη ⟨ἡ⟩[7] μία διαφορά. νῦν δ' ἐπειδὴ οὐκ ἔστιν,

---

[1] τις Y : om. vulg.　　　　　　[2] πρὸς τῷ Platt.
[3] seclusi. codices varia, ut videtur ; sic Bekker.
[4] οἷον . . . περίεργα seclusi.
[5] ἀλλ' . . . εἶδος seclusi.
[6] ἄνθρωπον vulg.　　　　　[7] ⟨ἡ⟩ Ogle.

90

" pale " and " dark ": neither " tame " nor " pale "
is a differentiation of " winged," but the beginning
of another line of differentiation, and can come in
here only *by accident*. Therefore, as I say, in dividing
we must distinguish the one original group forthwith
by numerous *differentiae* ; and then too the privative
terms will make valid *differentiae*, which they will
never do in the system of dichotomy.

Here are further considerations to show that it is
impossible to come at any of the particular species by
the method of dividing the group into two, as some
people have imagined.

Obviously it is impossible that one single *differentia*
is adequate for each of the particular species covered
by the division, whether you adopt as your *differentia*
the isolated term or the combination of terms [a] (for
this is intended by the continuity of the series of
differentiae throughout the division from the original
group, to indicate that the whole is a unity ; but, in
consequence of the form of the expression, the last
one comes to be considered as the sole *differentia*).
And it is evident that there cannot be more than one
such *differentia*; for the division proceeds steadily until
it reaches the ultimate *differentia*, and—supposing the
division is aiming at " Man "—this is either " cloven-
footed " alone, or else the whole combination, *e.g.*
if one combined " footed," " two-footed," " cloven-
footed." [b] If Man were merely a cloven-footed
animal, then this would be the one *differentia*, arrived
at by the right method. But as he is not merely

----

[a] *i.e.* the last term of any series, or all its terms together,
as he goes on to say. *Cf.* 643 b 15 f.

[b] This definition appears also in *Met.* 1037-1038.

644 a

ἀνάγκη πολλὰς εἶναι μὴ ὑπὸ μίαν διαίρεσιν. ἀλλὰ
μὴν πλείους γε τοῦ αὐτοῦ οὐκ ἔστιν ὑπὸ μίαν
10 διχοτομίαν εἶναι, ἀλλὰ μίαν κατὰ μίαν τελευτᾶν.
ὥστε ἀδύνατον ὁτιοῦν λαβεῖν τῶν καθ᾽ ἕκαστον
ζῴων δίχα διαιρουμένους.

IV. Ἀπορήσειε δ᾽ ἄν τις διὰ τί οὐκ ἄνωθεν ἑνὶ
ὀνόματι ἐμπεριλαβόντες ἅμα ἓν γένος ἄμφω προσ-
ηγόρευσαν οἱ ἄνθρωποι, ὃ περιέχει τά τε ἔνυδρα
15 καὶ τὰ πτηνὰ τῶν ζῴων· ἔστι γὰρ ἔνια πάθη
κοινὰ καὶ τούτοις [καὶ τοῖς ἄλλοις ζῴοις ἅπασιν].[1]
ἀλλ᾽ ὅμως ὀρθῶς διώρισται τοῦτον τὸν τρόπον.
ὅσα μὲν γὰρ διαφέρει τῶν γενῶν καθ᾽ ὑπεροχὴν καὶ
τῷ μᾶλλον καὶ ἧττον,[2] ταῦτα ὑπέζευκται ἑνὶ γένει,
ὅσα δ᾽ ἔχει τὸ ἀνάλογον, χωρίς· λέγω δ᾽ οἷον ὄρνις
20 ὄρνιθος διαφέρει τῷ μᾶλλον ἢ καθ᾽ ὑπεροχὴν (τὸ
μὲν γὰρ μακρόπτερον τὸ δὲ βραχύπτερον), ἰχθύες
δ᾽ ὄρνιθος τῷ ἀνάλογον (ὃ γὰρ ἐκείνῳ πτερόν, θα-
τέρῳ λεπίς). τοῦτο δὲ ποιεῖν ἐπὶ πᾶσιν οὐ ῥᾴδιον·
τὰ γὰρ πολλὰ ζῷα ἀνάλογον ταὐτὸ πέπονθεν.

Ἐπεὶ δ᾽ οὐσίαι μέν εἰσι τὰ ἔσχατα εἴδη, κατὰ
25 δὲ ταῦτα τὰ[3] τὸ εἶδος ἀδιάφορα (οἷον Σωκράτης,
Κορίσκος), ἀναγκαῖον ἢ τὰ καθόλου ὑπάρχοντα

─────────

[1] seclusi Ogle docente.
[2] sic Rackham : τὸ μᾶλλον καὶ τὸ (τὸ om. Y) ἧττον vulg.
[3] κατὰ δὲ ταῦτα τὰ Peck : ταῦτα δὲ κατὰ vulg.

─────────

[a] This paragraph has been corrupted by confusing interpolations, which I have bracketed in the Greek text and
omitted in the translation. With this passage cf. Met.
1037 b 27—1038 a 30.

[b] On this point see D'Arcy W. Thompson, Growth and
Form, esp. ch. 17, and the same author's paper Excess
and Defect ; or The Little More and the Little Less, in
Mind, xxxviii. (N.S.) 149, pp. 43-55. See also infra,
661 b 28 ff., 692 b 3 ff. ; and Introduction, p. 39.

that, it is necessary that there should be many *differentiae*, not under one line of division. And yet there cannot be more than one *differentia* for the same thing under one line of dichotomy : one line must end in one *differentia*. So it is impossible for those who follow the method of twofold division to arrive at any of the particular animals.[a]

IV. Some may find it puzzling that general usage has not combined the water-animals and the feathered animals into one higher group, and adopted one name to cover both, seeing that in fact these two groups have certain features in common. The answer is that in spite of this the present grouping is the right one ; because while groups that differ only " by excess " (that is, " by the more and less "[b]) are placed together in one group, those which differ so much that their characteristics can merely be called analogous are placed in separate groups. As an illustration : (*a*) one bird differs from another bird " by the more," or " by excess " : one bird's feathers are long, another's are short ; whereas (*b*) the difference between a Bird and a Fish is greater, and their correspondence is only by analogy : a fish has no feathers at all, but scales, which correspond to them. It is not easy to do this in all cases, for the corresponding analogous parts of most groups of animals are identical.

Now since the ultimate species are " real things," [c] Method. while within them are individuals which do not differ in species (as *e.g.* Socrates and Coriscus),[d] we shall have to choose (as I have pointed out)[e] between

---

[c] Lit. " substances."
[d] *i.e.* within the species " man."
[e] Above, at 639 a, b, etc.

πρότερον εἰπεῖν ἢ πολλάκις ταὐτὸν λέγειν, καθάπερ
εἴρηται. (τὰ δὲ καθόλου κοινά· τὰ γὰρ πλείοσιν
ὑπάρχοντα καθόλου λέγομεν.) ἀπορίαν δ᾽ ἔχει περὶ
πότερα δεῖ πραγματεύεσθαι. ᾗ μὲν γὰρ οὐσία τὸ
30 τῷ εἴδει ἄτομον, κράτιστον, εἴ τις δύναιτο, περὶ τῶν
καθ᾽ ἕκαστον καὶ ἀτόμων τῷ εἴδει θεωρεῖν χωρίς,
ὥσπερ περὶ ἀνθρώπου, οὕτω καὶ[1] περὶ ὄρνιθος, ⟨καὶ
μὴ περὶ ὁτουοῦν ὄρνιθος⟩ (ἔχει γὰρ εἴδη τὸ γένος
τοῦτο), ἀλλὰ περὶ τῶν ἀτόμων·[2] οἷον ἢ στρουθὸς ἢ
35 γέρανος ἤ τι τοιοῦτον. ᾗ δὲ συμβήσεται λέγειν
πολλάκις περὶ τοῦ αὐτοῦ πάθους διὰ τὸ κοινῇ
πλείοσιν ὑπάρχειν, ταύτῃ δ᾽ ἐστὶν ὑπάτοπον καὶ
644 b μακρὸν τὸ περὶ ἑκάστου λέγειν χωρίς. ἴσως μὲν
οὖν ὀρθῶς ἔχει τὰ μὲν κατὰ γένη κοινῇ λέγειν,
ὅσα λέγεται καλῶς ὡρισμένων τῶν ἀνθρώπων, καὶ
ἔχει τε μίαν φύσιν κοινὴν καὶ εἴδη ἐν αὐτοῖς[3] μὴ
5 πολὺ διεστῶτα, ὄρνις καὶ ἰχθύς, καὶ εἴ τι ἄλλο
ἐστὶν ἀνώνυμον μέν, τῷ γένει δ᾽ ὁμοῖα[4] περιέχει
τὰ ἐν αὐτῷ[5] εἴδη· ὅσα δὲ μὴ τοιαῦτα, καθ᾽
ἕκαστον, οἷον περὶ ἀνθρώπου καὶ εἴ τι τοιοῦτον
ἕτερόν ἐστιν.

Σχεδὸν δὲ τοῖς σχήμασι τῶν μορίων καὶ τοῦ
σώματος ὅλου, ἐὰν ὁμοιότητα ἔχωσιν, ὥρισται τὰ
γένη, οἷον τὸ τῶν ὀρνίθων γένος πρὸς αὐτὸ[6] πέ-

---

[1] καί] μὴ Bonitz.
[2] hunc locum correxi, Σ secutus ; ἔχει γὰρ εἴδη τὸ γένος
τοῦτο· ἀλλὰ περὶ ὁτουοῦν ὄρνιθος τῶν ἀτόμων, οἷον κτλ. vulg.
[3] αὐτοῖς vulg.: correxi.   [4] ὁμοίως vulg.: correxi.
[5] αὐτῷ vulg.: correxi.
[6] αὐτὸ Platt, fortasse Z¹: αὐτὸ Y: αὐτὰ Z², vulg.

describing first of all the general attributes of many species, and repeating the same thing many times over. (By "general" attributes I intend the "common" ones. That which belongs to many we call "general.") One may well hesitate whether of the two courses to follow. For, in so far as it is the specifically indivisible which is the "real thing," it would be best, if one could do it, to study separately the particular and specifically indivisible sorts, in the same way as one studies "Man," to do this with "Bird" too, that is, to study not just "Bird" in the mass, but—since "Bird" is a group which contains species—the indivisible species of it, e.g. Ostrich, Crane, and so on. Yet, on the other hand, this course is somewhat unreasonable and long-winded, because it makes us describe the same attributes time and again, as they happen to be common attributes of many species. So perhaps after all the right procedure is this : (a) So far as concerns the attributes of those groups which have been correctly marked off by popular usage—groups which possess one common nature apiece and contain in themselves species not far removed from one another, I mean Birds and Fishes and any other such group which though it may lack a popular name yet contains species generically similar—to describe the common attributes of each group all together ; and (b) with regard to those animals which are not covered by this, to describe the attributes of each of these by tself—e.g. those of Man, and of any other such species here may be.

Now it is practically by resemblance of the shapes of their parts, or of their whole body, that the groups are marked off from each other : as e.g. the groups

644 b

10 πονθε καὶ τὸ τῶν ἰχθύων καὶ τὰ μαλάκιά τε καὶ
τὰ ὄστρεια. τὰ γὰρ μόρια διαφέρουσι τούτων οὐ
τῇ ἀνάλογον ὁμοιότητι, οἷον ἐν ἀνθρώπῳ καὶ ἰχθύι
πέπονθεν ὀστοῦν πρὸς ἄκανθαν, ἀλλὰ μᾶλλον τοῖς
σωματικοῖς πάθεσιν, οἷον μεγέθει μικρότητι, μαλα-
15 κότητι σκληρότητι, λειότητι τραχύτητι καὶ τοῖς
τοιούτοις, ὅλως δὲ τῷ μᾶλλον καὶ ἧττον.

Πῶς μὲν οὖν ἀποδέχεσθαι δεῖ τὴν περὶ φύσεως
μέθοδον, καὶ τίνα τρόπον γίνοιτ᾽ ἂν ἡ θεωρία περὶ
αὐτῶν ὁδῷ καὶ ῥᾷστα, ἔτι δὲ περὶ διαιρέσεως, τίνα
τρόπον ἐνδέχεται μετιοῦσι λαμβάνειν χρησίμως, καὶ
20 διότι τὸ διχοτομεῖν τῇ μὲν ἀδύνατον τῇ δὲ κενόν,
εἴρηται. διωρισμένων δὲ τούτων περὶ τῶν ἐφεξῆς
λέγωμεν, ἀρχὴν τήνδε ποιησάμενοι.

V. Τῶν οὐσιῶν ὅσαι φύσει συνεστᾶσι, τὰς μὲν
⟨λέγομεν⟩¹ ἀγενήτους καὶ ἀφθάρτους εἶναι τὸν
ἅπαντα αἰῶνα, τὰς δὲ μετέχειν γενέσεως καὶ
25 φθορᾶς. συμβέβηκε δὲ περὶ μὲν ἐκείνας τιμίας
οὔσας καὶ θείας ἐλάττους ἡμῖν ὑπάρχειν θεωρίας
(καὶ γὰρ ἐξ ὧν ἄν τις σκέψαιτο περὶ αὐτῶν, καὶ
περὶ ὧν εἰδέναι ποθοῦμεν, παντελῶς ἐστὶν ὀλίγα τὰ
φανερὰ κατὰ τὴν αἴσθησιν), περὶ δὲ τῶν φθαρτῶν
φυτῶν τε καὶ ζῴων εὐποροῦμεν μᾶλλον πρὸς τὴν
30 γνῶσιν διὰ τὸ σύντροφον· πολλὰ γὰρ περὶ ἕκαστον
γένος λάβοι τις ἂν τῶν ὑπαρχόντων βουλόμενος
διαπονεῖν ἱκανῶς. ἔχει δ᾽ ἑκάτερα χάριν. τῶν μὲν
γὰρ εἰ καὶ κατὰ μικρὸν ἐφαπτόμεθα, ὅμως διὰ τὴν

---

¹ ⟨λέγομεν⟩ Peck.

ᵃ Lit., "softies." The group includes, roughly, the
cephalopod mollusca.
ᵇ Lit., "oysters" (bivalves).

Birds, Fishes, Cephalopods,[a] Testacea.[b] Within each of these groups, the parts do not differ so far that they correspond only by analogy (as a man's bone and a fish's spine) ; that is, they differ not structurally, but only in respect of bodily qualities, *e.g.* by being larger or smaller, softer or harder, smoother or rougher, and so forth, or, to put it generally, they differ " by the more and less."

We have now shown :

(1) how to test a method of Natural science ;
(2) what is the most systematic and easiest way of studying Natural science ;
(3) what is the most useful mode of Division for our present purpose ;
(4) why dichotomy is in one respect impossible and in another futile.

Now that we have made this beginning, and clearly distinguished these points, we may proceed.

V. Of the works of Nature there are, we hold, two kinds : those which are brought into being and perish, and those which are free from these processes throughout all ages. The latter are of the highest worth and are divine, but our opportunities for the study of them are somewhat scanty, since there is but little evidence available to our senses to enable us to consider them and all the things that we long to know about. We have better means of information, however, concerning the things that perish, that is to say, plants and animals, because we live among them ; and anyone who will but take enough trouble can learn much concerning every one of their kinds. Yet each of the two groups has its attractiveness. For although our grasp of the eternal things is but slight, nevertheless the joy which it brings is, by

A protreptic to the study of animals.

97

644 b

τιμιότητα τοῦ γνωρίζειν ἥδιον ἢ τὰ παρ' ἡμῖν
ἅπαντα, ὥσπερ καὶ τῶν ἐρωμένων τὸ τυχὸν καὶ
35 μικρὸν μόριον κατιδεῖν ἥδιόν ἐστιν ἢ πολλὰ ἕτερα
645 a καὶ μεγάλα δι' ἀκριβείας ἰδεῖν· τὰ δὲ διὰ τὸ μᾶλλον
καὶ πλείω γνωρίζειν αὐτῶν λαμβάνει τὴν τῆς ἐπι-
στήμης ὑπεροχήν, ἔτι δὲ διὰ τὸ πλησιαίτερα ἡμῶν
εἶναι καὶ τῆς φύσεως οἰκειότερα ἀντικαταλλάτ-
τεταί τι πρὸς τὴν περὶ τὰ θεῖα φιλοσοφίαν. ἐπεὶ
5 δὲ περὶ ἐκείνων διήλθομεν λέγοντες τὸ φαινόμενον
ἡμῖν, λοιπὸν περὶ τῆς ζωικῆς φύσεως εἰπεῖν, μηδὲν
παραλιπόντας εἰς δύναμιν μήτε ἀτιμότερον μήτε
τιμιώτερον. καὶ γὰρ ἐν τοῖς μὴ κεχαρισμένοις
αὐτῶν πρὸς τὴν αἴσθησιν κατὰ τὴν θεωρίαν ὅμως[1]
ἡ δημιουργήσασα φύσις ἀμηχάνους ἡδονὰς παρέχει
10 τοῖς δυναμένοις τὰς αἰτίας γνωρίζειν καὶ φύσει
φιλοσόφοις. καὶ γὰρ ἂν εἴη παράλογον καὶ ἄτοπον,
εἰ τὰς μὲν εἰκόνας αὐτῶν θεωροῦντες χαίρομεν ὅτι
τὴν δημιουργήσασαν τέχνην συνθεωροῦμεν, οἷον τὴν
γραφικὴν ἢ τὴν πλαστικήν, αὐτῶν δὲ τῶν φύσει
συνεστώτων μὴ μᾶλλον ἀγαπῶμεν τὴν θεωρίαν,
15 δυνάμενοί γε τὰς αἰτίας καθορᾶν. διὸ δεῖ μὴ
δυσχεραίνειν παιδικῶς τὴν περὶ τῶν ἀτιμοτέρων
ζῴων ἐπίσκεψιν· ἐν πᾶσι γὰρ τοῖς φυσικοῖς ἔνεστί

---

[1] ὅμως Bekker: ὁμοίως codd.

---

^a This passage, 645 a 6-15, is quoted by R. Boyle (*Of the
Usefulnesse of Naturall Philosophy*, 1663) both in Gaza's
Latin version and in an English translation, and he intro-
duces it thus : " And, methinks, *Aristotle* discourses very
Philosophically in that place, where passing from the consideration of the sublimist productions of Nature, to justifie
his diligence in recording the more homely Circumstances of
the History of Animals, he thus discourses." He also quotes

reason of their excellence and worth, greater than that of knowing all things that are here below ; just as the joy of a fleeting and partial glimpse of those whom we love is greater than that of an accurate view of other things, no matter how numerous or how great they are. But inasmuch as it is possible for us to obtain more and better information about things here on the earth, our knowledge of them has the advantage over the other ; and moreover, because they are nearer to us and more akin to our Nature, they are able to make up some of their leeway as against the philosophy which contemplates the things that are divine. Of " things divine " we have already treated and have set down our views concerning them ; so it now remains to speak of animals and their Nature. ^a So far as in us lies, we will not leave out any one of them, be it never so mean ; for though there are animals which have no attractiveness for the senses, yet for the eye of science, for the student who is naturally of a philosophic spirit and can discern the causes of things, Nature which fashioned them provides joys which cannot be measured. If we study mere likenesses of these things and take pleasure in so doing, because then we are contemplating the painter's or the carver's Art which fashioned them, and yet fail to delight much more in studying the works of Nature themselves, though we have the ability to discern the actual causes—that would be a strange absurdity indeed. Wherefore we must not betake ourselves to the consideration of the meaner animals with a bad grace, as though we were children ; since in all natural things there is somewhat of the mar-

the following passage, a 15-23, describing it as " that Judicious reasoning of *Aristotle.*"

τι θαυμαστόν· καὶ καθάπερ Ἡράκλειτος λέγεται
20 πρὸς τοὺς ξένους εἰπεῖν τοὺς βουλομένους ἐντυχεῖν
αὐτῷ, οἳ ἐπειδὴ προσιόντες εἶδον αὐτὸν θερόμενον
πρὸς τῷ ἰπνῷ ἔστησαν (ἐκέλευε γὰρ αὐτοὺς εἰσιέναι
θαρροῦντας· εἶναι γὰρ καὶ ἐνταῦθα θεούς), οὕτω καὶ
πρὸς τὴν ζήτησιν περὶ ἑκάστου τῶν ζῴων προσιέναι
δεῖ μὴ δυσωπούμενον, ὡς ἐν ἅπασιν ὄντος τινὸς
φυσικοῦ καὶ καλοῦ.

Τὸ γὰρ μὴ τυχόντως ἀλλ' ἕνεκά τινος ἐν τοῖς τῆς
25 φύσεως ἔργοις ἐστὶ καὶ μάλιστα· οὗ δ' ἕνεκα
συνέστηκεν ἢ γέγονε τέλους, τὴν τοῦ καλοῦ χώραν
εἴληφεν. εἰ δέ τις τὴν περὶ τῶν ἄλλων ζῴων
θεωρίαν ἄτιμον εἶναι νενόμικε, τὸν αὐτὸν τρόπον
οἴεσθαι χρὴ καὶ περὶ αὑτοῦ· οὐκ ἔστι γὰρ ἄνευ
πολλῆς δυσχερείας ἰδεῖν ἐξ ὧν συνέστηκε τὸ τῶν
30 ἀνθρώπων γένος, οἷον αἷμα, σάρκες, ὀστᾶ, φλέβες
καὶ τὰ τοιαῦτα μόρια. ὁμοίως τε δεῖ νομίζειν τὸν
περὶ οὑτινοσοῦν τῶν μορίων ἢ τῶν σκευῶν δια-
λεγόμενον μὴ περὶ τῆς ὕλης ποιεῖσθαι τὴν μνήμην,
μηδὲ ταύτης χάριν, ἀλλὰ τῆς ὅλης μορφῆς, οἷον καὶ
περὶ οἰκίας, ἀλλὰ μὴ πλίνθων καὶ πηλοῦ καὶ ξύλων·
35 καὶ τὸν περὶ φύσεως περὶ τῆς συνθέσεως καὶ τῆς
ὅλης οὐσίας, ἀλλὰ μὴ περὶ τούτων ἃ μὴ συμβαίνει
χωριζόμενά ποτε τῆς οὐσίας αὐτῶν.

---

[a] Or, with reference to another use of οὐσία, "which gives
them their being." Independent approaches to the position
that components are non-significant in isolation had been
made, *e.g.* by Anaxagoras, as a physical philosopher (see
my article in *C.Q.* xxv. 27 ff., 112 ff.), who held that "the
things (*i.e.* the constituent elements) in this world are not
separate one from another" (frag. 8, Diels, *Fragmente*[5],

vellous. There is a story which tells how some visitors once wished to meet Heracleitus, and when they entered and saw him in the kitchen, warming himself at the stove, they hesitated; but Heracleitus said, " Come in ; don't be afraid ; there are gods even here." In like manner, we ought not to hesitate nor to be abashed, but boldly to enter upon our researches concerning animals of every sort and kind, knowing that in not one of them is Nature or Beauty lacking.

I add " Beauty," because in the works of Nature purpose and not accident is predominant ; and the purpose or end for the sake of which those works have been constructed or formed has its place among what is beautiful. If, however, there is anyone who holds that the study of the animals is an unworthy pursuit, he ought to go further and hold the same opinion about the study of himself, for it is not possible without considerable disgust to look upon the blood, flesh, bones, blood-vessels, and suchlike parts of which the human body is constructed. In the same way, when the discussion turns upon any one of the parts or structures, we must not suppose that the lecturer is speaking of the material of them in itself and for its own sake ; he is speaking of the whole conformation. Just as in discussing a house, it is the whole figure and form of the house which concerns us, not merely the bricks and mortar and timber ; so in Natural science, it is the composite thing, the thing as a whole, which primarily concerns us, not the materials of it, which are not found apart from the thing itself whose materials they are.[a]

59 b 8) ; also from the logical point of view, as seen in Plato, *Theaetetus*, 201 e ff.

645 b Ἀναγκαῖον δὲ πρῶτον τὰ συμβεβηκότα διελεῖν περὶ ἕκαστον γένος, ὅσα καθ' αὑτὰ πᾶσιν ὑπάρχει τοῖς ζῴοις, μετὰ δὲ ταῦτα τὰς αἰτίας αὐτῶν πειρᾶσθαι διελεῖν. εἴρηται μὲν οὖν καὶ πρότερον ὅτι πολλὰ κοινὰ πολλοῖς ὑπάρχει τῶν ζῴων, τὰ μὲν ἁπλῶς (οἷον πόδες, πτερά, λεπίδες, καὶ πάθη δὴ τὸν αὐτὸν τρόπον τούτοις), τὰ δ' ἀνάλογον (λέγω δ' ἀνάλογον, ὅτι τοῖς μὲν ὑπάρχει πλεύμων, τοῖς δὲ πλεύμων μὲν οὔ, ὃ δὲ τοῖς ἔχουσι πλεύμονα, ἐκεί-
10 νοις ἕτερον ἀντὶ τούτου· καὶ τοῖς μὲν αἷμα, τοῖς δὲ τὸ ἀνάλογον τὴν αὐτὴν ἔχον δύναμιν ἥνπερ τοῖς ἐναίμοις τὸ αἷμα)· τὸ δὲ λέγειν χωρὶς περὶ ἑκάστων τῶν καθ' ἕκαστα, καὶ ἔμπροσθεν εἴπομεν ὅτι πολλάκις συμβήσεται ταὐτὰ λέγειν, ἐπειδὰν λέ-γωμεν περὶ πάντων τῶν ὑπαρχόντων· ὑπάρχει δὲ πολλοῖς ταὐτά. ταῦτα μὲν οὖν ταύτῃ διωρίσθω.

15 Ἐπεὶ δὲ τὸ μὲν ὄργανον πᾶν ἕνεκά του, τῶν δὲ τοῦ σώματος μορίων ἕκαστον ἕνεκά του, τὸ δ' οὗ ἕνεκα πρᾶξίς τις, φανερὸν ὅτι καὶ τὸ σύνολον σῶμα συνέστηκε πράξεώς τινος ἕνεκα πολυμεροῦς.[1] οὐ γὰρ ἡ πρίσις τοῦ πρίονος χάριν γέγονεν, ἀλλ' ὁ πρίων τῆς πρίσεως· χρῆσις γάρ τις ἡ πρίσις ἐστίν. ὥστε καὶ τὸ σῶμά πως τῆς ψυχῆς ἕνεκεν, καὶ τὰ
20 μόρια τῶν ἔργων πρὸς ἃ πέφυκεν ἕκαστον.

Λεκτέον ἄρα πρῶτον τὰς πράξεις τάς τε κοινὰς[2]

[1] πολυμεροῦς P: πλήρους vulg.: fortasse πολυμόρφου, cf. 646 b 15.

[2] πάντων post κοινὰς vulg.; delevi.

---

[a] Almost always used in the singular by Aristotle.

[b] By "blood" Aristotle means red blood only. "Blooded" and "bloodless" animals do not quite coincide with verte-brates and invertebrates; for there are some invertebrates which have red blood, *e.g.* molluscs (*Planorbis*), insect

First of all, our business must be to describe the Final summary of the Method. attributes found in each group; I mean those "essential" attributes which belong to all the animals, and after that to endeavour to describe the causes of them. It will be remembered that I have said already that there are many attributes which are common to many animals, either identically the same (*e.g.* organs like feet, feathers, and scales, and affections similarly), or else common by analogy only (*i.e.* some animals have a lung,[a] others have no lung but something else to correspond instead of it; again, some animals have blood, while others have its counterpart,[b] which in them has the same value as blood in the former). And I have pointed out above that to treat separately of all the particular species would mean continual repetition of the same things, if we are going to deal with all their attributes, as the same attributes are common to many animals. Such, then, are my views on this matter.

Now, as each of the parts of the body, like every other instrument, is for the sake of some purpose, viz. some action, it is evident that the body as a whole must exist for the sake of some complex action. Just as the saw is there for the sake of sawing and not sawing for the sake of the saw, because sawing is the using of the instrument, so in some way the body exists for the sake of the soul, and the parts of the body for the sake of those functions to which they are naturally adapted.

So first of all we must describe the actions (*a*)

larvae (*Chironomus*), worms (*Arenicola*). In other invertebrates the blood may be blue (Crustacea) or green (Sabellid worms), or there may be no respiratory pigment at all (most insects).

103

645 b

καὶ τὰς κατὰ γένος καὶ τὰς κατ᾽ εἶδος. λέγω δὲ κοινὰς μὲν αἳ πᾶσιν ὑπάρχουσι τοῖς ζῴοις, κατὰ γένος δὲ ὅσων παρ᾽ ἄλληλα τὰς διαφορὰς ὁρῶμεν 25 καθ᾽ ὑπεροχὴν οὔσας, οἷον ὄρνιθα λέγω κατὰ γένος, ἄνθρωπον δὲ κατ᾽ εἶδος, καὶ πᾶν ὃ κατὰ τὸν καθόλου λόγον μηδεμίαν ἔχει διαφοράν. τὰ μὲν γὰρ ἔχουσι τὸ κοινὸν κατ᾽ ἀναλογίαν, τὰ δὲ κατὰ γένος, τὰ δὲ κατ᾽ εἶδος.

Ὅσαι μὲν οὖν πράξεις ἄλλων ἕνεκα, δῆλον ὅτι καὶ ὧν αἱ πράξεις τὸν αὐτὸν τρόπον διεστᾶσιν 30 ὅνπερ αἱ πράξεις. ὁμοίως δὲ κἂν εἴ τινες πρότεραι καὶ τέλος ἑτέρων πράξεων τυγχάνουσιν οὖσαι, τὸν αὐτὸν ἕξει τρόπον καὶ τῶν μορίων ἕκαστον ὧν αἱ πράξεις αἱ τοιαῦται· καὶ τρίτον, ἃ τινῶν[1] ὄντων ἀναγκαῖον ὑπάρχειν. (λέγω δὲ πάθη καὶ πράξεις γένεσιν, αὔξησιν, ὀχείαν, ἐγρήγορσιν, ὕπνον, πο- 35 ρείαν, καὶ ὁπόσ᾽ ἄλλα τοιαῦτα τοῖς ζῴοις ὑπάρχει· μόρια δὲ λέγω ῥῖνα, ὀφθαλμὸν καὶ τὸ σύνολον 646 a πρόσωπον, ὧν ἕκαστον καλεῖται μέλος. ὁμοίως δὲ καὶ περὶ τῶν ἄλλων.)

Καὶ περὶ μὲν τοῦ τρόπου τῆς μεθόδου τοσαῦθ᾽ ἡμῖν εἰρήσθω· τὰς δ᾽ αἰτίας πειραθῶμεν εἰπεῖν περί τε τῶν κοινῶν καὶ τῶν ἰδίων, ἀρξάμενοι, καθάπερ διωρίσαμεν, πρῶτον ἀπὸ τῶν πρώτων.

---

[1] ἃ τινῶν Peck, cf. 677 a 18: ὧν vulg.: ἃ τούτων Ogle.

---

[a] See above, note on 644 a 17.
[b] Examples will occur during the course of the treatise.

which are common, and those which belong (*b*) to a group, or (*c*) to a species. By " common " I mean those that are present in all animals ; by " those which belong to a group " I mean those of animals whose differences we see to be differences " of excess " [a] in relation to one another : an example of this is the group Birds. Man is an example of a species ; so is every class which admits no differentiation of its general definition. These three sorts of common attributes are, respectively, (1) analogous, (2) generic, (3) specific.

Now it is evident that when one action is for the sake of another action, then the instruments which perform the two actions differ exactly as the two actions differ : and if one action is " prior " to another and is the " end " of that other action, then the part of the body to which it belongs will be " prior " to the part to which the other action belongs. There is also a third possibility, viz. that the action and its organ are there simply because the presence of others *necessarily* involves them.[b] (By affections and actions I mean Generation, Growth, Copulation, Waking, Sleep, Locomotion, and the other similar ones that are found in animals. Examples of parts are : Nose, Eye, Face ; each of these is named a " limb " or " member." And the same holds for the rest too.)

Let this suffice concerning the method of our inquiry, and let us now endeavour to describe the causes of all these things, particular as well as common ; and, according to the principles laid down, we will begin with the first ones first.

B

Ἐκ τίνων μὲν οὖν μορίων καὶ πόσων συνέστηκεν ἕκαστον τῶν ζῴων, ἐν ταῖς ἱστορίαις ταῖς
10 περὶ αὐτῶν δεδήλωται σαφέστερον· δι᾽ ἃς δ᾽ αἰτίας
ἕκαστον τούτων ἔχει τὸν τρόπον, ἐπισκεπτέον νῦν,
χωρίσαντας καθ᾽ αὑτὰ τῶν ἐν ταῖς ἱστορίαις εἰρημένων.

Τριῶν δ᾽ οὐσῶν τῶν συνθέσεων πρώτην μὲν ἄν
τις θείη τὴν ἐκ τῶν καλουμένων ὑπό τινων στοιχείων, οἷον γῆς, ἀέρος, ὕδατος, πυρός. ἔτι δὲ
15 βέλτιον ἴσως ἐκ τῶν δυνάμεων λέγειν, καὶ τούτων
οὐκ ἐξ ἁπασῶν, ἀλλ᾽ ὥσπερ ἐν ἑτέροις εἴρηται καὶ
πρότερον· ὑγρὸν γὰρ καὶ ξηρὸν καὶ θερμὸν καὶ
ψυχρὸν ὕλη τῶν συνθέτων σωμάτων ἐστίν, αἱ δ᾽
ἄλλαι διαφοραὶ ταύταις ἀκολουθοῦσιν, οἷον βάρος
καὶ κουφότης καὶ πυκνότης καὶ μανότης καὶ τρα-
20 χύτης καὶ λειότης καὶ τἆλλα τὰ τοιαῦτα πάθη τῶν
σωμάτων. δευτέρα δὲ σύστασις ἐκ τῶν πρώτων ἡ
τῶν ὁμοιομερῶν φύσις ἐν τοῖς ζῴοις ἐστίν, οἷον
ὀστοῦ καὶ σαρκὸς καὶ τῶν ἄλλων τῶν τοιούτων.

---

[a] For the threefold series cf. *De gen. an.* 714 a 9 ff. This
first " composition " seems to be intended to cover *non-organic* compounds.

[b] " Dynamis " here is clearly the pre-Aristotelian technical
term. See Introduction, p. 30.   [c] See *De gen. et corr.* chh. 2, 8.

[d] In some contexts, " fluid " and " solid " seem more

BOOK II

I HAVE already described with considerable detail in my *Researches upon Animals* what and how many are the parts of which the various animals are composed. We must now leave on one side what was said there, as our present task is to consider what are the *causes* through which each animal is as I there described it.

Three sorts of composition can be distinguished. (1) First of all [a] we may put composition out of the Elements (as some call them), viz. Earth, Air, Water, Fire; or perhaps it is better to say *dynameis* [b] instead of Elements—some of the *dynameis*, that is, not all, as I have stated previously elsewhere.[c] It is just these four, the fluid substance, the solid,[d] the hot, and the cold, which are the matter of composite bodies; and the other differences and qualities—such as heaviness lightness, firmness looseness, roughness smoothness, etc.—which composite bodies present are subsequent upon these. (2) The second sort of composition is the composition of the "uniform"[e] substances found in animals (such as bone, flesh, etc.). These also are composed out of the primary

appropriate; in others, "moist" and "dry" (the traditional renderings). Aristotle defines them at *De gen. and corr.* 329 b 30. See also below, 649 b 9. I have normally translated them "fluid" and "solid" throughout.

[e] "Uniform," "non-uniform"; see Introduction, p. 28.

τρίτη δὲ καὶ τελευταία κατ᾽ ἀριθμὸν ἡ τῶν ἀνομοιομερῶν, οἷον προσώπου καὶ χειρὸς καὶ τῶν
τοιούτων μορίων.

25 Ἐπεὶ δ᾽ ἐναντίως ἐπὶ τῆς γενέσεως ἔχει καὶ τῆς
οὐσίας—τὰ γὰρ ὕστερα τῇ γενέσει πρότερα τὴν
φύσιν ἐστί, καὶ πρῶτον τὸ τῇ γενέσει τελευταῖον
(οὐ γὰρ οἰκία πλίνθων ἕνεκέν ἐστι καὶ λίθων, ἀλλὰ
ταῦτα τῆς οἰκίας· ὁμοίως δὲ τοῦτ᾽ ἔχει καὶ περὶ τὴν
ἄλλην ὕλην· οὐ μόνον δὲ φανερὸν ὅτι τοῦτον ἔχει τὸν
30 τρόπον ἐκ τῆς ἐπαγωγῆς, ἀλλὰ καὶ κατὰ τὸν λόγον·
πᾶν γὰρ τὸ γινόμενον ἔκ τινος καὶ εἴς τι ποιεῖται
τὴν γένεσιν, καὶ ἀπ᾽ ἀρχῆς ἐπ᾽ ἀρχήν, ἀπὸ τῆς
πρώτης κινούσης καὶ ἐχούσης ἤδη τινὰ φύσιν ἐπὶ
τινα μορφὴν ἢ τοιοῦτον ἄλλο τέλος· ἄνθρωπος γὰρ
ἄνθρωπον καὶ φυτὸν γεννᾷ φυτὸν ἐκ τῆς περὶ
35 ἕκαστον ὑποκειμένης ὕλης)—τῷ μὲν οὖν χρόνῳ
646 b προτέραν τὴν ὕλην ἀναγκαῖον εἶναι καὶ τὴν γένεσιν,
τῷ λόγῳ δὲ τὴν οὐσίαν καὶ τὴν ἑκάστου μορφήν.
δῆλον δ᾽ ἂν λέγῃ τις τὸν λόγον τῆς γενέσεως· ὁ μὲν
γὰρ τῆς οἰκοδομήσεως λόγος ἔχει τὸν τῆς οἰκίας,
ὁ δὲ τῆς οἰκίας οὐκ ἔχει τὸν τῆς οἰκοδομήσεως.
5 ὁμοίως δὲ τοῦτο συμβέβηκε καὶ ἐπὶ τῶν ἄλλων.
ὥστε τὴν μὲν τῶν στοιχείων ὕλην ἀναγκαῖον εἶναι
τῶν ὁμοιομερῶν ἕνεκεν· ὕστερα γὰρ ἐκείνων ταῦτα

---

^a Or, " efficient."
^b Or, " in thought," " in conception."
^c Almost represented here by " definition."

substances. (3) The third and last is the composition of the " non-uniform " parts of the body, such as face, hand, and the like.

Now the order of things in the process of formation is the reverse of their real and essential order ; I mean that the later a thing comes in the formative process the earlier it comes in the order of Nature, and that which comes at the end of the process is at the beginning in the order of Nature. Just so bricks and stone come chronologically *before* the house, although the house is the purpose which they subserve, and not *vice versa*. And the same applies to materials of every kind. Thus the truth of my statement can be shown by induction ; but it can also be demonstrated logically, as follows. Everything which is in process of formation is in passage from one thing towards another thing, *i.e.* from one Cause towards another Cause ; in other words, it proceeds from a primary motive [a] Cause which to begin with possesses a definite nature, towards a Form or another such End. For example, a man begets a man and a plant begets a plant. These new individuals are made out of the substrate matter appropriate in each case. Thus, matter and the process of formation must come first in time, but logically [b] the real essence and the Form of the thing comes first. This is clear if we state the *logos* [c] of such a process. For example, the *logos* of the process of building includes the *logos* of a house, but that of a house does not include that of the process of building. And this holds good in all such cases. Hence we see that the matter, viz. the Elements, must exist for the sake of the uniform substances, because these come later in the process of formation than

τῇ γενέσει, τούτων δὲ τὰ ἀνομοιομερῆ. ταῦτα γὰρ
ἤδη τὸ τέλος ἔχει καὶ τὸ πέρας, ἐπὶ τοῦ τρίτου
λαβόντα τὴν σύστασιν ἀριθμοῦ, καθάπερ ἐπὶ πολλῶν
10 συμβαίνει τελειοῦσθαι τὰς γενέσεις.

Ἐξ ἀμφοτέρων μὲν οὖν τὰ ζῷα συνέστηκε τῶν
μορίων τούτων, ἀλλὰ τὰ ὁμοιομερῆ τῶν ἀνομοιομερῶν ἕνεκέν ἐστιν· ἐκείνων γὰρ ἔργα καὶ πράξεις
εἰσίν, οἷον ὀφθαλμοῦ καὶ μυκτῆρος καὶ τοῦ προσώπου παντὸς καὶ δακτύλου καὶ χειρὸς καὶ παντὸς
15 τοῦ βραχίονος. πολυμόρφων δὲ τῶν πράξεων καὶ
τῶν κινήσεων ὑπαρχουσῶν τοῖς ζῴοις ὅλοις τε καὶ
τοῖς μορίοις τοῖς τοιούτοις, ἀναγκαῖον ἐξ ὧν σύγ-
κεινται τὰς δυνάμεις ἀνομοίας ἔχειν· πρὸς μὲν γὰρ
τινα μαλακότης χρήσιμος πρὸς δέ τινα σκληρότης,
καὶ τὰ μὲν τάσιν ἔχειν δεῖ τὰ δὲ κάμψιν.

20 Τὰ μὲν οὖν ὁμοιομερῆ κατὰ μέρος διείληφε τὰς
δυνάμεις τὰς τοιαύτας (τὸ μὲν γὰρ αὐτῶν ἐστι
μαλακὸν τὸ δὲ σκληρόν, καὶ τὸ μὲν ὑγρὸν τὸ δὲ
ξηρόν, καὶ τὸ μὲν¹ γλίσχρον τὸ δὲ κραῦρον), τὰ
δ' ἀνομοιομερῆ κατὰ πολλὰς καὶ συγκειμένας
ἀλλήλαις· ἑτέρα γὰρ πρὸς τὸ πιέσαι τῇ χειρὶ χρή-
25 σιμος δύναμις καὶ πρὸς τὸ λαβεῖν. διόπερ ἐξ
ὀστῶν καὶ νεύρων καὶ σαρκὸς καὶ τῶν ἄλλων τῶν
τοιούτων συνεστήκασι τὰ ὀργανικὰ τῶν μορίων,
ἀλλ' οὐκ ἐκεῖνα ἐκ τούτων.

Ὡς μὲν οὖν ἕνεκά τινος διὰ ταύτην τὴν αἰτίαν
ἔχει περὶ τούτων τὸν εἰρημένον τρόπον, ἐπεὶ δὲ
ζητεῖται καὶ πῶς ἀναγκαῖον ἔχειν οὕτω, φανερὸν ὅτι
30 προϋπῆρχεν οὕτω πρὸς ἄλληλα ἔχοντα ἐξ ἀνάγκης

¹ τὸ μὲν PZ : om. vulg.

the Elements; just so the non-uniform parts come later than the uniform. The non-uniform parts, indeed, whose manner of composition is that of the third sort, have reached the goal and End of the whole process; and we often find that processes of formation reach their completion at this point.

Now animals are composed out of both of these two sorts of parts, uniform and non-uniform; the former, however, are for the sake of the latter, as it is to the latter that actions and operations belong (*e.g.* eye, nose, the face as a whole, finger, hand, the arm as a whole). And inasmuch as the actions and movements both of an animal as a whole and of its parts are manifold, the substances out of which these are composed must of necessity possess divers *dynameis*. Softness is is useful for some purposes, hardness for others; some parts must be able to stretch, some to bend.

In the uniform parts, then, such *dynameis* are found apportioned out separately: one of the parts, for instance, will be soft, another hard, while one is fluid, another solid; one viscous, another brittle. In the non-uniform parts, on the other hand, these *dynameis* are found in combination, not singly. For example, the hand needs one *dynamis* for the action of compressing and another for that of grasping. Hence it is that the instrumental parts of the body are composed of bones, sinews, flesh, and the rest of them, and not the other way round.

The Cause which I have just stated as controlling the relation between them is, of course, a Final Cause; but when we go on to inquire in what sense it is *necessary* that they should be related as they are, it becomes clear that they must of necessity have been thus related to each other from the beginning.

646 b

τὰ μὲν γὰρ ἀνομοιομερῆ ἐκ τῶν ὁμοιομερῶν ἐνδέχεται συνεστάναι, καὶ ἐκ πλειόνων καὶ ἑνός, οἷον
ἔνια τῶν σπλάγχνων· πολύμορφα γὰρ τοῖς σχήμασιν, ἐξ ὁμοιομεροῦς ὄντα σώματος ὡς εἰπεῖν
ἁπλῶς. τὰ δ' ὁμοιομερῆ ἐκ τούτων ἀδύνατον· τὸ
85 γὰρ ὁμοιομερὲς πόλλ' ἂν εἴη ἀνομοιομερῆ.

647 a Διὰ μὲν οὖν ταύτας τὰς αἰτίας τὰ μὲν ἁπλᾶ καὶ
ὁμοιομερῆ, τὰ δὲ σύνθετα καὶ ἀνομοιομερῆ τῶν
μορίων ἐν τοῖς ζῴοις ἐστίν.

Ὄντων δὲ τῶν μὲν ὀργανικῶν μερῶν τῶν δ'
αἰσθητηρίων ἐν τοῖς ζῴοις, τῶν μὲν ὀργανικῶν
5 ἕκαστον ἀνομοιομερές ἐστιν, ὥσπερ εἶπον πρότερον,
ἡ δ' αἴσθησις ἐγγίνεται πᾶσιν ἐν τοῖς ὁμοιομερέσιν
διὰ τὸ τῶν αἰσθήσεων ὁποιανοῦν ἑνός τινος εἶναι
γένους, καὶ τὸ αἰσθητήριον ἑκάστου δεκτικὸν εἶναι
τῶν αἰσθητῶν. πάσχει δὲ τὸ δυνάμει ὂν ὑπὸ τοῦ
ἐνεργείᾳ ὄντος, ὥστ' ἔστι τὸ αὐτὸ τῷ γένει, καὶ
10 ⟨εἰ⟩¹ ἐκεῖνο ἕν, καὶ τοῦτο ἕν, καὶ διὰ τοῦτο χεῖρα
μὲν ἢ πρόσωπον ἢ τῶν τοιούτων τι μορίων οὐδεὶς
ἐγχειρεῖ λέγειν τῶν φυσιολόγων τὸ μὲν εἶναι γῆν,
τὸ δ' ὕδωρ, τὸ δὲ πῦρ· τῶν δ' αἰσθητηρίων ἕκαστον

¹ ⟨εἰ⟩ Ogle.

ᵃ The translation " sense-organ " must not be taken to
imply that the part through which the sense functions is an

112

It is possible for the non-uniform parts to be constructed out of the uniform substances, either out of many of them, or out of one only. (Examples of the latter are furnished by certain of the viscera, which, although they are of manifold shapes and forms, yet for all practical purposes may be said to consist of one only of the uniform substances.) But it is impossible for the uniform substances to be constructed out of the non-uniform parts: for then we should have an uniform substance consisting of several non-uniform parts, which is absurd.

These, then, are the Causes owing to which some of the parts of animals are simple and uniform; while others are composite and non-uniform.

Now the parts can also be divided up into (a) instrumental parts and (b) sense-organs.[a] And we may say that each of the instrumental parts of the body, as I have stated earlier, is always non-uniform, while sensation in all cases takes place in parts that are uniform. The reasons why this is so are the following: The function of each of the senses is concerned with a single kind of sensible objects; and the sense-organ in each case **must** be such as can apprehend those objects. Now when one thing affects another, the thing which is affected must be *potentially* what the other is *actually*; so both are the same in kind, and therefore if the affecting thing is single, the affected one is single too. Hence we find that while with regard to the parts of the body such as hand, or face, none of the physiologers attempts to say that one of them is earth, and another water, and another fire; yet they do conjoin

" organ " in the stricter meaning of the word. " Organs " are normally " non-uniform," sense-organs are " uniform."

πρὸς ἕκαστον ἐπιζευγνύουσι τῶν σταιχείων, τὸ μὲν
ἀέρα φάσκοντες εἶναι, τὸ δὲ πῦρ.

Οὔσης δὲ τῆς αἰσθήσεως ἐν τοῖς ἁπλοῖς μέρεσιν
15 εὐλόγως μάλιστα συμβαίνει τὴν ἀφὴν ἐν ὁμοιομερεῖ
μὲν ἥκιστα δ' ἁπλῷ τῶν αἰσθητηρίων ἐγγίνεσθαι·
μάλιστα γὰρ αὕτη δοκεῖ πλειόνων εἶναι γενῶν, καὶ
πολλὰς ἔχειν ἐναντιώσεις τὸ ὑπὸ ταύτην αἰσθητόν,
θερμὸν ψυχρόν, ξηρὸν ὑγρὸν καὶ εἴ τι ἄλλο τοιοῦτον·
20 καὶ τὸ τούτων αἰσθητήριον, ἡ σὰρξ καὶ τὸ ταύτῃ
ἀνάλογον, σωματωδέστατόν ἐστι τῶν αἰσθητηρίων.
ἐπεὶ δ' ἀδύνατον εἶναι ζῷον ἄνευ αἰσθήσεως, καὶ
διὰ τοῦτο ἂν εἴη ἀναγκαῖον ἔχειν τοῖς ζῴοις ἔνια
μόρια ὁμοιομερῆ· ἡ μὲν γὰρ αἴσθησις ἐν τούτοις,
αἱ δὲ πράξεις διὰ τῶν ἀνομοιομερῶν ὑπάρχουσιν
αὐτοῖς.

25 Τῆς δ' αἰσθητικῆς δυνάμεως καὶ τῆς κινούσης
τὸ ζῷον καὶ τῆς θρεπτικῆς ἐν ταὐτῷ μορίῳ τοῦ
σώματος οὔσης, καθάπερ ἐν ἑτέροις εἴρηται πρότερον, ἀναγκαῖον τὸ ἔχον πρῶτον μόριον τὰς
τοιαύτας ἀρχάς, ᾗ μέν ἐστι δεκτικὸν πάντων τῶν
αἰσθητῶν, τῶν ἁπλῶν εἶναι μορίων, ᾗ δὲ κινητικὸν
30 καὶ πρακτικόν, τῶν ἀνομοιομερῶν. διόπερ ἐν μὲν
τοῖς ἀναίμοις ζῴοις τὸ ἀνάλογον, ἐν δὲ τοῖς ἐναίμοις
ἡ καρδία τοιοῦτόν ἐστιν· διαιρεῖται μὲν γὰρ εἰς
ὁμοιομερῆ καθάπερ τῶν ἄλλων σπλάγχνων ἕκαστον,
διὰ δὲ τὴν τοῦ σχήματος μορφὴν ἀνομοιομερές
ἐστιν. ταύτῃ δ' ἠκολούθηκε καὶ τῶν ἄλλων τῶν

---

each of the *sense*-organs with one of the elementary
substances, and they assert that this sense-organ is
air, this one fire.

Sensation thus takes place in the simple parts of
the body. The organ in which touch takes place is,
however, as we should expect, the least simple of all
the sense-organs, though of course like the others it
is uniform. This is evidently because the sense of
touch deals with more kinds of sense-objects than
one : and these objects may have several sorts of
oppositions in them, *e.g.* hot and cold, solid and fluid,
and the like. So the sense-organ which deals with
these—viz. the flesh, or its counterpart—is the most
corporeal of all the sense-organs. Another reason
we might adduce why animals must of necessity
possess some uniform parts at any rate, is that there
cannot be such a thing as an animal with no power
of sensation, and the seat of sensation is the uniform
parts. (The non-uniform parts supply the means for
the various activities, not for sensation.)

Further, since the faculties of sensation and of
motion and of nutrition are situated in one and the
same part of the body, as I stated in an earlier work,[a]
that part, which is the primary seat of these principles,
must of necessity be included not only among the
simple parts but also among the non-uniform parts—
the former in virtue of receiving all that is perceived
through the senses, the latter because it has to do
with motion and action. In blooded animals this
part is the heart, in bloodless animals the counterpart
of the heart, for the heart, like every one of the other
viscera, can be divided up into uniform pieces ; but
on the other hand it is non-uniform owing to its
shape and formation. Every one of the other so-

647 a

35 καλουμένων σπλάγχνων ἕκαστον· ἐκ τῆς αὐτῆς

647 b γὰρ ὕλης συνεστᾶσιν· αἱματικὴ γὰρ ἡ φύσις πάντων
αὐτῶν διὰ τὸ τὴν θέσιν ἔχειν ἐπὶ πόροις φλεβικοῖς
καὶ διαλήψεσιν. καθάπερ οὖν ῥέοντος ὕδατος ἰλύς,
τἆλλα σπλάγχνα τῆς διὰ τῶν φλεβῶν ῥύσεως τοῦ
αἵματος οἷον προχεύματά ἐστιν· ἡ δὲ καρδία, διὰ
5 τὸ τῶν φλεβῶν ἀρχὴ εἶναι καὶ ἔχειν ἐν αὐτῇ τὴν
δύναμιν τὴν δημιουργοῦσαν τὸ αἷμα πρώτην, εὔλογον ἐξ οἵας ἄρχεται[1] τροφῆς ἐκ τοιαύτης συνεστάναι καὶ αὐτήν.

Διότι μὲν οὖν αἱματικὰ τὴν μορφὴν τὰ σπλάγχνα
ἐστὶν εἴρηται, καὶ διότι τῇ μὲν ὁμοιομερῆ τῇ δ'
ἀνομοιομερῆ.

10 II. Τῶν δ' ὁμοιομερῶν μορίων ἐν τοῖς ζῴοις ἐστὶ
τὰ μὲν μαλακὰ καὶ ὑγρά, τὰ δὲ σκληρὰ καὶ στερεά,
ὑγρὰ μὲν ἢ ὅλως ἢ ἕως ἂν ᾖ ἐν τῇ φύσει, οἷον
αἷμα, ἰχώρ, πιμελή, στέαρ, μυελός, γονή, χολή,
γάλα ἐν τοῖς ἔχουσι, σάρξ, καὶ τὰ τούτοις ἀνάλογον·
15 οὐ γὰρ ἅπαντα τὰ ζῷα τούτων τῶν μορίων τέ-
τευχεν, ἀλλ' ἔνια τῶν ἀνάλογον τούτων τισίν. τὰ
δὲ ξηρὰ καὶ στερεὰ τῶν ὁμοιομερῶν ἐστίν, οἷον
ὀστοῦν, ἄκανθα, νεῦρον, φλέψ. καὶ γὰρ τῶν ὁμοιομερῶν ἡ διαίρεσις ἔχει διαφοράν· ἔστι γὰρ ὡς ἐνίων
τὸ μέρος ὁμώνυμον τῷ ὅλῳ, οἷον φλεβὸς φλέψ, ἔστι
20 δ' ὡς οὐχ ὁμώνυμον, ἀλλὰ προσώπου πρόσωπον
οὐδαμῶς.

---

[1] οἵας corr. in loco plurium litterarum Y : οἷ ας Z (ας Z² in
rasura). ἄρχεται (vel ἀρχή ἐστι) Peck, cf. 666 a 7, b 1, etc. :
δέχεται vulg.

called viscera follows suit. They are all composed of the same material, as they all have a sanguineous character, and this is because they are situated upon the channels of the blood-vessels and on the points of ramification. All these viscera (excluding the heart) may be compared to the mud which a running stream deposits; they are as it were deposits left by the current of blood in the blood-vessels. As for the heart itself, since it is the starting-point of the blood-vessels and contains the substance (*dynamis*) by which the blood is first fashioned, it is only to be expected that it will itself be composed out of that form of nutriment which it originates.

We have now stated why the viscera are sanguineous in formation, and why in one aspect they are uniform and in another non-uniform.

II. Of the uniform parts in animals, some are soft and fluid, some hard and firm. Some are permanently fluid, some are fluid only so long as they are in the living organism—*e.g.* blood, serum, lard, suet, marrow, semen, bile, milk (in the lactiferous species), flesh. (As these parts are of course not to be found in all animals, add to this list their counterparts.) Other of the uniform parts are solid and firm : examples are bone, fish-spine, sinew, blood-vessel. This division of the uniform parts admits a further distinction : There are some of them of which a portion has, in one sense, the same name as the whole (*e.g.* a portion of a blood-vessel has the name blood-vessel), and in another sense has not the same name. (In no sense is this the case with a non-uniform part ; for instance, a portion of a face cannot be called face at all.)

The uniform parts.

647 b

Πρῶτον μὲν οὖν καὶ τοῖς ὑγροῖς μορίοις καὶ τοῖς
ξηροῖς πολλοὶ τρόποι τῆς αἰτίας εἰσίν. τὰ μὲν γὰρ
ὡς ὕλη τῶν μερῶν τῶν ἀνομοιομερῶν ἐστιν (ἐκ
τούτων γὰρ συνέστηκεν ἕκαστον τῶν ὀργανικῶν
μερῶν, ἐξ ὀστῶν καὶ νεύρων καὶ σαρκῶν καὶ ἄλλων
25 τοιούτων συμβαλλομένων τὰ μὲν εἰς τὴν οὐσίαν τὰ
δ᾽ εἰς τὴν ἐργασίαν), τὰ δὲ τροφὴ τούτοις τῶν
ὑγρῶν ἐστί (πάντα γὰρ ἐξ ὑγροῦ λαμβάνει τὴν
αὔξησιν), τὰ δὲ περιττώματα συμβέβηκεν εἶναι
τούτων, οἷον τήν τε τῆς ξηρᾶς τροφῆς ὑπόστασιν
καὶ τὴν τῆς ὑγρᾶς τοῖς ἔχουσι κύστιν.

Αὐτῶν δὲ τούτων αἱ διαφοραὶ πρὸς ἄλληλα τοῦ
30 βελτίονος ἕνεκέν εἰσιν, οἷον τῶν τε ἄλλων καὶ
αἵματος πρὸς αἷμα· τὸ μὲν γὰρ λεπτότερον τὸ δὲ
παχύτερον καὶ τὸ μὲν καθαρώτερόν ἐστι τὸ δὲ
θολερώτερον, ἔτι δὲ τὸ μὲν ψυχρότερον τὸ δὲ θερμότερον, ἔν τε τοῖς μορίοις τοῦ ἑνὸς ζῴου (τὸ γὰρ
35 ἐν τοῖς ἄνω μέρεσι πρὸς τὰ κάτω μόρια διαφέρει
ταύταις ταῖς διαφοραῖς) καὶ ἑτέρῳ πρὸς ἕτερον.
648 a καὶ ὅλως τὰ μὲν ἔναιμα τῶν ζῴων ἐστί, τὰ δ᾽ ἀντὶ
τοῦ αἵματος ἔχει ἕτερόν τι μόριον τοιοῦτον.

Ἔστι δ᾽ ἰσχύος μὲν ποιητικώτερον τὸ παχύτερον
αἷμα καὶ θερμότερον, αἰσθητικώτερον δὲ καὶ νοερώ-
τερον τὸ λεπτότερον καὶ ψυχρότερον. τὴν αὐτὴν δ᾽
5 ἔχει διαφορὰν καὶ τὸ ἀνάλογον ὑπάρχον[1] πρὸς τὸ

---

[1] τὸ . . . ὑπάρχον P: τῶν . . . ὑπαρχόντων vulg.

---

[a] Or, "reason."
[b] See Introduction, pp. 32 ff.
[c] See Introduction, pp. 28 ff.
[d] With this passage compare Hippocrates, Περὶ διαίτης.
i. 35. See also below, 650 b 24 ff., and Introduction, pp.
37-39.

118

Now first of all there are many sorts of Cause [a] to which the existence of these uniform parts, both the fluid and the solid ones, is to be ascribed. Some of them act as the material for the non-uniform parts (*e.g.* each of the instrumental parts is composed of these uniform parts—bones, sinews, fleshes, and the like, which contribute either to its essence, or else towards the discharge of its proper function). Another group of the uniform parts—fluid ones—act as nutriment for the ones just mentioned, since everything that grows gets the material for its growth from what is fluid ; and yet a third group are residues [b] produced from the second group : examples, the excrement deposited from the solid nutriment and (in those animals which have a bladder) from the fluid nutriment.

Further, variations are found among different specimens of these uniform parts, and this is to subserve a good purpose. Blood is an excellent illustration. Blood can be thin or thick, clear or muddy, cold or warm ; and it can be different in different parts of the same animal : instances are known of animals in which the blood in the upper parts differs from that in the lower parts in respect of the characteristics just enumerated. And of course the blood of one animal differs from that of another. And there is the general division between the animals that have blood and those which instead of it have a part [c] which is similar to it though not actually blood.

The thicker and warmer the blood is, the more it makes for strength ; if it tends to be thin and cold, it is conducive to sensation and intelligence.[d] The same difference holds good with the counterpart of

119

αἷμα· διὸ καὶ μέλιτται καὶ ἄλλα τοιαῦτα ζῷα φρονιμώτερα τὴν φύσιν ἐστὶν ἐναίμων πολλῶν, καὶ τῶν
ἐναίμων τὰ ψυχρὸν ἔχοντα καὶ λεπτὸν αἷμα φρονιμώτερα τῶν ἐναντίων ἐστίν. ἄριστα δὲ τὰ θερμὸν
10 ἔχοντα καὶ λεπτὸν καὶ καθαρόν· ἅμα γὰρ πρός τ᾽
ἀνδρείαν τὰ τοιαῦτα καὶ πρὸς φρόνησιν ἔχει καλῶς.
διὸ καὶ τὰ ἄνω μόρια πρὸς τὰ κάτω ταύτην ἔχει
τὴν διαφοράν, καὶ πρὸς τὸ θῆλυ αὖ τὸ ἄρρεν, καὶ
τὰ δεξιὰ πρὸς τὰ ἀριστερὰ τοῦ σώματος.

Ὁμοίως δὲ καὶ περὶ τῶν ἄλλων καὶ τῶν τοιούτων
15 μορίων καὶ τῶν ἀνομοιομερῶν ὑποληπτέον ἔχειν
τὴν διαφοράν, τὰ μὲν πρὸς τὰ ἔργα καὶ τὴν οὐσίαν
ἑκάστῳ τῶν ζῴων, τὰ δὲ πρὸς τὸ βέλτιον ἢ χεῖρον,
οἷον ἐχόντων ὀφθαλμοὺς ἀμφοτέρων τὰ μέν ἐστι
σκληρόφθαλμα τὰ δ᾽ ὑγρόφθαλμα, καὶ τὰ μὲν οὐκ
ἔχει βλέφαρα τὰ δ᾽ ἔχει, πρὸς τὸ τὴν ὄψιν ἀκρι-
βεστέραν εἶναι.

20 Ὅτι δ᾽ ἀναγκαῖον ἔχειν ἢ αἷμα ἢ τὸ τούτῳ τὴν
αὐτὴν ἔχον φύσιν, καὶ τίς ἐστιν ἡ τοῦ αἵματος
φύσις, πρῶτον διελομένοις περὶ θερμοῦ καὶ ψυχροῦ,
οὕτω καὶ περὶ τούτου θεωρητέον τὰς αἰτίας. πολλῶν
γὰρ ἡ φύσις ἀνάγεται πρὸς ταύτας τὰς ἀρχάς, καὶ
25 πολλοὶ διαμφισβητοῦσι ποῖα θερμὰ καὶ ποῖα ψυχρὰ
τῶν ζῴων ἢ τῶν μορίων. ἔνιοι γὰρ τὰ ἔνυδρα τῶν
πεζῶν θερμότερά φασιν εἶναι, λέγοντες ὡς ἐπανισοῖ
τὴν ψυχρότητα τοῦ τόπου ἡ τῆς φύσεως αὐτῶν

---

^a This sentiment, which at first sight appears to go against
the Aristotelian teleology, is supported by actual instances,
e.g. the horns of the backward-grazing oxen (659 a 19) and
of the deer (663 a 11) and the talons of certain birds (694 a 20).

blood in other creatures : and thus we can explain
why bees and other similar creatures are of a more
intelligent nature than many animals that have
blood in them ; and among the latter class, why
some (viz. those whose blood is cold and thin) are
more intelligent than others. Best of all are those
animals whose blood is hot and also thin and clear ;
they stand well both for courage and for intelligence.
Consequently, too, the upper parts of the body have
this pre-eminence over the lower parts ; the male over
the female ; and the right side of the body over the
left.

What applies to the blood applies as well to the
other uniform parts and also to the non-uniform
parts ; similar variations occur. And it must be
supposed that these variations either have some reference to the activities of the creatures and to their
essential nature, or else bring them some advantage
or disadvantage.[a] For example, the eyes of some
creatures are hard in substance, of others, fluid ;
some have eyelids, others have not. In both cases
the difference is for the sake of greater accuracy of
vision.

Before we can go on to consider the reasons why all
animals must of necessity have blood in them or something which possesses the same nature, and also what
the nature of blood itself is, we must first come to
some decision about hot and cold. The nature of many
things is to be referred back to these two principles,
and there is much dispute about which animals and
which parts of animals are hot and which are cold.
Some assert that water-animals are hotter than land-
animals, and they allege that the creatures' natural
heat makes up for the coldness of their habitat.

121

648 a

θερμότης, καὶ τὰ ἄναιμα τῶν ἐναίμων καὶ τὰ θήλεα
τῶν ἀρρένων, οἷον Παρμενίδης τὰς γυναῖκας τῶν
30 ἀνδρῶν θερμοτέρας εἶναί φησι καὶ ἕτεροί τινες ὡς
διὰ τὴν θερμότητα καὶ πολυαιμούσαις γινομένων
τῶν γυναικείων, Ἐμπεδοκλῆς δὲ τοὐναντίον· ἔτι δ'
αἷμα καὶ χολὴν οἱ μὲν θερμὸν ὁποτερονοῦν εἶναί
φασιν αὐτῶν, οἱ δὲ ψυχρόν. εἰ δ' ἔχει τοσαύτην
τὸ θερμὸν καὶ τὸ ψυχρὸν ἀμφισβήτησιν, τί χρὴ
35 περὶ τῶν ἄλλων ὑπολαβεῖν; ταῦτα γὰρ ἡμῖν ἐν-
αργέστατα τῶν περὶ τὴν αἴσθησιν.

Ἔοικε δὲ διὰ τὸ πολλαχῶς λέγεσθαι τὸ θερμό-
648 b τερον ταῦτα συμβαίνειν· ἕκαστος γὰρ δοκεῖ τι
λέγειν τἀναντία λέγων. διὸ δεῖ μὴ λανθάνειν πῶς
δεῖ τῶν φύσει συνεστώτων τὰ μὲν θερμὰ λέγειν τὰ
δὲ ψυχρὰ καὶ τὰ μὲν ξηρὰ τὰ δ' ὑγρά, ἐπεὶ ὅτι γ'
αἴτια ταῦτα σχεδὸν καὶ θανάτου καὶ ζωῆς ἔοικεν
5 εἶναι φανερόν, ἔτι δ' ὕπνου καὶ ἐγρηγόρσεως καὶ
ἀκμῆς καὶ γήρως καὶ νόσου καὶ ὑγιείας, ἀλλ' οὐ
τραχύτητες καὶ λειότητες οὐδὲ βαρύτητες καὶ κου-
φότητες οὐδ' ἄλλο τῶν τοιούτων οὐδὲν ὡς εἰπεῖν.
καὶ τοῦτ' εὐλόγως συμβέβηκεν· καθάπερ γὰρ ἐν
ἑτέροις εἴρηται πρότερον, ἀρχαὶ τῶν φυσικῶν
10 στοιχείων αὗταί εἰσι, θερμὸν καὶ ψυχρὸν καὶ
ξηρὸν καὶ ὑγρόν.

Πότερον οὖν ἁπλῶς λέγεται τὸ θερμὸν ἢ πλεοναχῶς; δεῖ δὴ λαβεῖν τί ἔργον τοῦ θερμοτέρου, ἢ

ᵃ See above, 646 a 15, and note.

Further, it is asserted that bloodless animals are hotter than those that have blood ; and that females are hotter than males. Parmenides and others, for instance, assert that women are hotter than men on the ground of the menstrual flow, which they say is due to their heat and the abundance of their blood. Empedocles, however, maintains the opposite opinion. Again, some say that blood is hot and bile cold, others that bile is hot and blood cold. And if there is so much dispute about the hot and the cold, which after all are the most distinct of the things which affect our senses, what line are we to take about the rest of them ?

Now it looks as if the difficulty is due to the term "hotter" being used in more senses than one, as there seems to be something in what each of these writers says, though their statements are contradictory. Hence we must permit no ambiguity in our application of the descriptions "hot" and "cold," "solid" and "fluid" to the substances that are found produced by nature. It is surely sufficiently established that these four principles (and not to any appreciable extent roughness, smoothness, heaviness, lightness, or any such things) are practically the causes controlling life and death, not to mention sleep and waking, prime and age, disease and health. And this, after all, is but reasonable, because (as I have said previously in another work) these four—hot, cold, solid, fluid—are the principles of the physical Elements.[a]

Let us consider, then, whether the term "hot" has one sense or several. To decide this point, we must find out what is the particular effect which a body has in virtue of being hotter than another, or, if there are several such effects, how many there are.

The primary substances : (1) "hot" and "cold."

123

648 b

πόσα, εἰ πλείω. ἕνα μὲν δὴ τρόπον λέγεται μᾶλλον
θερμὸν ὑφ' οὗ μᾶλλον θερμαίνεται τὸ ἁπτόμενον,
15 ἄλλως δὲ τὸ μᾶλλον αἴσθησιν ἐμποιοῦν ἐν τῷ
θιγγάνειν, καὶ τοῦτ', ἐὰν μετὰ λύπης. ἔστι δ' ὅτε
δοκεῖ τοῦτ' εἶναι ψεῦδος· ἐνίοτε γὰρ ἡ ἕξις αἰτία
τοῦ ἀλγεῖν αἰσθανομένοις. ἔτι τὸ τηκτικώτερον τοῦ
τηκτοῦ καὶ τοῦ καυστοῦ καυστικώτερον. ἔτι ἐὰν
ᾖ τὸ μὲν πλέον τὸ δ' ἔλαττον τὸ αὐτό, τὸ πλέον τοῦ
20 ἐλάττονος θερμότερον. πρὸς δὲ τούτοις δυοῖν τὸ
μὴ ταχέως ψυχόμενον ἀλλὰ βραδέως θερμότερον,
καὶ τὸ θᾶττον θερμαινόμενον τοῦ θερμαινομένου
βραδέως θερμότερον εἶναι τὴν φύσιν φαμέν, ὡς τὸ
μὲν ἐναντίον ὅτι πόρρω, τὸ δ' ὅμοιον ὅτι ἐγγύς.
λέγεται μὲν οὖν εἰ μὴ πλεοναχῶς, ἀλλὰ τοσαυταχῶς
25 ἕτερον ἑτέρου θερμότερον· τούτους δὲ τοὺς τρόπους
ἀδύνατον ὑπάρχειν τῷ αὐτῷ πάντας· θερμαίνει μὲν
γὰρ μᾶλλον τὸ ζέον ὕδωρ τῆς φλογός, καίει δὲ καὶ
τήκει τὸ καυστὸν καὶ τηκτὸν ἡ φλόξ, τὸ δ' ὕδωρ
οὐδέν. ἔτι θερμότερον μὲν τὸ ζέον ὕδωρ ἢ πῦρ
ὀλίγον, ψύχεται δὲ καὶ θᾶττον καὶ μᾶλλον τὸ θερμὸν
30 ὕδωρ μικροῦ πυρός· οὐ γὰρ γίνεται ψυχρὸν πῦρ,
ὕδωρ δὲ γίνεται πᾶν. ἔτι θερμότερον μὲν κατὰ τὴν
ἁφὴν τὸ ζέον ὕδωρ, ψύχεται δὲ θᾶττον καὶ πήγνυται
τοῦ ἐλαίου. ἔτι τὸ αἷμα κατὰ μὲν τὴν ἁφὴν θερμότερον ὕδατος καὶ ἐλαίου, πήγνυται δὲ θᾶττον. ἔτι
λίθοι καὶ σίδηρος καὶ τὰ τοιαῦτα θερμαίνεται μὲν
35 βραδύτερον ὕδατος, καίει δὲ θερμανθέντα μᾶλλον.
πρὸς δὲ τούτοις τῶν λεγομένων θερμῶν τὰ μὲν

ᵃ Alluding, perhaps, to the expansion due to heat.

A is said to be " hotter " than *B* (1) if that which comes into contact with it is heated more by it than by *B*. (2) If it produces a more violent sensation when touched, and especially if the sensation is accompanied by pain. (The latter is not always a true indication, since sometimes the pain is due to the condition of the percipient.) (3) If it is a better melting or burning agent. (4) If it is of the same composition as *B*, but greater in bulk,ᵃ it is said to be " hotter " than *B*, and in addition (5) if it cools more slowly than *B*, or warms up more quickly : in both these cases we call the thing " hotter " in its nature—as we call one thing " contrary " to another when it is far removed from it, and " like " it when it is near it. But although the senses in which one thing is said to be " hotter " than another are certainly as many as this, if not more, yet no one thing can be " hotter " in all of these ways at once. For instance, boiling water can impart heat more effectively than flame ; but flame is able to cause burning and melting, whereas water is not. Again, boiling water is hotter than a small fire, but the hot water will cool off more quickly and more thoroughly than the small fire, since fire does not become cold, but all water does. Again, boiling water is hotter to the touch than oil, yet it cools and solidifies more quickly. And again, blood is warmer to the touch than either water or oil, yet it congeals more quickly. Again, stone and iron and such substances get hot more slowly than water, but once they are hot they burn other things more than water can. In addition to all this there is another distinction to be made among the things that are called " hot " : in some of them the

125

649 a ἀλλοτρίαν ἔχει τὴν θερμότητα τὰ δ' οἰκείαν, διαφέρει δὲ τὸ θερμὸν εἶναι οὕτως ἢ ἐκείνως πλεῖστον,
ἐγγὺς γὰρ τοῦ κατὰ συμβεβηκὸς εἶναι θερμὸν ἀλλὰ
μὴ καθ' αὑτὸ θάτερον αὐτῶν· ὥσπερ ἂν εἴ τις λέγοι,
ε εἰ συμβεβηκὸς εἴη τῷ πυρέττοντι εἶναι μουσικῷ,
τὸν μουσικὸν εἶναι θερμότερον ἢ τὸν μεθ' ὑγιείας
θερμόν. ἐπεὶ δ' ἐστὶ τὸ μὲν καθ' αὑτὸ θερμὸν τὸ
δὲ κατὰ συμβεβηκός, ψύχεται μὲν βραδύτερον τὸ
καθ' αὑτό, θερμαίνει δὲ μᾶλλον πολλάκις τὴν αἴσθησιν τὸ κατὰ συμβεβηκός· καὶ πάλιν καίει μὲν
10 μᾶλλον τὸ καθ' αὑτὸ θερμόν, οἷον ἡ φλὸξ τοῦ
ὕδατος τοῦ ζέοντος, θερμαίνει δὲ κατὰ τὴν ἁφὴν τὸ
ζέον μᾶλλον, τὸ κατὰ συμβεβηκὸς θερμόν. ὥστε
φανερὸν ὅτι τὸ κρῖναι δυοῖν πότερον θερμότερον οὐχ
ἁπλοῦν· ὡδὶ μὲν γὰρ τόδε ἔσται θερμότερον, ὡδὶ δὲ
15 θάτερον. ἔνια δὲ τῶν τοιούτων οὐδ' ἔστιν ἁπλῶς
εἰπεῖν ὅτι θερμὸν ἢ μὴ θερμόν· ὁ μὲν γάρ ποτε
τυγχάνει ὂν τὸ ὑποκείμενον οὐ θερμόν, συνδυαζό-
μενον δὲ θερμόν, οἷον εἴ τις θεῖτο ὄνομα ὕδατι ἢ
σιδήρῳ θερμῷ· τοῦτον γὰρ τὸν τρόπον τὸ αἷμα
20 θερμόν ἐστιν. καὶ ποιεῖ δὲ φανερὸν ἐν τοῖς τοιούτοις ὅτι τὸ ψυχρὸν φύσις τις ἀλλ' οὐ στέρησίς ἐστιν,
ἐν ὅσοις τὸ ὑποκείμενον κατὰ πάθος θερμόν ἐστιν.
τάχα δὲ καὶ ἡ τοῦ πυρὸς φύσις, εἰ ἔτυχε, τοιαύτη
τις ἐστίν· ἴσως γὰρ τὸ ὑποκείμενόν ἐστιν ἢ καπνὸς
ἢ ἄνθραξ, ὧν τὸ μὲν ἀεὶ θερμόν (ἀναθυμίασις γὰρ ὁ
καπνός), ὁ δ' ἄνθραξ ἀποσβεσθεὶς ψυχρός. ἔλαιον
δὲ καὶ πεύκη γένοιτ' ἂν ψυχρά. ἔχει δὲ θερμότητα

ᵃ That is, " blood " is really " hot x," and the " x " is no
more hot of its own nature than the " water " in " hot water."
Cf. 649 b 21 ff., and Torstrik, *Rh. Mus.* xii. 161 ff.

ᵇ Perhaps a reference to the resin which is in firwood or is
obtained from it.

heat is their own; in others it has been derived from without. And there is a very great difference between these two ways of being hot, because one of them comes near to being hot " by accident " and not hot " of itself"; as is obvious, supposing anyone were to assert, if a fever-patient were " by accident " a man of culture, that the man of culture is hotter than a man whose heat is due to his perfect health. Thus some things are hot " of themselves " and some hot " by accident," and though the former cool more slowly, the latter are in many cases hotter in their effect upon the senses. Again, the former have a greater power of burning : *e.g.* a flame burns you more than boiling water, yet the boiling water, which is hot only " by accident," causes a stronger sensation of heat if you touch it. From this it is plain that it is no simple matter to decide which of two things is the hotter. The first will be hotter in one way, and the second in another. In some cases of this sort it is actually impossible to say simply that a thing is hot or is not hot. I mean cases in which the substratum in its permanent nature is not hot, but when coupled ⟨with heat⟩ is hot; as if we were to give a special name to hot water or hot iron : that is the mode in which blood is hot.[a] These cases, in which the substratum is hot merely through some external influence, make it clear that cold is not just a privation but a real thing in itself. Perhaps even fire may be an instance of this kind. It may be that its substratum is smoke or charcoal: and, though smoke is always hot because it is an exhalation, charcoal when it goes out is cold. In the same way oil and firwood [b] become cold. Further, practically all

649 a

25 καὶ τὰ πυρωθέντα πάντα σχεδόν, οἷον κονία καὶ
τέφρα, καὶ τὰ ὑποστήματα τῶν ζῴων, καὶ τῶν
περιττωμάτων ἡ χολή, τῷ ἐμπεπυρεῦσθαι καὶ
ἐγκαταλελεῖφθαί τι ἐν αὐτοῖς θερμόν. ἄλλον δὲ
τρόπον θερμὰ[1] πεύκη καὶ τὰ πίονα, τῷ ταχὺ μεταβάλλειν εἰς ἐνέργειαν πυρός.

30 Δοκεῖ δὲ τὸ θερμὸν καὶ πηγνύναι καὶ τήκειν. ὅσα
μὲν οὖν ὕδατος μόνον, ταῦτα πήγνυσι τὸ ψυχρόν,
ὅσα δὲ γῆς, τὸ πῦρ· καὶ τῶν θερμῶν πήγνυται ὑπὸ
ψυχροῦ ταχὺ μὲν ὅσα γῆς μᾶλλον καὶ ἀλύτως,
λυτῶς δ’ ὅσα ὕδατος. ἀλλὰ περὶ μὲν τούτων ἐν
ἑτέροις διώρισται σαφέστερον, ποῖα τὰ πηκτά, καὶ
πήγνυται διὰ τίνας αἰτίας.

35 Τὸ δὲ τί θερμὸν καὶ ποῖον θερμότερον ἐπειδὴ
649 b λέγεται πλεοναχῶς, οὐ τὸν αὐτὸν τρόπον ὑπάρξει
πᾶσιν, ἀλλὰ προσδιοριστέον ὅτι καθ’ αὑτὸ μὲν τόδε,
κατὰ συμβεβηκὸς δὲ πολλάκις θάτερον,[2] ἔτι δὲ
δυνάμει μὲν τοδί, τοδὶ δὲ κατ’ ἐνέργειαν, καὶ τόνδε
μὲν τὸν τρόπον τοδί, τῷ μᾶλλον τὴν ἁφὴν θερ-
5 μαίνειν, τοδὶ δὲ τῷ φλόγα ποιεῖν καὶ πυροῦν.
λεγομένου δὲ τοῦ θερμοῦ πολλαχῆς, ἀκολουθήσει
δῆλον ὅτι καὶ τὸ ψυχρὸν κατὰ τὸν αὐτὸν λόγον.

Καὶ περὶ μὲν θερμοῦ καὶ ψυχροῦ καὶ τῆς
ὑπεροχῆς αὐτῶν διωρίσθω τὸν τρόπον τοῦτον.

III. Ἐχόμενον δὲ καὶ περὶ ξηροῦ καὶ ὑγροῦ διελ-
10 θεῖν ἀκολούθως τοῖς εἰρημένοις. λέγεται δὲ ταῦτα

---

[1] θερμὰ Peck : θερμὸν vulg.
[2] πολλάκις θάτερον] num τἆλλο θερμότερον?

---

a See *Meteor.* 382 b 31 ff., 388 b 10 ff.
b Probably the text should be altered to read : " *B* hotter
by accident."
c See note on 646 a 16, and Introd. p. 32.

things that have passed through a process of combustion have heat in them, such as cinder, ash, the
excrement of animals, and bile (an instance of a
residue). These have passed through fire and some
heat is left behind in them. Firwood and fatty
substances are hot in another way : they can quickly
change into the actuality of fire.

We must recognize that " the hot " can cause both
congealing and melting. Things that consist of
water only are solidified by the cold, those that consist of earth, by fire. Again, hot things are solidified
by cold : those that consist chiefly of earth solidify
quickly, and the product cannot be dissolved again ;
those that consist chiefly of water can be dissolved
after solidification. I have dealt more particularly
with these matters in another work,[a] where I have
stated what things can be solidified, and the causes
that are responsible for it.

So, in view of the fact that there are numerous
senses in which a thing is said to be " hot " or " hotter,"
the same meaning will not apply to all instances, but
we must specify further, and say that A is hotter
" of itself," B perhaps " by accident "[b] ; and again
that C is hotter potentially, D actually ; and we
must also say in what way the thing's heat manifests
itself : e.g. E causes a greater sensation of heat when
touched ; F causes flame and sets things on fire.
And of course, if " the hot " is used in all these
senses, there will be an equal variety of senses
attaching to " the cold."

This will suffice for our examination of the terms
" hot " and " cold," " hotter " and " colder."

III. It follows on naturally after this to discuss (b) "solid "
" the solid " and " the fluid "[c] on similar lines. and " fluid. "

129

649 b

πλεοναχῶς, οἷον τὰ μὲν δυνάμει τὰ δ' ἐνεργείᾳ.
κρύσταλλος γὰρ καὶ πᾶν τὸ πεπηγὸς ὑγρὸν λέγεται
ξηρὰ¹ μὲν ἐνεργείᾳ καὶ κατὰ συμβεβηκός, ὄντα
δυνάμει καὶ καθ' αὑτὰ ὑγρά, γῆ δὲ καὶ τέφρα καὶ
15 τὰ τοιαῦτα μιχθέντα ὑγρῷ ἐνεργείᾳ μὲν ὑγρὰ καὶ
κατὰ συμβεβηκός, καθ' αὑτὰ δὲ καὶ δυνάμει ξηρά·
διακριθέντα δὲ ταῦτα τὰ μὲν ὕδατος ἀναπληστικὰ
καὶ ἐνεργείᾳ καὶ δυνάμει ὑγρά, τὰ δὲ γῆς ἅπαντα
ξηρά, καὶ τὸ κυρίως καὶ ἁπλῶς ξηρὸν τοῦτον
μάλιστα λέγεται τὸν τρόπον. ὁμοίως δὲ καὶ θάτερα
20 τὰ ὑγρὰ κατὰ τὸν αὐτὸν λόγον ἔχει τὸ κυρίως καὶ
ἁπλῶς, καὶ ἐπὶ θερμῶν καὶ ψυχρῶν. τούτων δὲ
διωρισμένων φανερὸν ὅτι τὸ αἷμα ὡδὶ μέν ἐστι
θερμόν [οἷον τί² ἦν αὐτῷ τὸ αἵματι εἶναι;]· καθάπερ
γὰρ³ εἰ ὀνόματί τινι⁴ σημαίνοιμεν τὸ ζέον ὕδωρ,
οὕτω λέγεται· τὸ δ' ὑποκείμενον καὶ ὅ ποτε ὂν
25 αἷμά ἐστιν, οὐ θερμόν· καὶ καθ' αὑτὸ ἔστι μὲν ὡς
θερμόν ἐστιν, ἔστι δ' ὡς οὔ· ἐν μὲν γὰρ τῷ
λόγῳ ὑπάρξει αὐτοῦ ἡ θερμότης, ὥσπερ ἐν τῷ
τοῦ λευκοῦ ἀνθρώπου τὸ λευκόν· ᾗ δὲ κατὰ πάθος,
τὸ αἷμα οὐ καθ' αὑτὸ θερμόν.⁵

Ὁμοίως δὲ καὶ περὶ ξηροῦ καὶ ὑγροῦ. διὸ καὶ

---

¹ ξηρὰ Peck : ξηρὸν vulg.
² οἷόν τι Bekker. haec, signo interrog. adscr., seclusi.
³ γὰρ Z : om. vulg.
⁴ ὀνόματί τινι PSUZ² : ὀνόματί τι EY : ὀνόματι vulg.
⁵ ll. 22-29 interpunctionem correxi.

ᵃ *i.e.* they assume the shape of the receptacle into which
they are put.

These terms are used in several senses. *E.g.* " solid " and " fluid " may mean either potentially solid and fluid or actually solid and fluid. Ice and other congealed fluids are said to be solid actually and by accident, though in themselves and potentially they are fluid. On the other hand, earth and ash and the like, when they have been mixed with a fluid, are fluid actually and by accident, but potentially and in themselves they are solid. When these mixtures have been resolved again into their components, we have on the one hand the watery constituents, which are anaplestic,[a] and fluid actually as well as potentially, and on the other hand the earthy components which are all solid : and these are the cases where the term " solid " is applicable most properly and absolutely. In the same way, only those things which are actually as well as potentially fluid, or hot, or cold, are such in the proper and absolute sense of the terms. Bearing this distinction in mind, we see it is plain that in one way blood is hot [*e.g.* what is the essential definition of blood ?], for the term " blood " is used just as the term for " boiling water " would be, if we had a special name to denote that ; but in another way, *i.e.* in respect of its permanent substratum, blood is not hot. This means that in one respect blood is essentially hot, and in another respect is not. Heat will be included in the *logos* of blood, just as fairness is included in the *logos* of a fair man, and in this way blood is essentially hot ; but in so far as it is hot owing to external influence, blood is not essentially hot.

A similar argument would hold with regard to the solid and the fluid. And that is why some of these

649 b

ἐν τῇ φύσει τῶν τοιούτων τὰ μὲν θερμὰ καὶ ὑγρά,
30 χωριζόμενα δὲ πήγνυται καὶ ψυχρὰ φαίνεται, οἷον
τὸ αἷμα, τὰ δὲ θερμὰ καὶ πάχος ἔχοντα καθάπερ ἡ
χολή, χωριζόμενα δ' ἐκ τῆς φύσεως τῶν ἐχόντων
τοὐναντίον πάσχει· ψύχεται γὰρ καὶ ὑγραίνεται· τὸ
μὲν γὰρ αἷμα ξηραίνεται μᾶλλον, ὑγραίνεται δ' ἡ
ξανθὴ χολή. τὸ δὲ μᾶλλον καὶ ἧττον μετέχειν τῶν
35 ἀντικειμένων ὡς ὑπάρχον[1] δεῖ τιθέναι τούτοις.

650 a

Πῶς μὲν οὖν θερμὸν καὶ πῶς ὑγρόν, καὶ πῶς
τῶν ἐναντίων ἡ φύσις τοῦ αἵματος κεκοινώνηκεν,
εἴρηται σχεδόν.

Ἐπεὶ δ' ἀνάγκη πᾶν τὸ αὐξανόμενον λαμβάνειν
τροφήν, ἡ δὲ τροφὴ πᾶσιν ἐξ ὑγροῦ καὶ ξηροῦ, καὶ
5 τούτων ἡ πέψις γίνεται καὶ ἡ μεταβολὴ διὰ τῆς τοῦ
θερμοῦ δυνάμεως, καὶ τὰ ζῷα πάντα καὶ τὰ φυτά,
κἂν εἰ μὴ δι' ἄλλην αἰτίαν, ἀλλὰ διὰ ταύτην ἀναγκαῖον ἔχειν ἀρχὴν θερμοῦ φυσικήν. [καὶ ταύτην
ὥσπερ] αἱ ⟨δ'⟩[2] ἐργασίαι τῆς τροφῆς πλειόνων εἰσὶ
μορίων· ἡ μὲν γὰρ πρώτη φανερὰ τοῖς ζῴοις
10 λειτουργία διὰ τοῦ στόματος οὖσα καὶ τῶν ἐν
τούτῳ μορίων, ὅσων ἡ τροφὴ δεῖται διαιρέσεως.
ἀλλ' αὕτη μὲν οὐδεμιᾶς αἰτία πέψεως, ἀλλ' εὐπεψίας μᾶλλον· ἡ γὰρ εἰς μικρὰ διαίρεσις τῆς
τροφῆς ῥάω ποιεῖ τῷ θερμῷ τὴν ἐργασίαν· ἡ δὲ τῆς
ἄνω καὶ τῆς κάτω κοιλίας ἤδη μετὰ θερμότητος

---

[1] ὑπάρχον Peck : ὑπάρχοντα vulg.
[2] καὶ ταύτην ὥσπερ seclusi, ⟨δ'⟩ supplevi : καὶ ταύτην
⟨πλείοσι μορίοις ἐνυπάρχουσαν⟩ Camus.

---

[a] See above, note on 644 a 17.
[b] See Introduction, p. 34.
[c] Lit. " the *dynamis* of the hot substance," perhaps here
something more than a mere periphrasis for " the hot sub-

substances while in the living organism are hot and fluid, but when separated from it congeal and are observed to be cold, as blood does; others, like yellow bile, are hot and of a thick consistency while in the organism, but when separated from it undergo a change in the opposite direction and become cool and fluid. Blood becomes more solid, yellow bile becomes fluid. And we must assume that " more and less " [a] participation in opposite characteristics is a property of these substances.

We have now pretty well explained in what way blood is hot, in what way it is fluid, and in what way it participates in opposite characteristics.

Everything that grows must of necessity take food. This food is always supplied by fluid and solid matter, and the concoction [b] and transformation of these is effected by the agency of heat.[c] Hence, apart from other reasons, this would be a sufficient one for holding that of necessity all animals and plants must have in them a natural source of heat; though there are several parts which exert action upon the food. In the case of those animals whose food needs to be broken up, the first duty clearly belongs to the mouth and the parts in the mouth. But this operation does nothing whatever towards causing concoction: it merely enables the concoction to turn out successfully; because when the food has been broken up into small pieces the action of the heat upon it is rendered easier. The natural heat comes into play in the upper and in the lower gut,

stance," as emphasizing its proper and specific natural character, which makes it a particularly good agent for effecting concoction. See Introduction, pp. 30-32.

650 a

15 φυσικῆς ποιεῖται τὴν πέψιν. ὥσπερ δὲ καὶ τὸ
στόμα τῆς ἀκατεργάστου τροφῆς πόρος ἐστί, καὶ
τὸ συνεχὲς αὐτῷ μόριον ὃ καλοῦσιν οἰσοφάγον,
ὅσα τῶν ζῴων ἔχει τοῦτο τὸ μόριον, ἕως εἰς
τὴν κοιλίαν, οὕτω καὶ ἄλλους δεῖ πόρους[1] εἶναι, δι᾽
20 ὧν ἅπαν λήψεται τὸ σῶμα τὴν τροφήν, ὥσπερ
ἐκ φάτνης, ἐκ τῆς κοιλίας καὶ τῆς τῶν ἐντέρων
φύσεως. τὰ μὲν γὰρ φυτὰ λαμβάνει τὴν τροφὴν
κατειργασμένην ἐκ τῆς γῆς ταῖς ῥίζαις (διὸ καὶ
περίττωμα οὐ γίνεται τοῖς φυτοῖς· τῇ γὰρ γῇ καὶ
τῇ ἐν αὐτῇ θερμότητι χρῆται ὥσπερ κοιλίᾳ), τὰ δὲ
ζῷα πάντα μὲν σχεδόν, τὰ δὲ πορευτικὰ φανερῶς,
25 οἷον γῆν ἐν αὑτοῖς ἔχει τὸ τῆς κοιλίας κύτος, ἐξ
ἧς, ὥσπερ ἐκεῖνα ταῖς ῥίζαις, ταῦτα δεῖ τινι τὴν
τροφὴν λαμβάνειν, ἕως τὸ τῆς ἐχομένης πέψεως
λάβῃ τέλος. ἡ μὲν γὰρ τοῦ στόματος ἐργασία παρα-
δίδωσι τῇ κοιλίᾳ, παρὰ δὲ ταύτης ἕτερον ἀναγκαῖον
λαμβάνειν, ὅπερ συμβέβηκεν· αἱ γὰρ φλέβες κατα-
30 τείνονται διὰ τοῦ μεσεντερίου παράπαν, κάτωθεν
ἀρξάμεναι μέχρι τῆς κοιλίας. δεῖ δὲ ταῦτα θεωρεῖν
ἔκ τε τῶν ἀνατομῶν καὶ τῆς φυσικῆς ἱστορίας.

Ἐπεὶ δὲ πάσης τροφῆς ἐστί τι δεκτικὸν καὶ τῶν
γινομένων περιττωμάτων, αἱ δὲ φλέβες οἷον ἀγγεῖον
αἵματός εἰσι, φανερὸν ὅτι τὸ αἷμα ἡ τελευταία
35 τροφὴ τοῖς ζῴοις τοῖς ἐναίμοις ἐστί, τοῖς δ᾽ ἀναίμοις

---

[1] ἄλλους δεῖ πόρους Peck : ἄλλας ἀρχὰς δεῖ πλείους vulg.

---

[a] Cf. Shakespeare, Coriolanus I. i. 133-152.
[b] The membrane to which the intestines are attached.
[c] Dissections (or Anatomy) is a treatise which has not
survived.

134

which effect the concoction of the food by its aid. And, just as the mouth (and in some animals the so-called oesophagus too which is continuous with it) is the passage for the as yet untreated food, and conveys it to the stomach ; so there must be other passages through which as from a manger the body as a whole may receive its food from the stomach and from the system of the intestines.[a] Plants get their food from the earth by their roots ; and since it is already treated and prepared no residue is produced by plants—they use the earth and the heat in it instead of a stomach, whereas practically all animals, and unmistakably those that move about from place to place, have a stomach, or bag,—as it were an earth inside them—and in order to get the food out of this, so that finally after the successive stages of concoction it may reach its completion, they must have some instrument corresponding to the roots of a plant. The mouth, then, having done its duty by the food, passes it on to the stomach, and there must of necessity be another part to receive it in its turn from the stomach. This duty is undertaken by the blood-vessels, which begin at the bottom of the mesentery,[b] and extend throughout the length of it right up to the stomach. These matters should be studied in the *Dissections*[c] and my treatise on *Natural History*.[d]

We see then that there is a receptacle for the food at each of its stages, and also for the residues that are produced ; and as the blood-vessels are a sort of container for the blood, it is plain that the blood (or its counterpart) is the final form of that food in living

[d] The *Natural History*, otherwise *History of Animals* or *Researches upon Animals*. See 495 b 19 ff., 514 b 10 ff.

650 a

τὸ ἀνάλογον. καὶ διὰ τοῦτο μὴ λαμβάνουσί τε
650 b τροφὴν ὑπολείπει τοῦτο καὶ λαμβάνουσιν αὐξάνεται,
καὶ χρηστῆς μὲν οὔσης ὑγιεινόν, φαύλης δὲ φαῦλον.
ὅτι μὲν οὖν τὸ αἷμα τροφῆς ἕνεκεν ὑπάρχει τοῖς
ἐναίμοις, φανερὸν ἐκ τούτων καὶ τῶν τοιούτων. καὶ
γὰρ διὰ τοῦτο θιγγανόμενον αἴσθησιν οὐ ποιεῖ
5 (ὥσπερ οὐδ᾽ ἄλλο τῶν περιττωμάτων οὐδέν, οὐδ᾽ ἡ
τροφὴ) καθάπερ σάρξ·[1] αὕτη γὰρ θιγγανομένη ποιεῖ
αἴσθησιν. οὐ γὰρ συνεχές ἐστι τὸ αἷμα ταύτῃ οὐδὲ
συμπεφυκός, ἀλλ᾽ οἷον ἐν ἀγγείῳ τυγχάνει κείμενον
ἔν τε τῇ καρδίᾳ καὶ ταῖς φλεψίν. ὃν δὲ τρόπον
10 λαμβάνει ἐξ αὐτοῦ τὰ μόρια τὴν αὔξησιν, ἔτι δὲ
περὶ τροφῆς ὅλως, ἐν τοῖς περὶ γενέσεως καὶ ἐν
ἑτέροις οἰκειότερόν ἐστι διελθεῖν. νῦν δ᾽ ἐπὶ
τοσοῦτον εἰρήσθω (τοσοῦτον γὰρ χρήσιμον), ὅτι τὸ
αἷμα τροφῆς ἕνεκα καὶ τροφῆς τῶν μορίων ἐστίν.

IV. Τὰς δὲ καλουμένας ἶνας τὸ μὲν ἔχει αἷμα
15 τὸ δ᾽ οὐκ ἔχει, οἷον τὸ τῶν ἐλάφων καὶ προκῶν.
διόπερ οὐ πήγνυται τὸ τοιοῦτον αἷμα· τοῦ γὰρ
αἵματος τὸ μὲν ὑδατῶδες μᾶλλόν[2] ἐστι, διὸ καὶ οὐ
πήγνυται, τὸ δὲ γεῶδες πήγνυται ουνεξατμίζοντος
τοῦ ὑγροῦ· αἱ δ᾽ ἶνες γῆς εἰσιν.

Συμβαίνει δ᾽ ἔνιά γε καὶ γλαφυρωτέραν ἔχειν
20 τὴν διάνοιαν τῶν τοιούτων, οὐ διὰ τὴν ψυχρότητα
τοῦ αἵματος, ἀλλὰ διὰ τὴν λεπτότητα μᾶλλον καὶ

---

[1] ll. 4 f., interpunctionem correxit Cornford.
[2] μᾶλλον Z : μᾶλλον ψυχρόν vulg.

---

[a] In the Second Book. Also in *De gen. et corr.*
[b] With the sentiments of the following passage and its
terminology (" more intelligent," " soul," " blend," etc.)
compare the very interesting passage in Hippocrates, Περὶ
διαίτης, i. 35. *Cf.* 648 a 3.

creatures. This explains why the blood diminishes in quantity when no food is taken and increases when it is ; and why, when the food is good, the blood is healthy, when bad, poor. These and similar considerations make it clear that the purpose of the blood in living creatures is to provide them with nourishment ; and also why it is that when the blood is touched it yields no sensation, as flesh does when it is touched. Indeed, none of the residues yields any sensation either, nor does the nourishment. This difference of behaviour is because the blood is not continuous with the flesh nor conjoined to it organically : it just stands in the heart and in the blood-vessels like water in a jar. A description of the way in which the parts of the body derive their growth from the blood, and the discussion of nourishment in general, comes more appropriately in the treatise on *Generation*[a] and elsewhere. For the present it is enough to have said that the purpose of the blood is to provide nourishment, that is to say, nourishment for the parts of the body. So much and no more is pertinent to our present inquiry.

IV. The blood of some animals contains what are called fibres ; the blood of others (*e.g.* the deer and the gazelle) does not. Blood which lacks fibres does not congeal, for the following reason. Part of the blood is of a more watery nature, and therefore does not congeal ; while the other part, which is earthy, congeals as the fluid part evaporates off. The fibres are this earthy part.

The uniform parts: Blood.

Now some of the animals whose blood is watery have a specially subtle intelligence.[b] This is due not to the coldness of their blood, but to its greater thin-

650 b

διὰ τὸ καθαροὶ εἶναι· τὸ γὰρ γεῶδες οὐδέτερον ἔχει
τούτων. εὐκινητοτέραν γὰρ ἔχουσι τὴν αἴσθησιν τὰ
λεπτοτέραν ἔχοντα τὴν ὑγρότητα καὶ καθαρωτέραν.
διὰ γὰρ τοῦτο καὶ τῶν ἀναίμων ἔνια συνετωτέραν ἔχει
25 τὴν ψυχὴν ἐνίων ἐναίμων, καθάπερ εἴρηται πρότερον,
οἷον ἡ μέλιττα καὶ τὸ γένος τὸ τῶν μυρμήκων κἂν
εἴ τι ἕτερον τοιοῦτόν ἐστιν. δειλότερα δὲ τὰ λίαν
ὑδατώδη. ὁ γὰρ φόβος καταψύχει· προωδοποίηται
οὖν τῷ πάθει τὰ τοιαύτην ἔχοντα τὴν ἐν τῇ καρδίᾳ
30 κρᾶσιν· τὸ γὰρ ὕδωρ τῷ ψυχρῷ πηκτόν ἐστιν. διὸ
καὶ τἆλλα τὰ ἄναιμα δειλότερα τῶν ἐναίμων ἐστὶν
ὡς ἁπλῶς εἰπεῖν, καὶ ἀκινητίζει τε φοβούμενα καὶ
προΐεται περιττώματα καὶ μεταβάλλει ἔνια τὰς
χρόας αὐτῶν. τὰ δὲ πολλὰς ἔχοντα λίαν ἶνας καὶ
παχείας γεωδέστερα τὴν φύσιν ἐστὶ καὶ θυμώδη τὸ
35 ἦθος καὶ ἐκστατικὰ διὰ τὸν θυμόν. θερμότητος
γὰρ ποιητικὸν ὁ θυμός, τὰ δὲ στερεὰ θερμανθέντα
651 a μᾶλλον θερμαίνει τῶν ὑγρῶν· αἱ δ' ἶνες στερεὸν καὶ
γεῶδες, ὥστε γίνονται οἷον πυρίαι ἐν τῷ αἵματι
καὶ ζέσιν ποιοῦσιν ἐν τοῖς θυμοῖς. διὸ οἱ ταῦροι καὶ
οἱ κάπροι θυμώδεις καὶ ἐκστατικοί· τὸ γὰρ αἷμα
τούτων ἰνωδέστατον, καὶ τό γε τοῦ ταύρου τάχιστα
5 πήγνυται πάντων. ἐξαιρουμένων δὲ τούτων τῶν
ἰνῶν οὐ πήγνυται τὸ αἷμα· καθάπερ γὰρ ἐκ πηλοῦ
εἴ τις ἐξέλοι τὸ γεῶδες οὐ πήγνυται τὸ ὕδωρ, οὕτω
καὶ τὸ αἷμα· αἱ γὰρ ἶνες γῆς. μὴ ἐξαιρουμένων

---

^a At 648 a 2 ff.
^b For the connexion between fear and cold cf. 667 a 16,
692 a 22 ff., and *Rhetoric*, 1389 b 30.

ness and clarity, neither of which characteristics belongs to the earthy substance ; and an animal which has the thinner and clearer sort of fluid in it has also a more mobile faculty of sensation. This is why, as I said before,[a] some of the bloodless creatures have a more intelligent Soul than some of the blooded ones ; e.g. the bee and the ants and such insects. Those, however, that have excessively watery blood are somewhat timorous. This is because water is congealed by cold ; and coldness also accompanies fear ; therefore in those creatures whose heart contains a predominantly watery blend, the way is already prepared for a timorous disposition.[b] This, too, is why, generally speaking, the bloodless creatures are more timorous than the blooded ones and why they stand motionless when they are frightened and discharge their residues and (in some cases) change their colour. On the other side, there are the animals that have specially plentiful and thick fibres in their blood ; these are of an earthier nature, and are of a passionate temperament and liable to outbursts of passion. Passion produces heat ; and solids, when they have been heated, give off more heat than fluids. So the fibres, which are solid and earthy, become as it were embers inside the blood and cause it to boil up when the fits of passion come on. That is why bulls and boars are so liable to these fits of passion. Their blood is very fibrous ; indeed, that of the bull is the quickest of all to congeal. But just as when the earthy matter is taken out of mud, the water which remains does not congeal ; so when the fibres, which consist of earth, are taken out of the blood, it no longer congeals. If they are

δὲ πήγνυται, οἷον ὑγρὰ γῆ ὑπὸ ψύχους· τοῦ γὰρ
θερμοῦ ὑπὸ τοῦ ψυχροῦ ἐκθλιβομένου συνεξατμίζει
10 τὸ ὑγρόν, καθάπερ εἴρηται πρότερον, καὶ πήγνυται
οὐχ ὑπὸ θερμοῦ ἀλλ' ὑπὸ ψυχροῦ ξηραινόμενον. ἐν
δὲ τοῖς σώμασιν ὑγρόν ἐστι διὰ τὴν θερμότητα τὴν
ἐν τοῖς ζῴοις.

Πολλῶν δ' ἐστὶν αἰτία ἡ τοῦ αἵματος φύσις καὶ
κατὰ τὸ ἦθος τοῖς ζῴοις καὶ κατὰ τὴν αἴσθησιν,
εὐλόγως· ὕλη γάρ ἐστι παντὸς τοῦ σώματος· ἡ γὰρ
15 τροφὴ ὕλη, τὸ δ' αἷμα ἡ ἐσχάτη τροφή. πολλὴν
οὖν ποιεῖ διαφορὰν θερμὸν ὂν καὶ ψυχρὸν καὶ λεπτὸν
καὶ παχὺ καὶ θολερὸν καὶ καθαρόν. ἰχὼρ δ' ἐστὶ
τὸ ὑδατῶδες τοῦ αἵματος διὰ τὸ μήπω πεπέφθαι ἢ
διεφθάρθαι, ὥστε ὁ μὲν ἐξ ἀνάγκης ἰχώρ, ὁ δ'
αἵματος χάριν ἐστίν.

20 V. Πιμελὴ δὲ καὶ στέαρ διαφέρουσι μὲν ἀλλήλων
κατὰ τὴν τοῦ αἵματος διαφοράν. ἔστι γὰρ ἑκάτερον αὐτῶν αἷμα πεπεμμένον δι' εὐτροφίαν, καὶ τὸ
μὴ καταναλισκόμενον εἰς τὸ σαρκῶδες μόριον τῶν
ζῴων, εὔπεπτον δὲ καὶ εὐτραφές. δηλοῖ δὲ τὸ
25 λιπαρὸν αὐτῶν· τῶν γὰρ ὑγρῶν τὸ λιπαρὸν κοινὸν
ἀέρος καὶ πυρός ἐστιν. διὰ τοῦτο οὐδὲν ἔχει τῶν
ἀναίμων οὔτε πιμελὴν οὔτε στέαρ, ὅτι οὐδ' αἷμα.
τῶν δ' ἐναίμων τὰ μὲν σωματῶδες ἔχοντα τὸ αἷμα
στέαρ ἔχει μᾶλλον. τὸ γὰρ στέαρ γεῶδές ἐστι, διὸ

---

^a As it were, the " raw " material.

^b I have used the terms " lard " and " suet " rather than
" soft fat " and " hard fat " because they represent more
closely the distinction made by Aristotle. The difference
between them is now known to be less fundamental, and is

not taken out, it does congeal, as moist earth does under the influence of cold : the cold expels the heat and makes the fluid evaporate, as has been said before ; so it is due to the solidifying effect of the cold, and not of the hot, that what remains becomes congealed. And while it is in the body the blood is fluid on account of the heat which is there.

There are many points both in regard to the temperament of animals and their power of sensation which are controlled by the character of the blood. This is what we should expect : for the blood is the material [a] of which the whole body consists—material in the case of living creatures being nourishment, and blood is the final form which the nourishment assumes. For this reason a great deal depends upon whether the blood be hot, cold, thin, thick, muddy, or clear. Serum is the watery part of blood ; and it is watery either because it has not yet undergone concoction or because it has been already corrupted ; consequently some of the serum is the result of a *necessary* process, and some is there for the *purpose* of producing blood.

V. The difference between lard and suet [b] is parallel to a difference in the blood. They both consist of blood that has been concocted as the result of plentiful nourishment ; that is, the surplus blood that is not used up to nourish the fleshy parts of the animal, but is well concocted and well nourished. (This point is proved by their greasiness, for grease in fluids is a combination of Air and Fire.) This explains why there is no lard or suet in any of the bloodless animals. And among the others, those whose blood is denser tend to contain suet rather than lard. Suet

due to varying proportions of unsaturated triglycerides and the lengths of the carbon chains.

141

651 a

πήγνυται καθάπερ καὶ τὸ αἷμα τὸ ἰνῶδες καὶ αὐτὸ
καὶ οἱ ζωμοὶ οἱ τοιοῦτοι· ὀλίγον γὰρ ἔχει ὕδατος,
30 τὸ δὲ πολὺ γῆς. διὸ τὰ μὴ ἀμφώδοντα ἀλλὰ
κερατώδη στέαρ ἔχει. φανερὰ δ' ἡ φύσις αὐτῶν
τοῦ τοιούτου στοιχείου πλήρης οὖσα τῷ κερατώδης
εἶναι καὶ ἀστραγάλους ἔχειν· ἅπαντα γὰρ ξηρὰ καὶ
γεηρὰ τὴν φύσιν ἐστίν. τὰ δ' ἀμφώδοντα καὶ
35 ἀκέρατα καὶ πολυσχιδῆ πιμελὴν ἔχει ἀντὶ στέατος,
ἣ οὐ πήγνυται οὐδὲ θρύπτεται ξηραινομένη διὰ τὸ
μὴ εἶναι γεώδη τὴν φύσιν αὐτῆς.

Μέτρια μὲν οὖν ταῦτα ὄντα ἐν τοῖς μορίοις τῶν
651 b ζῴων ὠφελεῖ (πρὸς μὲν γὰρ αἴσθησιν οὐκ ἐμποδίζει,
πρὸς δ' ὑγίειαν καὶ δύναμιν ἔχει βοήθειαν), ὑπερ-
βάλλοντα δὲ τῷ πλήθει φθείρει καὶ βλάπτει. εἰ
γὰρ πᾶν γένοιτο τὸ σῶμα πιμελὴ καὶ στέαρ, ἀπό-
λοιτ' ἄν. ζῷον μὲν γάρ ἐστι κατὰ τὸ αἰσθητικὸν
5 μόριον, ἡ δὲ σὰρξ καὶ τὸ ἀνάλογον αἰσθητικόν· τὸ
δ' αἷμα, ὥσπερ εἴρηται καὶ πρότερον, οὐκ ἔχει
αἴσθησιν, διὸ οὐδὲ πιμελὴ οὐδὲ στέαρ· αἷμα γὰρ
πεπεμμένον ἐστίν. ὥστ' εἰ πᾶν γένοιτο τὸ σῶμα
τοιοῦτον, οὐκ ἂν ἔχοι οὐδεμίαν αἴσθησιν. διὸ καὶ
γηράσκει ταχέως τὰ λίαν πίονα· ὀλίγαιμα γὰρ ἅτε εἰς
15 τὴν πιότητα ἀναλισκομένου τοῦ αἵματος, τὰ δ' ὀλίγαιμα ἤδη προωδοποίηται πρὸς τὴν φθοράν· ἡ γὰρ
φθορὰ ὀλιγαιμία τις ἐστί, καὶ τὸ ὀλίγαιμον[1] παθητικὸν καὶ ὑπὸ ψυχροῦ τοῦ τυχόντος καὶ ὑπὸ θερμοῦ.

_____
[1] sic Th. : *animal pauci sanguinis* Σ: ὀλίγον vulg.

142

is of an earthy character; it contains but little water against a large proportion of earth; so it congeals just as fibrous blood and broths do. So too the animals which have horns but have teeth in one jaw only contain suet. And it is clear that their natural constitution is full of this element (earth) from the fact that they have horns and huckle-bones, for they are all of them solid and earthy in constitution. On the other hand, the animals which have incisor teeth in both jaws and have toes (not uncloven hoofs), but no horns, contain lard instead of suet. Lard neither congeals nor splits up into small pieces when it dries, owing to the fact that it is not earthy.

Lard and suet when present in the parts of animals in moderate quantities are beneficial: they do not hinder the action of the senses, and they contribute towards the health and strength of the body. But when the amount of them is excessive they are destructive and injurious. This is shown by the consideration that if the whole body were to become lard and suet, it would perish. The *sine qua non* of a living creature is its sensory part, which is flesh or its counterpart; and since, as I have said before, blood is not sensitive, neither lard nor suet, which are just concocted blood, is sensitive. Therefore, if the whole body were to become either of these, it would have no sensation whatever. For this reason, too, unduly fat animals age quickly: their blood gets used up to produce fat, so there is very little of it left; and anything that has but little blood is well on the road to decay. In fact, decay is just a form of blood-deficiency; and an animal deficient in blood is easily susceptible to the effects of accidental cold and

651 b

καὶ ἀγονώτερα δὴ τὰ πίονά ἐστι διὰ τὴν αὐτὴν
αἰτίαν· ὃ γὰρ ἔδει ἐκ τοῦ αἵματος εἰς τὴν γονὴν
15 ἰέναι καὶ τὸ σπέρμα, τοῦτ' εἰς τὴν πιμελὴν ἀναλίσκεται καὶ τὸ στέαρ· πεττόμενον γὰρ τὸ αἷμα
γίνεται ταῦτα, ὥστε ἢ ὅλως οὐ γίνεται περίττωμα
αὐτοῖς οὐδὲν ἢ ὀλίγον.

Καὶ περὶ μὲν αἵματος καὶ ἰχῶρος καὶ πιμελῆς
καὶ στέατος, τί τέ ἐστιν ἕκαστον αὐτῶν καὶ διὰ
τίνας αἰτίας, εἴρηται.

20 VI. Ἔστι δὲ καὶ ὁ μυελὸς αἵματός τις φύσις, καὶ
οὐχ ὥσπερ οἴονταί τινες, τῆς γονῆς σπερματικὴ
δύναμις. δηλοῖ δ' ἐν τοῖς νέοις πάμπαν· ἅτε γὰρ
ἐξ αἵματος συνεστώτων τῶν μορίων καὶ τῆς τροφῆς
οὔσης τοῖς ἐμβρύοις αἵματος, καὶ ἐν τοῖς ὀστοῖς ὁ
25 μυελὸς αἱματώδης ἐστίν· αὐξανομένων δὲ καὶ πετ-
τομένων, καθάπερ καὶ τὰ μόρια μεταβάλλει καὶ τὰ
σπλάγχνα τὰς χρόας (ὑπερβολῇ γὰρ αἱματῶδες καὶ
τῶν σπλάγχνων ἕκαστόν ἐστιν ἔτι νέων ὄντων),
οὕτω καὶ ὁ μυελός.

Καὶ τῶν μὲν πιμελωδῶν λιπαρὸς καὶ πιμελῇ
ὅμοιος, ὅσοις δὲ μὴ πιμελῇ ὅμοιον[1] ἀλλὰ στέαρ
30 γίνεται τὸ αἷμα πεττόμενον, τούτοις δὲ στεατώδης.
διὸ τοῖς μὲν κερατοφόροις καὶ μὴ ἀμφώδουσι
στεατώδης, τοῖς δ' ἀμφώδουσι καὶ πολυσχιδέσι
πιμελώδης. (ἥκιστα δὲ τοιοῦτος ὁ ῥαχίτης ἐστὶ
μυελὸς διὰ τὸ δεῖν αὐτὸν εἶναι συνεχῆ καὶ διέχειν
διὰ πάσης τῆς ῥάχεως διῃρημένης κατὰ τοὺς
35 σφονδύλους· λιπαρὸς δ' ὢν ἢ στεατώδης οὐκ ἂν
ὁμοίως ἦν συνεχής, ἀλλ' ἢ θραυστὸς ἢ ὑγρός.)

[1] ὅμοιον Z[1] : ὅμοιος alii.

---

[a] *e.g.* secretion of semen. See above, on 647 b 27.
[b] Plato, *Timaeus*, 73 c.

heat. The same cause is responsible for the comparative sterility of fat animals : that part of the blood which ought to go to form semen and seed gets used up in forming lard and suet, which are formed by the concoction of blood. Hence in fat animals there is either no residue [a] at all, or else very little.

I have now spoken of blood, serum, lard and suet, describing the nature and the Causes of each of them.

VI. Marrow, again. is really a form of blood, and not, Marrow. as some [b] think, the same as the seminal substance [c] of the seed. This is proved by the case of very young animals. In the embryo, the parts are composed out of blood and its nourishment is blood ; so it is not surprising that the marrow in the bones has a blood-like appearance. As they grow and become mature, [d] the marrow changes its colour just like the other parts [e] of the body and the viscera, which while the creature is young all have a blood-like appearance owing to the large quantity of blood in them.

Animals which contain lard have greasy marrow, like lard ; those whose concocted blood produces not a substance like lard but suet have suety marrow. Hence, in the horned animals which have teeth in one jaw only the marrow is suety, and in the animals that have teeth in both jaws and are polydactylous it is like lard. (The spinal marrow cannot possibly be of this nature because it has to be continuous and to pass without a break right through the whole spine which is divided into separate vertebrae ; and if it were fatty or suety it could not hold together as well as it does, but it would be either brittle or fluid.)

---

[c] *Dynamis.* See Introduction, pp. 30 ff. and note on 646 a 14.
　　　　[d] Lit. " are concocted."
　[e] A good instance of Aristotle's usage of the term " part."

Ἔνια δ' οὐκ ἔχει τῶν ζῴων ὡς ἀξίως εἰπεῖν
μυελόν, ὅσων τὰ ὀστᾶ ἰσχυρὰ καὶ πυκνά, οἷον τὰ
652 a τοῦ λέοντος· τούτου γὰρ τὰ ὀστᾶ, διὰ τὸ πάμπαν
ἄσημον ἔχειν, δοκεῖ οὐκ ἔχειν ὅλως μυελόν. ἐπεὶ
δὲ τὴν μὲν τῶν ὀστῶν ἀνάγκη φύσιν ὑπάρχειν τοῖς
ζῴοις ἢ τὸ ἀνάλογον τοῖς ὀστοῖς, οἷον τοῖς ἐνύδροις
5 τὴν ἄκανθαν, ἀναγκαῖον ἐνίοις ὑπάρχειν καὶ μυελόν,
ἐμπεριλαμβανομένης τῆς τροφῆς ἐξ ἧς γίνεται τὰ
ὀστᾶ. ὅτι δ' ἡ τροφὴ πᾶσιν αἷμα, εἴρηται πρότερον. εὐλόγως δὲ καὶ στεατώδεις οἱ μυελοὶ καὶ
πιμελώδεις εἰσίν· διὰ γὰρ τὴν ἀλέαν τὴν γινομένην
ὑπὸ τοῦ περιέχεσθαι τοῖς ὀστοῖς πέττεται τὸ αἷμα,
10 ἡ δὲ καθ' αὑτὸ πέψις αἵματος στέαρ καὶ πιμελή
ἐστιν. καὶ ἐν τοῖς δὴ τὰ ὀστᾶ πυκνὰ ἔχουσι καὶ
ἰσχυρὰ εὐλόγως ἐν τοῖς μὲν οὐκ ἔνεστι, τοῖς δ'
ὀλίγος[1] ἔνεστιν· εἰς γὰρ τὰ ὀστᾶ ἀναλίσκεται ἡ
τροφή.

Ἐν δὲ τοῖς μὴ ἔχουσιν ὀστᾶ ἀλλ' ἄκανθαν ὁ
ῥαχίτης μόνος ἐστὶ μυελός· ὀλίγαιμά τε γὰρ φύσει
15 ὑπάρχει ὄντα, καὶ κοίλη ἄκανθα μόνον ἡ τῆς ῥά-
χεώς ἐστιν. διὸ ἐν ταύτῃ ἐγγίνεται· μόνη τε γὰρ
ἔχει χώραν, καὶ μόνη δεῖται συνδέσμου διὰ τὰς
διαλήψεις. διὸ καὶ ὁ ἐνταῦθα μυελός, ὥσπερ
εἴρηται, ἀλλοιότερός ἐστιν· διὰ τὸ ἀντὶ περόνης

---

[1] ὀλίγοις per errorem Bekker.

Some animals have no marrow worth mentioning :
these are they whose bones are strong and close-
textured : for instance, the Lion, whose bones contain so insignificant an amount of marrow that they
look as if they contained none at all.   Now in view of
the fact that the bodies of animals must have in them
either bones or the counterpart of bones (*e.g.* the
spines in water-animals), it follows of necessity that
some of them must contain marrow as well, due to the
enclosing of the nourishment out of which the bones
are formed.   Now we have stated already that the
nourishment of all the parts of the body is blood.
And it is quite reasonable that the various sorts of
marrow should be suety and lardy ; because the
blood undergoes concoction owing to the heat produced by its being surrounded by bone, and the
product of blood when it undergoes concoction by
itself is suet and lard.   And also, of the animals that
have strong, close-textured bones, some have no
marrow, others have but little, and this is reasonable
too, because the nourishment gets used up to supply
the substance of the bones themselves.

In those animals that have no bones but spine
instead, the backbone contains the only marrow they
possess.   It is the nature of these creatures to have
but a small amount of blood, and their only hollow
spine is that of the backbone.   Therefore the marrow
is formed in it—indeed, it is the only bone where
there is room for the marrow, and the only one
which requires something to connect it together,
owing to its being divided up into segments.
This also explains why the marrow here is (as
I have already said) somewhat different from the
marrow elsewhere.   It has to serve as a fastening,

147

γὰρ γίνεσθαι γλίσχρος, καὶ νευρώδης ἐστὶν ἵν᾽
ἔχῃ τάσιν.

20 Διὰ τί μὲν οὖν μυελὸν ἔχει τὰ ζῷα τὰ ἔχοντα
μυελόν, εἴρηται· καὶ τί ἐστιν ὁ μυελός, ἐκ τούτων
φανερόν, ὅτι τῆς αἱματικῆς τροφῆς τῆς εἰς ὀστᾶ
καὶ ἄκανθαν μεριζομένης ἐστὶ τὸ ἐμπεριλαμβανό-
μενον περίττωμα πεφθέν.

VII. Περὶ δ᾽ ἐγκεφάλου σχεδόν ἐστιν ἐχόμενον
25 εἰπεῖν· πολλοῖς γὰρ καὶ ὁ ἐγκέφαλος δοκεῖ μυελὸς
εἶναι καὶ ἀρχὴ τοῦ μυελοῦ διὰ τὸ συνεχῆ τὸν
ῥαχίτην αὐτῷ ὁρᾶν μυελόν. ἔστι δὲ πᾶν τοὐναντίον
αὐτῷ τὴν φύσιν ὡς εἰπεῖν· ὁ μὲν γὰρ ἐγκέφαλος
ψυχρότατον τῶν ἐν τῷ σώματι μορίων, ὁ δὲ μυελὸς
θερμὸς τὴν φύσιν· δηλοῖ δ᾽ ἡ λιπαρότης αὐτοῦ καὶ
30 τὸ πῖον. διὸ καὶ συνεχὴς ὁ ῥαχίτης τῷ ἐγκεφάλῳ
ἐστίν· ἀεὶ γὰρ ἡ φύσις μηχανᾶται πρὸς τὴν ἑκάστου
ὑπερβολὴν βοήθειαν τὴν τοῦ ἐναντίου παρεδρίαν, ἵνα
ἀνισάζῃ τὴν θατέρου ὑπερβολὴν θάτερον. ὅτι μὲν
οὖν ὁ μυελὸς θερμός[1] ἐστι, δῆλον ἐκ πολλῶν. ἡ δὲ
35 τοῦ ἐγκεφάλου ψυχρότης φανερὰ μὲν καὶ κατὰ τὴν
θίξιν, ἔτι δ᾽ ἀναιμότατον τῶν ὑγρῶν τῶν ἐν τῷ
σώματι πάντων (οὐδ᾽ ὁτιοῦν γὰρ αἵματος ἔχει ἐν
652 b αὑτῷ) καὶ αὐχμηρότατον. ἔστι δ᾽ οὔτε περίττωμα
οὔτε τῶν συνεχῶν μορίων, ἀλλὰ ἴδιος ἡ φύσις, καὶ
εὐλόγως τοιαύτη. ὅτι μὲν οὖν οὐκ ἔχει συνέχειαν
οὐδεμίαν πρὸς τὰ αἰσθητικὰ μόρια, δῆλον μὲν καὶ
5 διὰ τῆς ὄψεως, ἔτι δὲ μᾶλλον τῷ μηδεμίαν ποιεῖν
αἴσθησιν θιγγανόμενος, ὥσπερ οὐδὲ τὸ αἷμα οὐδὲ τὸ
περίττωμα τῶν ζῴων.

[1] θερμός PZ : θερμόν vulg.

148

and so it is sticky ; and it is sinewy too so that it can stretch.

We have now explained why marrow is present in certain animals. We have also made clear what marrow is. The surplus of the blood-like nourishment which is distributed to the bones and spine gets enclosed within them, and after it has undergone concoction then it is marrow.

VII. The brain is the next subject on our list. It comes appropriately after the marrow, as many think that the brain is really marrow [a] and is the source of the marrow, because, as observation shows, the spinal marrow is continuous with the brain. As a matter of fact, however, the two are quite opposite in nature. The brain is the coldest of all the parts in the body, whereas the marrow is hot, as is shown by the fact that it is greasy and fat. And that is the real reason why the spinal marrow is continuous with the brain. Nature is always contriving to set next to anything that is excessive a reinforcement of the opposite substance, so that the one may level out the excess of the other. Now there are many indications that the marrow is hot ; and the coldness of the brain is shown not only by its being cold to the touch, but also by its being the driest of all the fluid parts of the body and the one that has the least blood in it—in fact, it has none at all. It is, however, not a residue, nor is it to be classed among the parts that are continuous. It is peculiar in its nature, and this after all is but reasonable. Inspection shows that the brain has no continuity with the sensory parts, but this is shown still more unmistakably by the fact that like the blood and the residue of animals it produces no sensation when it is touched.

Brain.

[a] *Cf.* Plato, *Timaeus* 75 c, d.

652 b

Ὑπάρχει δὲ τοῖς ζῴοις πρὸς τὴν τῆς φύσεως
ὅλης σωτηρίαν. οἱ μὲν γὰρ τοῦ ζῴου τὴν ψυχὴν
τιθέασι πῦρ ἢ τοιαύτην τινὰ δύναμιν, φορτικῶς
τιθέντες· βέλτιον δ' ἴσως φάναι ἐν τοιούτῳ τινὶ
10 σώματι συνεστάναι. τούτου δ' αἴτιον ὅτι τοῖς τῆς
ψυχῆς ἔργοις ὑπηρετικώτατον τῶν σωμάτων τὸ
θερμόν ἐστιν· τὸ τρέφειν γὰρ καὶ κινεῖν ψυχῆς
ἔργον ἐστί, ταῦτα δὲ διὰ ταύτης μάλιστα γίνεται
τῆς δυνάμεως. ὅμοιον οὖν τὸ τὴν ψυχὴν εἶναι
φάναι πῦρ καὶ τὸ πρίονα ἢ τρύπανον τὸν τέκτονα
15 ἢ τὴν τεκτονικήν, ὅτι τὸ ἔργον περαίνεται ἐγγὺς
ἀλλήλων οὖσιν. ὅτι μὲν οὖν θερμότητος τὰ ζῷα
μετέχειν ἀναγκαῖον, δῆλον ἐκ τούτων· ἐπεὶ δ'
ἅπαντα δεῖται τῆς ἐναντίας ῥοπῆς, ἵνα τυγχάνῃ τοῦ
μετρίου καὶ τοῦ μέσου (τὴν γὰρ οὐσίαν ἔχει τοῦτο
καὶ τὸν λόγον, τῶν δ' ἄκρων ἑκάτερον οὐκ ἔχει
20 χωρίς), διὰ ταύτην τὴν αἰτίαν πρὸς τὸν τῆς καρδίας
τόπον καὶ τὴν ἐν αὐτῇ θερμότητα μεμηχάνηται τὸν
ἐγκέφαλον ἡ φύσις, καὶ τούτου χάριν ὑπάρχει τοῦτο
τὸ μόριον τοῖς ζῴοις, τὴν φύσιν ἔχον κοινὴν ὕδατος
καὶ γῆς, καὶ διὰ τοῦτο τὰ ⟨μὲν⟩[1] ἔναιμα ἔχει πάντα
ἐγκέφαλον, τῶν δ' ἄλλων οὐδὲν ὡς εἰπεῖν, πλὴν ὅτι
25 κατὰ τὸ ἀνάλογον, οἷον ὁ πολύπους· ὀλιγόθερμα γὰρ
πάντα διὰ τὴν ἀναιμίαν.

Ὁ μὲν οὖν ἐγκέφαλος εὔκρατον ποιεῖ τὴν ἐν τῇ
καρδίᾳ θερμότητα καὶ ζέσιν· ἵνα δὲ καὶ τοῦτο τὸ
μόριον τυγχάνῃ μετρίας θερμότητος, ἀφ' ἑκατέρας
τῆς φλεβός, τῆς τε μεγάλης καὶ τῆς καλουμένης
30 ἀορτῆς, τελευτῶσιν αἱ φλέβες εἰς τὴν μήνιγγα τὴν

---

[1] ⟨μὲν⟩ Rackham.

[a] e.g. Democritus ; see Aristotle, De anima, 403 b 31.
[b] Or, " proportion."

150

The brain is present in order to preserve the animal organism as a whole. Some [a] maintain that the Soul of an animal is Fire or some such substance. This is a crude way of putting it; and might be improved upon by saying that the Soul subsists in some body of a fiery nature. The reason for this is that the hot substance is the most serviceable of all for the activities of the Soul, since one of the activities of the Soul is to nourish; another is to cause motion; and these are most readily effected by means of this substance (viz. the hot). So to say that the Soul is fire is like saying that the craftsman, or his craft, is the saw or the auger which he uses, on the ground that the activity is performed while the two are near together. From what we have said this at any rate is clear: animals must of necessity have in them a certain amount of heat. Now, everything needs something to counterbalance it, so that it may achieve moderation and the mean; for it is the mean, and not either of the extremes apart, which has reality and rationality.[b] For this cause nature has contrived the brain to counterbalance the region of the heart and the heat in it; and that is why animals have a brain, the composition of which is a combination of Water and Earth. Hence, although all blooded animals have a brain, practically none of the others has (unless it be just a counterpart, as in the case of the Octopus), for since they lack blood they have but little heat.

The brain, then, makes the heat and the boiling in the heart well blent and tempered; yet in order that the brain may still have a moderate heat, blood-vessels run from the great Blood-vessel and what is known as the Aorta, till they reach the membrane

151

652 b

περὶ τὸν ἐγκέφαλον. πρὸς δὲ τὸ τῇ θερμότητι μὴ
βλάπτειν, ἀντὶ μὲν μεγάλων ⟨καὶ⟩[1] ὀλίγων πυκναὶ
καὶ λεπταὶ φλέβες περιέχουσιν αὐτόν, ἀντὶ δὲ θολεροῦ[2] καὶ παχέος αἵματος λεπτὸν καὶ καθαρόν. διὸ
καὶ τὰ ῥεύματα τοῖς σώμασιν ἐκ τῆς κεφαλῆς ἐστι
35 τὴν ἀρχήν, ὅσοις ἂν ᾖ τὰ περὶ τὸν ἐγκέφαλον
ψυχρότερα τῆς συμμέτρου κράσεως· ἀναθυμιω-
653 a μένης γὰρ διὰ τῶν φλεβῶν ἄνω τῆς τροφῆς τὸ
περίττωμα ψυχόμενον διὰ τὴν τοῦ τόπου τούτου
δύναμιν ῥεύματα ποιεῖ φλέγματος καὶ ἰχῶρος.
δεῖ δὲ λαβεῖν, ὡς μεγάλῳ παρεικάζοντα μικρόν,
ὁμοίως συμβαίνειν ὥσπερ τὴν τῶν ὑετῶν γένεσιν·
5 ἀναθυμιωμένης γὰρ ἐκ τῆς γῆς τῆς ἀτμίδος καὶ
φερομένης ὑπὸ τοῦ θερμοῦ πρὸς τὸν ἄνω τόπον,
ὅταν ἐν τῷ ὑπὲρ τῆς γῆς γένηται ἀέρι ὄντι ψυχρῷ,
συνίσταται πάλιν εἰς ὕδωρ διὰ τὴν ψύξιν καὶ ῥεῖ
κάτω πρὸς τὴν γῆν. ἀλλὰ περὶ μὲν τούτων ἐν ταῖς
τῶν νόσων ἀρχαῖς ἁρμόττει λέγειν, ἐφ' ὅσον τῆς φυ-
10 σικῆς φιλοσοφίας ἐστὶν εἰπεῖν περὶ αὐτῶν.

Ποιεῖ δὲ καὶ τὸν ὕπνον τοῖς ζῴοις τοῦτο τὸ
μόριον τοῖς ἔχουσιν ἐγκέφαλον, τοῖς δὲ μὴ ἔχουσι
τὸ ἀνάλογον· καταψῦχον γὰρ τὴν ἀπὸ τῆς τροφῆς
τοῦ αἵματος ἐπίρρυσιν (ἢ καὶ διά τινας ὁμοίας
αἰτίας ἄλλας), βαρύνει τε τὸν τόπον (διὸ τὴν κεφαλὴν
15 καρηβαροῦσιν οἱ ὑπνώσσοντες) καὶ κάτω ποιεῖ τὸ
θερμὸν ὑποφεύγειν μετὰ τοῦ αἵματος. διὸ πλεῖον
ἀθροιζόμενον ἐπὶ τὸν κάτω τόπον ἀπεργάζεται τὸν
ὕπνον, καὶ τὸ δύνασθαι ἑστάναι ὀρθὰ ἀφαιρεῖται
ὅσα τῶν ζῴων ὀρθὰ τὴν φύσιν ἐστί, τῶν δ' ἄλλων

---

[1] ⟨καὶ⟩ Rackham.
[2] θολεροῦ coni. Buss. (*turbidi* Σ): πολλοῦ vulg.

which surrounds the brain. And in order to prevent injury being done through heat, the blood-vessels surrounding it are not few and large but small and multitudinous; and the blood is not muddy and thick but thin and clear. This also explains why fluxes begin in the head; they occur when the parts around the brain are colder than the rightly-proportioned blend.[a] What happens is that, as the nourishment exhales upwards through the blood-vessels, the residue from it becomes cooled owing to the specific nature of the brain, and produces fluxes of phlegm and serum. And we should be justified in maintaining that this process resembles, on a small scale, the one which produces rain-showers. Damp vapour exhales up from the earth and is carried into the upper regions by the heat; and when it reaches the cold air up aloft, it condenses back again into water owing to the cold, and pours down towards the earth. However, so far as Natural Philosophy is concerned with these matters, the proper place to speak of them is in the *Origins of Diseases.*[b]

Furthermore, it is the brain (or, if there is no brain, its counterpart) which produces sleep in animals. It cools the onflow of blood which comes from the food (or else is due to other causes of the same sort), and weighs down the part where it is (that is why when a person is sleepy his head is weighed down), and causes the hot substance to escape below together with the blood. Hence, the blood accumulates unduly in the lower region of the body and produces sleep; at the same time it takes away from those animals whose nature is to stand upright the power to do so, and the others it prevents from

[a] See p. 38. [b] No such treatise exists.

653 a

τὴν ὀρθότητα τῆς κεφαλῆς· περὶ ὧν εἴρηται καθ'
20 αὐτὰ ἔν τε τοῖς περὶ αἰσθήσεως καὶ περὶ ὕπνου
διωρισμένοις.

"Ὅτι δ' ἐστὶν ὁ ἐγκέφαλος κοινὸς ὕδατος καὶ
γῆς, δηλοῖ τὸ συμβαῖνον περὶ αὐτόν· ἑψόμενος γὰρ
γίνεται ξηρὸς καὶ σκληρός, καὶ λείπεται τὸ γεῶδες
ἐξατμισθέντος τοῦ ὕδατος ὑπὸ τῆς θερμότητος,
ὥσπερ τὰ τῶν χεδρόπων ἑψήματα καὶ τῶν ἄλλων
25 καρπῶν, διὰ τὸ γῆς εἶναι τὸ πλεῖστον μέρος, ἐξ-
ιόντος τοῦ μιχθέντος ὑγροῦ· καὶ γὰρ ταῦτα γίνεται
σκληρὰ καὶ γεηρὰ πάμπαν.

"Ἔχει δὲ τῶν ζῴων ἐγκέφαλον πλεῖστον ἄνθρωπος
ὡς κατὰ μέγεθος, καὶ τῶν ἀνθρώπων οἱ ἄρρενες τῶν
θηλειῶν· καὶ γὰρ τὸν περὶ τὴν καρδίαν καὶ τὸν
30 πλεύμονα τόπον θερμότατον καὶ ἐναιμότατον. διὸ
καὶ μόνον ἐστὶ τῶν ζῴων ὀρθόν· ἡ γὰρ τοῦ θερμοῦ
φύσις ἐνισχύουσα ποιεῖ τὴν αὔξησιν ἀπὸ τοῦ μέσου
κατὰ τὴν αὑτῆς φοράν. πρὸς οὖν πολλὴν θερμότητα
ἀντίκειται πλείων ὑγρότης καὶ ψυχρότης, καὶ διὰ τὸ
πλῆθος ὀψιαίτατα πήγνυται τὸ περὶ τὴν κεφαλὴν
35 ὀστοῦν, ὃ καλοῦσι βρέγμα τινές, διὰ τὸ πολὺν
χρόνον τὸ θερμὸν ἀπατμίζειν· τῶν δ' ἄλλων οὐδενὶ
τοῦτο συμβαίνει τῶν ἐναίμων ζῴων. καὶ ῥαφὰς δὲ
653 b πλείστας ἔχει περὶ τὴν κεφαλήν, καὶ τὸ ἄρρεν
πλείους τῶν θηλειῶν, διὰ τὴν αὐτὴν αἰτίαν, ὅπως ὁ
τόπος εὔπνους ᾖ, καὶ μᾶλλον ὁ πλείων ἐγκέφαλος·
ὑγραινόμενος γὰρ ἢ ξηραινόμενος μᾶλλον οὐ ποιήσει
τὸ αὑτοῦ ἔργον, ἀλλ' ἢ οὐ ψύξει ἢ πήξει, ὥστε

---

[a] See *De somno*, 455 b 28 ff., especially 456 b 17 ff.
[b] The cranial bone, which covers the anterior fontanelle.

holding their heads upright. These matters have been spoken of separately in the treatises on *Sensation* and on *Sleep*.[a]

I said the brain is compounded of Water and Earth. This is shown by what happens when it is boiled. Then it becomes solid and hard : the earthy substance is left behind after the Water has evaporated owing to the heat. It is just what happens when pulse and other forms of fruit are boiled ; they also get hard and earthy altogether, because the greater part of them is earth, and the fluid mixed with it departs when they are boiled.

Of all the animals, man has the largest brain for his size ; and men have a larger brain than women. In both cases the largeness is due to there being a great deal of heat and blood in the region around the heart and the lung. This too explains why man is the only animal that stands upright. As the hot substance prevails in the body it induces growth, beginning from the centre along its own line of travel. It is against great heat, then, that a large supply of fluid and cold is provided. This bulk of moisture is also the reason why the bone that surrounds the brain (called by some the *bregma*)[b] is the last of all to solidify ; the hot substance takes a long time to evaporate it off. This phenomenon does not occur in any other of the blooded animals. Again, man has more sutures in the skull than any other animal, and males have more than females. The size of the brain is the reason for this also ; it is to secure ventilation, and the larger the brain, the more ventilation it requires. If the brain becomes unduly fluid or unduly solid, it will not perform its proper function, but will either fail to cool the blood or else

155

5 νόσους καὶ παρανοίας ποιεῖν καὶ θανάτους· τὸ γὰρ
ἐν τῇ καρδίᾳ θερμὸν καὶ ἡ ἀρχὴ συμπαθέστατόν
ἐστι καὶ ταχεῖαν ποιεῖται τὴν αἴσθησιν μεταβάλ-
λοντός τι καὶ πάσχοντος τοῦ περὶ τὸν ἐγκέφαλον
αἵματος.

Περὶ μὲν οὖν τῶν συμφύτων τοῖς ζῴοις ὑγρῶν
10 σχεδὸν εἴρηται περὶ πάντων· τῶν δ' ὑστερογενῶν
τά τε περιττώματα τῆς τροφῆς ἐστί, τό τε τῆς
κύστεως ὑπόστημα καὶ τὸ τῆς κοιλίας, καὶ παρὰ
ταῦτα γονὴ καὶ γάλα τοῖς πεφυκόσιν ἔχειν ἕκαστα
τούτων. τὰ μὲν οὖν τῆς τροφῆς περιττώματα περὶ
τὴν τῆς τροφῆς σκέψιν καὶ θεωρίαν οἰκείους ἔχει
15 τοὺς λόγους, τίσι τε τῶν ζῴων ὑπάρχει καὶ διὰ
τίνας αἰτίας, τὰ δὲ περὶ σπέρματος καὶ γάλακτος ἐν
τοῖς περὶ γενέσεως· τὸ μὲν γὰρ ἀρχὴ γενέσεως
αὐτῶν ἐστι, τὸ δὲ χάριν γενέσεως.

VIII. Περὶ δὲ τῶν ἄλλων μορίων τῶν ὁμοιο-
20 μερῶν σκεπτέον, καὶ πρῶτον περὶ σαρκὸς ἐν τοῖς
ἔχουσι σάρκας, ἐν δὲ τοῖς ἄλλοις τὸ ἀνάλογον· τοῦτο
γὰρ ἀρχὴ καὶ σῶμα καθ' αὑτὸ τῶν ζῴων ἐστίν.
δῆλον δὲ καὶ κατὰ τὸν λόγον· τὸ γὰρ ζῷον ὁρι-
ζόμεθα τῷ ἔχειν αἴσθησιν, πρῶτον δὲ τὴν πρώτην·
αὕτη δ' ἐστὶν ἁφή, ταύτης δ' αἰσθητήριον τὸ τοιοῦ-
25 τον μόριόν ἐστιν, ἤτοι τὸ πρῶτον, ὥσπερ ἡ κόρη

ᵃ At *De gen. an.* 722 a, 776 a 15 ff.

will make it set fast, thus producing various forms of disease, madness, and death. Indeed, the heat that is in the heart, being the source, is extremely responsive to any influence upon it; and if the blood which surrounds the brain undergoes any change or any other affection, then this heat at once becomes sensitive of it.

We may now claim to have considered all the fluids which are present in animal bodies from their very earliest stages. There are others which are first produced only at some later stage, and among these we must reckon the residues of the nourishment— that is to say, the deposits from the bladder and from the gut; and also semen, and milk; these make their appearance according to the species and sex of the animal concerned. Discussion of the residues of the nourishment will come in appropriately during our general consideration and examination of nourishment; we shall then show in what animals they occur, and why they do so. Semen, which gives rise to generation, and milk, which exists on account of generation, we shall deal with in the treatise on *Generation*.[a]

VIII. We must now go on to consider the rest of the uniform parts. Let us take first of all Flesh (and, where Flesh is absent, its counterpart), for this is to animals both a principle and a body in itself. Its primacy can also be logically shown, as follows. We define an animal as something that has the power of sensation, and chiefly the primary sensation, which is touch; and the organ through which this sensation is effected is the flesh (or its counterpart). And flesh is either its primary organ (comparable to the pupil in the case of sight), or else it is the organ and

Flesh and Bone.

157

653 b

τῆς ὄψεως, ἢ τὸ δι' οὗ συνειλημμένον, ὥσπερ ἂν εἴ
τις προσλάβοι τῇ κόρῃ τὸ διαφανὲς πᾶν. ἐπὶ μὲν
οὖν τῶν ἄλλων αἰσθήσεων ἀδύνατόν τε καὶ οὐδὲν
προὔργου τοῦτ' ἦν ποιῆσαι τῇ φύσει, τὸ δ' ἁπτικὸν
ἐξ ἀνάγκης· μόνον γὰρ ἢ μάλιστα τοῦτ' ἐστὶ σωμα-
30 τῶδες τῶν αἰσθητηρίων. κατὰ δὲ τὴν αἴσθησιν
φανερὸν πάντα τἆλλα τούτου χάριν ὄντα, λέγω δ'
οἷον ὀστᾶ καὶ δέρμα καὶ νεῦρα καὶ φλέβες, ἔτι δὲ
τρίχες καὶ τὸ τῶν ὀνύχων γένος, καὶ εἴ τι τοιοῦτον
ἕτερόν ἐστιν. ἡ μὲν γὰρ τῶν ὀστῶν φύσις σωτηρίας ἕνεκεν μεμηχάνηται ⟨τοῦ⟩[1] μαλακοῦ, σκληρὰ
35 τὴν φύσιν οὖσα, ἐν τοῖς ἔχουσιν ὀστᾶ· ἐν δὲ τοῖς
μὴ ἔχουσι, τὸ ἀνάλογον, οἷον ἐν τοῖς ἰχθύσι τοῖς
μὲν ἄκανθα τοῖς δὲ χόνδρος.

Τὰ μὲν οὖν ἔχει τῶν ζῴων ἐντὸς τὴν τοιαύτην
654 a βοήθειαν, ἔνια δὲ τῶν ἀναίμων ἐκτός, ὥσπερ τῶν τε
μαλακοστράκων ἕκαστον, οἷον καρκίνοι καὶ τὸ τῶν
καράβων γένος, καὶ τὸ τῶν ὀστρακοδέρμων ὡσαύτως, οἷον τὰ καλούμενα ὄστρεα· πᾶσι γὰρ τούτοις
5 τὸ μὲν σαρκῶδες ἐντός, τὸ δὲ συνέχον καὶ φυλάττον
ἐκτὸς τὸ γεῶδές ἐστιν· πρὸς γὰρ τῇ φυλακῇ τῆς
συνεχείας, τῷ ἔχειν ὀλίγον αὐτῶν τὴν φύσιν θερμὸν
ἀναίμων ὄντων, οἷον πνιγεύς τις περικείμενον τὸ
ὄστρακον φυλάττει τὸ ἐμπεπυρευμένον θερμόν. ἡ
δὲ χελώνη καὶ τὸ τῶν ἐμύδων γένος ὁμοίως ἔχειν

[1] ⟨τοῦ⟩ Ogle.

---

[a] Apparently because the objects with which it deals are
more " corporeal " than those of the other senses—it has
to be in bodily contact with them.

[b] As apart from *a priori* reasoning.

[c] Sometimes, as here, " counterpart " could be represented
by the modern term " analogue."

[d] Lit., " the soft-shelled creatures."

the medium of the sensation combined in one (comparable to the pupil *plus* the whole of the transparent medium in the case of sight). Now not only was it pointless, it was impossible for Nature to make such a combination in the case of the other senses ; with touch, however, it was due to *necessity*, since its sense-organ is the only one which is corporeal— or at least it is definitely the most corporeal one.[a] It is also clear from our actual experience in sensation[b] that all the other parts exist for the sake of the organ of touch (the flesh). In these I include the bones, the skin, the sinews, the blood-vessels ; also the hair, nails of every sort and kind, and the like. The bones, for instance, which are hard in substance, have been devised for the preservation of the soft parts. The same is true of the counterpart[c] of the bones in other creatures : two examples in species of fish are spine and cartilage.

Now with some animals this hard supporting substance is situated inside the body, with others (some of the bloodless ones) it is outside. It is outside in the case of all the Crustacea[d] (*e.g.* the Crabs and the group of Crayfish), and the group of Testacea[e] too, *e.g.* those that are known as Oysters. All these have their fleshy part inside, and the earthy part which holds it together and protects it is outside—outside, because it performs an additional function as well : since these creatures are bloodless, they possess but little heat, and the shell acts like a *couvre-feu* ; it encloses the faintly burning heat and protects it. Another quite different group of creatures, the Turtles and the group of freshwater

[e] Lit., " the shell-skinned creatures."   " Testacea " is the nearest modern term.   See Introduction, p. 23.

654 a

δοκεῖ τούτοις, ἕτερον ὂν γένος τούτων. τὰ δ'
10 ἔντομα τῶν ζῴων καὶ τὰ μαλάκια τούτοις τ'
ἐναντίως καὶ αὑτοῖς ἀντικειμένως συνέστηκεν· οὐδὲν
γὰρ ὀστῶδες ἔχειν ἔοικεν οὐδὲ γεηρὸν ἀποκεκρι-
μένον, ὅ τι καὶ ἄξιον εἰπεῖν, ἀλλὰ τὰ μὲν μαλάκια
σχεδὸν ὅλα σαρκώδη καὶ μαλακά, πρὸς δὲ τὸ μὴ
εὔφθαρτον εἶναι τὸ σῶμα αὐτῶν, καθάπερ τὰ
15 σαρκώδη, μεταξὺ σαρκὸς καὶ νεύρου τὴν φύσιν ἔχει.
μαλακὸν μὲν γὰρ ὥσπερ σάρξ ἐστιν, ἔχει δὲ τάσιν
ὥσπερ νεῦρον· τὴν δὲ σχίσιν ἔχει τῆς σαρκὸς οὐ
κατ' εὐθυωρίαν ἀλλὰ κατὰ κύκλους διαιρετήν·
οὕτως γὰρ [ἂν]¹ ἔχον χρησιμώτατον ἂν εἴη² πρὸς τὴν
20 ἰσχύν. ὑπάρχει δ' ἐν αὐτοῖς καὶ τὸ ἀνάλογον ταῖς
τῶν ἰχθύων ἀκάνθαις, οἷον ἐν μὲν ταῖς σηπίαις τὸ
καλούμενον σηπίον, ἐν δὲ ταῖς τευθίσι τὸ καλούμενον ξίφος. τὸ³ δ' αὖ τῶν πολυπόδων ⟨γένος⟩⁴
τοιοῦτον οὐδὲν ἔχει διὰ τὸ μικρὸν ἔχειν τὸ κύτος
τὴν καλουμένην κεφαλήν, θάτερα δ' εὐμήκη. διὸ
πρὸς τὴν ὀρθότητα αὐτῶν καὶ τὴν ἀκαμψίαν ὑπ-
25 έγραψε ταῦτα ἡ φύσις, ὥσπερ τῶν ἐναίμων τοῖς
μὲν ὀστοῦν τοῖς δ' ἄκανθαν. τὰ δ' ἔντομα τούτοις
τ' ἐναντίως ἔχει καὶ τοῖς ἐναίμοις, καθάπερ εἴπομεν·
οὐδὲν γὰρ ἀφωρισμένον ἔχει σκληρόν, τὸ δὲ μαλακόν, ἀλλ' ὅλον τὸ σῶμα σκληρόν, σκληρότητα δὲ
τοιαύτην, ὀστοῦ μὲν σαρκωδεστέραν, σαρκὸς δ'

---

¹ [ἂν] seclusi. ² χρησιμα´τατα εἴη SU.
³ τὸ Platt: τὰ vulg. ⁴ ⟨γένος⟩ Platt.

Tortoises, are apparently in like case. On the other
hand, the Insects and the Cephalopods are differently constructed from these, as well as being
different from each other. Not only, as it appears,
have they no bony part, but they have practically
no earthy part at all distinct from the rest of the
body. The Cephalopods are almost wholly soft
and fleshy, yet in order to prevent their bodies
from being easily destructible as fleshy structures are, the substance of which they are formed
is intermediate between flesh and sinew, having the
softness of flesh and the elasticity of sinew. When
it is split up, it breaks as flesh does, that is, not
longitudinally but into circular portions. The reason
for this seems to be that such a structure secures
the greatest strength. There is found also in these
creatures the counterpart of the spinous bones of
fishes ; examples are : the " pounce " (*os sepiae*) of
the cuttlefish, and the " pen " (*gladius*) of the
calamaries. Nothing of this sort, however, appears
in the Octopuses : this is because in them what is
called the " head " forms but a small sac, whereas
in the cuttlefish and calamaries the " head " is of
considerable length. So we see that, in order to
secure that they should be straight and inflexible,
nature prescribed for them this hard support, just
as she gave to the blooded creatures bones or spines.
Quite a different contrivance obtains in the Insects—
different both from the Cephalopods and from the
blooded creatures, as has already been stated. In
the Insects we do not find the clear-cut distinction
of hard parts and soft ; here, the whole body is hard,
yet its hardness is such that it is more fleshlike than

654 a

30 ὀστωδεστέραν καὶ γεωδεστέραν, πρὸς τὸ μὴ εὐ-
διαίρετον εἶναι τὸ σῶμα αὐτῶν.

IX. Ἔχει δ' ὁμοίως ἥ τε τῶν ὀστῶν καὶ ἡ
τῶν φλεβῶν φύσις. ἑκατέρα γὰρ αὐτῶν ἀφ' ἑνὸς
ἠργμένη συνεχής ἐστι, καὶ οὔτ' ὀστοῦν ἐστιν αὐτὸ
35 καθ' αὑτὸ οὐδέν, ἀλλ' ἢ μόριον ὡς συνεχοῦς ἢ
ἁπτόμενον καὶ προσδεδεμένον, ἵνα χρῆται ἡ φύσις
654 b καὶ ὡς ἑνὶ καὶ συνεχεῖ καὶ ὡς δυσὶ καὶ διῃρημένοις
πρὸς τὴν κάμψιν. ὁμοίως δὲ καὶ φλὲψ οὐδεμία
αὐτὴ καθ' αὑτήν ἐστιν, ἀλλὰ πᾶσαι μόριον μιᾶς
εἰσιν. ὀστοῦν τε γὰρ εἴ τι κεχωρισμένον ἦν, τό τ'
5 ἔργον οὐκ ἂν ἐποίει οὗ χάριν ἡ τῶν ὀστῶν ἐστι
φύσις (οὔτε γὰρ ἂν κάμψεως ἦν αἴτιον οὔτ' ὀρθό-
τητος οὐδεμιᾶς μὴ συνεχὲς ὂν ἀλλὰ διαλεῖπον), ἔτι
τ' ἔβλαπτεν ἂν ὥσπερ ἄκανθά τις ἢ βέλος ἐνὸν ταῖς
σαρξίν. εἴτε φλὲψ ἦν τις κεχωρισμένη καὶ μὴ
συνεχὴς πρὸς τὴν ἀρχήν, οὐκ ἂν ἔσωζε τὸ ἐν αὐτῇ
10 αἷμα· ἡ γὰρ ἀπ' ἐκείνης θερμότης κωλύει πήγνυ-
σθαι, φαίνεται δὲ καὶ σηπόμενον τὸ χωριζόμενον.
ἀρχὴ δὲ τῶν μὲν φλεβῶν ἡ καρδία, τῶν δ' ὀστῶν ἡ
καλουμένη ῥάχις τοῖς ἔχουσιν ὀστᾶ πᾶσιν, ἀφ' ἧς
συνεχὴς ἡ τῶν ἄλλων ὀστῶν ἐστι φύσις· ἡ γὰρ τὸ
μῆκος καὶ τὴν ὀρθότητα συνέχουσα τῶν ζῴων ἡ
15 ῥάχις ἐστίν. ἐπεὶ δ' ἀνάγκη κινουμένου τοῦ ζῴου
162

bone is and more bony and earthy than flesh. The purpose of this is to ensure that the body shall not easily break up.

IX. The system of the bones is similar to that of the blood-vessels : each is a connected system beginning from one point. There is no such thing as a bone by itself in isolation ; every bone is either actually part of the connected scheme, or else is attached to it and so is in contact with it. This enables Nature to use any couple of bones either as a single connected piece, or, when flexion is required, as two distinct pieces. In like manner, there is no such thing as a blood-vessel by itself in isolation : they are all of them parts of one blood-vessel. An isolated bone could never discharge the function for which all bones exist ; for, being discontinuous and disconnected from the rest, it could never serve as the means either for bending or for straightening a limb ; but worse than that, it would be a source of harm, like a thorn or an arrow sticking in the flesh. Similarly, if we imagine a blood-vessel isolated and not connected with the source of them all, it could never keep the blood within it in a proper condition, since it is the heat which comes from that source which prevents the blood from congealing, as is shown by the putrefaction of blood when separated from it. This source of the blood-vessels is of course the heart, and the corresponding source of the bones in all bony species is what is called the backbone. The system of the bones is a connected whole, starting from the backbone, since the backbone connects together the length of the animal's body and holds it straight. Now although this backbone is a unity because it is connected together, it

κάμπτεσθαι τὸ σῶμα, μία μὲν διὰ τὴν συνέχειάν
ἐστι, πολυμερὴς δὲ τῇ διαιρέσει τῶν σπονδύλων.
ἐκ δὲ ταύτης τοῖς ἔχουσι κῶλα συνεχῆ [πρὸς αὐτὴν]
τὰ τούτων ὀστᾶ [τῶν ἁρμονιῶν] ἐστιν· τὰ¹ μὲν [ἔχει
τὰ κῶλα κάμψιν συνδεδεμένα τοῖς² νεύροις, καὶ] τῶν
20 ἐσχάτων συναρμοττόντων, τοῦ μὲν ὄντος κοίλου
τοῦ δὲ περιφεροῦς, ἢ καὶ ἀμφοτέρων κοίλων, ἐν
μέσῳ δὲ περιειληφότων, οἷον γόμφον, ἀστράγαλον,
ἵνα γίνηται κάμψις καὶ ἔκτασις (ἄλλως γὰρ ἢ ὅλως
ἀδύνατον, ἢ οὐ καλῶς ἂν ἐποίουν τὴν τοιαύτην κίνησιν)· ἔνια δ' αὐτῶν ὁμοίαν ἔχοντα τὴν ἀρχὴν τὴν
25 θατέρου τῇ τελευτῇ θατέρου [συνδέδεται νεύροις]·³
καὶ χονδρώδη δὲ μόρια μεταξὺ τῶν κάμψεών
ἐστιν,⁴ οἷον στοιβή, πρὸς τὸ ἄλληλα μὴ τρίβειν.

Περὶ δὲ τὰ ὀστᾶ αἱ σάρκες περιπεφύκασι,
προσειλημμέναι λεπτοῖς καὶ ἰνώδεσι δεσμοῖς· ὧν
ἕνεκεν τὸ τῶν ὀστῶν ἐστι γένος. ὥσπερ γὰρ οἱ
30 πλάττοντες ἐκ πηλοῦ ζῷον ἤ τινος ἄλλης ὑγρᾶς
συστάσεως ὑφιστᾶσι τῶν στερεῶν τι σωμάτων, εἶθ'
οὕτω περιπλάττουσι, τὸν αὐτὸν τρόπον ἡ φύσις
δεδημιούργηκεν ἐκ τῶν σαρκῶν τὸ ζῷον. τοῖς
μὲν οὖν ἄλλοις ὑπεστιν ὀστᾶ τοῖς σαρκώδεσι μορίοις, τοῖς μὲν κινουμένοις διὰ κάμψιν τούτου
35 χάριν, τοῖς δ' ἀκινήτοις φυλακῆς ἕνεκεν, οἷον αἱ
655 a συγκλείουσαι πλευραὶ τὸ στῆθος σωτηρίας χάριν

---

¹ τὰ Peck : τὰς Z : ᾗ vulg. : ὀστᾶ τῶν μορίων ἐστιν· τὰς μὲν
(ᾗ μὲν vulg.) ἔχει τὰ κῶλα καὶ κάμψιν Ζ.
² τοῖς SU : τε vulg. : γε ΕΥ.
³ ll. 16-25 : hunc locum correxi, partim Σ et Albertum
secutus. vid. p. 46. fortasse et ἐπεὶ δ' ἀνάγκη . . . σπονδύλων (ll. 14-16) secludenda.
⁴ εἰσιν vulg.

is also a thing of many parts because of its division
into vertebrae, since the body must be able to bend
while the animal is in motion. And the bones of the
various limbs (in those animals which have them)
are connected with this backbone, from which they
originate. Some of them have extremities which fit
on to each other: either (*a*) one is hollow and the
other rounded, or (*b*) both are hollow and hold a
huckle-bone between them (as it might be a bolt),
to admit of bending and extension, since these
movements would be quite impossible or at any rate
unsatisfactory without such an arrangement. (*c*)
There are some joints in which the adjacent ends of
the two bones are similar in shape; [these are bound
together by sinews,] and there are pieces of cartilage
inserted in between them, like a pad, to prevent
them from rubbing against each other.[a]

Now the whole system of the bones exists to subserve the fleshy parts of the body, which have their
place around the bones and are attached to them by
thin fibrous threads. Modellers who set out to mould
an animal out of clay or some other plastic substance
begin first of all with a hard and solid core and mould
their figure round it. Nature's method has been the
same in fashioning animals out of flesh. With one
exception, all the fleshy parts have a core of bone:
for the parts that move and bend, this is present as
a means for enabling the limb to bend; for those
that do not move, it serves as a protection: an
example of this are the ribs, enclosing the chest,
which are a means of protection for the viscera in

---

[a] The text of this paragraph has been confused by a
number of interpolations, most of which I have omitted in
translating.

655 a

τῶν περὶ τὴν καρδίαν σπλάγχνων· τὰ δὲ περὶ τὴν
κοιλίαν ἀνόστεα πᾶσιν, ὅπως μὴ κωλύῃ τὴν ἀνοίδησιν τὴν ἀπὸ τῆς τροφῆς γινομένην τοῖς ζῴοις
ἐξ ἀνάγκης καὶ τοῖς θήλεσι τὴν ἐν αὐτοῖς τῶν ἐμβρύων αὔξησιν.

5 Τὰ μὲν οὖν ζῳοτόκα τῶν ζῴων καὶ ἐν αὐτοῖς καὶ
ἐκτὸς παραπλησίαν ἔχει τὴν τῶν ὀστῶν δύναμιν καὶ
ἰσχυράν. πολὺ γὰρ μείζω πάντα τὰ τοιαῦτα τῶν
μὴ ζῳοτόκων ὡς κατὰ λόγον εἰπεῖν τῶν σωμάτων·
ἐνιαχοῦ γὰρ πολλὰ γίνεται μεγάλα τῶν ζῳοτόκων,
10 οἷον ἐν Λιβύῃ καὶ τοῖς τόποις τοῖς θερμοῖς καὶ τοῖς
ξηροῖς. τοῖς δὲ μεγάλοις ἰσχυροτέρων δεῖ τῶν
ὑπερεισμάτων καὶ μειζόνων καὶ σκληροτέρων, καὶ
τούτων αὐτῶν τοῖς βιαστικωτέροις. διὸ τὰ τῶν
ἀρρένων σκληρότερα ἢ τὰ τῶν θηλειῶν, καὶ τὰ τῶν
σαρκοφάγων (ἡ τροφὴ γὰρ διὰ μάχης τούτοις),
ὥσπερ τὰ τοῦ λέοντος· οὕτω γὰρ ἔχει ταῦτα
15 σκληρὰν τὴν φύσιν ὥστ᾿ ἐξάπτεσθαι τυπτομένων
καθάπερ ἐκ λίθων πῦρ. ἔχει δὲ καὶ ὁ δελφὶς οὐκ
ἀκάνθας ἀλλ᾿ ὀστᾶ· ζῳοτόκος γάρ ἐστιν.

Τοῖς δ᾿ ἐναίμοις μὲν μὴ ζῳοτόκοις δὲ παρ-
αλλάττει κατὰ μικρὸν ἡ φύσις, οἷον τοῖς ὄρνισιν
ὀστᾶ μέν, ἀσθενέστερα δέ. τῶν δ᾿ ἰχθύων τοῖς μὲν
20 ᾠοτόκοις ἄκανθα, καὶ τοῖς ὄφεσιν ἀκανθώδης ἐστὶν
ἡ τῶν ὀστῶν φύσις, πλὴν τοῖς λίαν μεγάλοις· τούτοις δέ, δι᾿ ἅπερ καὶ τοῖς ζῳοτόκοις, πρὸς τὴν
ἰσχὺν ἰσχυροτέρων δεῖ τῶν στερεωμάτων. τὰ δὲ
καλούμενα σελάχη χονδράκανθα τὴν φύσιν ἐστίν·
ὑγροτέραν τε γὰρ ἀναγκαῖον αὐτῶν εἶναι τὴν κί-

─────────

^a Cartilaginous fishes, including the sharks.

the region of the heart. The exception is the parts near the belly, which in all animals are boneless. The purpose of this is that the swelling which takes place *of necessity* after the receipt of nourishment may not be hampered, and (in females) to prevent any interference with the growth of the fetus.

The nature of the bones is similar in all viviparous animals (that is, internally viviparous as well as externally) ; and as the Vivipara are much larger proportionately in bodily size than other animals, their bones are strong. In some places many of these animals grow to a great size, as for example in Libya and other hot dry countries. These large animals need stronger and bigger and harder supports, especially those of them that are particularly violent in their habits. Hence, the bones of males are harder than the bones of females, and those of carnivorous animals than those of herbivorous, because the carnivorous have to fight for their food. An example is the Lion : it has such hard bones that when they are struck fire is kindled as it is from stones. Note that the Dolphin, being viviparous, has bones like the other viviparous creatures, and not fish-spines.

In the creatures which though blooded are not viviparous Nature has made a series of graduated changes : for example, birds have bones, but they are weaker than the bones of the Vivipara. The oviparous fishes have fish-spine, not bone ; and the serpents have bone whose nature is that of fish-spine ; except the very large species, and they have bones, because (just like the Vivipara) if their bodies are to be strong the solid framework of them must be stronger. The creatures called Selachia [a] have spines made of cartilage. This is because their movement

355 a

25 νησιν, ὥστε δεῖ καὶ τὴν τῶν ἐρεισμάτων μὴ κραῦρον εἶναι ἀλλὰ μαλακωτέραν, καὶ τὸ γεῶδες εἰς
τὸ δέρμα πᾶν ἀνήλωκεν ἡ φύσις· ἅμα δὲ τὴν αὐτὴν
ὑπεροχὴν εἰς πολλοὺς τόπους ἀδυνατεῖ διανέμειν ἡ
φύσις. ἔνεστι δὲ καὶ ἐν τοῖς ζῳοτόκοις πολλὰ τῶν
ὀστῶν χονδρώδη, ἐν ὅσοις συμφέρει μαλακὸν εἶναι
30 καὶ μυξῶδες[1] τὸ στερεὸν διὰ τὴν σάρκα τὴν περι-
κειμένην, οἷον συμβέβηκε περί τε τὰ ὦτα καὶ
τοὺς μυκτῆρας· θραύεται γὰρ τὰ κραῦρα ταχέως
ἐν τοῖς ἀπέχουσιν. ἡ δὲ φύσις ἡ αὐτὴ χόνδρου
καὶ ὀστοῦ ἐστι, διαφέρει δὲ τῷ μᾶλλον καὶ ἧττον·
διὸ καὶ οὐδέτερον αὐξάνεται ἀποκοπέν.

35 Οἱ μὲν οὖν ἐν τοῖς πεζοῖς ἀμύελοι χόνδροι κεχω-
ρισμένῳ μυελῷ· τὸ γὰρ χωριζόμενον εἰς ἅπαν
μεμιγμένον μαλακὴν ποιεῖ καὶ μυξώδη[2] τὴν τοῦ
χόνδρου σύστασιν. ἐν δὲ τοῖς σελάχεσιν ἡ ῥάχις
355 b χονδρώδης μέν ἐστιν, ἔχει δὲ μυελόν· ἀντ᾽ ὀστοῦ
γὰρ αὐτοῖς ὑπάρχει τοῦτο τὸ μόριον.

Σύνεγγυς δὲ κατὰ τὴν ἁφήν ἐστι τοῖς ὀστοῖς καὶ
τὰ τοιάδε τῶν μορίων, οἷον ὄνυχές τε καὶ ὁπλαὶ καὶ
χηλαὶ καὶ κέρατα καὶ ῥύγχη τὰ τῶν ὀρνίθων. πάντα
5 δὲ ταῦτα βοηθείας ἔχουσι χάριν [τὰ ζῷα][3]· τὰ γὰρ ἐξ
αὐτῶν συνεστηκότα ὅλα καὶ συνώνυμα τοῖς μορίοις,
οἷον ὁπλή τε ὅλη καὶ κέρας ὅλον, μεμηχάνηται πρὸς
τὴν σωτηρίαν ἑκάστοις. ἐν τούτῳ δὲ τῷ γένει καὶ

---

[1] ζυμῶδες Z.   [2] ζυμώδη EPSZ.
[3] [τὰ ζῷα] secludit Rackham.

---

[a] Cf. the " law of organic equivalents."
[b] See note on 644 a 17.

has to be somewhat supple, and accordingly the supporting framework of their bodies must be somewhat pliable, not brittle. In addition, Nature cannot allot the same plentiful supply of any one substance to many different parts of the body ;[a] and in the case of the Selachia she has used up all the available earthy substance in constructing their skin. In the Vivipara too there are many instances of cartilaginous bones : they are found where it is an advantage that the solid framework should be pliable and glutinous for the benefit of the flesh that surrounds them. This applies to the ears and the nostrils. Such projecting parts quickly get broken if they are brittle. Cartilage and bone are the same in kind and differ only by " the more and less "[b] ; so neither of them continues to grow when it has been cut out of the living organism.

The cartilages of land-animals contain no marrow—that is, no marrow existing as a separate thing. What in ordinary bones is separable is here mixed in with the body of the cartilage and gives it its pliable and glutinous character. In the Selachia, however, although the backbone is cartilaginous it contains marrow, because it stands to these creatures in place of a bone.

The following substances or " parts " resemble bones very closely as regards their feel : the various sorts of nail ; hoof and talon ; horn, and beak. All these substances are present for the sake of self-defence. This is shown by the fact that the complete structures which are made out of them and bear the same names—e.g. the complete hoof, or horn—have been contrived in each case by Nature for the creature's self-preservation. We must reckon the teeth in this

169

655 b

ἡ τῶν ὀδόντων ἐστὶ φύσις, τοῖς μὲν ὑπάρχουσα
10 πρὸς ἓν ἔργον τὴν τῆς τροφῆς ἐργασίαν, τοῖς δὲ
πρός τε τοῦτο καὶ πρὸς ἀλκήν, οἷον τοῖς καρχαρ-
όδουσι καὶ χαυλιόδουσι πᾶσιν. ἐξ ἀνάγκης δὲ
πάντα ταῦτα γεώδη καὶ στερεὰν ἔχει τὴν φύσιν·
ὅπλου γὰρ αὕτη δύναμις. διὸ καὶ πάντα τὰ τοιαῦτα
μᾶλλον ἐν τοῖς τετράποσιν ὑπάρχει ζῴοις τῶν
15 ζῳοτόκων, διὰ τὸ γεωδεστέραν ἔχειν πάντα τὴν
σύστασιν ἢ τὸ τῶν ἀνθρώπων γένος. ἀλλὰ καὶ
περὶ τούτων καὶ τῶν ἐχομένων, οἷον δέρματος καὶ
κύστεως[1] καὶ ὑμένος καὶ τριχῶν καὶ πτερῶν καὶ
τῶν ἀνάλογον τούτοις καὶ εἴ τι τοιοῦτόν ἐστι μέρος,
ὕστερον ἅμα τοῖς ἀνομοιομερέσι θεωρητέον τὴν
20 αἰτίαν αὐτῶν, καὶ τίνος ἕνεκεν ὑπάρχει τοῖς ζῴοις
ἕκαστον· ἐκ τῶν ἔργων γὰρ γνωρίζειν, ὥσπερ
κἀκεῖνα, καὶ ταῦτα ἀναγκαῖον ἂν εἴη. ἀλλ' ὅτι
συνώνυμα τοῖς ὅλοις τὰ μέρη, τὴν τάξιν ἀπέλαβεν
ἐν τοῖς ὁμοιομερέσι νῦν. εἰσὶ δ' ἀρχαὶ πάντων
τούτων τό τε ὀστοῦν καὶ ἡ σάρξ. ἔτι δὲ περὶ
γονῆς καὶ γάλακτος ἀπελίπομεν ἐν τῇ περὶ τῶν
25 ὑγρῶν καὶ ὁμοιομερῶν θεωρίᾳ· τοῖς γὰρ περὶ
γενέσεως λόγοις ἁρμόττουσαν ἔχει τὴν σκέψιν· τὸ
μὲν γὰρ αὐτῶν ἀρχὴ τὸ δὲ τροφὴ τῶν γινομένων
ἐστίν.

X. Νῦν δὲ λέγωμεν οἷον ἀπ' ἀρχῆς πάλιν, ἀρξάμενοι πρῶτον ἀπὸ τῶν πρώτων. πᾶσι γὰρ τοῖς

---

[1] σκύτεος Buss. (σκύτεως EY).

class too. In some creatures teeth are present to discharge one function only—viz. mastication; in others they are a means of force as well (*e.g.* sawlike teeth and tusks). All these parts are of necessity earthy and solid in character; that is the proper sort of substance for a weapon. So there is a tendency for all parts of this sort to appear in the four-footed Vivipara more extensively than in man, because the former all have more earthy matter in their constitution. We shall, however, consider these substances, and the other kindred ones such as skin, bladder, membrane, hair, feather, and the counterparts of them, and all such parts, when we come to deal with the non-uniform parts. Then also we shall consider the Causes of them and for what purpose each of them is present in animal bodies; since it is true to say, of both sets of things, that our knowledge of them must be derived from a study of the functions which they discharge. The reason why we have just been taking them with the uniform substances and out of their proper order is that in them the name of the complete structure is the same as that of a portion of it, and also because the sources and principles of them all are bone and flesh. We also left out all mention of semen and milk when we were considering the fluid uniform substances. As semen is the source of the things that are generated and milk is the food that feeds them, the proper place to discuss these is in the treatise dealing with *Generation*.

X. We may now make what is practically a fresh beginning. We will begin first of all with the things that come first in importance. *The non-uniform parts.*

30 ζῴοις τοῖς τελείοις[1] δύο τὰ ἀναγκαιότατα μόριά
ἐστιν, ᾗ τε δέχονται τὴν τροφὴν καὶ ᾗ τὸ περίττωμα
ἀφιᾶσιν[2]· οὔτε γὰρ εἶναι οὔτε αὐξάνεσθαι ἐνδέχεται
ἄνευ τροφῆς. (τὰ μὲν οὖν φυτά—καὶ γὰρ ταῦτα ζῆν
φαμεν—τοῦ μὲν ἀχρήστου περιττώματος οὐκ ἔχει
35 τόπον· ἐκ τῆς γῆς γὰρ λαμβάνει πεπεμμένην τὴν
τροφήν, ἀντὶ δὲ τούτου προΐεται τὰ σπέρματα καὶ
τοὺς καρπούς.) τρίτον δὲ μέρος ἐν πᾶσίν ἐστι τὸ
τούτων μέσον, ἐν ᾧ ἡ ἀρχή ἐστιν ἡ τῆς ζωῆς. ἡ
656 a μὲν οὖν τῶν φυτῶν φύσις οὖσα μόνιμος οὐ πολυ-
ειδής ἐστι τῶν ἀνομοιομερῶν· πρὸς γὰρ ὀλίγας
πράξεις ὀλίγων ὀργάνων ἡ χρῆσις· διὸ θεωρητέον
καθ᾽ αὑτὰ περὶ τῆς ἰδέας αὐτῶν. τὰ δὲ πρὸς τῷ
5 ζῆν αἴσθησιν ἔχοντα πολυμορφοτέραν ἔχει τὴν
ἰδέαν, καὶ τούτων ἕτερα πρὸ ἑτέρων μᾶλλον, καὶ
πολυχουστέραν ὅσων μὴ μόνον τοῦ ζῆν ἀλλὰ καὶ
τοῦ εὖ ζῆν ἡ φύσις μετείληφεν. τοιοῦτο δ᾽ ἐστὶ τὸ
τῶν ἀνθρώπων γένος· ἢ γὰρ μόνον μετέχει τοῦ
θείου τῶν ἡμῖν γνωρίμων ζῴων, ἢ μάλιστα πάντων.
10 ὥστε διά τε τοῦτο, καὶ διὰ τὸ γνώριμον εἶναι
μάλιστ᾽ αὐτοῦ τὴν τῶν ἔξωθεν μορίων μορφήν,
περὶ τούτου λεκτέον πρῶτον. εὐθὺς γὰρ καὶ τὰ
φύσει μόρια κατὰ φύσιν ἔχει τούτῳ μόνῳ, καὶ τὸ

---

[1] τοῖς τελείοις Peck : τοῖς γε τ. Ogle : καὶ τελειουμένοις καὶ
τελείοις Platt : καὶ τελείοις vulg.
[2] ἀφιᾶσιν SUY : ἀφήσουσιν alii.

---

[a] These three parts of the "perfect" animals are again
referred to at *De juv. et sen.* 468 a 13 ff. At *De gen. an.*

An animal can neither exist nor grow without food. Therefore in all living creatures of perfect formation *a* there are two parts most necessary above all : one by which food is taken in and the other by which residues are eliminated. (Plants—which also we include under the head of living things—have, it is true, no place for the useless residue, but this is because their food, which they get out of the earth, is already concocted before it enters them, and instead of this residue they yield their fruit and seeds.) And in all creatures there is a third part intermediate between these indispensable two, and this is the seat of the source and principle of life. Plants, again, are so made as to remain in one place, and thus they do not exhibit a great variety of non-uniform substances; they have few actions to perform, and therefore but few organs are needed to perform them. For this reason we must consider plants and their formations separately. But with creatures that not only live but also have the power of sensation, the formations are more varied, and there is more diversity in some than in others, the greatest variety being found in those creatures which in addition to living have the capability of living the good life, as man has. Man is the only one of the animals known to us who has something of the divine in him, or if there are others, he has most. This is one reason why we ought to speak about man first, and another is that the shape of his external parts is better known than that of other animals. Another and obvious reason is that in man and in man alone do the natural parts appear in their natural situation : the

733 b 1 and 737 b 16, 26, the " perfect " animals are the viviparous ones. For the " most highly finished " animals see 666 a 28.

656 a

τούτου ἄνω πρὸς τὸ τοῦ ὅλου ἔχει ἄνω· μόνον γὰρ
ὀρθόν ἐστι τῶν ζῴων ἄνθρωπος.

Τὸ μὲν οὖν ἔχειν τὴν κεφαλὴν ἄσαρκον ἐκ τῶν
15 περὶ τὸν ἐγκέφαλον εἰρημένων ἀναγκαῖον συμβέβηκεν. οὐ γὰρ ὥσπερ τινές λέγουσιν, ὅτι εἰ
σαρκώδης ἦν, μακροβιώτερον ἂν ἦν τὸ γένος,
ἀλλ᾿ εὐαισθησίας ἕνεκεν ἄσαρκον εἶναί φασιν·
αἰσθάνεσθαι μὲν γὰρ τῷ ἐγκεφάλῳ, τὴν δ᾿ αἴσθησιν
οὐ προσίεσθαι τὰ μόρια τὰ σαρκώδη λίαν. τούτων
20 δ᾿ οὐδέτερόν ἐστιν ἀληθές, ἀλλὰ πολύσαρκος μὲν
ὁ τόπος ὢν ὁ περὶ τὸν ἐγκέφαλον τοὐναντίον ἂν
ἀπειργάζετο οὗ ἕνεκα ὑπάρχει τοῖς ζῴοις ὁ ἐγκέφαλος (οὐ γὰρ ἂν ἐδύνατο καταψύχειν ἀλεαίνων
αὐτὸς λίαν), τῶν τ᾿ αἰσθήσεων οὐκ αἴτιος οὐδεμιᾶς,
ὅς γε ἀναίσθητος καὶ αὐτός ἐστιν ὥσπερ ὁτιοῦν
25 τῶν περιττωμάτων. ἀλλ᾿ οὐχ εὑρίσκοντες διὰ
τίνα αἰτίαν ἔνιαι τῶν αἰσθήσεων ἐν τῇ κεφαλῇ
τοῖς ζῴοις εἰσί, τοῦτο δ᾿ ὁρῶντες ἰδιαίτερον ὂν
τῶν ἄλλων μορίων, ἐκ συλλογισμοῦ πρὸς ἄλληλα
συνδυάζουσιν. ὅτι μὲν οὖν ἀρχὴ τῶν αἰσθήσεών
ἐστιν ὁ περὶ τὴν καρδίαν τόπος, διώρισται πρότερον ἐν τοῖς περὶ αἰσθήσεως, καὶ διότι αἱ μὲν δύο
30 φανερῶς ἠρτημέναι πρὸς τὴν καρδίαν εἰσίν, ἥ τε
τῶν ἁπτῶν καὶ ἡ τῶν χυμῶν, τῶν δὲ τριῶν ἡ μὲν
τῆς ὀσφρήσεως μέση, ἀκοὴ δὲ καὶ ὄψις μάλιστ᾿ ἐν
τῇ κεφαλῇ διὰ τὴν τῶν αἰσθητηρίων φύσιν εἰσί, καὶ

ᵃ See the identical phrase in *De resp.* 477 a 22.
ᵇ Cf. Plato, *Timaeus* 75 A–C.

upper part of man is placed towards the upper part
of the universe.[a] In other words, man is the only
animal that stands upright.

In man, the head is lacking in flesh, and this follows
of necessity from what we have said about the brain.
Some [b] say (erroneously) that if the head abounded
with flesh mankind's lifespan would be longer than
it is, and they explain the absence of flesh as on purpose to facilitate sensation, their view being that the
brain is the organ of sensation, and that sensation
cannot penetrate parts that are too fleshy. Neither
of these assertions is true. The truth is that if
the part surrounding the brain were fleshy, the
effect of the brain would be the very reverse
of that for which it is intended : it would be
unable to cool the rest of the body because it would
be too hot itself. And, of course, the brain is not
responsible for any of the sensations at all ; it has no
more power of sensation than any of the residues.
People adopt these erroneous views because they are
unable to discover the reason why some of the senses
are placed in the head ; but they see that the head
is a somewhat unusual part, compared with the rest,
so they put two and two together and argue that
the brain is the seat of sensation. The correct view,
that the seat and source of sensation is the region
of the heart, has already been set forth in the treatise
Of Sensation,[c] where also I show why it is that two of
the senses, touch and taste, are evidently connected
to the heart ; of the remaining three, smell is placed
between the other two, hearing and sight, and these
are practically always located in the head : this is
owing to the nature of the organs through which

[a] *De sensu*, 438 b 25 ff.

656 a

τούτων ἡ ὄψις πᾶσιν· ἐπεὶ ἥ γ' ἀκοὴ καὶ ἡ ὄσφρησις
85 ἐπὶ τῶν ἰχθύων καὶ τῶν τοιούτων ποιεῖ τὸ λεγόμενον φανερόν· ἀκούουσι μὲν γὰρ καὶ ὀσφραίνονται,
αἰσθητήριον δ' οὐδὲν ἔχουσι φανερὸν ἐν τῇ κεφαλῇ
τούτων τῶν αἰσθητῶν.¹ ἡ δ' ὄψις πᾶσι τοῖς ἔχουσιν
656 b εὐλόγως ἐστὶ περὶ τὸν ἐγκέφαλον· ὁ μὲν γὰρ ὑγρὸς
καὶ ψυχρός, ἡ δ' ὕδωρ τὴν φύσιν ἐστίν· τοῦτο γὰρ
τῶν διαφανῶν εὐφυλακτότατόν ἐστιν. ἔτι δὲ τὰς
ἀκριβεστέρας τῶν αἰσθήσεων διὰ τῶν καθαρωτέρων
ἐχόντων τὸ αἷμα μορίων ἀναγκαῖον ἀκριβεστέρας
5 γίνεσθαι· ἐκκόπτει γὰρ ἡ τῆς ἐν τῷ αἵματι θερμότητος κίνησις τὴν αἰσθητικὴν ἐνέργειαν· διὰ
ταύτας τὰς αἰτίας ἐν τῇ κεφαλῇ τούτων τὰ αἰσθη-
τήριά ἐστιν.

Οὐ μόνον δ' ἐστὶ τὸ ἔμπροσθεν ἄσαρκον, ἀλλὰ τὸ
ὄπισθεν τῆς κεφαλῆς, διὰ τὸ πᾶσι τοῖς ἔχουσιν
αὐτὴν ὀρθότατον δεῖν εἶναι τοῦτο τὸ μόριον· οὐδὲν
10 γὰρ ὀρθοῦσθαι δύναται φορτίον ἔχον, ἣν δ' ἂν
τοιοῦτον, εἰ σεσαρκωμένην εἶχε τὴν κεφαλήν. ᾗ καὶ
δῆλον ὅτι οὐ τῆς τοῦ ἐγκεφάλου αἰσθήσεως χάριν
ἄσαρκος ἡ κεφαλή ἐστιν· τὸ γὰρ ὄπισθεν οὐκ ἔχει
ἐγκέφαλον, ἄσαρκον δ' ὁμοίως.

Ἔχει δὲ καὶ τὴν ἀκοὴν εὐλόγως ἔνια τῶν ζῴων
15 ἐν τῷ τόπῳ τῷ περὶ τὴν κεφαλήν· τὸ γὰρ κενὸν
καλούμενον ἀέρος πλῆρές ἐστι, τὸ δὲ τῆς ἀκοῆς
αἰσθητήριον ἀέρος εἶναί φαμεν.

---

¹ (ἐπεὶ . . . αἰσθητῶν) Cook Wilson, qui et ⟨οὐ⟩ post
λεγόμενον, l. 35.

they operate. Sight is always located there. The case of hearing and smell in fishes and the like shows that the opinion I maintain is patently correct. These creatures hear and smell, although they have no obvious and visible organs for these senses in the head. As for sight, it is reasonable enough that when present it should always be located near the brain, for the brain is fluid and cold, and the sense-organ of sight is identical in its nature with water, which of all transparent substances is the easiest to keep confined. Again, those senses which are intended for more precise work than the others must necessarily receive greater precision by being situated in parts where the blood is specially pure, since the movement of the heat in the blood ousts the activity appropriate to sensation. These are the reasons why the organs of these senses are placed in the head.

Now the back of the head is free from fleshiness as well as the front. This is because the head is the part which all animals that possess one have to hold as upright as possible. Nothing that carries a burden can raise itself upright, and the head would be burdened if it were well covered with flesh. And this is another reason to show that the lack of flesh on the head is not for the purpose of enabling the brain to function in sensation. There is no brain in the back of the head, although the back has no more flesh on it than the front.

Some animals have their organ of hearing as well as of sight located in the region of the head. This is well explained on our view, which is that the organ of hearing is of air. The space in the head called the *vacuum* is full of air.

656 b

’Εκ μὲν οὖν τῶν ὀφθαλμῶν οἱ πόροι φέρουσιν εἰς
τὰς περὶ τὸν ἐγκέφαλον φλέβας· πάλιν δ’ ἐκ τῶν
ὤτων ὡσαύτως πόρος εἰς τοὔπισθεν συνάπτει.

20 [’Εστι δ’ οὔτ’ ἄναιμον οὐδὲν αἰσθητικὸν οὔτε τὸ
αἷμα, ἀλλὰ τῶν ἐκ τούτου τι. διόπερ οὐδὲν ἐν
τοῖς ἐναίμοις ἄναιμον αἰσθητικόν, οὐδ’ αὐτὸ τὸ
αἷμα[1]· οὐδὲν γὰρ τῶν ζῴων μόριον.][2]

’Εχει δ’ ἐν τῷ ἔμπροσθεν τὸν ἐγκέφαλον πάντα
τὰ ἔχοντα τοῦτο τὸ μόριον, διὰ τὸ ἔμπροσθεν
εἶναι ἐφ’ ὃ αἰσθάνεται, τὴν δ’ αἴσθησιν ἀπὸ τῆς
25 καρδίας, ταύτην δ’ εἶναι ἐν τοῖς ἔμπροσθεν, καὶ
τὸ αἰσθάνεσθαι διὰ τῶν ἐναίμων γίνεσθαι μορίων,
φλεβῶν δ’ εἶναι κενὸν τὸ ὄπισθεν κύτος. τέτακται
δὲ τὸν τρόπον τοῦτον τὰ αἰσθητήρια τῇ φύσει
καλῶς, τὰ μὲν τῆς ἀκοῆς ἐπὶ μέσης τῆς περιφερείας
(ἀκούει γὰρ οὐ μόνον κατ’ εὐθυωρίαν ἀλλὰ πάν-
80 τοθεν), ἡ δ’ ὄψις εἰς τὸ ἔμπροσθεν (ὁρᾷ γὰρ κατ’
εὐθυωρίαν, ἡ δὲ κίνησις εἰς τὸ ἔμπροσθεν, προορᾶν
δὲ δεῖ ἐφ’ ὃ ἡ κίνησις). ἡ δὲ τῆς ὀσφρήσεως
μεταξὺ τῶν ὀμμάτων εὐλόγως. διπλοῦν μὲν γάρ
ἐστιν ἕκαστον τῶν αἰσθητηρίων διὰ τὸ διπλοῦν
εἶναι τὸ σῶμα, τὸ μὲν δεξιὸν τὸ δ’ ἀριστερόν. ἐπὶ
85 μὲν οὖν τῆς ἀφῆς τοῦτ’ ἄδηλον· τούτου δ’ αἴτιον
ὅτι οὐκ ἔστι τὸ πρῶτον αἰσθητήριον ἡ σὰρξ καὶ τὸ
τοιοῦτον μόριον, ἀλλ’ ἐντός. ἐπὶ δὲ τῆς γλώττης
ἧττον μέν, μᾶλλον δ’ ἢ ἐπὶ τῆς ἀφῆς· ἔστι γὰρ οἷον

---

[1] οὐδ’ αὐτὸ τὸ αἷμα om. E.
[2] ll. 19-22 seclusi (20-22 Ogle): partim ex 666 a 16 trans-
lata.

---

[a] This passage seems to be a note on a remark which comes
a few lines below, and should probably be omitted from the
text. Part of it is taken from 666 a 16.

Passages (or channels) run from the eyes to the blood-vessels that are round the brain. And, again, a passage runs from the ears and connects to the back of the brain.

[No bloodless part is capable of sensation, nor indeed is the blood itself. It is the parts which are made out of blood that have this faculty. Hence, in the blooded animals, no bloodless part is capable of sensation, nor indeed is the blood itself, for it is no part of animals.] [a]

The brain, whenever there is one, is in the forepart of the head. This is (a) because all acts of sensation take place in a forward direction; (b) because the heart, from which sensation has its origin, is in the forepart of the body; and (c) because the process of sensation depends upon parts that have blood in them, whereas the sac at the back of the head contains no blood-vessels at all. In fact, Nature has located the sense-organs in a very satisfactory manner. The ears are half-way round the circumference of the head, because they are to hear sounds from all directions alike and not only from straight before them. The eyes face front : this is because sight is along one straight line, and we must be able to see along the line in which we are moving, which is directly forward. The nostrils are between the eyes, and this is quite reasonable. Each of the sense-organs is double, because the body itself is double : it has a right side and a left side. It must be admitted that this duality is not at all clear in the case of touch : this is because the primary sense-organ of touch is not the flesh or a corresponding part, but something internal. With the tongue the duality is not very clear, but more so than with touch.

657 a ἁφή τις καὶ αὕτη ἡ αἴσθησις. ὅμως δὲ δῆλον καὶ
ἐπὶ ταύτης· φαίνεται γὰρ ἐσχισμένη. ἐπὶ δὲ τῶν
ἄλλων αἰσθητηρίων φανερωτέρως ἐστὶν ἡ αἴσθησις
διμερής· ὦτά τε γὰρ δύο καὶ ὄμματα καὶ ἡ τῶν
μυκτήρων δύναμις διφυής ἐστιν. ἄλλον οὖν ἂν
5 τρόπον κειμένη καὶ διεσπασμένη, καθάπερ ἡ τῆς
ἀκοῆς, οὐκ ἂν ἐποίει τὸ αὑτῆς ἔργον, οὐδὲ τὸ
μόριον ἐν ᾧ ἐστίν· διὰ γὰρ τῆς ἀναπνοῆς ἡ αἴσθησις
τοῖς ἔχουσι μυκτῆρας, τοῦτο δὲ τὸ μόριον κατὰ
μέσον καὶ ἐν τοῖς ἔμπροσθέν ἐστιν. διόπερ εἰς
μέσον τῶν τριῶν αἰσθητηρίων συνήγαγεν ἡ φύσις
10 τοὺς μυκτῆρας, οἷον ἐπὶ στάθμην θεῖσα μίαν ἐπὶ τὴν
τῆς ἀναπνοῆς κίνησιν.

Καλῶς δὲ καὶ τοῖς ἄλλοις ἔχει ταῦτα τὰ αἰσθητήρια ζῴοις πρὸς τὴν ἰδίαν φύσιν ἑκάστῳ. XI.
τὰ μὲν γὰρ τετράποδα ἀπηρτημένα ἔχει τὰ ὦτα καὶ
ἄνωθεν τῶν ὀμμάτων, ὡς δόξειεν ἄν, οὐκ ἔχει δέ,
15 ἀλλὰ φαίνεται διὰ τὸ μὴ ὀρθὰ εἶναι τὰ ζῷα ἀλλὰ
κύπτειν. οὕτω δὲ τὸ πλεῖστον κινουμένων χρήσιμα
μετεωρότερά τ᾽ ὄντα καὶ κινούμενα· δέχεται γὰρ
στρεφόμενα πάντοθεν τοὺς ψόφους μᾶλλον.

XII. Οἱ δ᾽ ὄρνιθες τοὺς πόρους μόνον ἔχουσι διὰ
τὴν τοῦ δέρματος σκληρότητα καὶ τὸ ἔχειν μὴ
20 τρίχας ἀλλὰ πτερωτὰ εἶναι· οὐκ οὖν ἔχει τοιαύτην
ὕλην ἐξ ἧς ἂν ἔπλασε τὰ ὦτα. ὁμοίως δὲ καὶ τῶν

---

[a] Aristotle seems to refer here to the forked tongues of
certain animals. See 660 b 7 ff.

(Taste, in fact, is itself, as it were, a sort of touch.)
The duality is plain, however, even with this sense,
for it is seen to be divided.ᵃ With the other senses,
the organ is more evidently parted into two : there
are two ears and two eyes, and two passages for the
nostrils in the nose. The sense of smell, if it had
been otherwise placed—separated into two, that is,
like the sense of hearing—would not have been
able to perform its proper function ; nor would
that part of the body in which it is situated, since
in animals which have nostrils, the sensation of
smell is effected by means of inspiration, and this
part is at the front and in the middle. This is
why Nature has brought the nostrils together in a
straight line and made them the central of the three
sense-organs in the head, located where the motion
of in-breathing takes place.

In the other animals as well as in man these sense-
organs are very satisfactorily arranged as required
by the peculiar nature of each animal. XI. For
instance, the quadrupeds have ears that stand out free
from the head, and they are higher than the eyes—
or appear to be, although this is not really so : it is
an illusion due to the fact that these animals are not
upright but stand on all fours. And as they are
usually in this posture when in motion, it is useful for
them to have their ears well up in the air, and also
movable : this enables them to be turned round and
pick up sounds better from all directions.

XII. Birds have the auditory passages only,
owing to the hardness of their skin, and because
they have feathers instead of hair, which means that
they have not got the right material for forming
ears. The same argument applies to those oviparous

181

657 a

τετραπόδων τὰ ᾠοτόκα καὶ φολιδωτά· ὁ γὰρ αὐτὸς
ἁρμόσει καὶ ἐπ᾽ ἐκείνων λόγος. ἔχει δὲ καὶ ἡ
φώκη τῶν ζῳοτόκων οὐκ ὦτα ἀλλὰ πόρους ἀκοῆς,
διὰ τὸ πεπηρωμένον εἶναι τετράπουν.

25 XIII. Καὶ οἱ μὲν ἄνθρωποι καὶ οἱ ὄρνιθες καὶ τὰ
ζῳοτόκα καὶ τὰ ᾠοτόκα τῶν τετραπόδων φυλακὴν
ἔχουσι τῆς ὄψεως, τὰ μὲν ζῳοτόκα βλέφαρα δύο,
οἷς καὶ σκαρδαμύττουσι, τῶν δ᾽ ὀρνίθων ἄλλοι τε
καὶ οἱ βαρεῖς καὶ τὰ ᾠοτόκα τῶν τετραπόδων τῇ
30 κάτω βλεφαρίδι μύουσιν· σκαρδαμύττουσι δ᾽ οἱ
ὄρνιθες ἐκ τῶν κανθῶν ὑμένι. τοῦ μὲν οὖν φυλακὴν
ἔχειν αἴτιον τὸ ὑγρὰ τὰ ὄμματα εἶναι ἵνα ὀξὺ
βλέπωσι [τοῦτον τὸν τρόπον ὑπὸ τῆς φύσεως][1]·
σκληρόδερμα γὰρ ὄντα ἀβλαβέστερα μὲν ἂν ἦν
ὑπὸ τῶν ἔξωθεν προσπιπτόντων, οὐκ ὀξυωπὰ δέ.
τούτου μὲν οὖν[2] ἕνεκα λεπτὸν τὸ δέρμα τὸ περὶ
35 τὴν κόρην ἐστί, τῆς δὲ σωτηρίας χάριν τὰ βλέφαρα·
καὶ διὰ τοῦτο σκαρδαμύττει τε πάντα καὶ μάλιστ᾽
ἄνθρωπος, πάντα μὲν ὅπως τὰ προσπίπτοντα τοῖς
657 b βλεφάροις κωλύωσι (καὶ τοῦτο οὐκ ἐκ προαιρέσεως,
ἀλλ᾽ ἡ φύσις ἐποίησε), πλειστάκις δ᾽ ὁ ἄνθρωπος
διὰ τὸ λεπτοδερμότατος εἶναι.

Ἡ δὲ βλεφαρίς ἐστι δέρματι περιειλημμένη· διὸ
καὶ οὐ συμφύεται οὔτε βλεφαρὶς οὔτ᾽ ἀκροποσθία,
ὅτι ἄνευ σαρκὸς δέρματά ἐστιν.

5 Τῶν δ᾽ ὀρνίθων ὅσοι τῇ κάτω βλεφαρίδι μύουσι,
καὶ τὰ ᾠοτόκα τῶν τετραπόδων, διὰ τὴν σκληρό-

---

[1] om. Z[1].   [2] τούτου μὲν οὖν] τοῦ μὲν οὖν εὖ EPZ.

---

[a] Or, " imperfectly developed." *Cf.* Bk. III. ch. viii.

182

quadrupeds which have horny scales. One viviparous animal, the Seal, has no ears but only auditory
passages ; but this is because, though a quadruped,
it is deformed.[a]

XIII. Man, the Birds, and the Quadrupeds (both Eyes.
viviparous and oviparous) have a protective covering
for their eyes. The viviparous quadrupeds have
two eyelids to each eye (which also enable them
to blink) ; some of the birds, especially the heavily
built ones, and the oviparous quadrupeds, when
they close their eyes, do so with the lower eyelid ;
birds, however, can blink, with the aid of a membrane that comes out of the corner of the eye. The
reason for the existence of these protective coverings is that the eye is fluid in order to ensure
keenness of vision. If the eye had been constructed with a hard skin it would of course have
been less liable to injury by impact from without,
but its vision would have been duller. For this
cause the skin round the pupil is left thin and fine,
and the safety of the eye is ensured by the addition
of the eyelids. The movement of the eyelids known
as blinking is a natural and instinctive one, not
dependent on the will, and its object is to prevent
things from getting into the eyes. All animals
that have eyelids do it, but human beings blink
most of all, because they have the thinnest and
finest skin.

Now the eyelid is encased with skin ; and that is
why, like the tip of the foreskin, it will not unite
again once it has been cut, because both of them
are skin and contain no flesh.

We said just now that some birds and the oviparous quadrupeds close the eye with the lower

τητα τοῦ δέρματος τοῦ περὶ τὴν κεφαλὴν οὕτω
μύουσιν. οἱ μὲν γὰρ βαρεῖς τῶν πτερωτῶν διὰ τὸ
μὴ πτητικοὶ εἶναι τὴν τῶν πτερῶν αὔξησιν εἰς τὴν
τοῦ δέρματος παχύτητα τετραμμένην ἔχουσιν. διὸ
10 καὶ οὗτοι μὲν τῷ κάτω βλεφάρῳ μύουσι, περι-
στεραὶ δὲ καὶ τὰ τοιαῦτα ἀμφοῖν. τὰ δὲ τετράποδα
τῶν ᾠοτόκων φολιδωτά ἐστιν· ταῦτα δὲ σκληρότερα πάντα τριχός, ὥστε καὶ τὰ δέρματα τοῦ
δέρματος. τὸ μὲν οὖν περὶ τὴν κεφαλὴν σκληρόν
ἐστιν αὐτοῖς, διόπερ οὐκ ἔχει βλέφαρον ἐκεῖθεν,
15 τὸ δὲ κάτωθεν σαρκῶδες, ὥστ' ἔχειν τὸ βλέφαρον
λεπτότητα καὶ τάσιν.

Σκαρδαμύττουσι δ' οἱ βαρεῖς ὄρνιθες τούτῳ μὲν
οὔ, τῷ δ' ὑμένι, διὰ τὸ βραδεῖαν εἶναι τὴν τούτου
κίνησιν, δεῖν δὲ ταχεῖαν γίνεσθαι, ὁ δ' ὑμὴν τοιοῦτον. ἐκ δὲ τοῦ κανθοῦ τοῦ παρὰ τοὺς μυκτῆρας
20 σκαρδαμύττουσιν, ὅτι βέλτιον ἀπ' ἀρχῆς μιᾶς τὴν
φύσιν εἶναι αὐτῶν, οὗτοι δ' ἔχουσιν ἀρχὴν τὴν πρὸς
τὸν μυκτῆρα πρόσφυσιν· καὶ τὸ πρόσθιον ἀρχὴ τοῦ
πλαγίου μᾶλλον.

Τὰ δὲ τετράποδα καὶ ᾠοτόκα οὐ σκαρδαμύττει
ὁμοίως, ὅτι οὐδ' ὑγρὰν αὐτοῖς ἀναγκαῖον ἔχειν καὶ
ἀκριβῆ τὴν ὄψιν ἐπιγείοις οὖσιν· τοῖς δ' ὄρνισιν
25 ἀναγκαῖον, πόρρωθεν γὰρ ἡ χρῆσις τῆς ὄψεως. διὸ
καὶ τὰ γαμψώνυχα μὲν ὀξυωπά (ἄνωθεν γὰρ αὐτοῖς
ἡ θεωρία τῆς τροφῆς, διὸ καὶ ἀναπέτονται ταῦτα
μάλιστα τῶν ὀρνέων εἰς ὕψος), τὰ δ' ἐπίγεια καὶ
μὴ πτητικά, οἷον ἀλεκτρυόνες καὶ τὰ τοιαῦτα,

eyelid only. This is due to the hardness of the skin which surrounds the head. (*a*) The heavily built birds are not great fliers, and so the material which would have supplied growth for the wings has been diverted, resulting in thickness of the skin. These creatures, then, use only the bottom eyelid to cover the eye ; whereas pigeons and such use both eyelids. (*b*) With regard to the oviparous quadrupeds : As the horny scales with which they are covered are in every case harder than hair, so their skin also is harder than ordinary skin. And as the skin on their heads is hard, they can have no upper eyelid ; but lower down the skin has some flesh with it, and so they have a lower eyelid that is thin and extensible.

Now the heavily built birds blink not with this lower eyelid, because its motion is slow, but with the membrane above mentioned, whose motion is swift, as is requisite. This blinking or nictitating begins at the corner of the eye nearest the nostrils, because it is better that the membranes should have one place of origin rather than two, and in these birds this is where the eye and nostril are conjoined ; also, the front is more a place of origin than the side.

The oviparous quadrupeds do not blink in this way, because, unlike birds, which have to use their eyes over great distances, they go upon the ground, and therefore there is no need for them to have fluid eyes or great accuracy of sight. The crook-taloned birds are sharp-sighted, for they view their prey from above, and that also explains why they fly to a greater height than other birds. The birds that remain on the ground, however, and do not fly much (*e.g.* barn-door fowls and the like) are

657 b

οὐκ ὀξυωπά· οὐδὲν γὰρ αὐτὰ κατεπείγει πρὸς
τὸν βίον.

30 Οἱ δ' ἰχθύες καὶ τὰ ἔντομα καὶ τὰ σκληρόδερμα
διαφέροντα μὲν ἔχουσι τὰ ὄμματα, βλέφαρον δ'
οὐδὲν αὐτῶν ἔχει. τὰ μὲν γὰρ σκληρόδερμα ὅλως
οὐκ ἔχει· ἡ δὲ τοῦ βλεφάρου χρῆσις ταχεῖαν τὴν¹
δερματικὴν ἔχει ἐργασίαν· ἀλλ' ἀντὶ ταύτης τῆς
35 φυλακῆς πάντα σκληρόφθαλμά ἐστιν, οἷον βλέποντα
διὰ τοῦ βλεφάρου προσπεφυκότος. ἐπεὶ δ' ἀναγκαῖον διὰ τὴν σκληρότητα ἀμβλύτερον βλέπειν,
κινουμένους ἐποίησεν ἡ φύσις τοὺς ὀφθαλμοὺς τοῖς
658 a ἐντόμοις, καὶ μᾶλλον ἔτι τοῖς σκληροδέρμοις, ὥσπερ
ἔνια τῶν τετραπόδων τὰ ὦτα, ὅπως ὀξύτερον βλέπῃ
στρέφοντα πρὸς τὸ φῶς καὶ δεχόμενα τὴν αὐγήν.
οἱ δ' ἰχθύες ὑγρόφθαλμοι μέν εἰσιν· ἀναγκαία γὰρ
5 τοῖς πολλὴν ποιουμένοις κίνησιν ἡ τῆς ὄψεως ἐκ
πολλοῦ χρῆσις. τοῖς μὲν οὖν πεζοῖς ὁ ἀὴρ εὐ-
δίοπτος· ἐκείνοις δ' ἐπεὶ τὸ ὕδωρ πρὸς μὲν τὸ ὀξὺ
βλέπειν ἐναντίον, οὐκ ἔχει δὲ πολλὰ τὰ προσκρούσ-
ματα πρὸς τὴν ὄψιν ὥσπερ ὁ ἀήρ, διὰ μὲν τοῦτ'
οὐκ ἔχει βλέφαρον (οὐδὲν γὰρ ἡ φύσις ποιεῖ μάτην),
10 πρὸς δὲ τὴν παχύτητα τοῦ ὕδατος ὑγρόφθαλμοί
εἰσιν.

XIV. Βλεφαρίδας δ' ἐπὶ τῶν βλεφάρων ἔχουσιν
ὅσα τρίχας ἔχουσιν, ὄρνιθες δὲ καὶ τῶν φολιδωτῶν
οὐδέν· οὐ γὰρ ἔχουσι τρίχας. περὶ γὰρ τοῦ στρου-
θοῦ τοῦ Λιβυκοῦ τὴν αἰτίαν ὕστερον ἐροῦμεν· τοῦτο

---

¹ τὴν Ogle: καὶ vulg.: τὴν ante ἐργασίαν vulg., om. SU.

not sharp-sighted, since there is no urgent necessity for it in their kind of life.

Many differences in the eye itself are found among the Fishes, the Insects and the hard-skinned Crustacea, though not one of them has eyelids. In the hard-skinned Crustacea there cannot be an eyelid at all, for the action of an eyelid depends upon swift working of the skin. To compensate for the lack of this protection, all these creatures have hard eyes : it is as though the eyelid were all of a piece with the eyeball, and the creature looked through the lid as well. But since the vision is bound to be dimmed by this hardness of the eye, Nature has given the Insects (and even more noticeably the Crustacea) movable eyes, just as she has given some quadrupeds movable ears ; this is to enable them to turn towards the light and catch its rays and so to quicken their vision. Fish have fluid eyes for the following reason. They move about a good deal and have to use their sight over long distances. Now when land-animals do this, they are looking through air, which is highly transparent ; but fish move about in water, which is inimical to sharpness of vision ; so to counteract its opacity their eyes are fluid in composition. At the same time, water contains far fewer objects to strike against the eyes than the air does ; hence fish need no eyelids, and because Nature never makes anything without a purpose, they have none.

XIV. Those animals that have hair on their body have eyelashes on their eyelids : the others (birds and the creatures with horny scales) have none. There is one exception to this rule : the Libyan ostrich, which has eyelashes. The cause of this

658 a

15 γὰρ ἔχει βλεφαρίδας τὸ ζῷον. καὶ τῶν ἐχόντων
τρίχας ἐπ' ἀμφότερα οἱ ἄνθρωποι μόνον ἔχουσιν.
τὰ γὰρ τετράποδα τῶν ζῴων ἐν τοῖς ὑπτίοις οὐκ
ἔχει τρίχας, ἀλλ' ἐν τοῖς πρανέσι μᾶλλον· οἱ δ'
ἄνθρωποι τοὐναντίον ἐν τοῖς ὑπτίοις μᾶλλον ἢ ἐν
τοῖς πρανέσιν. σκέπης γὰρ χάριν αἱ τρίχες ὑπάρχουσι τοῖς ἔχουσιν· τοῖς μὲν οὖν τετράποσι τὰ
20 πρανῆ δεῖται μᾶλλον τῆς σκέπης, τὰ δὲ πρόσθια
τιμιώτερα μέν, ἀλλὰ λεάζει διὰ τὴν κάμψιν· τοῖς
δ' ἀνθρώποις ἐπεὶ ἐξ ἴσου διὰ τὴν ὀρθότητα τὰ
πρόσθια τοῖς ὀπισθίοις, τοῖς τιμιωτέροις ὑπέγραψεν
ἡ φύσις τὴν βοήθειαν· ἀεὶ γὰρ ἐκ τῶν ἐνδεχομένων αἰτία τοῦ βελτίονός ἐστιν. καὶ διὰ τοῦτο
25 τῶν τετραπόδων οὐθὲν οὔτε βλεφαρίδα ἔχει τὴν
κάτωθεν, ἀλλ' ὑπὸ τοῦτο τὸ βλέφαρον ἐνίοις παρα-
φύονται μαναὶ τρίχες, οὔτ' ἐν ταῖς μασχάλαις οὔτ'
ἐπὶ τῆς ἥβης, ὥσπερ τοῖς ἀνθρώποις· ἀλλ' ἀντὶ
τούτων τὰ μὲν καθ' ὅλον τὸ σῶμα πρανές[1] δεδά-
30 συνται ταῖς θριξίν, οἷον τὸ τῶν κυνῶν γένος, τὰ δὲ
λοφιὰν ἔχει, καθάπερ ἵπποι καὶ τὰ τοιαῦτα τῶν
ζῴων, τὰ δὲ χαίτην, ὥσπερ ὁ ἄρρην λέων. ἔτι δ'
ὅσα κέρκους ἔχει μῆκος ἐχούσας, καὶ ταύτας ἐπι-
κεκόσμηκεν ἡ φύσις θριξί, τοῖς μὲν μικρὸν ἔχουσι
τὸν στόλον μακραῖς, ὥσπερ τοῖς ἵπποις, τοῖς δὲ
35 μακρὸν βραχείαις, καὶ κατὰ τὴν τοῦ ἄλλου σώματος
φύσιν· πανταχοῦ γὰρ ἀποδίδωσι λαβοῦσα ἑτέρωθεν
πρὸς ἄλλο μόριον. ὅσοις δὲ τὸ σῶμα δασὺ λίαν
658 b πεποίηκε, τούτοις ἐνδεῶς ἔχει τὰ περὶ τὴν κέρκον,
οἷον ἐπὶ τῶν ἄρκτων συμβέβηκεν.

---

[1] πρανὲς delet Platt.

188

will be explained later.[a] Man is the only animal which has eyelashes on both lids. Why is this? The quadrupeds tend to have more hair on their backs than on the underside of the body; but in man the reverse is true. The purpose of hair is to give protection; and as the quadrupeds go on all fours, they need more protection on their backs; so they have no hair on their front, although the front is the nobler of the two sides. Man goes upright, and so there is nothing to choose as regards his need of protection between front and back. Therefore Nature has prescribed the protection for the nobler side, the front—an example of how, out of given conditions, she is always the cause of that which is the better. This, then, is why none of the quadrupeds has lower eyelashes (though some have a few scattered hairs growing on the lower eyelid), or hair in the axillae or on the pubes, as man has. Instead of this, some of them have thick hair all over the back part of[b] their body (*e.g.* dogs), some of them have a mane (*e.g.* horses and such), others a flowing mane, like the male lion. Again, if an animal has a tail of any length, Nature decks that with hair too; long hair for tails with a short stem (*e.g.* horses), short hair for tails with a long stem. This, however, is not independent of the general condition of the whole animal, for Nature gives something to one part of the body only after she has taken it from another part. So when she has made an animal's body extremely hairy, we find that there is not much hair about the tail. An example of this is the Bears.

[a] See 697 b 13 ff.
[b] Platt deletes " the back part of."

658 b

Τὴν δὲ κεφαλὴν ἄνθρωπός ἐστι τῶν ζῴων δασύ-
τατον, ἐξ ἀνάγκης μὲν διὰ τὴν ὑγρότητα τοῦ
ἐγκεφάλου καὶ διὰ τὰς ῥαφάς (ὅπου γὰρ ὑγρὸν καὶ
5 θερμὸν πλεῖστον, ἐνταῦθ᾽ ἀναγκαῖον πλείστην εἶναι
τὴν ἔκφυσιν), ἕνεκεν δὲ βοηθείας, ὅπως σκεπάζωσι
φυλάττουσαι τὰς ὑπερβολὰς τοῦ τε ψύχους καὶ τῆς
ἀλέας. πλεῖστος δ᾽ ὢν καὶ ὑγρότατος ὁ τῶν
ἀνθρώπων ἐγκέφαλος πλείστης καὶ τῆς φυλακῆς
δεῖται· τὸ γὰρ ὑγρότατον καὶ ζεῖ καὶ ψύχεται
10 μάλιστα, τὸ δ᾽ ἐναντίως ἔχον ἀπαθέστερόν ἐστιν.

Ἀλλὰ περὶ μὲν τούτων παρεκβῆναι συμβέβηκεν
ἐχομένοις τῆς περὶ τὰς βλεφαρίδας αἰτίας, διὰ τὴν
συγγένειαν αὐτῶν, ὥστε περὶ τῶν λοιπῶν ἐν τοῖς
οἰκείοις καιροῖς ἀποδοτέον τὴν μνείαν.

XV. Αἱ δ᾽ ὀφρύες καὶ αἱ βλεφαρίδες ἀμφότεραι
15 βοηθείας χάριν εἰσίν, αἱ μὲν ὀφρύες τῶν κατα-
βαινόντων ὑγρῶν, ὅπως ἀποστέγωσιν οἷον ἀπογεί-
σωμα τῶν ἀπὸ τῆς κεφαλῆς ὑγρῶν, αἱ δὲ βλεφαρίδες
τῶν πρὸς τὰ ὄμματα προσπιπτόντων ἕνεκεν, οἷον
τὰ χαρακώματα ποιοῦσί τινες πρὸ τῶν ἐργμάτων.[1]
εἰσὶ δ᾽ αἱ μὲν ὀφρύες ἐπὶ συνθέσει ὀστῶν, διὸ καὶ
20 δασύνονται πολλοῖς ἀπογηράσκουσιν οὕτως ὥστε
δεῖσθαι κουρᾶς· αἱ δὲ βλεφαρίδες ἐπὶ πέρατι
φλεβίων, ᾗ γὰρ τὸ δέρμα περαίνει, καὶ τὰ φλέβια

---

[1] ἐργμάτων scripsi : ἐργμάτων Bekker : ἐρυμάτων editores.

---

[a] This is one of the passages fastened upon by Bacon in
his tirade against the importation of final causes into physics,
Adv. of Learning (publ. 1605), ii. pp. 29, 30 : " This I finde
done not onely by *Plato*, who euer ancreth vppon that shoare,
but by *Aristotle*, *Galen*, and others, who do vsually likewise
fall vppon these flatts *of discoursing causes* ; For to say *that
the haires of the Eye-liddes are for a quic-sette and fence about*

Man has the hairiest head of all the animals. This is (*a*) due to *necessity*, because the brain is fluid, and the skull has many sutures ; and a large outgrowth necessarily occurs where there is a large amount of fluid and hot substance. But also (*b*) it is *on purpose* to give protection ; that is, the hair affords shelter both from excessive cold and from excessive heat. The human brain is the biggest and the most fluid of all brains ; therefore it needs the greatest amount of protection. A very fluid thing is very liable both to violent heating and violent cooling, while substances of an opposite nature are less liable to such affections.

This, however, is a digression. We were led into it because the subject was connected with our investigation of the cause of eyelashes. Anything further that there is to be said about it will be said in its proper place.

XV. Both eyebrows and eyelashes exist to afford protection to the eyes : the eyebrows, like the eaves of a house, are to protect the eyes from the fluids that run down from the head ; the eyelashes are like the palisades which are sometimes put up in front of an enclosure ; their purpose is to keep out things that try to get in.[a] However, the eyebrows are placed where two bones join (which is why they often get so thick in old age that they have to be cut); and the eyelashes are placed at the ends of small blood-vessels, which have to stop where the skin itself comes to

the Sight . . . and the like, is well inquired & collected in Metaphisicke, but in Phisicke they are impertinent." But there is no incompatibility, p. 33, " For the cause rendred *that the haires about the Eye-liddes are for the safeguard of the sight,* doth not impugne the cause rendred, *that Pilositie is incident to Orifices of Moisture.*" See also Xen. *Mem.* i. 4. 6.

πέρας ἔχει τοῦ μήκους· ὥστ᾽ ἀναγκαῖον διὰ τὴν
ἀπιοῦσαν ἰκμάδα σωματικὴν οὖσαν, ἂν μή τι τῆς
φύσεως ἔργον ἐμποδίσῃ πρὸς ἄλλην χρῆσιν, καὶ
25 διὰ τὴν τοιαύτην αἰτίαν ἐξ ἀνάγκης ἐν τοῖς τόποις
τούτοις γίνεσθαι τρίχας.

XVI. Τοῖς μὲν οὖν ἄλλοις ζῴοις τοῖς τετράποσι
καὶ ζῳοτόκοις οὐ πόρρω τρόπον τινὰ διέστηκεν
ἀλλήλων τὸ τῆς ὀσφρήσεως αἰσθητήριον, ἀλλ᾽ ὅσα
30 μὲν ἔχει προμήκεις εἰς στενὸν ἀπηγμένας τὰς
σιαγόνας, ἐν τῷ καλουμένῳ ῥύγχει καὶ τὸ τῶν
μυκτήρων ἐνυπάρχει μόριον κατὰ τὸν ἐνδεχόμενον
τρόπον, τοῖς δ᾽ ἄλλοις μᾶλλον διηρθρωμένον ἐστὶ
πρὸς τὰς σιαγόνας. ὁ δ᾽ ἐλέφας ἰδιαίτατον ἔχει
τοῦτο τὸ μόριον τῶν ἄλλων ζῴων· τό τε γὰρ
35 μέγεθος καὶ τὴν δύναμιν[a] ἔχει περιττήν. μυκτὴρ
γάρ ἐστιν ᾧ τὴν τροφὴν προσάγεται, καθάπερ χειρὶ
659 a χρώμενος, πρὸς τὸ στόμα, τήν τε ξηρὰν καὶ τὴν
ὑγράν, καὶ τὰ δένδρα περιελίττων ἀνασπᾷ, καὶ
χρῆται καθάπερ ἂν εἰ χειρί. τὴν γὰρ φύσιν ἑλῶδές
ἅμα τὸ ζῷόν ἐστι καὶ πεζόν, ὥστ᾽ ἐπεὶ τὴν τροφὴν
ἐξ ὑγροῦ συνέβαινεν ἔχειν, ἀναπνεῖν δ᾽ ἀναγκαῖον
5 πεζὸν ὂν καὶ ἔναιμον, καὶ μὴ ταχεῖαν ποιεῖσθαι τὴν
μεταβολὴν ἐκ τοῦ ὑγροῦ πρὸς τὸ ξηρόν, καθάπερ
ἔνια τῶν ζῳοτόκων καὶ ἐναίμων καὶ ἀναπνεόντων,
τὸ γὰρ μέγεθος ὂν ὑπερβάλλον, ἀναγκαῖον ὁμοίως
ἦν χρῆσθαι τῷ ὑγρῷ ὥσπερ καὶ τῇ γῇ. οἷον οὖν
τοῖς κολυμβηταῖς ἔνιοι πρὸς τὴν ἀναπνοὴν ὄργανα
10 πορίζονται, ἵνα πολὺν χρόνον ἐν τῇ θαλάττῃ μέ-
νοντες ἕλκωσιν ἔξωθεν τοῦ ὑγροῦ διὰ τοῦ ὀργάνου
τὸν ἀέρα, τοιοῦτον ἡ φύσις τὸ τοῦ μυκτῆρος μέγεθος ἐποίησε τοῖς ἐλέφασιν. διόπερ ἀναπνέουσιν

---

[a] Or " strength."

an end. Thus, owing to the fact that the moisture which comes off is corporeal in composition, hair must be formed at these places even on account of a *necessary* cause such as this, unless some function of Nature impedes by diverting the moisture to another use.

XVI. The general run of viviparous quadrupeds differ very little among themselves as regards the organ of smell. The following variations occur, however. Those animals whose jaws project forward and become gradually narrower, forming what is called a snout, have the organ of smell in their snout—this being the only possibility ; in the others, the jaws and nostrils are more definitely separated. The elephant's nose is unique owing to its enormous size and its extraordinary character.[a] By means of his nose, as if it were a hand, the elephant conveys his food, both solid and fluid, to his mouth ; by means of it he tears up trees, by winding it round them. In fact, he uses it for all purposes as if it were a hand. This is because the elephant has a double character : he is a land-animal, but he also lives in swamps. He has to get his food from the water ; yet he has to breathe, because he is a land-animal and has blood ; owing to his enormous size, however, he cannot transfer himself quickly from the water on to the land, as do quite a number of blooded viviparous animals that breathe ; hence he has to be equally at home on land and in the water. Some divers, when they go down into the sea, provide themselves with a breathing-machine, by means of which they can inhale the air from above the surface while they remain for a long time in the water. Nature has provided the elephant with something of this sort by giving him a long nose. If ever the

ἄραντες ἄνω διὰ τοῦ ὕδατος τὸν μυκτῆρα, ἄν ποτε
ποιῶνται δι' ὑγροῦ τὴν πορείαν· καθάπερ γὰρ
15 εἴπομεν, μυκτήρ ἐστιν ἡ προβοσκὶς τοῖς ἐλέφασιν.
ἐπεὶ δ' ἀδύνατον ἦν εἶναι τὸν μυκτῆρα τοιοῦτον μὴ
μαλακὸν ὄντα μηδὲ κάμπτεσθαι δυνάμενον (ἐνεπό-
διζε γὰρ ἂν τῷ μήκει πρὸς τὸ λαβεῖν τὴν θύραθεν
τροφήν, καθάπερ φασὶ τὰ κέρατα τοῖς ὀπισθονόμοις
20 βουσίν· καὶ γὰρ ἐκείνους νέμεσθαί φασιν ὑπο-
χωροῦντας παλιμπυγηδόν)—ὑπάρξαντος οὖν τοιούτου τοῦ μυκτῆρος, ἡ φύσις παρακαταχρῆται, καθάπερ εἴωθεν, ἐπὶ πλείονα τοῖς αὐτοῖς μορίοις, ἀντὶ
τῆς τῶν προσθίων ποδῶν χρείας. τούτους γὰρ τὰ
πολυδάκτυλα τῶν τετραπόδων ἀντὶ χειρῶν ἔχουσιν,
25 ἀλλ' οὐ μόνον ἔνεχ' ὑποστάσεως τοῦ βάρους· οἱ δ'
ἐλέφαντες τῶν πολυδακτύλων εἰσί, καὶ οὔτε διχά-
λους ἔχουσιν οὔτε μώνυχας τοὺς πόδας· ἐπεὶ δὲ τὸ
μέγεθος πολὺ καὶ τὸ βάρος τὸ τοῦ σώματος, διὰ
τοῦτο μόνον ἐρείσματός εἰσι χάριν, καὶ διὰ τὴν
βραδυτῆτα καὶ τὴν ἀφυΐαν τῆς κάμψεως οὐ χρήσιμοι[1] πρὸς ἄλλο οὐθέν.

30 Διὰ μὲν οὖν τὴν ἀναπνοὴν ἔχει μυκτῆρα, καθάπερ
καὶ τῶν ἄλλων ἕκαστον τῶν ἐχόντων πλεύμονα
ζῴων, διὰ δὲ τὴν ἐν τῷ ὑγρῷ διατριβὴν καὶ τὴν
βραδυτῆτα τῆς ἐκεῖθεν μεταβολῆς δυνάμενον ἑλίτ-
τεσθαι καὶ μακρόν· ἀφηρημένης δὲ τῆς τῶν ποδῶν
35 χρήσεως, καὶ ἡ φύσις, ὥσπερ εἴπομεν, καταχρῆται
καὶ πρὸς τὴν ἀπὸ τῶν ποδῶν γινομένην ἂν βοήθειαν
τούτῳ τῷ μορίῳ.

659 b   Οἱ δ' ὄρνιθες καὶ οἱ ὄφεις καὶ ὅσα ἄλλ' ἔναιμα

---

[1] χρήσιμοι Rackham : χρήσιμον vulg.

elephant has to make his way through deep water, he will put his trunk up to the surface and breathe through it. This is possible, because, as I have said already, the trunk is really a nostril. Now it would have been impossible for the nostril to be put to all these uses if it had not been soft and able to bend ; for then by its very length it would have prevented the animal from getting its food, just as they say the horns of the " backward-grazing " oxen do, forcing them to walk backwards as they feed.[a] So the trunk is soft and pliable ; and in consequence Nature, as usual, takes advantage of this to make it discharge an extra function beside its original one : it has to serve instead of forefeet. Now in polydactylous quadrupeds the forefeet are there to serve as hands, not merely in order to support the weight of the animal ; but elephants (which must be included under this class of animals, because they have neither a solid hoof nor a cloven one) are so large and so heavy that their forefeet can serve only as supports ; and indeed they are no good for anything else because they move so slowly and are quite unsuited for bending.

So the elephant's nostril is there, in the first place, to enable him to breathe (as in all animals that have a lung) ; and also it is lengthened and able to coil itself round things because the elephant spends much of his time in the water and cannot quickly emerge upon land. And as his forefeet are not available for the normal function, Nature, as we said, presses the trunk into service to supply what should have been forthcoming from the feet.

The Birds and Serpents and the quadrupeds which

---

[a] See above, on 648 a 16. This is from Herodotus, iv. 183.

659 b

καὶ ᾠοτόκα[1] τῶν τετραπόδων, τοὺς μὲν πόρους
ἔχουσι τῶν μυκτήρων πρὸ τοῦ στόματος, ὥστε δ'
εἰπεῖν μυκτῆρας, εἰ μὴ διὰ τὸ ἔργον, οὐκ ἔχουσι
φανερῶς διηρθρωμένους· ἀλλ' ἥ γ' ὄρνις ὥστε
5 μηθὲν ἂν εἰπεῖν ἔχει[2] ῥῖνας. τοῦτο δὲ συμβέβηκεν,
ὅτι ἀντὶ σιαγόνων ἔχει τὸ καλούμενον ῥύγχος. αἰτία
δὲ τούτων ἡ φύσις ἡ τῶν ὀρνίθων συνεστηκυῖα
τοῦτον τὸν τρόπον. δίπουν γάρ ἐστι καὶ πτερυ-
γωτόν, ὥστ' ἀνάγκη μικρὸν τὸ βάρος ἔχειν τὸ τοῦ
αὐχένος καὶ τὸ τῆς κεφαλῆς, ὥσπερ καὶ τὸ στῆθος
10 στενόν· ὅπως μὲν οὖν ᾖ χρήσιμον πρός τε τὴν
ἀλκὴν καὶ διὰ τὴν τροφήν, ὀστῶδες ἔχουσι τὸ
ῥύγχος, στενὸν δὲ διὰ τὴν μικρότητα τῆς κεφαλῆς.
ἐν δὲ τῷ ῥύγχει τοὺς πόρους ἔχουσι τῆς ὀσφρήσεως,
μυκτῆρας δ' ἔχειν ἀδύνατον.

Περὶ δὲ τῶν ἄλλων ζῴων τῶν μὴ ἀναπνεόντων
15 εἴρηται πρότερον δι' ἣν αἰτίαν οὐκ ἔχουσι μυκτῆρας, ἀλλὰ τὰ μὲν διὰ τῶν βραγχίων, τὰ δὲ διὰ
τοῦ αὐλοῦ, τὰ δ' ἔντομα διὰ τοῦ ὑποζώματος
αἰσθάνονται τῶν ὀσμῶν, καὶ πάντα τῷ συμφύτῳ
πνεύματι τοῦ σώματος ᾧπερ[3] ⟨καὶ⟩[4] κινεῖται· τοῦτο
δ' ὑπάρχει φύσει πᾶσι καὶ οὐ θύραθεν ἐπείσακτόν
ἐστιν.

20 Ὑπὸ δὲ τοὺς μυκτῆρας ἡ τῶν χειλῶν ἐστι φύσις
τοῖς ἔχουσι τῶν ἐναίμων ὀδόντας. τοῖς γὰρ ὄρνισι,
καθάπερ εἴπομεν, διὰ τὴν τροφὴν καὶ τὴν ἀλκὴν τὸ
ῥύγχος ὀστῶδές ἐστιν· συνῆκται γὰρ εἰς ἓν ἀντ'
ὀδόντων καὶ χειλῶν, ὥσπερ ἂν εἴ τις ἀφελὼν
25 ἀνθρώπου τὰ χείλη καὶ συμφύσας τοὺς ἄνωθεν

---

[1] ᾠοτόκα Z, vulg. : ζῳοτόκα EPSUY.
[2] ἔχει S : ἔχειν vulg.

like them are blooded and oviparous, have their nostril-passages in front of the mouth : but they have nothing which except for its function can be called nostrils—nothing distinctly articulated. A bird, at any rate, one might say has no nose at all. The reason for this is that its beak really replaces jaws. And this is because of the natural structure of birds. A bird is a winged biped ; hence its head and its neck must be light in weight, and its breast must be narrow ; and it has a beak, which (a) is made out of bony material, so that it will serve as a weapon as well as for the uptake of food, and (b) is narrow, owing to the small size of the head. It has the passages for smell in this beak, but it is impossible for it to have nostrils there.

We have spoken already about the animals that do not breathe, and shown why they have no nostrils : some of them smell by means of the gills, some through a blow-hole ; while the insects smell through the middle part of the body. All of them smell, as all of them move, by means of the connate *pneuma* [a] of their bodies, which is not introduced from without, but is present in all of them by nature.

In all blooded animals that have teeth, the lips have Lips. their place below the nostrils. (As stated already, birds have a bony beak for getting food and for defence ; and this is as it were teeth and lips run into one. The nature of the beak can be illustrated thus. Supposing, in a human being, that the lips were removed, and all the upper teeth were welded to-

[a] *Cf. De somno et vig.* 455 b 34 ff. For a full account of Σύμφυτον Πνεῦμα see *G.A.* (Loeb edn.), pp. 576 ff.

---

[3] ὥπερ SUZ² : ὥσπερ vulg.   [4] ⟨καὶ⟩ Peck.

659 b

ὀδόντας χωρὶς καὶ τοὺς κάτωθεν προαγάγοι μῆκος
ποιήσας ἀμφοτέρωθεν εἰς στενόν· εἴη γὰρ ἂν τοῦτο
ἤδη ῥύγχος ὀρνιθῶδες. τοῖς μὲν οὖν ἄλλοις ζῴοις
πρὸς σωτηρίαν τῶν ὀδόντων ἡ τῶν χειλῶν φύσις
ἐστὶ καὶ πρὸς φυλακήν, διόπερ ὡς ἐκείνων μετ-
30 έχουσι τοῦ ἀκριβῶς καὶ καλῶς ἢ τοὐναντίον, οὕτω
καὶ τοῦ διηρθρῶσθαι τοῦτο τὸ μόριον ἔχουσιν· οἱ δ'
ἄνθρωποι μαλακὰ καὶ σαρκώδη καὶ δυνάμενα χωρί-
ζεσθαι, φυλακῆς θ' ἕνεκα τῶν ὀδόντων ὥσπερ καὶ
τὰ ἄλλα, καὶ μᾶλλον ἔτι διὰ τὸ εὖ· πρὸς γὰρ τὸ
χρῆσθαι τῷ λόγῳ καὶ ταῦτα. ὥσπερ γὰρ τὴν
35 γλῶτταν οὐχ ὁμοίαν τοῖς ἄλλοις ἐποίησεν ἡ φύσις,
πρὸς ἐργασίας δύο καταχρησαμένη, καθάπερ
660 a εἴπομεν ποιεῖν αὐτὴν ἐπὶ πολλῶν, τὴν μὲν γλῶτταν
τῶν τε χυμῶν ἕνεκεν καὶ τοῦ λόγου, τὰ δὲ χείλη
τούτου θ' ἕνεκεν καὶ τῆς τῶν ὀδόντων φυλακῆς. ὁ
μὲν γὰρ λόγος ὁ διὰ τῆς φωνῆς ἐκ τῶν γραμμάτων
5 σύγκειται, τῆς δὲ γλώττης μὴ τοιαύτης οὔσης μηδὲ
τῶν χειλῶν ὑγρῶν οὐκ ἂν ἦν φθέγγεσθαι τὰ πλεῖστα
τῶν γραμμάτων· τὰ μὲν γὰρ τῆς γλώττης εἰσὶ
προσβολαί, τὰ δὲ συμβολαὶ τῶν χειλῶν. ποίας δὲ
ταῦτα καὶ πόσας καὶ τίνας ἔχει διαφοράς, δεῖ
πυνθάνεσθαι παρὰ τῶν μετρικῶν.

Ἀνάγκη δ' ἦν εὐθὺς ἀκολουθῆσαι τούτων τῶν
10 μορίων ἑκάτερον πρὸς τὴν εἰρημένην χρῆσιν εὐεργὰ
καὶ τοιαύτην ἔχοντα τὴν φύσιν· διὸ σάρκινα. μα-
λακωτάτη δ' ἡ σὰρξ ἡ τῶν ἀνθρώπων ὑπῆρχεν.
τοῦτο δὲ διὰ τὸ αἰσθητικώτατον εἶναι τῶν ζῴων
τὴν διὰ τῆς ἁφῆς αἴσθησιν.

198

gether, and similarly all the bottom teeth, and then each set were extended in a forward direction, and made to taper : this would result in a beak such as birds have.) In all animals except man the lips are intended to preserve and to protect the teeth ; hence we find that the distinctness of formation in the lips is directly proportionate to the nicety and exactitude of formation in the teeth. In man the lips are soft and fleshy and can be separated. Their purpose is (as in other animals) to protect the teeth ; but— still more important—they subserve a *good* purpose, inasmuch as they are among the parts that make speech possible. This double function of the human lips, to facilitate speech as well as to protect the teeth, may be compared with that of the human tongue, which is unlike that of any other animal, and is used by Nature for two functions (a device of hers which we have often noted), (*a*) to perceive the various tastes, and (*b*) to be the means of speech. Now vocal speech consists of combinations of the various letters or sounds, some of which are produced by an impact of the tongue, others by closing the lips ; and if the lips were not supple, or if the tongue were other than it is, the greater part of these could not possibly be pronounced. For further particulars about the various differences between these sounds you must consult the authorities on Metre.

It was *necessary*, however, from the start that each of these two parts should be adapted and well-fitted for their function as stated above ; therefore their nature had to be suitable thereto, and that is why they are made of flesh. Human flesh is the softest kind of flesh there is ; and this is because man's sense of touch is much more delicate than that of any other creature.

XVII. Ὑπὸ δὲ τὸν οὐρανὸν ἐν τῷ στόματι ἡ
15 γλῶττα τοῖς ζῴοις ἐστί, τοῖς μὲν πεζοῖς σχεδὸν
ὁμοίως πᾶσι, τοῖς δ᾽ ἄλλοις ἀνομοίως καὶ αὐτοῖς
πρὸς αὑτὰ καὶ πρὸς τὰ πεζὰ τῶν ζῴων. ὁ μὲν οὖν
ἄνθρωπος ἀπολελυμένην τε μάλιστα τὴν γλῶτταν
καὶ πλατεῖαν καὶ μαλακωτάτην ἔχει,[1] ὅπως πρὸς
ἀμφοτέρας ᾖ τὰς ἐργασίας χρήσιμος, πρός τε τὴν
20 τῶν χυμῶν αἴσθησιν (ὁ γὰρ ἄνθρωπος εὐαισθητό-
τατος τῶν ἄλλων ζῴων, καὶ ἡ μαλακὴ γλῶττα
⟨αἰσθητικωτάτη⟩[2]· ἁπτικωτάτη γάρ, ἡ δὲ γεῦσις ἁφή
τίς ἐστιν), καὶ πρὸς τὴν τῶν γραμμάτων διάρθρωσιν
καὶ πρὸς τὸν λόγον ἡ μαλακὴ καὶ πλατεῖα χρή-
σιμος· συστέλλειν γὰρ καὶ προβάλλειν παντοδαπὴ
25 τοιαύτη οὖσα καὶ ἀπολελυμένη μάλιστ᾽ ἂν δύναιτο.
δηλοῖ δ᾽ ὅσοις μὴ λίαν ἀπολέλυται· ψελλίζονται
γὰρ καὶ τραυλίζουσι, τοῦτο δ᾽ ἐστὶν ἔνδεια τῶν
γραμμάτων.

Ἔν τε τῷ πλατεῖαν εἶναι καὶ τὸ στενήν ἐστιν·
ἐν γὰρ τῷ μεγάλῳ καὶ τὸ μικρόν, ἐν δὲ τῷ μικρῷ
τὸ μέγα οὐκ ἔστιν. διὸ καὶ τῶν ὀρνίθων οἱ μάλιστα
30 φθεγγόμενοι γράμματα πλατυγλωττότεροι τῶν ἄλλων εἰσίν. τὰ δ᾽ ἔναιμα καὶ ζῳοτόκα τῶν τετραπόδων βραχεῖαν τῆς φωνῆς ἔχει διάρθρωσιν·
σκληράν τε γὰρ καὶ οὐκ ἀπολελυμένην ἔχουσι
καὶ παχεῖαν τὴν γλῶτταν. τῶν δ᾽ ὀρνίθων ἔνιοι
πολύφωνοι, καὶ πλατυτέραν οἱ γαμψώνυχοι ἔχουσιν.
35 πολύφωνοι δ᾽ οἱ μικρότεροι. καὶ χρῶνται τῇ
γλώττῃ καὶ πρὸς ἑρμηνείαν ἀλλήλοις πάντες μέν
660 b ἕτεροι δὲ τῶν ἑτέρων μᾶλλον, ὥστ᾽ ἐπ᾽ ἐνίων καὶ

---

[1] καὶ μαλ. ἔχει post τε vulg.; transposui.
[2] αἰσθητικωτάτη supplevi.

XVII. Under the vaulted roof of the mouth is Tongue. placed the tongue, and it is practically the same in all land-animals; but there are variations in the other groups, whose tongues are as a whole different from those of land-animals and also different among themselves. The human tongue is the freest, the broadest, and the softest of all : this is to enable it to fulfil both its functions. On the one hand, it has to perceive all the various tastes, for man has the most delicate senses of all the animals, and a soft tongue is the most sensitive, because it is the most responsive to touch, and taste is a sort of touch. It has, also, to articulate the various sounds and to produce speech, and for this a tongue which is soft and broad is admirably suited, because it can roll back and dart forward in all directions ; and herein too its freedom and looseness assists it. This is shown by the case of those whose tongues are slightly tied : their speech is indistinct and lisping, which is due to the fact that they cannot produce all the sounds.

A tongue which is broad can also become narrow, on the principle that the great includes the small, but not *vice versa*. That is why the clearest talkers, even among birds, are those which have the broadest tongues. On the other hand, the blooded viviparous quadrupeds have a limited vocal articulation ; it is because their tongues are hard and thick and not sufficiently loose. Some birds— the smaller sorts—have a large variety of notes. The crook-taloned birds have fairly broad tongues. All birds use their tongues as a means of communication with other birds, and some to a very considerable extent, so much so that it is probable that in

μάθησιν εἶναι δοκεῖν παρ' ἀλλήλων· εἴρηται δὲ περὶ
αὐτῶν ἐν ταῖς ἱστορίαις ταῖς περὶ τῶν ζώων.

Τῶν δὲ πεζῶν καὶ ᾠοτόκων καὶ ἐναίμων πρὸς
μὲν τὴν τῆς φωνῆς ἐργασίαν ἄχρηστον τὰ πολλὰ
5 τὴν γλῶτταν ἔχει καὶ προσδεδεμένην καὶ σκληράν,
πρὸς δὲ τὴν τῶν χυμῶν γεῦσιν οἵ τ' ὄφεις καὶ οἱ
σαῦροι μακρὰν καὶ δικρόαν ἔχουσιν, οἱ μὲν ὄφεις
οὕτω μακρὰν ὥστ' ἐκτείνεσθαι ἐκ μικροῦ ἐπὶ πολύ,
δικρόαν δὲ καὶ τὸ ἄκρον λεπτὸν καὶ τριχῶδες διὰ
τὴν λιχνείαν τῆς φύσεως· διπλῆν γὰρ τὴν ἡδονὴν
10 κτᾶται τῶν χυμῶν, ὥσπερ διπλῆν ἔχοντα τὴν τῆς
γεύσεως αἴσθησιν.

Ἔχει δὲ καὶ τὰ μὴ ἔναιμα τῶν ζῴων τὸ αἰσθητικὸν τῶν χυμῶν μόριον καὶ τὰ ἔναιμα πάντα· καὶ
γὰρ ὅσα μὴ δοκεῖ τοῖς πολλοῖς ἔχειν, οἷον ἔνιοι τῶν
ἰχθύων, καὶ οὗτοι τρόπον τινὰ γλίσχρον ἔχουσι, καὶ
15 σχεδὸν παραπλησίοις τοῖς ποταμίοις κροκοδείλοις.
οὐ φαίνονται δ' οἱ πλεῖστοι αὐτῶν ἔχειν διά τιν'
αἰτίαν εὔλογον· ἀκανθώδης τε γάρ ἐστιν ὁ τόπος
τοῦ στόματος πᾶσι τοῖς τοιούτοις, καὶ διὰ τὸ
μικρὸν χρόνον εἶναι τὴν αἴσθησιν τοῖς ἐνύδροις τῶν
χυμῶν, ὥσπερ καὶ ἡ χρῆσις αὐτῆς βραχεῖα, οὕτω
20 βραχεῖαν ἔχουσιν αὐτῆς καὶ τὴν διάρθρωσιν. ταχεῖα
δ' ἡ δίοδος εἰς τὴν κοιλίαν διὰ τὸ μὴ οἷόν τ' εἶναι
διατρίβειν ἐκχυμίζοντας· παρεμπίπτοι γὰρ ἂν τὸ
ὕδωρ. ὥστ' ἐὰν μή τις τὸ στόμα ἐπικλίνῃ, μὴ
φαίνεσθαι ἀφεστηκὸς τοῦτο τὸ μόριον. ἀκανθώδης
δ' ἐστὶν οὗτος ὁ τόπος· σύγκειται γὰρ ἐκ τῆς
25 συμφαύσεως τῶν βραγχίων, ὧν ἡ φύσις ἀκανθώδης
ἐστίν.

ᵃ See *Hist. An.* 504 b 1, 536 a 20 ff., 597 b 26, 608 a 17.

some cases information is actually conveyed from one bird to another. I have spoken of these in the *Researches upon Animals.*[a]

The tongue is useless for the purpose of speech in most of the oviparous and blooded land-animals because it is fastened down and is hard ; but it is very useful for the purpose of taste, *e.g.* in the serpents and lizards, which have long, forked tongues. Serpents' tongues are very long, but can be rolled into a small compass and then extended to a great distance ; they are also forked, and the tips of them are fine and hairy, owing to their having such inordinate appetites ; by this means the serpents get a double pleasure out of what they taste, owing to their possessing as it were a double organ for this sense.

Even some of the bloodless animals have an organ for perceiving tastes ; and of course all the blooded animals have one, including those which most people would say had not, *e.g.*, certain of the fishes, which have a paltry sort of tongue, very like what the river-crocodiles have. Most of these creatures look as if they had no tongue, and there is good reason for this. (1) All animals of this sort have spinous mouths ; (2) the time which water-animals have for perceiving tastes is short ; hence, since the use of this sense is short, so is the articulation of its organ. The reason why their food passes very quickly into the stomach is because they cannot spend much time sucking out its juices, otherwise the water would get in as well. So unless you pull the mouth well open, you will not be able to see that the tongue is a separate projection. The inside of the mouth is spinous, because it is formed by the juxtaposition of the gills which are of a spinous nature.

Τοῖς δὲ κροκοδείλοις συμβάλλεταί τι πρὸς τὴν
τοῦ μορίου τούτου ἀναπηρίαν καὶ τὸ τὴν σιαγόνα
τὴν κάτω ἀκίνητον ἔχειν. ἔστι μὲν γὰρ ἡ γλῶττα
τῇ κάτω συμφυής, οἱ δ᾽ ἔχουσιν ὥσπερ ἀνάπαλιν
τὴν ἄνω κάτω· τοῖς γὰρ ἄλλοις ἡ ἄνω ἀκίνητος.
30 πρὸς μὲν οὖν τῇ ἄνω οὐκ ἔχουσι τὴν γλῶτταν, ὅτι
ἐναντίως ἂν ἔχοι πρὸς τὴν τῆς τροφῆς εἴσοδον, πρὸς
δὲ τῇ κάτω, ὅτι ὥσπερ μετακειμένη ἡ ἄνω ἐστίν.
ἔτι δὲ καὶ συμβέβηκεν αὐτῷ πεζῷ ὄντι ζῆν ἰχθύων
βίον, ὥστε καὶ διὰ τοῦτο ἀναγκαῖον ἀδιάρθρωτον
αὐτὸν ἔχειν τοῦτο τὸ μόριον.

35 Τὸν δ᾽ οὐρανὸν σαρκώδη πολλοὶ καὶ τῶν ἰχθύων
ἔχουσι, καὶ τῶν ποταμίων ἔνιοι σφόδρα σαρκώδη
καὶ μαλακόν, οἷον οἱ καλούμενοι κυπρῖνοι, ὥστε
661 a δοκεῖν τοῖς μὴ σκοποῦσιν ἀκριβῶς γλῶτταν ἔχειν
ταύτην. οἱ δ᾽ ἰχθύες διὰ τὴν εἰρημένην αἰτίαν
ἔχουσι μὲν οὐ σαφῆ δ᾽ ἔχουσι τὴν διάρθρωσιν τῆς
γλώττης. ἐπεὶ δὲ [τῆς τροφῆς χάριν][1] καὶ τῶν
5 χυμῶν αἴσθησις ἔνεστι μὲν τῷ γλωττοειδεῖ μορίῳ,
οὐ παντὶ[2] δ᾽ ὁμοίως ἀλλὰ τῷ ἄκρῳ μάλιστα, διὰ
τοῦτο τοῖς ἰχθύσι τοῦτ᾽ ἀφώρισται μόνον.

Ἐπιθυμίαν δ᾽ ἔχει τροφῆς τὰ ζῷα πάντα ὡς
ἔχοντα αἴσθησιν τῆς ἡδονῆς τῆς γινομένης ἐκ τῆς
τροφῆς· ἡ γὰρ ἐπιθυμία τοῦ ἡδέος ἐστίν. ἀλλὰ τὸ
μόριον οὐχ ὅμοιον τοῦτο πᾶσιν, ᾧ τὴν αἴσθησιν
10 ποιοῦνται τῆς τροφῆς, ἀλλὰ τοῖς μὲν ἀπολελυμένον
τοῖς δὲ προσπεφυκός, ὅσοις μηδὲν ἔργον ὑπάρχει

---

[1] [τῆς τροφῆς χάριν] praecedentium interpretationem seclusi,
cetera correxi: τῆς ἐν τοῖς χυμοῖς ἐστιν ἡ αἴσθησις (εἰς αἴσθησιν
Z) τὸ μὲν (μὲν τὸ EYZ) γλωττοειδὲς ἔχει (ἔχει om. Z) μόριον
vulg.   [2] παντὶ Z: πάντη vulg.

Among the factors which contribute to the deformity of the crocodile's tongue is the immobility of its lower jaw, to which the tongue is naturally joined. We must remember, however, that the crocodile's jaws are topsy-turvy ; the bottom one is on top and the top one below ; this is clearly so, because in other animals the top jaw is the immovable one. The tongue is not fixed to the upper jaw (as one might expect it to be) because it would get in the way of the food as it entered the mouth, but to the lower one, which is really the upper one in the wrong place. Furthermore, although the crocodile is a land-animal, his manner of life is that of a fish, and this is another reason why he must have a tongue that is not distinctly articulated.

Many fish, however, have a *fleshy* roof to their mouths. In some of the fresh-water fish—*e.g.* those known as Cyprinoi—it is very fleshy and soft, so that casual observers think it is a tongue. In fish, however, for the reason already given, the tongue, though articulated, is not distinctly so ; yet, inasmuch as the power also of perceiving tastes resides in the tongue-like organ, though not in the whole of it equally but chiefly in the tip, therefore on this account in fish the tip only is separate from the jaw.

Now all animals are able to perceive the pleasant taste which is derived from food, and so they have a desire for food, because desire aims at getting that which is pleasant. The part, however, by which this perception or sensation of the food takes place, is not identical in all of them, for some have a tongue which moves freely and loosely, others (which have no vocal functions) have a tongue that is fastened down.

φωνῆς, καὶ τοῖς μὲν σκληρὸν τοῖς δὲ μαλακὸν
ἢ σαρκῶδες. διὸ καὶ τοῖς μαλακοστράκοις, οἷον
καράβοις καὶ τοῖς τοιούτοις, ἐντὸς ὑπάρχει τι τοῦ
15 στόματος τοιοῦτον, καὶ τοῖς μαλακίοις, οἷον σηπίαις
καὶ πολύποσιν. τῶν δ᾽ ἐντόμων ζῴων ἔνια μὲν
ἐντὸς ἔχει τὸ τοιοῦτον μόριον, οἷον τὸ τῶν μυρμήκων γένος, ὡσαύτως δὲ καὶ τῶν ὀστρακοδέρμων
πολλά· τὰ δ᾽ ἐκτός, οἷον κέντρον, σομφὸν δὲ τὴν
φύσιν καὶ κοῖλον, ὥσθ᾽ ἅμα τούτῳ καὶ γεύεσθαι καὶ
20 τὴν τροφὴν ἀνασπᾶν. δῆλον δὲ τοῦτο ἐπί τε μυιῶν
καὶ μελιττῶν καὶ πάντων τῶν τοιούτων, ἔτι δ᾽
ἐπ᾽ ἐνίων τῶν ὀστρακοδέρμων· ταῖς γὰρ πορφύραις
τοσαύτην ἔχει δύναμιν τοῦτο τὸ μόριον ὥστε καὶ
τῶν κογχυλίων διατρυπῶσι τὸ ὄστρακον, οἷον τῶν
στρόμβων οἷς δελεάζουσιν αὐτάς. ἔτι δ᾽ οἵ τ᾽
οἶστροι καὶ οἱ μύωπες οἱ μὲν τὰ τῶν ἀνθρώπων
25 οἱ δὲ καὶ τὰ τῶν ἄλλων ζῴων δέρματα διαιροῦσιν.
ἐν μὲν οὖν τούτοις τοῖς ζῴοις ἡ γλῶττα τοιαύτη
τὴν φύσιν ἐστίν, ὥσπερ ἀντιστρόφως ἔχουσα τῷ
μυκτῆρι τῷ τῶν ἐλεφάντων· καὶ γὰρ ἐκείνοις πρὸς
βοήθειαν ὁ μυκτήρ, καὶ τούτοις ἡ γλῶττα ἀντὶ
κέντρου ἐστίν. ἐπὶ δὲ τῶν ἄλλων ζῴων ἡ γλῶττα
30 πάντων ἐστὶν οἵανπερ εἴπομεν.

ᵃ Under this name Aristotle probably includes several
species of Purpura and Murex. Tyrian purple (6, 6′ dibrom-

Some again have a hard tongue ; others a soft or fleshy one. So we find that even the Crustacea—*e.g.* the Crayfish and such—have a tongue-like object inside the mouth, and so have the Cephalopods—*e.g.* the Sepias and the Octopuses. Of the Insects, some have this organ inside the mouth (*e.g.* the Ants), and so have many of the Testacea. Others have it outside, as though it were a sting, in which case it is spongy and hollow, and so they can use it both for tasting and for drawing up their food. Clear examples of this are flies and bees and all such creatures, and also some of the Testacea. In the Purpurae,[a] for instance, this " tongue " has such strength that they can actually bore through the shells of shellfish with it, including those of the spiral snails which are used as baits for them. Also, there are among the gadflies and cattle-flies creatures that can pierce through the skin of the human body, and some can actually puncture animal hides as well. Tongues of this sort, we may say, are on a par with the elephant's nose ; in their tongue these creatures have a useful sting just as the elephant has a handy implement in his trunk.

In all other animals the tongue conforms to the description we have given.

indigo) is obtained from *Murex brandaris.* For the boring powers of these creatures' tongues see the reference for *Purpura lapillus* given by Ogle (Forbes and Hanley, *Brit. Mollusca,* iii. 385).

Γ

Ἐχόμενον δὲ τῶν εἰρημένων ἡ τῶν ὀδόντων
35 ἐστὶ φύσις τοῖς ζῴοις, καὶ τὸ στόμα τὸ περι-
εχόμενον ὑπὸ τούτων καὶ συνεστηκὸς ἐκ τούτων.

661 b   Τοῖς μὲν οὖν ἄλλοις ἡ τῶν ὀδόντων φύσις κοινὴ
μὲν ἐπὶ τὴν τῆς τροφῆς ἐργασίαν ὑπάρχει, χωρὶς
δὲ κατὰ γένη τοῖς μὲν ἀλκῆς χάριν, καὶ ταύτης διῃρημένης, ἐπί τε τὸ ποιεῖν καὶ τὸ μὴ πάσχειν·
5 τὰ μὲν γὰρ ἀμφοῖν ἕνεκεν ἔχει, καὶ τοῦ μὴ παθεῖν
καὶ τοῦ ποιεῖν, οἷον ὅσα σαρκοφάγα τῶν ἀγρίων
τὴν φύσιν ἐστίν, τὰ δὲ βοηθείας χάριν, ὥσπερ
πολλὰ τῶν ἀγρίων καὶ τῶν ἡμέρων.

Ὁ δ' ἄνθρωπος πρός τε τὴν κοινὴν χρῆσιν καλῶς
ἔχει πεφυκότας· τοὺς μὲν προσθίους ὀξεῖς, ἵνα
διαιρῶσι, τοὺς δὲ γομφίους πλατεῖς, ἵνα λεαίνωσιν·
10 ὁρίζουσι δ' ἑκατέρους οἱ κυνόδοντες, μέσοι τὴν
φύσιν ἀμφοτέρων ὄντες· τό τε γὰρ μέσον ἀμφοτέρων
μετέχει τῶν ἄκρων, οἵ τε κυνόδοντες τῇ μὲν
ὀξεῖς τῇ δὲ πλατεῖς εἰσιν· ὁμοίως δὲ καὶ ἐπὶ
τῶν ἄλλων ζῴων, ὅσα μὴ πάντας ἔχουσιν ὀξεῖς
—μάλιστα δὲ καὶ τούτους τοιούτους καὶ τοσούτους πρὸς τὴν διάλεκτον· πολλὰ γὰρ πρὸς τὴν

208

BOOK III

THE subject which follows naturally after our previous remarks is that of the Teeth. We shall also speak about the Mouth, for this is bounded by the teeth and is really formed by them.

In the lower animals teeth have one common function, namely, mastication ; but they have additional functions in different groups of animals. In some they are present to serve as weapons, offensive and defensive, for there are animals which have them both for offence and defence (*e.g.* the wild carnivora) ; others (including many animals both wild and domesticated) have them for purposes of assistance.

Human teeth too are admirably adapted for the common purpose that all teeth subserve : the front ones are sharp, to bite up the food; the molars are broad and flat, to grind it small ; and on the border between the two are the dog-teeth whose nature is intermediate between the two : and just as a mean shares the nature of both its extremes, so the dog-teeth are broad in one part and sharp in another. Thus the provision is similar to that of the other animals, except those whose teeth are all sharp ; but in man even these sharp teeth, in respect of character and number, are adapted chiefly for the purposes of speech, since the

209

681 b

15 γένεσιν τῶν γραμμάτων οἱ πρόσθιοι τῶν ὀδόντων
συμβάλλονται.

Ἔνια δὲ τῶν ζῴων, ὥσπερ εἴπομεν, τροφῆς χάριν
ἔχει μόνον. ὅσα δὲ καὶ πρὸς βοήθειάν τε καὶ πρὸς
ἀλκήν, τὰ μὲν χαυλιόδοντας ἔχει, καθάπερ ὗς, τὰ
δ' ὀξεῖς καὶ ἐπαλλάττοντας, ὅθεν καρχαρόδοντα
20 καλεῖται. ἐπεὶ γὰρ ἐν τοῖς ὀδοῦσιν ἡ ἰσχὺς αὐτῶν,
τοῦτο δὲ γίνοιτ' ἂν διὰ τὴν ὀξύτητα, οἱ χρήσιμοι
πρὸς τὴν ἀλκὴν ἐναλλὰξ ἐμπίπτουσιν, ὅπως μὴ
ἀμβλύνωνται τριβόμενοι πρὸς ἀλλήλους. οὐδὲν δὲ
τῶν ζῴων ἐστὶν ἅμα καρχαρόδουν καὶ χαυλιόδουν,
διὰ τὸ μηδὲν μάτην ποιεῖν τὴν φύσιν μηδὲ περί-
25 εργον· ἔστι δὲ τῶν μὲν διὰ πληγῆς ἡ βοήθεια,
τῶν δὲ διὰ δήγματος. διόπερ αἱ θήλειαι τῶν ὑῶν
δάκνουσιν· οὐ γὰρ ἔχουσι χαυλιόδοντας.

(Καθόλου δὲ χρεών τι λαβεῖν, ὃ καὶ ἐπὶ τούτων
καὶ ἐπὶ πολλῶν τῶν ὕστερον λεχθησομένων ἔσται
χρήσιμον. τῶν τε γὰρ πρὸς ἀλκήν τε καὶ βοήθειαν
30 ὀργανικῶν μορίων ἕκαστα ἀποδίδωσιν ἡ φύσις τοῖς
δυναμένοις χρῆσθαι μόνοις ἢ μᾶλλον, μάλιστα δὲ
τῷ μάλιστα, οἷον κέντρον, πλῆκτρον, κέρατα,
χαυλιόδοντας καὶ εἴ τι τοιοῦτον ἕτερον. ἐπεὶ δὲ τὸ
ἄρρεν ἰσχυρότερον καὶ θυμικώτερον, τὰ μὲν μόνα
τὰ δὲ μᾶλλον ἔχει τὰ τοιαῦτα τῶν μορίων. ὅσα
35 μὲν γὰρ ἀναγκαῖον καὶ τοῖς θήλεσιν ἔχειν, οἷον τὰ
πρὸς τὴν τροφήν, ἔχουσι μὲν ἧττον δ' ἔχουσιν, ὅσα
δὲ πρὸς μηδὲν τῶν ἀναγκαίων, οὐκ ἔχουσιν. καὶ

---

* See note on 644 a 17.

210

front teeth contribute a great deal to the formation of the sounds.

As we have said, the teeth of some of the animals have one function only, to break up the food. Of those animals whose teeth serve also as a defence and as weapons, some (like the Swine) have tusks, some have sharp interlocking teeth, and are called "saw-toothed" as a result. The strength of these latter animals lies in their teeth, and sharpness is the means of securing this; so the teeth which are serviceable as weapons are arranged to fit in side by side when the jaws are closed to prevent them from rubbing against each other and becoming blunt. No animal has saw-teeth as well as tusks; for Nature never does anything without purpose or makes anything superfluously. These teeth are used in self-defence by biting; tusks by striking. This explains why sows bite: they have no tusks.

(At this point we should make a generalization, which will help us both in our study of the foregoing cases and of many that are to follow. Nature allots defensive and offensive organs only to those creatures which can make use of them, or allots them "in a greater degree," [a] and "in the greatest degree" to the animal which can use them to the greatest extent. This applies to stings, spurs, horns, tusks, and the rest. Example: Males are stronger than females and more spirited; hence sometimes the male of a species has one of these parts and the female has none, sometimes the male has it "in a greater degree." Parts which are necessary for the female as well as for the male, as for instance those needed for feeding, are of course present though "in a less degree"; but those which serve no necessary end are not

"The more and the less."

662 a διὰ τοῦτο τῶν ἐλάφων οἱ μὲν ἄρρενες ἔχουσι
κέρατα, αἱ δὲ θήλειαι οὐκ ἔχουσιν. διαφέρει δὲ
καὶ τὰ κέρατα τῶν θηλειῶν βοῶν καὶ τῶν ταύρων·
ὁμοίως δὲ καὶ ἐν τοῖς προβάτοις. καὶ πλῆκτρα
5 τῶν ἀρρένων ἐχόντων αἱ πολλαὶ τῶν θηλειῶν οὐκ
ἔχουσιν. ὡς δ' αὔτως ἔχει τοῦτο καὶ ἐπὶ τῶν
ἄλλων τῶν τοιούτων.)

Οἱ δ' ἰχθύες πάντες εἰσὶ καρχαρόδοντες, πλὴν
τοῦ ἑνὸς τοῦ καλουμένου σκάρου· πολλοὶ δ' ἔχουσι
καὶ ἐν ταῖς γλώτταις ὀδόντας καὶ ἐν τοῖς οὐρανοῖς.
τούτου δ' αἴτιον ὅτι ἀναγκαῖον ἐν ὑγροῖς οὖσι
10 παρεισδέχεσθαι τὸ ὑγρὸν ἅμα τῇ τροφῇ, καὶ τοῦτο
ταχέως ἐκπέμπειν. οὐ γὰρ ἐνδέχεται λεαίνοντας
διατρίβειν· εἰσρέοι γὰρ ἂν τὸ ὑγρὸν εἰς τὰς κοιλίας.
διὰ τοῦτο πάντες εἰσὶν ὀξεῖς πρὸς τὴν διαίρεσιν
μόνον, καὶ[1] πολλοὶ καὶ πολλαχῇ, ἵνα ἀντὶ τοῦ
λεαίνειν εἰς πολλὰ κερματίζωσι τῷ πλήθει. γαμψοὶ
15 δὲ διὰ τὸ τὴν ἀλκὴν σχεδὸν ἅπασαν αὐτοῖς διὰ
τούτων εἶναι.

Ἔχει δὲ καὶ τὴν τοῦ στόματος φύσιν τὰ ζῷα
τούτων τε τῶν ἔργων ἕνεκα καὶ ἔτι τῆς ἀναπνοῆς,
ὅσα ἀναπνεῖ τῶν ζῴων καὶ καταψύχεται θύραθεν.
ἡ γὰρ φύσις αὐτὴ καθ' αὑτήν, ὥσπερ εἴπομεν, τοῖς
20 κοινοῖς πάντων μορίοις εἰς πολλὰ τῶν ἰδίων καταχρῆται, οἷον καὶ ἐπὶ τοῦ στόματος ἡ μὲν τροφὴ
πάντων κοινόν, ἡ δ' ἀλκὴ τινῶν ἴδιον καὶ ὁ λόγος
ἑτέρων, ἔτι δὲ τὸ ἀναπνεῖν οὐ πάντων κοινόν. ἡ δὲ

---

[1] sic P: διαίρεσιν. πάλιν καὶ vulg.

---

[a] Probably the parrot-fish. *Cf.* 675 a 3.

present. Thus, stags have horns, does have not. Thus, too, cows' horns are different from bulls' horns, and ewes' from rams'. In many species the males have spurs while the females have not. And so with the other such parts.)

All fishes are saw-toothed except one species, the Scarus.[a] Many of them have teeth on their tongues and in the roof of the mouth. This is because as they live in the water they cannot help letting some of it in as they take in their food, and they have to get it out again as quickly as possible. If they failed to do so, and spent time grinding the food small, the water would run down into their gut. So all their teeth are sharp and intended only for cutting up the food. Further, they are numerous and placed all over the mouth; so by reason of their multitude they can reduce the food into tiny pieces, and this takes the place of the grinding process. They are also curved; this is because practically the whole of a fish's offensive force is concentrated in its teeth.

The mouth, too, is present in animals on purpose Mouth. to fulfil these same offices, but it has also a further purpose, at any rate in those animals which breathe and are cooled from without—namely, to effect respiration. As we said earlier, Nature will often quite spontaneously take some part that is common to all animals and press it into service for some specialized purpose. Thus, the mouth is common to all animals, and its normal and universal function has to do with food : but sometimes it has an extra function, peculiar to some species only : in some it is a weapon, in others a means of speech ; or more generally, though not universally, it serves for respiration. Nature has

662 a

φύσις ἅπαντα συνήγαγεν εἰς ἕν, ποιοῦσα διαφορὰν
αὐτοῦ τοῦ μορίου πρὸς τὰς τῆς ἐργασίας διαφοράς.
25 διὸ τὰ μέν ἐστι συστομώτερα, τὰ δὲ μεγαλόστομα.
ὅσα μὲν γὰρ τροφῆς καὶ ἀναπνοῆς καὶ λόγου χάριν,
συστομώτερα, τῶν δὲ βοηθείας χάριν τὰ μὲν
καρχαρόδοντα πάντα ἀνερρωγότα· οὔσης γὰρ
αὐτοῖς τῆς ἀλκῆς ἐν τοῖς δήγμασι χρήσιμον τὸ
μεγάλην εἶναι τὴν ἀνάπτυξιν τοῦ στόματος· πλείοσι
30 γὰρ καὶ κατὰ μεῖζον δήξεται, ὅσονπερ ἂν ἐπὶ τὸ
πλέον ἀνερρώγῃ τὸ στόμα. ἔχουσι δὲ καὶ τῶν
ἰχθύων οἱ δηκτικοὶ καὶ σαρκοφάγοι τοιοῦτον στόμα,
οἱ δὲ μὴ σαρκοφάγοι μύουροι· τοιοῦτον γὰρ αὐτοῖς
χρήσιμον, ἐκεῖνο δὲ ἄχρηστον.

Τοῖς δ᾽ ὄρνισίν ἐστι τὸ καλούμενον ῥύγχος στόμα·
35 τοῦτο γὰρ ἀντὶ χειλῶν καὶ ὀδόντων ἔχουσιν. δια-
662 b φέρει δὲ τοῦτο κατὰ τὰς χρήσεις καὶ τὰς βοηθείας.
τὰ μὲν γὰρ γαμψώνυχα καλούμενα διὰ τὸ σαρκο-
φαγεῖν καὶ μηδενὶ τρέφεσθαι καρπῷ γαμψὸν ἔχει τὸ
ῥύγχος ἅπαντα· χρήσιμον γὰρ πρὸς τὸ κρατεῖν καὶ
βιαστικώτερον τοιοῦτο πεφυκός. ἡ δ᾽ ἀλκὴ ἐν
5 τούτῳ τε καὶ τοῖς ὄνυξι· διὸ καὶ τοὺς ὄνυχας
γαμψοτέρους ἔχουσιν. τῶν δ᾽ ἄλλων ἑκάστῳ πρὸς
τὸν βίον χρήσιμόν ἐστι τὸ ῥύγχος, οἷον τοῖς μὲν
δρυοκόποις ἰσχυρὸν καὶ σκληρόν, καὶ κόρακι καὶ
κορακώδεσι, τοῖς δὲ μικροῖς γλαφυρὸν πρὸς τὰς
συλλογὰς τῶν καρπῶν καὶ τὰς λήψεις τῶν ζω-
10 δαρίων. ὅσα δὲ ποηφάγα καὶ ὅσα παρ᾽ ἕλη ζῇ,

brought all these functions together under one part, whose formation she varies in the different species to suit its various duties. That is why the animals which use their mouths for feeding, respiration and speaking have rather narrow mouths, while those that use them for self-defence have wide and gaping mouths. All the saw-toothed creatures have these wide mouths, for their method of attack is biting, and therefore they find it an advantage to have a mouth that will open wide ; and the wider it opens the greater the space the bite will enclose, and the greater the number of teeth brought into action. Biting and carnivorous fishes have mouths of this sort ; in the non-carnivorous ones it is on a tapering snout, and this suits their habits, whereas a gaping mouth would be useless.

In birds, the mouth appears in the form of a beak, Beak. which serves them instead of lips and teeth. Various sorts of beak are found, to suit the various uses including defensive purposes to which it is put. All of the birds known as crook-taloned have a curved beak, because they feed on flesh and take no vegetable food : a beak of this form is useful to them in mastering their prey, as being more adapted for the exertion of force. Their beak, then, is one weapon of offence, and their claws are another ; that is why their claws are exceptionally curved. Every bird has a beak which is serviceable for its particular mode of life. The woodpeckers, for instance, have a strong, hard beak ; so have crows, and other birds closely related to them ; small birds, on the other hand, have a finely constructed beak, for picking up seeds and catching minute animals. Birds that feed on herbage and that live by marshes (*e.g.* swimmers and

καθάπερ τὰ πλωτὰ καὶ στεγανόποδα, τὰ μὲν ἄλλον
τρόπον χρήσιμον ἔχει τὸ ῥύγχος, τὰ δὲ πλατύρυγχα
αὐτῶν ἐστιν· τοιούτῳ γὰρ ὄντι ῥᾳδίως δύναται
ὀρύσσειν, ὥσπερ καὶ τῶν τετραπόδων τὸ τῆς ὑός·
καὶ γὰρ αὕτη ῥιζοφάγος. ἔτι δ᾽ ἔχουσι καὶ τὰ
15 ῥιζοφάγα τῶν ὀρνέων καὶ τῶν ὁμοιοβίων ἔνια τὰ
ἄκρα τοῦ ῥύγχους κεχαραγμένα· ποηφάγοις γὰρ
τούτοις οὖσι ποιεῖ ῥᾳδίως.

Περὶ μὲν οὖν τῶν ἄλλων μορίων τῶν ἐν τῇ
κεφαλῇ σχεδὸν εἴρηται, τῶν δ᾽ ἀνθρώπων καλεῖται
τὸ μεταξὺ τῆς κεφαλῆς καὶ τοῦ αὐχένος πρόσωπον,
20 ἀπὸ τῆς πράξεως αὐτῆς ὀνομασθέν, ὡς ἔοικεν· διὰ
γὰρ τὸ μόνον ὀρθὸν εἶναι τῶν ζῴων μόνον πρόσ-
ωθεν ὄπωπε καὶ τὴν φωνὴν εἰς τὸ πρόσω διαπέμπει.

II. Περὶ δὲ κεράτων λεκτέον· καὶ γὰρ ταῦτα
πέφυκε τοῖς ἔχουσιν ἐν τῇ κεφαλῇ. ἔχει δ᾽ οὐδὲν
25 μὴ ζῳοτόκον. καθ᾽ ὁμοιότητα δὲ καὶ μεταφορὰν
λέγεται καὶ ἑτέρων τινῶν κέρατα· ἀλλ᾽ οὐδενὶ
αὐτῶν τὸ ἔργον τοῦ κέρατος ὑπάρχει. βοηθείας
γὰρ καὶ ἀλκῆς χάριν ἔχουσι τὰ ζῳοτόκα, ὃ τῶν
ἄλλων τῶν λεγομένων ἔχειν κέρας οὐδενὶ συμ-
βέβηκεν· οὐδὲν γὰρ χρῆται τοῖς κέρασιν οὔτ᾽
30 ἀμυνόμενον οὔτε πρὸς τὸ κρατεῖν, ἅπερ ἰσχύος
ἐστὶν ἔργα. ὅσα μὲν οὖν πολυσχιδῆ τῶν ζῴων,
οὐδὲν ἔχει κέρας. τούτου δ᾽ αἴτιον ὅτι τὸ μὲν
κέρας βοηθείας αἴτιόν ἐστι, τοῖς δὲ πολυσχιδέσιν
ὑπάρχουσιν ἕτεραι βοήθειαι· δέδωκε γὰρ ἡ φύσις
τοῖς μὲν ὄνυχας τοῖς δ᾽ ὀδόντας μαχητικούς, τοῖς

---

[a] Under this heading all the Mammalia known to Aristotle

web-footed birds) have a beak adapted for their mode of life, a special instance of which is the broad beak, which enables them to dig for roots easily, just as the broad snout of the pig enables it to dig—an example of a root-eating quadruped. These root-eating birds and other birds of similar habits sometimes have sharp points at the end of the beak. This enables them to deal easily with the herbaceous food which they take.

We have now, I think, spoken of practically all the parts that have their place in the head; but in man, the portion of the body between the head and the neck is called the *Prosōpon* (Face), a name derived, no doubt, from the function it performs. Man, the only animal that stands upright, is the only one that looks straight before him (*prosōthen opōpe*) or sends forth his voice straight before him (*prosō, opa*).

II. We still have to speak of Horns: these also, Horns when present, grow out of the head. Horns are found only in the Vivipara; though some other creatures have what are called horns, owing to their resemblance to real horns. None of these so-called horns, however, performs the function proper to horns. The reason why the Vivipara have horns is for the sake of self-defence and attack, and this is not true of any of these other creatures, since none of them uses its " horns " for such feats of strength either defensively or offensively. The polydactylous animals,[a] moreover, have no horns, because they possess other means of defence. Nature has given them claws or teeth to fight with, or some other part capable of

are included, except ruminants, solid-hoofed animals, and Cetacea.

217

862 b
35 δ' ἄλλο τι μόριον ἱκανὸν ἀμύνειν. τῶν δὲ διχάλων
663 a τὰ μὲν πολλὰ κέρατα ἔχει πρὸς ἀλκήν, καὶ τῶν
μωνύχων ἔνια, τὰ δὲ καὶ πρὸς βοήθειαν, ὅσοις[1] μὴ
δέδωκεν ἡ φύσις ἄλλην ἀλκὴν πρὸς σωτηρίαν, οἷον
ταχυτῆτα σώματος, καθάπερ τοῖς ἵπποις βεβοήθη-
κεν, ἢ μέγεθος, ὥσπερ ταῖς καμήλοις· καὶ γὰρ
5 μεγέθους ὑπερβολὴ τὴν ἀπὸ τῶν ἄλλων ζῴων
φθορὰν ἱκανὴ κωλύειν, ὅπερ συμβέβηκε ταῖς καμή-
λοις, ἔτι δὲ μᾶλλον τοῖς ἐλέφασιν. τὰ δὲ χαυλι-
όδοντα, ὥσπερ καὶ τὸ τῶν ὑῶν γένος, δίχαλον ⟨ὄν⟩.[2]

   Ὅσοις δ' ἄχρηστος πέφυκεν ἡ τῶν κεράτων
ἐξοχή, τούτοις προστέθεικεν ἑτέραν βοήθειαν ἡ
10 φύσις, οἷον ταῖς μὲν ἐλάφοις τάχος (τὸ γὰρ μέγεθος αὐτῶν καὶ τὸ πολυσχιδὲς μᾶλλον βλάπτει ἢ
ὠφελεῖ), καὶ βουβάλοις δὲ καὶ δορκάσι (πρὸς ἔνια
μὲν γὰρ ἀνθιστάμενα τοῖς κέρασιν ἀμύνονται, τὰ δὲ
θηριώδη καὶ μάχιμα ἀποφεύγουσι), τοῖς δὲ βονάσοις
15 (καὶ γὰρ τούτοις γαμψὰ τὰ κέρατα πέφυκε πρὸς
ἄλληλα) τὴν τοῦ περιττώματος ἄφεσιν· τούτῳ γὰρ
ἀμύνεται φοβηθέντα· καὶ ταύτῃ δὲ τῇ προέσει δια-
σῴζεται ἕτερα. ἅμα δ' ἱκανὰς καὶ πλείους βοηθείας
οὐ δέδωκεν ἡ φύσις τοῖς αὐτοῖς.

   Ἔστι δὲ τὰ πλεῖστα τῶν κερατοφόρων δίχαλα,
λέγεται δὲ καὶ μώνυχον, ὃν καλοῦσιν Ἰνδικὸν ὄνον.
20 Τὰ μὲν οὖν πλεῖστα, καθάπερ καὶ τὸ σῶμα
διῄρηται τῶν ζῴων οἷς ποιεῖται τὴν κίνησιν, δεξιὸν
καὶ ἀριστερόν, καὶ κέρατα δύο πέφυκεν ἔχειν διὰ

---

[1] δὲ post ὅσοις vulg. : del. Platt, Thurot.
[2] ⟨ὄν⟩ Ogle.

---

[a] Cf. above, on 648 a 16.
[b] The European bison.
[c] This is probably the Indian Rhinoceros. This account

218

rendering adequate defence. Most of the cloven-hoofed animals, and some of the solid-hoofed, have horns, as weapons of offence; some have horns for self-defence, as those animals which have not been given means of safety and self-defence of a different order—the speed, for instance, which Nature has given to horses, or the enormous size which camels have (and elephants even more), which is sufficient to prevent them from being destroyed by other animals. Some, however, have tusks, for instance swine, although they are cloven-hoofed.

In some animals the horns are a useless appendage,[a] and to these Nature has given an additional means of defence. Deer have been given speed (because the size of their horns and the numerous branches are more of a nuisance to them than a help). So have the antelopes and the gazelles, which, although they withstand some attackers and defend themselves with their horns, run away from really fierce fighters. The Bonasus,[b] whose horns curve inwards to meet each other, protects itself when frightened by the discharge of its excrement. There are other animals that protect themselves in the same way. Nature, however, has not given more than one adequate means of protection to any one animal.

Most of the horned animals are cloven-hoofed, though there is said to be one that is solid-hoofed, the Indian Ass,[c] as it is called.

The great majority of horned animals have two horns, just as, in respect of the parts by which its movement is effected, the body is divided into two—the right and the left. And the

of it comes from the *Indica* of Ktesias of Knidos, quoted in Photius's *Bibliotheca*, lxxii. pp. 48 b 19 (Bekker) foll.

τὴν αὐτὴν αἰτίαν[1]· ἔστι δὲ καὶ μονοκέρατα, οἷον ὅ
τ' ὄρυξ καὶ ὁ Ἰνδικὸς καλούμενος ὄνος. ἔστι δ' ὁ
μὲν ὄρυξ δίχαλον, ὁ δ' ὄνος μώνυχον. ἔχει δὲ τὰ
25 μονοκέρατα τὸ κέρας ἐν τῷ μέσῳ τῆς κεφαλῆς·
οὕτω γὰρ ἑκάτερον τῶν μερῶν μάλιστ' ἂν ἔχοι
κέρας ἕν· τὸ γὰρ μέσον ὁμοίως κοινὸν ἀμφοτέρων
τῶν ἐσχάτων. εὐλόγως δ' ἂν δόξειε μονόκερων
εἶναι τὸ μώνυχον τοῦ διχάλου μᾶλλον· ὁπλὴ γὰρ
καὶ χηλὴ τὴν αὐτὴν ἔχει κέρατι φύσιν, ὥσθ' ἅμα
30 καὶ τοῖς αὐτοῖς ἡ σχίσις γίνεται τῶν ὁπλῶν καὶ
τῶν κεράτων. ἔτι δ' ἡ σχίσις καὶ τὸ δίχαλον κατ'
ἔλλειψιν τῆς φύσεώς ἐστιν, ὥστ' εὐλόγως τοῖς
μωνύχοις ἐν ταῖς ὁπλαῖς δοῦσα τὴν ὑπεροχὴν
ἡ φύσις ἄνωθεν ἀφεῖλε καὶ μονόκερων ἐποίησεν.

Ὀρθῶς δὲ καὶ τὸ ἐπὶ τῆς κεφαλῆς ποιῆσαι τὴν
35 τῶν κεράτων φύσιν, ἀλλὰ μὴ καθάπερ ὁ Αἰσώπου
Μῶμος διαμέμφεται τὸν ταῦρον ὅτι οὐκ ἐπὶ τοῖς
663 b ὤμοις ἔχει τὰ κέρατα, ὅθεν τὰς πληγὰς ἐποίειτ'
ἂν ἰσχυροτάτας, ἀλλ' ἐπὶ τοῦ ἀσθενεστάτου μέρους
τῆς κεφαλῆς. οὐ γὰρ ὀξὺ βλέπων ὁ Μῶμος ταῦτ'
ἐπετίμησεν. ὥσπερ γὰρ καὶ εἰ ἑτέρωθί που τοῦ
5 σώματος κέρατα ἐπεφύκει, βάρος ἂν παρεῖχεν ἄλλως οὐδὲν ὄντα χρήσιμα κἂν ἐμπόδια τῶν ἔργων
πολλοῖς ἦν, οὕτω καὶ ἐπὶ τῶν ὤμων πεφυκότα. οὐ
γὰρ μόνον χρὴ σκοπεῖν πόθεν ἰσχυρότεραι αἱ πλη-
γαί, ἀλλὰ καὶ πόθεν πορρώτεραι· ὥστ' ἐπεὶ χεῖρας
μὲν οὐκ ἔχουσιν, ἐπὶ δὲ τῶν ποδῶν ἀδύνατον, ἐν δὲ

---

[1] αὐτὴν αἰτίαν Peck : αἰτίαν ταύτην vulg.

* See Babrius, *Myth. Aesop.* lix. 8-10.

reason in both cases is the same. There are, however, some animals that have one horn only, *e.g.*
the Oryx (whose hoof is cloven) and the " Indian
Ass " (whose hoof is solid). These creatures have
their horn in the middle of the head : this is the
nearest approximation to letting each side have its
own horn, because the middle is common equally to
both extremes. Now it is quite reasonable that the
one horn should go with the solid hoof rather than
with the cloven hoof, because hoof is identical in
nature with horn, and we should expect to find
divided hoofs and divided horns together in the same
animal. Again, division of the hoof is really due to
deficiency of material, so it is reasonable that as
Nature has used more material in the hoofs of the
solid-hoofed animals, she has taken something away
from the upper parts and made one horn only.

Again, Nature acted aright in placing the horns
on the head. Momus in Aesop's fable *a* is quite
wrong when he finds fault with the bull for having
his horns on the head, which is the weakest part of
all, instead of on the shoulders, which, he says,
would have enabled them to deliver the strongest
possible blow. Such a criticism shows Momus's
lack of perspicacity. If the horns had been placed
on the shoulders, as indeed on any other part of the
body, they would have been a dead weight, and
would have been no assistance but rather a hindrance
to many of the animal's activities. And besides,
strength of stroke is not the only point to be considered : width of range is equally important.
Where could the horns have been placed to secure
this ? It would have been impossible to have them
on the feet ; knees with horns on them would have

663 b

τοῖς γόνασιν ὄντα τὴν κάμψιν ἐκώλυεν ἄν, ἀναγ-
10 καῖον ὥσπερ νῦν ἔχουσιν, ἐπὶ τῆς κεφαλῆς ἔχειν.
ἅμα δὲ καὶ πρὸς τὰς ἄλλας κινήσεις τοῦ σώματος
ἀνεμπόδιστα πέφυκεν οὕτω μάλιστα.

Ἔστι δὲ τὰ κέρατα δι᾽ ὅλου στερεὰ τοῖς ἐλάφοις
μόνοις, καὶ ἀποβάλλει μόνον, ἕνεκεν μὲν ὠφελείας
κουφιζόμενον, ἐξ ἀνάγκης δὲ διὰ τὸ βάρος. τῶν δ᾽
15 ἄλλων τὰ κέρατα μέχρι τινὸς κοῖλα, τὰ δ᾽ ἄκρα
στερεὰ διὰ τὸ πρὸς τὰς πληγὰς τοῦτ᾽ εἶναι χρήσιμον. ὅπως δὲ μηδὲ τὸ κοῖλον ἀσθενὲς ᾖ ὅ[1]
πέφυκεν ἐκ τοῦ δέρματος, ἐν τούτῳ[2] ἐνήρμοσται
⟨τὸ⟩[3] στερεὸν ἐκ τῶν ὀστῶν· οὕτω γὰρ καὶ τὰ
κέρατα ἔχοντα πρὸς ἀλκήν τε χρησιμώτατ᾽ ἐστὶ[4]
20 καὶ πρὸς τὸν ἄλλον βίον ἀνοχλότατα.

Τίνος μὲν οὖν ἕνεκεν ἡ τῶν κεράτων φύσις,
εἴρηται, καὶ διὰ τίν᾽ αἰτίαν τὰ μὲν ἔχουσι τοιαῦτα
τὰ δ᾽ οὐκ ἔχουσιν·

Πῶς δὲ τῆς ἀναγκαίας φύσεως ἐχούσης τοῖς
ὑπάρχουσιν ἐξ ἀνάγκης ἡ κατὰ τὸν λόγον φύσις
ἕνεκά του κατακέχρηται, λέγωμεν.

25 Πρῶτον μὲν οὖν τὸ σωματῶδες καὶ γεῶδες πλεῖον
ὑπάρχει τοῖς μείζοσι τῶν ζῴων, κερατοφόρον δὲ
μικρὸν πάμπαν οὐδὲν ἴσμεν· ἐλάχιστον γάρ ἐστι τῶν
γνωριζομένων δορκάς. δεῖ δὲ τὴν φύσιν θεωρεῖν
εἰς τὰ πολλὰ βλέποντα· ἢ γὰρ ἐν τῷ παντὶ ἢ ὡς ἐπὶ
τὸ πολὺ τὸ κατὰ φύσιν ἐστίν. τὸ δ᾽ ὀστῶδες ἐν

---

[1] ὅ Peck, cf. *Hist. An.* 500 a 8: οὐ vulg., om. EPΥ: οὐ
suprascr. Z (v. p. 46).   [2] τούτῳ Peck: τούτῳ δ᾽ vulg.
[3] ⟨τὸ⟩ Peck: cf. *Hist. An.* 500 a 9.
[4] ἐστὶ Platt: εἶναι vulg.: εἴη ἂν Thurot.

---

[a] For the contrast between "necessary nature" and

222

been unable to bend ; and the bull has no hands ; so they had to be where they are—on the head. And being there, they offer the least possible hindrance to the movements of the body in general.

Deer alone have horns that are solid throughout ; and deer alone shed their horns : this is done (a) on purpose to get the advantage of the extra lightness, (b) of necessity, owing to the weight of the horns. In other animals the horns are hollow up to a certain distance, but the tips are solid because solid tips are an advantage when striking. And to prevent undue weakness even in the hollow part, which grows out from the skin, the solid piece which is fitted into it comes up from the bones. In this way the horns are rendered most serviceable for offensive purposes and least hampering during the rest of the time.

This completes our statement of the *purpose* for which horns exist and the reason why some animals have them and some have not.

We must now describe the character of that " necessary nature," owing to which certain things are present *of necessity*, things which have been used by " rational nature " to subserve a " purpose." [a]

To begin with, then : the larger the animal, the greater the quantity of corporeal or earthy matter there is in it. We know no really small horned animal—the smallest known one is the gazelle. (To study Nature we have to consider the majority of cases, for it is either in what is universal or what happens in the majority of cases that Nature's ways are to be found. Now all the bone in animals'

" rational nature " see above 640 b 8-29, 641 a 25 ff., 642 a 1 ff., and *cf. G.A.* (Loeb edn.), Introd. § 14.

663 b

30 τοῖς σώμασι τῶν ζῴων γεῶδες ὑπάρχει· διὸ καὶ
πλεῖστον ἐν τοῖς μεγίστοις ὡς ἐπὶ τὸ πολὺ βλέ-
ψαντας εἰπεῖν. τὴν γοῦν τοιούτου σώματος περιτ-
τωματικὴν ὑπερβολὴν ἐν τοῖς μείζοσι τῶν ζῴων
ὑπάρχουσαν ἐπὶ βοήθειαν καὶ τὸ συμφέρον καταχρῆται ἡ φύσις, καὶ τὴν ῥέουσαν ἐξ ἀνάγκης εἰς τὸν
35 ἄνω τόπον τοῖς μὲν εἰς ὀδόντας καὶ χαυλιόδοντας
ἀπένειμε, τοῖς δ᾽ εἰς κέρατα. διὸ τῶν κερατοφόρων οὐδέν ἐστιν ἄμφωδον· ἄνω γὰρ οὐκ ἔχει τοὺς
664 a προσθίους ὀδόντας· ἀφελοῦσα γὰρ ἐντεῦθεν ἡ φύσις
τοῖς κέρασι προσέθηκε, καὶ ἡ διδομένη τροφὴ εἰς
τοὺς ὀδόντας τούτους εἰς τὴν τῶν κεράτων αὔξησιν
ἀναλίσκεται. τοῦ δὲ τὰς θηλείας ἐλάφους κέρατα
μὲν μὴ ἔχειν, περὶ δὲ τοὺς ὀδόντας ὁμοίως τοῖς
5 ἄρρεσιν, αἴτιον τὸ τὴν αὐτὴν εἶναι φύσιν ἀμφοῖν
καὶ κερατοφόρον, ἀφῄρηται δὲ τὰ κέρατα ταῖς
θηλείαις διὰ τὸ χρήσιμα μὲν μὴ εἶναι μηδὲ τοῖς
ἄρρεσιν, βλάπτεσθαι δ᾽ ἧσσον διὰ τὴν ἰσχύν.

Τῶν δ᾽ ἄλλων ζῴων ὅσοις μὴ εἰς κέρατα ἀποκρίνεται τὸ τοιοῦτον μόριον τοῦ σώματος, ἐνίοις
10 μὲν τῶν ὀδόντων αὐτῶν ἐπηύξησε τὸ μέγεθος κοινῇ
πάντων, ἐνίοις δὲ χαυλιόδοντας ὥσπερ κέρατα ἐκ
τῶν γνάθων ἐποίησεν.

Περὶ μὲν οὖν τῶν ἐν τῇ κεφαλῇ μορίων ταύτῃ
διωρίσθω

III. Ὑπὸ δὲ τὴν κεφαλὴν ὁ αὐχὴν πεφυκώς ἐστι
τοῖς ἔχουσιν αὐχένα τῶν ζῴων. οὐ γὰρ πάντα
15 τοῦτο τὸ μόριον ἔχει, ἀλλὰ μόνα τὰ ἔχοντα ὧν

---

^a *i.e.* constituent substance. See on 648 a 2.

224

bodies consists of earthy matter; so if we consider the majority of cases, we can say that there is most earthy matter in the biggest animals.) At any rate, in the larger animals there is present a surplus of this corporeal or earthy matter, produced as a residue, and this Nature makes use of and turns to advantage to provide them with means of defence. That portion of it which by necessity courses upwards she allots to form teeth and tusks in some animals, and to form horns in others. And we can see from this why no horned animal has incisor teeth in both jaws, but only in the bottom jaw. Nature has taken away from the teeth to add to the horns; so that the nourishment which would normally be supplied to the upper teeth is here used to grow the horns. Why is it, then, that female deer, although they have no horns, are no better off for teeth than the male deer? The answer is: Both of them are, by nature, horned animals; but the females have lost their horns because they would be not only useless but dangerous. The horns are indeed of no more use to the males, but they are less dangerous because the males are stronger.

Thus in some animals this " part "[a] of the body is secreted for the formation of horns; in others, however, it causes a general increase in the size of the teeth, and in others again it produces tusks, which are like horns springing out of the jaws instead of the head.

We have now dealt with the " parts " that appertain to the head.

III. The place of the neck, when there is one, is below the head. I say " when there is one," because only those animals have this part which also have

Of the Neck: the Oesophagus.

664 a

χάριν ὁ αὐχὴν πέφυκεν· ταῦτα δ' ἐστὶν ὅ τε φάρυγξ
καὶ ὁ καλούμενος οἰσοφάγος.

Ὁ μὲν οὖν φάρυγξ τοῦ πνεύματος ἕνεκεν πέφυκεν·
διὰ τούτου γὰρ εἰσάγεται τὸ πνεῦμα τὰ ζῷα καὶ
ἐκπέμπει ἀναπνέοντα καὶ ἐκπνέοντα. διὸ τὰ μὴ
20 ἔχοντα πλεύμονα οὐκ ἔχουσιν οὐδ' αὐχένα, οἷον
τὸ τῶν ἰχθύων γένος. ὁ δ' οἰσοφάγος ἐστὶ δι' οὗ
ἡ τροφὴ πορεύεται εἰς τὴν κοιλίαν· ὥσθ' ὅσα μὴ
ἔχει αὐχένα, οὐδ' οἰσοφάγον ἐπιδήλως ἔχουσιν.
οὐκ ἀναγκαῖον δ' ἔχειν τὸν οἰσοφάγον τῆς τροφῆς
ἕνεκεν· οὐθὲν γὰρ παρασκευάζει πρὸς αὐτήν. ἔτι
25 δὲ μετὰ τὴν τοῦ στόματος θέσιν ἐνδέχεται κεῖσθαι
τὴν κοιλίαν εὐθέως, τὸν δὲ πλεύμονα οὐκ ἐνδέχεται.
δεῖ γὰρ εἶναί τινα κοινὸν οἷον αὐλῶνα, δι' οὗ με-
ριεῖται τὸ πνεῦμα κατὰ τὰς ἀρτηρίας εἰς τὰς
σύριγγας, διμερῆ ὄντα[1]· καὶ κάλλιστ' ἂν οὕτως
ἀποτελοῖ τὴν ἀναπνοὴν καὶ ἐκπνοήν. τοῦ δ' ὀρ-
30 γάνου τοῦ περὶ τὴν ἀναπνοὴν ἐξ ἀνάγκης ἔχοντος
μῆκος, ἀναγκαῖον τὸν οἰσοφάγον εἶναι μεταξὺ τοῦ
στόματος καὶ τῆς κοιλίας. ἔστι δ' ὁ μὲν οἰσοφάγος
σαρκώδης, ἔχων νευρώδη τάσιν, νευρώδης μέν,
ὅπως ἔχῃ διάτασιν εἰσιούσης τῆς τροφῆς, σαρκώδης
35 δέ, ὅπως μαλακὸς ᾖ καὶ ἐνδιδῷ καὶ μὴ βλάπτηται
τραχυνόμενος ὑπὸ τῶν κατιόντων.

Ἡ δὲ καλουμένη φάρυγξ καὶ ἀρτηρία συνέστηκεν
664 b ἐκ χονδρώδους σώματος· οὐ γὰρ μόνον ἀναπνοῆς
ἕνεκέν ἐστιν ἀλλὰ καὶ φωνῆς, δεῖ δὲ τὸ ψοφήσειν
μέλλον λεῖον εἶναι καὶ στερεότητα ἔχειν. κεῖται δ'
ἔμπροσθεν ἡ ἀρτηρία τοῦ οἰσοφάγου, καίπερ ἐμ-
ποδίζουσα αὐτὸν περὶ τὴν ὑποδοχὴν τῆς τροφῆς·
5 ἐὰν γάρ τι παρεισρυῇ ξηρὸν ἢ ὑγρὸν εἰς τὴν ἀρτη-

---

[1] διμερῆ ὄντα Peck : διμερὴς ὢν vulg. : διμεροῦς ὄντος Th.

those parts that the neck subserves—viz. the larynx and the oesophagus, as it is called.

The larynx is present for the sake of the breath : when animals breathe in and out, the breath passes through the larynx. Thus creatures which have no lung (*e.g.* fish) have no neck either. The oesophagus is the passage by which the food makes its way to the stomach ; so those that have no neck have no distinct oesophagus. So far as food is concerned, however, an oesophagus is not necessary, since it exerts no action upon the food ; and there is really no reason why the stomach should not be placed immediately next the mouth. The lung, however, could not be so placed, because some sort of tube must be present, common to both lungs, and divided into two, by which the breath is divided along the bronchial tubes into the air-tubes : this is the best method for securing good breathing, both in and out. This respiratory organ, then, of necessity, is of some length ; and this necessitates the presence of an oesophagus, to connect the mouth to the stomach. Now the oesophagus is fleshy, and it can also be extended like a sinew. It is sinewy so that it can stretch as the food enters in ; and it is fleshy so that it may be soft and yielding and not be damaged by the food grating on it as it goes down.

What are called the larynx and windpipe are constructed of cartilaginous substance, since the purpose they serve includes speech as well as respiration ; and an instrument that is to produce sound must be smooth and firm. The windpipe is situated in front of the oesophagus, although it causes it some hindrance when food is being admitted—as when a piece of food, no matter whether solid or fluid, gets

Larynx and windpipe.

664 b

ρίαν, πνιγμοὺς καὶ πόνους καὶ βῆχας χαλεπὰς
ἐμποιεῖ. ὃ δὴ καὶ θαυμάσειεν ἄν τις τῶν λεγόντων
ὡς ταύτῃ τὸ ποτὸν δέχεται τὸ ζῷον· συμβαίνει γὰρ
φανερῶς τὰ λεχθέντα πᾶσιν οἷς ἂν παραρρυῇ τι τῆς
10 τροφῆς. πολλαχῇ δὲ γελοῖον φαίνεται τὸ λέγειν ὡς
ταύτῃ τὸ ποτὸν εἰσδέχεται τὰ ζῷα. πόρος γὰρ
οὐδείς ἐστιν εἰς τὴν κοιλίαν ἀπὸ τοῦ πλεύμονος,
ὥσπερ ἐκ τοῦ στόματος ὁρῶμεν τὸν οἰσοφάγον.
ἔτι δ᾽ ἐν τοῖς ἐμέτοις καὶ ναυτίαις οὐκ ἄδηλον πόθεν
τὸ ὑγρὸν φαίνεται πορευόμενον. δῆλον δὲ καὶ ὅτι
15 οὐκ εὐθέως εἰς τὴν κύστιν συλλέγεται τὸ ὑγρόν,
ἀλλ᾽ εἰς τὴν κοιλίαν πρότερον· τὰ γὰρ τῆς κοιλίας
περιττώματα φαίνεται χρωματίζειν ἡ ἰλὺς τοῦ μέ-
λανος οἴνου· συμβέβηκε δὲ τοῦτο πολλάκις φανερὸν
καὶ ἐπὶ τῶν εἰς τὴν κοιλίαν τραυμάτων. ἀλλὰ γὰρ
ἴσως εὔηθες τὸ τοὺς εὐήθεις τῶν λόγων λίαν
ἐξετάζειν.

20 Ἡ δ᾽ ἀρτηρία τῷ διακεῖσθαι, καθάπερ εἴπομεν,
ἐν τῷ πρόσθεν ὑπὸ τῆς τροφῆς ἐνοχλεῖται· ἀλλ᾽ ἡ
φύσις πρὸς τοῦτο μεμηχάνηται τὴν ἐπιγλωττίδα.
ταύτην δ᾽ οὐκ ἔχουσιν ἅπαντα τὰ ζῳοτοκοῦντα,[1]
ἀλλ᾽ ὅσα πλεύμονα ἔχει καὶ τὸ δέρμα τριχωτόν, καὶ
25 μὴ φολιδωτὰ μηδὲ πτερωτὰ πέφυκεν. τούτοις δ᾽
ἀντὶ τῆς ἐπιγλωττίδος συνάγεται καὶ διοίγεται ὁ
φάρυγξ ὅνπερ τρόπον ἐκείνοις· ἐπιβάλλει τε καὶ
ἀναπτύσσεται, τοῦ ⟨μὲν⟩[2] πνεύματος τῇ εἰσόδῳ τε
καὶ ἐξόδῳ ἀναπτυσσόμενος, τῆς δὲ τροφῆς εἰσ-

---

[1] ζῳοτοκοῦντα] ζῷα τὰ ἔναιμα Ogle.
[2] ⟨μὲν⟩ supplevi et interpunctionem hic correxi.

into the windpipe by mistake, and causes a great deal
of choking and distress and violent coughing. This
sort of thing occurs and can be observed whenever a
piece of food goes the wrong way ; yet they must be
mysteries to those who hold that animals take in their
drink by way of the windpipe. [a] And there are many
counts on which we can show that this is a ridiculous
opinion to hold. (a) There is no passage leading from
the lung into the stomach, such as the oesophagus,
which, as we can see, leads thither from the mouth.
And again, (b) there is no doubt where the fluid discharge comes from in cases of vomiting and sea-sick-
ness. (c) It is plain, too, that the fluid matter which
we take does not collect immediately in the bladder,
but goes first into the stomach. This is shown by
the fact that the dregs of dark wine affect the colour of the residual discharge from the stomach ; and
this colouring has often been observed in cases where
the stomach has been wounded. Still, perhaps it is
silly to be too minute in discussing these silly theories.

The windpipe, as we have said, is situated in front, Epiglottis.
and therefore is interfered with by the food. To deal
with this difficulty, Nature has contrived the epiglottis. Not all Vivipara [b] have this, but only those
which have a lung, and a hairy skin, and are not
covered with horny scales or feathers. Those that
are so covered have, to serve instead of the epiglottis,
a larynx which closes and opens, just as the epiglottis
does in the others ; it comes down and lifts up again :
it lifts up during the entrance and exit of the breath,
and subsides while food is being taken, to prevent

[a] See *e.g.* Plato, *Timaeus* 70 c 7, and Taylor *ad loc.*
[b] Ogle changes the text here to read " blooded animals,"
which brings the statement nearer the truth.

664 b

ιούσης ἐπιπτυσσόμενος, ἵνα μηθὲν παραρρυῇ πρὸς[1]
30 τὴν ἀρτηρίαν. ἐὰν δέ τι πλημμεληθῇ παρὰ τὴν
τοιαύτην κίνησιν καὶ προσφερομένης τῆς τροφῆς
ἀναπνεύσῃ τις, βῆχας καὶ πνιγμοὺς ποιεῖ, καθάπερ
εἴρηται. οὕτω δὲ καλῶς μεμηχάνηται καὶ ἡ ταύτης
καὶ ἡ τῆς γλώττης κίνησις, ὥστε τῆς τροφῆς ἐν μὲν
τῷ στόματι λεαινομένης, παρ' αὐτὴν δὲ διιούσης,
35 τὴν μὲν ὀλιγάκις ὑπὸ τοὺς ὀδόντας πίπτειν, εἰς δὲ
τὴν ἀρτηρίαν σπάνιόν τι παραρρεῖν.

665 a Οὐκ ἔχει δὲ τὰ λεχθέντα ζῷα τὴν ἐπιγλωττίδα
διὰ τὸ ξηρὰς εἶναι τὰς σάρκας αὐτῶν καὶ τὸ δέρμα
σκληρόν, ὥστ' οὐκ ἂν εὐκίνητον ἦν τὸ τοιοῦτον
μόριον αὐτοῖς ἐκ τοιαύτης σαρκὸς καὶ ἐκ τοιούτου
δέρματος συνεστηκός, ἀλλ' αὐτῆς τῆς ἀρτηρίας
5 τῶν ἐσχάτων θᾶσσον ἐγίνετ' ἂν ἡ συναγωγὴ τῆς ἐκ
τῆς οἰκείας σαρκὸς ἐπιγλωττίδος, ἣν ἔχουσι τὰ
τριχωτά.

  Δι' ἣν μὲν οὖν αἰτίαν τὰ μὲν ἔχει τῶν ζῴων τὰ δ'
οὐκ ἔχει, ταῦτ' εἰρήσθω, καὶ διότι τῆς ἀρτηρίας τὴν
φαυλότητα τῆς θέσεως ἰάτρευκεν ἡ φύσις, μηχανη-
σαμένη τὴν καλουμένην ἐπιγλωττίδα. κεῖται δ'
10 ἔμπροσθεν ἡ φάρυγξ τοῦ οἰσοφάγου ἐξ ἀνάγκης. ἡ
μὲν γὰρ καρδία ἐν τοῖς ἔμπροσθεν καὶ ἐν μέσῳ
κεῖται, ἐν ᾗ τὴν ἀρχήν φαμεν τῆς ζωῆς καὶ πάσης
κινήσεώς τε καὶ αἰσθήσεως (ἐπὶ τὸ καλούμενον γὰρ
ἔμπροσθεν ἡ αἴσθησις καὶ ἡ κίνησις· αὐτῷ γὰρ τῷ
15 λόγῳ τούτῳ διώρισται τὸ ἔμπροσθεν καὶ ὄπισθεν),
ὁ δὲ πλεύμων κεῖται οὗ ἡ καρδία καὶ περὶ ταύτην,
ἡ δ' ἀναπνοὴ διά τε τοῦτον[2] καὶ διὰ τὴν ἀρχὴν τὴν
ἐν τῇ καρδίᾳ ἐνυπάρχουσαν. ἡ δ' ἀναπνοὴ γίνεται
τοῖς ζῴοις διὰ τῆς ἀρτηρίας· ὥστ' ἐπεὶ τὴν καρδίαν

[1] πρὸς PZ: παρὰ vulg.

anything coming in by mistake into the windpipe. If there is any error in this movement, or if you breathe in while you are taking food, coughing and choking results, as I have said. But the movement of the epiglottis and of the tongue has been so neatly contrived that while the food is being masticated in the mouth and is passing over the epiglottis, the tongue seldom gets in the way of the teeth, and hardly ever does any food slip into the windpipe.

I mentioned some animals that have no epiglottis. This is because their flesh is dry and their skin hard; and thus if they had one, it would not move easily, because it would have to be made out of constituents of this sort. It is quicker to contract the edges of the windpipe itself than it would be to close an epiglottis, if, as in the hairy creatures, it were made out of the same sort of flesh as the rest of their bodies.

This will suffice to show why some animals have an epiglottis and some not; how Nature has contrived it so as to remedy the unsatisfactory position of the windpipe in front of the oesophagus. Still, the windpipe is bound by necessity to be in this position for the following reason. The heart is situated in the middle of the body and in the fore part of it; and in the heart, we hold, is the principle of life and of all movement and sensation. Both of these activities take place in the direction we call forwards: that is the very principle which constitutes the distinction between before and behind. The lung is situated in the region of the heart, and surrounding it. Now breathing takes place for the sake of the lung and the principle which is situated in the heart: and the breath passes through the windpipe. So, since the

---

² τοῦτον SUY : τοῦτο vulg.

665 a

ἐν τοῖς ἔμπροσθεν πρώτην ἀναγκαῖον κεῖσθαι, καὶ
20 τὸν φάρυγγα καὶ τὴν ἀρτηρίαν πρότερον ἀναγκαῖον
κεῖσθαι τοῦ οἰσοφάγου· τὰ μὲν γὰρ πρὸς τὸν
πλεύμονα τείνει καὶ τὴν καρδίαν, ὁ δ᾽ εἰς τὴν
κοιλίαν. ὅλως δ᾽ ἀεὶ τὸ βέλτιον καὶ τιμιώτερον,
ὅπου μηδὲν μεῖζον ἕτερον ἐμποδίζει, τοῦ μὲν
ἄνω καὶ κάτω ἐν τοῖς μᾶλλόν ἐστιν ἄνω, τοῦ δ᾽
25 ἔμπροσθεν καὶ ὄπισθεν ἐν τοῖς ἔμπροσθεν, τοῦ
δεξιοῦ δὲ καὶ ἀριστεροῦ ἐν τοῖς δεξιοῖς.

Καὶ περὶ μὲν αὐχένος τε καὶ οἰσοφάγου καὶ
ἀρτηρίας εἴρηται, ἑπόμενον δ᾽ ἐστὶ περὶ σπλάγχνων
εἰπεῖν.

IV. Ταῦτα δ᾽ ἐστὶν ἴδια τῶν ἐναίμων, καὶ τοῖς
30 μὲν ἅπανθ᾽ ὑπάρχει, τοῖς δ᾽ οὐχ ὑπάρχει. τῶν δ᾽
ἀναίμων οὐδὲν ἔχει σπλάγχνον. Δημόκριτος δ᾽
ἔοικεν οὐ καλῶς διαλαβεῖν περὶ αὐτῶν, εἴπερ ᾠήθη
διὰ μικρότητα τῶν ἀναίμων ζῴων ἄδηλα εἶναι
ταῦτα. συνισταμένων γὰρ εὐθέως τῶν ἐναίμων καὶ
πάμπαν ὄντων μικρῶν ἔνδηλα γίνεται καρδία τε καὶ
35 ἧπαρ· φαίνεται γὰρ ἐν μὲν τοῖς ᾠοῖς ἐνίοτε τριταίοις
665 b οὖσι στιγμῆς ἔχοντα μέγεθος, πάμμικρα δὲ καὶ ἐν
τοῖς ἐκβολίμοις τῶν ἐμβρύων. ἔτι δ᾽ ὥσπερ τῶν ἐκτὸς μορίων οὐ πᾶσι τῶν αὐτῶν χρῆσις, ἀλλ᾽ ἑκάστοις ἰδίᾳ πεπόρισται πρός τε τοὺς βίους καὶ τὰς
5 κινήσεις, οὕτω καὶ τὰ ἐντὸς ἄλλα πέφυκεν ἄλλοις.

Τὰ δὲ σπλάγχνα τῶν αἱματικῶν ἐστιν ἴδια, διὸ
καὶ συνέστηκεν αὐτῶν ἕκαστον ἐξ αἱματικῆς ὕλης.
δῆλον δ᾽ ἐν τοῖς νεογνοῖς τούτων· αἱματωδέστερα
γὰρ καὶ μέγιστα κατὰ λόγον διὰ τὸ εἶναι τὸ εἶδος

ᵃ Limited by Aristotle to blood-like viscera only.

heart must of necessity be situated in the front place of all, both the larynx and the windpipe, which lead to the lung and the heart, must of necessity be situated in front of the oesophagus which leads merely to the stomach. Speaking generally, unless some greater object interferes, that which is better and more honourable tends to be above rather than below, in front rather than at the back, and on the right side rather than on the left.

We have now spoken of the neck, the oesophagus, and the windpipe, and our next topic is the viscera.

IV. Only blooded animals have viscera.[a] Some, but not all, have a complete set of them. As no bloodless animals have them, Democritus must have been wrong in his ideas on this point, if he really supposed that the viscera in bloodless creatures are invisible owing to the smallness of the creatures themselves. Against this we can put the fact that the heart and the liver are visible in blooded animals as soon as they are formed at all, that is, when they are quite small: in eggs they are visible, just about the size of a point, sometimes as early as the third day, and very small ones are visible in aborted embryos. Further, just as each animal is equipped with those external parts which are necessary to it for its manner of life and its motion, and no two animals require exactly the same ones, so it is with the internal parts : they vary in the various animals.

Viscera, then, are peculiar to the blooded animals, and that is why each one of the viscera is formed of blood-like material. This is clearly to be seen in the new-born offspring of blooded animals ; in them the viscera are more blood-like, and at their largest in

τῆς ὕλης καὶ τὸ πλῆθος ἐμφανέστατον κατὰ τὴν
10 πρώτην σύστασιν. καρδία μὲν οὖν ἅπασιν ὑπάρχει
τοῖς ἐναίμοις· δι' ἣν δ' αἰτίαν, εἴρηται καὶ πρότερον.
αἷμα μὲν γὰρ ἔχειν τοῖς ἐναίμοις δῆλον ὡς ἀναγκαῖον, ὑγροῦ δ' ὄντος τοῦ αἵματος ἀναγκαῖον ἀγγεῖον ὑπάρχειν, ἐφ' ὃ δὴ καὶ φαίνεται μεμηχανῆσθαι
τὰς φλέβας ἡ φύσις· ἀρχὴν δὲ τούτων ἀναγκαῖον
15 εἶναι μίαν (ὅπου γὰρ ἐνδέχεται, μίαν βέλτιον ἢ
πολλάς), ἡ δὲ καρδία τῶν φλεβῶν ἀρχή· φαίνονται
γὰρ ἐκ ταύτης οὖσαι[1] καὶ οὐ διὰ ταύτης, καὶ ἡ
φύσις αὐτῆς φλεβώδης ὡς ὁμογενοῦς οὔσης. ἔχει
δὲ καὶ ἡ θέσις αὐτῆς ἀρχικὴν χώραν· περὶ μέσον
γάρ, μᾶλλον δ' ἐν τῷ ἄνω ἢ κάτω καὶ ἔμπροσθεν ἢ
20 ὄπισθεν· ἐν τοῖς γὰρ τιμιωτέροις τὸ τιμιώτερον
καθίδρυκεν ἡ φύσις, οὗ μή τι κωλύει μεῖζον. ἐμφανέστατον δὲ τὸ λεχθέν ἐστιν ἐπὶ τῶν ἀνθρώπων, βούλεται δὲ καὶ ἐν τοῖς ἄλλοις ὁμολόγως ἐν
μέσῳ κεῖσθαι τοῦ ἀναγκαίου σώματος, τούτου δὲ
πέρας ᾗ τὰ περιττώματα ἀποκρίνεται· τὰ δὲ κῶλα
25 πέφυκεν ἄλλοις ἄλλως, καὶ οὐκ ἔστι τῶν πρὸς
τὸ ζῆν ἀναγκαίων, διὸ καὶ ἀφαιρουμένων ζῶσιν·
δῆλον δ' ὡς οὐδὲ προστιθέμενα φθείρει.
    Οἱ δ' ἐν τῇ κεφαλῇ λέγοντες τὴν ἀρχὴν τῶν
φλεβῶν οὐκ ὀρθῶς ὑπέλαβον. πρῶτον μὲν γὰρ
πολλὰς ἀρχὰς καὶ διεσπασμένας[2] ποιοῦσιν, εἶτ' ἐν

---

[1] ἰοῦσαι Z.   [2] διεσπαρμένας ESUYZ.

[a] The first observer after Aristotle to realize the disparity
in the relative sizes of the parts with time was Leonardo da
Vinci (A.D. 1452–1518).

proportion[a] : this is because the nature of the material and its bulk are especially obvious at the first stage of a creature's formation. The heart is present in all blooded animals, and the reason for this has been already stated : It is obviously necessary for all blooded creatures to have blood, and as blood is a fluid, there must of necessity be a vessel to hold it, and it is evidently for this purpose that Nature has contrived the blood-vessels. And these blood-vessels must have a source—one source (one is always better than many where it is possible), and this source is the heart. This is certain, because the blood-vessels come out of the heart and do not pass through it ; and again, the heart is homogeneous and in character identical with the blood-vessels. Further-more, the place in which it is set is the place of primacy and governance. It is in a central position, and rather in the upper part of the body than the lower, and in front rather than at the back ; Nature always gives the more honourable place to the more honourable part, unless something more important prevents it. What I have just said is seen most clearly in the case of man, yet in other animals the heart tends in a similar way to be in the centre of the " necessary body," *i.e.* the portion of it which is terminated by the vent where the residues are discharged. The limbs vary in the various animals, and cannot be reckoned among the parts that are " necessary " for life, which is why animals can lose them and still remain alive ; and obviously they could have limbs added to them without being killed.

Those who suppose that the source of the blood-vessels is in the head are wrong, because : (1) this involves holding that there are many sources,

665 b
30 τόπῳ ψυχρῷ. δηλοῖ δὲ δύσριγος ὤν, ὁ δὲ περὶ τὴν
καρδίαν τοὐναντίον. ὥσπερ δ' ἐλέχθη, διὰ μὲν τῶν
ἄλλων σπλάγχνων διέχουσιν αἱ φλέβες, διὰ δὲ τῆς
καρδίας οὐ διατείνει φλέψ· ὅθεν καὶ δῆλον ὅτι
μόριον καὶ ἀρχὴ τῶν φλεβῶν ἐστὶν ἡ καρδία. καὶ
τοῦτ' εὐλόγως· μέσον γὰρ τὸ τῆς καρδίας ἐστὶ
35 σῶμα πυκνὸν καὶ κοῖλον πεφυκός, ἔτι δὲ πλῆρες
666 a αἵματος ὡς τῶν φλεβῶν ἐντεῦθεν ἠργμένων, κοῖλον
μὲν πρὸς τὴν ὑποδοχὴν τοῦ αἵματος, πυκνὸν δὲ
πρὸς τὸ φυλάσσειν τὴν ἀρχὴν τῆς θερμότητος. ἐν
ταύτῃ γὰρ μόνῃ τῶν σπλάγχνων καὶ τοῦ σώματος
5 αἷμα ἄνευ φλεβῶν ἐστι, τῶν δ' ἄλλων μορίων
ἕκαστον ἐν ταῖς φλεψὶν ἔχει τὸ αἷμα. καὶ τοῦτ' εὐ-
λόγως· ἐκ τῆς καρδίας γὰρ ἐποχετεύεται [καὶ]¹ εἰς
τὰς φλέβας, εἰς δὲ τὴν καρδίαν οὐκ ἄλλοθεν· αὕτη
γάρ ἐστιν ἀρχὴ καὶ πηγὴ τοῦ αἵματος ἢ ὑποδοχὴ
πρώτη. ἐκ τῶν ἀνατομῶν δὲ κατάδηλα μᾶλλον
10 ταῦτα, καὶ ἐκ τῶν γενέσεων· εὐθέως γάρ ἐστιν
ἔναιμος πρώτη γινομένη τῶν μορίων ἁπάντων. ἔτι
δ' αἱ κινήσεις τῶν ἡδέων καὶ τῶν λυπηρῶν καὶ
ὅλως πάσης αἰσθήσεως ἐντεῦθεν ἀρχόμεναι φαίνονται καὶ πρὸς ταύτην περαίνουσαι. οὕτω δ' ἔχει
καὶ κατὰ τὸν λόγον, ἀρχὴν γὰρ εἶναι δεῖ μίαν, ὅπου
15 ἐνδέχεται· εὐφυέστατος δὲ τῶν τόπων ὁ μέσος, ἐν
γὰρ τὸ μέσον καὶ ἐπὶ πᾶν ἐφικτὸν ὁμοίως ἢ παραπλησίως. ἔτι δ' ἐπεὶ οὔτε τῶν ἀναίμων οὐθὲν

¹ καὶ om. Z.

ᵃ Or "traverse." The connotation of this term seems to vary.

236

scattered about ; and (2) it involves placing them in a cold region (its intolerance of cold proves this). The region round the heart, on the other hand, is warm. And (3) as has been said already, the blood-vessels run all through *a* the other viscera, whereas none passes through the heart ; which clearly shows that the heart forms part of the blood-vessels and is their source. Which is reasonable enough ; since the centre of the heart is a body of dense and hollow structure, and this is full of blood ; it is hollow to form a receptacle for the blood ; dense to guard the source of heat ; and the store of blood is obviously there because that is the starting-point of the blood-vessels. In none other of the viscera and in no other part of the body is there blood and yet no blood-vessels ; in each of the other parts the blood is contained in blood-vessels. And this too is reasonable, as the blood is conveyed and conducted away from the heart into the blood-vessels, whereas none is thus conveyed into the heart from elsewhere, for the heart is itself the source and spring of the blood, or the first receptacle of it. All this, however, is more clearly brought out in *Dissections* and *Formative Processes*, where it is shown that the heart is the first of all the parts to be formed and has blood in it straightway. Further, all motions of sensation, including those produced by what is pleasant and painful, undoubtedly begin in the heart and have their final ending there. This is in accord with reason ; since, wherever possible, there must be one source only ; and the best situation for that is the centre, because there is only one centre, and the centre is equally (or nearly equally) accessible from every direction. Again, as every bloodless part, and the

666 a

αἰσθητικὸν οὔτε τὸ αἷμα, δῆλον ὡς τὸ πρῶτον ἔχον
ὡς ἐν ἀγγείῳ δ' ἔχον ἀναγκαῖον εἶναι τὴν ἀρχήν.
   Οὐ μόνον δὲ κατὰ τὸν λόγον οὕτως ἔχειν φαίνεται,
20 ἀλλὰ καὶ κατὰ τὴν αἴσθησιν. ἐν γὰρ τοῖς ἐμβρύοις
εὐθέως ἡ καρδία φαίνεται κινουμένη τῶν μορίων
καθάπερ εἰ ζῷον, ὡς ἀρχὴ τῆς φύσεως τοῖς ἐναίμοις
οὖσα. μαρτύριον δὲ τῶν εἰρημένων καὶ τὸ πᾶσι
τοῖς ἐναίμοις ὑπάρχειν αὐτήν· ἀναγκαῖον γὰρ αὐτοῖς
25 ἔχειν τὴν ἀρχὴν τοῦ αἵματος. ὑπάρχει δὲ καὶ τὸ
ἧπαρ πᾶσι τοῖς ἐναίμοις· ἀλλ' οὐθεὶς ἂν ἀξιώσειεν
αὐτὸ ἀρχὴν εἶναι οὔτε τοῦ ὅλου σώματος οὔτε τοῦ
αἵματος· κεῖται γὰρ οὐδαμῶς πρὸς ἀρχοειδῆ θέσιν,
ἔχει δ' ὥσπερ ἀντίζυγον ἐν τοῖς μάλιστ' ἀπηκριβω-
μένοις τὸν σπλῆνα. ἔτι δ' ὑποδοχὴν αἵματος οὐκ
30 ἔχει ἐν ἑαυτῷ καθάπερ ἡ καρδία, ἀλλ' ὥσπερ τὰ
λοιπὰ ἐν φλεβί. ἔτι δὲ τείνει δι' αὐτοῦ φλέψ, δι'[1]
ἐκείνης δ' οὐδεμία· πασῶν γὰρ τῶν φλεβῶν ἐκ τῆς
καρδίας αἱ ἀρχαί. ἐπεὶ οὖν ἀνάγκη μὲν θάτερον
τούτων ἀρχὴν εἶναι, μή ἐστι δὲ τὸ ἧπαρ, ἀνάγκη
τὴν καρδίαν εἶναι καὶ τοῦ αἵματος ἀρχήν. τὸ μὲν
35 γὰρ ζῷον αἰσθήσει ὥρισται, αἰσθητικὸν δὲ πρῶτον
τὸ πρῶτον ἔναιμον, τοιοῦτον δ' ἡ καρδία· καὶ γὰρ
666 b ἀρχὴ τοῦ αἵματος καὶ ἔναιμον πρῶτον.

   Ἔστι δ' αὐτῆς τὸ ἄκρον ὀξὺ καὶ στερεώτερον,
      [1] δι' Th.: ἐξ vulg.; mox ἐκείνου EUYZ.

---

[a] *Cor primum vivens ultimum moriens* : cf. *De gen. an.*
741 b 15 ff., and Ebstein & al., *Mitt. z. Gesch. der Medizin u.
Naturw.*, 1920, 19, 102, 219, 305.   [b] See 655 b 29, n.

blood itself as well, is without sensation, it is clear that the part where the blood is present first, *and* which holds it as in a receptacle, must of necessity be the source.

This reasoning is supported by the evidence of the senses. In embryos, as soon as they are formed, the heart can be seen moving before any of the other parts, just like a living creature [a]; which shows that it is the source of their nature in all blooded animals. Another piece of evidence to support this is that all blooded creatures have a heart : why ? because they are bound to have a source for their blood. All blooded creatures, it is true, have a liver too ; but no one would care to maintain that the liver is the source either of the blood or of the whole body, because it is nowhere near the place of primacy and governance, and, also, in the most highly finished [b] animals it has something to counterbalance it, as it were, viz. the spleen. Again, the liver has no receptacle for blood in itself as the heart has : like the rest of the viscera, it keeps its blood in a blood-vessel. Again, a blood-vessel runs all through it, whereas no blood-vessel runs through the heart : all blood-vessels have their source from the heart and begin there. Since, therefore, of necessity the source must be one of these two, the heart or the liver, and as it is not the liver, it must of necessity be the heart which is the source of the blood just as it is of the rest. An animal is defined by the fact that it possesses sensation : and the part of the body to have sensation first is the part that has blood in it first—in other words, the heart, which is the source of the blood and the first part to have it.

The apex of the heart is sharp and more solid than

κεῖται δὲ πρὸς τῷ στήθει καὶ ὅλως ἐν τοῖς πρόσθεν
τοῦ σώματος πρὸς τὸ μὴ καταψύχεσθαι αὐτό· πᾶσι
5 γὰρ ἀσαρκότερον τὸ στῆθος, τὰ δὲ πρανῆ σαρκω-
δέστερα, διὸ πολλὴν ἔχει σκέπην τὸ θερμὸν κατὰ
τὸν νῶτον. ἔστι δ' ἡ καρδία τοῖς μὲν ἄλλοις ζῴοις
κατὰ μέσον τοῦ στηθικοῦ τόπου, τοῖς δ' ἀνθρώποις
μικρὸν εἰς τὰ εὐώνυμα παρεκκλίνουσα πρὸς τὸ
ἀνισοῦν τὴν κατάψυξιν τῶν ἀριστερῶν· μάλιστα γὰρ
10 τῶν ἄλλων ζῴων ἄνθρωπος ἔχει κατεψυγμένα τὰ
ἀριστερά. ὅτι δὲ καὶ ἐν τοῖς ἰχθύσιν ὁμοίως ἡ
καρδία κεῖται, πρότερον εἴρηται, καὶ διότι φαίνεται
ἀνομοίως. ἔχει δὲ πρὸς τὴν κεφαλὴν τὸ ὀξύ· ἔστι
δ' αὕτη τὸ πρόσθεν, ἐπὶ ταύτην γὰρ ἡ κίνησις.

Ἔχει δὲ καὶ νεύρων πλῆθος ἡ καρδία, καὶ τοῦτ'
15 εὐλόγως· ἀπὸ ταύτης γὰρ αἱ κινήσεις, περαίνονται
δὲ διὰ τοῦ ἕλκειν καὶ ἀνιέναι· δεῖ οὖν τοιαύτης
ὑπηρεσίας καὶ ἰσχύος. ἡ δὲ καρδία, καθάπερ
εἴπομεν καὶ πρότερον, οἷον ζῷόν τι πέφυκεν ἐν
τοῖς ἔχουσιν.

Ἔστι δ' ἀνόστεος πάντων ὅσα καὶ ἡμεῖς τεθεά-
μεθα, πλὴν τῶν ἵππων καὶ γένους τινὸς βοῶν·
20 τούτοις δὲ διὰ τὸ μέγεθος οἷον ἐρείσματος χάριν
ὀστοῦν ὕπεστι, καθάπερ καὶ τοῖς ὅλοις σώμασιν.

Κοιλίας δ' ἔχουσιν αἱ μὲν τῶν μεγάλων ζῴων
τρεῖς, αἱ δὲ τῶν ἐλασσόνων δύο, μίαν δὲ πᾶσαι· δι'
ἣν δ' αἰτίαν, εἴρηται. δεῖ γὰρ εἶναι τόπον τινὰ τῆς

ᵃ At *De respir.* 478 b 3. And see the next note.
ᵇ Instead of towards the breast. The meaning of this
passage is made clear by *Hist. An.* 507 a 2 ff. In all animals,
says Aristotle, the " apex " of the heart points forwards, and
in most animals " forwards " is towards the breast. Fishes
240

the rest, and it lies towards the breast, and altogether in the fore part of the body so as to prevent it from getting cooled : for in all animals the breast has comparatively little flesh on it, while the back is well supplied and so gives the heat of the body ample protection on that side. In animals other than man the heart is in the centre of the region of the breast ; in man it inclines slightly to the left side so as to counteract the cooling there, for in man the left side is much colder than in other creatures. I have said already that the placing of the heart is the same in fishes as in other animals, though it appears to be different, together with the reasons *a* for the apparent difference. In fishes its apex is turned towards the head *b* ; but in them the head is "forwards," because the head is in the line of direction in which they move.

The heart has in it an abundance of sinews, which is reasonable enough, as the motions of the body have their origin there ; and as these are performed by contraction and relaxation, the heart needs the sinews to serve it and to give it strength. We have said already that the heart is like a living creature inside the body that contains it.

In all cases that we have examined the heart is boneless, except in horses and a certain kind of ox. In these, owing to its great size, the heart has a bone for a support, just as the whole body is supported by bones.

In the large animals, the heart has three cavities, in the smaller ones, two only ; and in no species has it less than one. The reason for this has been given : there

appear to be an exception to this rule, but only because in them " forwards " is towards the head.

666 b

καρδίας καὶ ὑποδοχὴν τοῦ πρώτου αἵματος. (ὅτι
25 δὲ πρῶτον ἐν τῇ καρδίᾳ γίνεται τὸ αἷμα, πολλάκις
εἰρήκαμεν.) διὰ δὲ¹ τὸ τὰς ἀρχηγοὺς φλέβας δύο
εἶναι, τήν τε μεγάλην καλουμένην καὶ τὴν ἀορτήν,
ἑκατέρας δ᾽² οὔσης ἀρχῆς τῶν φλεβῶν, καὶ διαφορὰς ἐχουσῶν, περὶ ὧν ὕστερον ἐροῦμεν, βέλτιον
καὶ τὰς ἀρχὰς αὐτῶν κεχωρίσθαι· τοῦτο δ᾽ ἂν εἴη
30 διφυοῦς ὄντος τοῦ αἵματος καὶ κεχωρισμένου.
διόπερ ἐν οἷς ἐνδέχεται, δύ᾽ εἰσὶν ὑποδοχαί. ἐνδέχεται δ᾽ ἐν τοῖς μεγάλοις· τούτων γὰρ ἔχουσι καὶ
αἱ καρδίαι μέγεθος. ἔτι δὲ βέλτιον τρεῖς εἶναι τὰς
κοιλίας, ὅπως ᾖ μία ἀρχὴ κοινή· τὸ δὲ μέσον καὶ
περιττὸν ἀρχή· ὥστε μεγέθους δεῖ μείζονος αὐταῖς
35 ἀεί, διόπερ αἱ μέγισται τρεῖς ἔχουσι μόναι.

667 a

Τούτων δὲ πλεῖστον μὲν αἷμα καὶ θερμότατον
ἔχουσιν αἱ δεξιαί (διὸ καὶ τῶν μερῶν θερμότερα τὰ
δεξιά), ἐλάχιστον δὲ καὶ ψυχρότερον αἱ ἀριστεραί,
μέσον δ᾽ αἱ μέσαι τῷ πλήθει καὶ θερμότητι, καθα-
ρώτατον δέ· δεῖ γὰρ τὴν ἀρχὴν ὅτι μάλιστ᾽ ἠρεμεῖν,
5 τοιαύτη δ᾽ ἂν εἴη καθαροῦ τοῦ αἵματος ὄντος, τῷ
πλήθει δὲ καὶ θερμότητι μέσου.

Ἔχουσι δὲ καὶ διάρθρωσίν τινα αἱ καρδίαι παραπλησίαν ταῖς ῥαφαῖς. οὐκ εἰσὶ δὲ συναφεῖς ὡς
τινος ἐκ πλειόνων συνθέτου, ἀλλὰ καθάπερ εἴπομεν,
διαρθρώσει μᾶλλον. εἰσὶ δὲ τῶν μὲν αἰσθητικῶν
10 ἀρθρωδέστεραι, τῶν δὲ νωθροτέρων ἀναρθρότεραι,

¹ διὰ δὲ ESUYZ: διὰ vulg.
² δ᾽ Peck : γὰρ vulg., om. Ogle.

must be some place in the heart which will be a receptacle for the blood when first formed. (As we have stated several times, blood is first formed in the heart.) Now there are two chief blood-vessels, the so-called Great Blood-vessel, and the Aorta; each of these is the source of other blood-vessels; and the two differ from each other (this will be discussed later); hence it is better for them to have separate sources. This result can be obtained by having two separate supplies of blood, and thus we find two receptacles wherever this is possible, as in the larger animals which of course have large hearts. But it is better still to have three cavities, and then there is an odd one in the middle which can be a common source for the other two; since, however, this requires the heart to be particularly large, only the very largest hearts have three cavities.

Of these cavities it is the right-hand one which contains the most blood and the hottest (that is why the right side of the body is hotter than the left); the left-hand cavity contains least blood, and it is colder. The blood in the middle cavity is intermediate both in amount and heat, although it is the purest of them all; this is because the source must remain as calm as possible, and this is secured when the blood is pure, and intermediate in its amount and heat.

The heart has also a sort of articulation, which resembles the sutures of the skull. By this I do not mean to say that the heart is a composite thing, consisting of several parts joined together, but an articulated whole, as I said. This articulation is more distinct in animals whose sensation is keen, and less distinct in the duller ones, such as swine. There are

667 a

καθάπερ αἱ τῶν ὑῶν. αἱ δὲ διαφοραὶ τῆς καρδίας
κατὰ μέγεθός τε καὶ μικρότητα καὶ σκληρότητά τε
καὶ μαλακότητα τείνουσί πη καὶ πρὸς τὰ ἤθη· τὰ
μὲν γὰρ ἀναίσθητα σκληρὰν ἔχει τὴν καρδίαν καὶ
15 πυκνήν, τὰ δ' αἰσθητικὰ μαλακωτέραν, καὶ τὰ μὲν
μεγάλας ἔχοντα τὰς καρδίας δειλά, τὰ δ' ἐλάσσους
καὶ μέσας θαρραλεώτερα (τὸ γὰρ συμβαῖνον πάθος
ὑπὸ τοῦ φοβεῖσθαι προϋπάρχει τούτοις διὰ τὸ μὴ
ἀνάλογον ἔχειν τὸ θερμὸν τῇ καρδίᾳ, μικρὸν δ' ὂν
ἐν μεγάλοις ἀμαυροῦσθαι, καὶ τὸ αἷμα ψυχρότερον
20 εἶναι). μεγάλας δὲ τὰς καρδίας ἔχουσι λαγώς,
ἔλαφος, μῦς, ὕαινα, ὄνος, πάρδαλις,[1] γαλῆ, καὶ
τἆλλα σχεδὸν πάνθ' ὅσα φανερῶς δειλὰ ἢ διὰ
φόβον κακοῦργα.

Παραπλησίως δὲ καὶ ἐπὶ τῶν φλεβῶν καὶ ἐπὶ
τῶν κοιλιῶν ἔχει· ψυχραὶ γὰρ αἱ μεγάλαι φλέβες
25 καὶ κοιλίαι. ὥσπερ γὰρ ἐν μικρῷ καὶ ἐν μεγάλῳ
οἰκήματι τὸ ἴσον πῦρ ἧσσον ἐν τοῖς μείζοσι θερμαίνει, οὕτω κἂν τούτοις τὸ θερμόν· ἀγγεῖα γὰρ
καὶ ἡ φλὲψ καὶ ἡ κοιλία. ἔτι δ' αἱ ἀλλότριαι κινήσεις ἕκαστον τῶν θερμῶν καταψύχουσιν, ἐν δὲ ταῖς
εὐρυχωρεστέραις τὸ πνεῦμα πλεῖον καὶ ἐνισχύει
30 μᾶλλον· διὸ τῶν μεγαλοκοιλίων οὐδὲν οὐδὲ τῶν
μεγαλοφλέβων πῖόν ἐστι κατὰ σάρκα, ἀλλὰ πάντα
ἢ τὰ πλεῖστα τῶν τοιούτων ἀδηλόφλεβα καὶ μικρο-
κοίλια φαίνεται.

Μόνον δὲ τῶν σπλάγχνων καὶ ὅλως τῶν ἐν τῷ

[1] πάρδαλις] δορκαλίς Platt.

other differences in the heart; some hearts are large, some small, some are hard, some soft; and these tend by some means to influence the creature's temperament. Illustrations of this are: animals whose powers of sensation are small have hearts that are hard and dense, those whose sensation is keen have softer ones; and those with large hearts are cowardly, those with small or moderate-sized ones, courageous (this is because in the former class the affection which is normally produced by fear is present to begin with,[a] as their heat is not proportionate to the size of their heart, but is small and therefore hardly noticeable in the enormous space that it occupies; so that their blood is comparatively cold). The following creatures have large hearts: the hare, the deer, the mouse, the hyena, the ass, the leopard, the marten, and practically all other animals whose cowardice is either outright or else betrayed by their mischievous behaviour.

Similar conditions obtain in the blood-vessels and the cavities of the heart: if they are large, they are cold. The effect of the same-sized fire is less in a large room than in a small one; and the same applies to the heat in these receptacles, the blood-vessels and the cavities. Further, extraneous motions have a cooling effect upon hot things; and the more roomy a receptacle is, the greater the amount of air (or *pneuma*) in it and the stronger its effect. Thus we find that no animal which has large cavities or large blood-vessels has fat flesh, and conversely, that all (or most) fat animals have indistinguishable blood-vessels and small cavities.

The heart is the only one of the viscera—indeed

---

[a] *Cf.* 650 b 27. See also 692 a 20.

667 a

σώματι μορίων ἡ καρδία χαλεπὸν πάθος οὐδὲν
ὑποφέρει, καὶ τοῦτ' εὐλόγως· φθειρομένης γὰρ τῆς
35 ἀρχῆς οὐκ ἔστιν ἐξ οὗ γένοιτ' ἂν βοήθεια τοῖς
667 b ἄλλοις ἐκ ταύτης ἠρτημένοις. σημεῖον δὲ τοῦ
μηθὲν ἐπιδέχεσθαι πάθος τὴν καρδίαν τὸ ἐν μηδενὶ
τῶν θυομένων ἱερείων ὦφθαι τοιοῦτον πάθος περὶ
αὐτὴν ὥσπερ ἐπὶ τῶν ἄλλων σπλάγχνων. οἵ τε
5 γὰρ νεφροὶ πολλάκις φαίνονται λίθων μεστοὶ καὶ
φυμάτων καὶ δοθιήνων καὶ τὸ ἧπαρ, ὡσαύτως δὲ
καὶ ὁ πλεύμων, μάλιστα δ' ὁ σπλήν. πολλὰ δὲ καὶ
ἕτερα παθήματα συμβαίνοντα περὶ αὐτὰ φαίνεται,
ἥκιστα δὲ τοῦ μὲν πλεύμονος περὶ τὴν ἀρτηρίαν,
τοῦ δ' ἥπατος περὶ τὴν σύναψιν τῇ μεγάλῃ φλεβί,
10 καὶ τοῦτ' εὐλόγως· ταύτῃ γὰρ μάλιστα κοινωνοῦσι
τῇ καρδίᾳ. ὅσα δὲ διὰ νόσον καὶ τοιαῦτα πάθη
φαίνεται τελευτῶντα τῶν ζῴων, τούτοις ἀνατεμνο-
μένοις φαίνεται περὶ τὴν καρδίαν νοσώδη πάθη.

Καὶ περὶ μὲν τῆς καρδίας, ποία τις, καὶ τίνος
ἕνεκεν καὶ διὰ τίν' αἰτίαν ὑπάρχει τοῖς ἔχουσιν,
τοσαῦτ' εἰρήσθω.

15 V. Ἑπόμενον δ' ἂν εἴη περὶ τῶν φλεβῶν εἰπεῖν,
τῆς τε μεγάλης καὶ τῆς ἀορτῆς· αὗται γὰρ ἐκ τῆς
καρδίας πρῶται δέχονται τὸ αἷμα, αἱ δὲ λοιπαὶ
τούτων ἀποφυάδες εἰσίν. ὅτι μὲν οὖν τοῦ αἵματος
χάριν εἰσί, πρότερον εἴρηται· τό τε γὰρ ὑγρὸν ἅπαν
20 ἀγγείου δεῖται, καὶ τὸ φλεβῶν γένος ἀγγεῖον, τὸ δ'

the only part in the whole body—which cannot withstand any serious affection. This is readily understood : the other parts depend upon the heart, and when this source itself is ailing, there is no place whence they can obtain succour. A proof that the heart cannot put up with any affection is this : Never has the heart in a sacrificial victim been observed to be affected in the way that the other viscera sometimes are. Very often the kidneys are found to be full of stones, growths, and small abscesses ; so is the liver, and the lung, and especially the spleen. Many other affections are observed in these organs ; but in the lung they occur least often in that portion which is nearest the windpipe, and in the liver in that portion which is nearest its junction with the Great Blood-vessel. This is readily understood : those are the places where they are most closely in communication with the heart. Those animals, however, which die as the result of disease, and affections such as I have mentioned, when cut open are seen to have diseased affections of the heart.

We have now spoken of the heart : we have said what its nature is, what purpose it serves, and why it is present ; and that will suffice.

V. I suppose that the next subject for us to discuss is the Blood-vessels, that is, the Great Blood-vessel and the Aorta. It is these into which the blood goes first after it leaves the heart, and the other blood-vessels are merely branches from these. We have already said that these blood-vessels are present for the sake of the blood : fluid substances always need a receptacle, and the blood-vessels generally are the receptacles which hold the blood. We may

Blood-vessels.

αἷμα ἐν ταύταις· διότι δὲ δύο καὶ ἀπὸ μιᾶς ἀρχῆς καθ᾽ ἅπαν τὸ σῶμα διατείνουσι, λέγωμεν.

Τοῦ μὲν οὖν εἰς μίαν ἀρχὴν συντελεῖν καὶ ἀπὸ μιᾶς αἴτιον τὸ μίαν ἔχειν πάντα τὴν αἰσθητικὴν ψυχὴν ἐνεργείᾳ, ὥστε καὶ τὸ μόριον ἓν τὸ ταύτην ἔχον πρώτως (ἐν μὲν τοῖς ἐναίμοις κατὰ δύναμιν
25 καὶ κατ᾽ ἐνέργειαν, τῶν δ᾽ ἀναίμων ἐνίοις κατ᾽ ἐνέργειαν μόνον), διὸ καὶ τὴν τοῦ θερμοῦ ἀρχὴν ἀναγκαῖον ἐν τῷ αὐτῷ τόπῳ εἶναι· αὕτη δ᾽ ἐστὶν αἰτία καὶ τῷ αἵματι τῆς ὑγρότητος καὶ τῆς θερμότητος. διὰ μὲν οὖν τὸ ἐν ἑνὶ εἶναι μορίῳ τὴν αἰσθητικὴν ἀρχὴν καὶ τὴν τῆς θερμότητος καὶ ἡ
30 τοῦ αἵματος ἀπὸ μιᾶς ἐστιν ἀρχῆς, διὰ δὲ τὴν τοῦ αἵματος ἑνότητα καὶ ἡ τῶν φλεβῶν ἀπὸ μιᾶς.

Δύο δ᾽ εἰσὶ διὰ τὸ τὰ σώματα εἶναι διμερῆ τῶν ἐναίμων καὶ πορευτικῶν· ἐν πᾶσι γὰρ τούτοις διώρισται τὸ ἔμπροσθεν καὶ τὸ ὄπισθεν καὶ τὸ δεξιὸν καὶ τὸ ἀριστερὸν καὶ τὸ ἄνω καὶ τὸ κάτω.
35 ὅσῳ δὲ τιμιώτερον καὶ ἡγεμονικώτερον τὸ ἔμ-
668 a προσθεν τοῦ ὄπισθεν, τοσούτῳ καὶ ἡ μεγάλη φλὲψ τῆς ἀορτῆς· ἡ μὲν γὰρ ἐν τοῖς ἔμπροσθεν, ἡ δ᾽ ἐν τοῖς ὄπισθεν κεῖται, καὶ τὴν μὲν ἅπαντ᾽ ἔχει τὰ ἔναιμα φανερῶς, τὴν δ᾽ ἔνια μὲν ἀμυδρῶς ἔνια δ᾽ ἀφανῶς.

Τοῦ δ᾽ εἰς τὸ πᾶν διαδεδόσθαι τὸ σῶμα τὰς
5 φλέβας αἴτιον τὸ παντὸς εἶναι τοῦ σώματος ὕλην τὸ αἷμα, τοῖς δ᾽ ἀναίμοις τὸ ἀνάλογον, ταῦτα δ᾽ ἐν

---

now go on to explain why there are two of these blood-vessels, why they begin from a single source, and why they extend all over the body.

The reason why finally they both coincide in one source and also begin from one source is this. The sensory Soul is, in all animals, one *actually*; therefore the part which primarily contains this Soul is also one (one *potentially* as well as *actually* in the blooded animals, but in some of the bloodless animals it is only *actually* one [a]), and for this reason the source of heat also must of necessity be in the selfsame place. But this concerns the blood, for this source is the cause of the blood's heat and fluidity. Thus we see that because the source of sensation and the source of heat are in one and the same part, the blood must originate from one source too ; and because there is this one origin of the blood, the blood-vessels also must originate from one source.

The blood-vessels are, however, two in number, because the bodies of the blooded creatures that move about are bilateral : we can distinguish in all of them front and back, right and left, upper and lower. And just as the fore part is more honourable and more suited to rule than the back part, so is the Great Blood-vessel pre-eminent over the Aorta. The Great Blood-vessel lies in front, while the Aorta is at the back. All blooded creatures have a Great Blood-vessel, plainly visible ; but in some of them the Aorta is indistinct and in others it cannot be detected.

The reason why the blood-vessels are distributed all over the body is that blood (and in bloodless creatures, its counterpart) is the material out of which the whole body is constructed, and blood-vessels (and their counterparts) are the channels in

868 a

φλεβὶ καὶ τῷ ἀνάλογον κεῖσθαι. πῶς μὲν οὖν
τρέφεται τὰ ζῷα καὶ ἐκ τίνος καὶ τίνα τρόπον
ἀναλαμβάνουσιν ἐκ τῆς κοιλίας ἐν τοῖς περὶ γενέσεως λόγοις μᾶλλον ἁρμόζει σκοπεῖν καὶ λέγειν.
10 [Συνισταμένων δὲ τῶν μορίων ἐκ τοῦ αἵματος,
καθάπερ εἴπομεν, εὐλόγως ἡ τῶν φλεβῶν ῥύσις
διὰ παντὸς τοῦ σώματος πέφυκεν· δεῖ γὰρ καὶ τὸ
αἷμα διὰ παντὸς καὶ παρὰ πᾶν εἶναι, εἴπερ τῶν μορίων ἕκαστον ἐκ τούτου συνέστηκεν.][1]

Ἔοικε δ' ὥσπερ ἔν τε τοῖς κήποις αἱ ὑδραγωγίαι
15 κατασκευάζονται ἀπὸ μιᾶς ἀρχῆς καὶ πηγῆς εἰς
πολλοὺς ὀχετοὺς καὶ ἄλλους ἀεὶ πρὸς τὸ πάντῃ
μεταδιδόναι, καὶ ἐν ταῖς οἰκοδομίαις παρὰ πᾶσαν
τὴν τῶν θεμελίων ὑπογραφὴν λίθοι παραβέβληνται,
διὰ τὸ τὰ μὲν κηπευόμενα φύεσθαι ἐκ τοῦ ὕδατος,
τοὺς δὲ θεμελίους ἐκ τῶν λίθων οἰκοδομεῖσθαι, τὸν
20 αὐτὸν τρόπον καὶ ἡ φύσις τὸ αἷμα διὰ παντὸς
ὠχέτευκε τοῦ σώματος, ἐπειδὴ παντὸς ὕλη πέφυκε
τοῦτο. γίνεται δὲ κατάδηλον ἐν τοῖς μάλιστα κατα-
λελεπτυσμένοις· οὐθὲν γὰρ ἄλλο φαίνεται παρὰ τὰς
φλέβας, καθάπερ ἐπὶ τῶν ἀμπελίνων τε καὶ συκίνων
25 φύλλων καὶ ὅσ' ἄλλα τοιαῦτα· καὶ γὰρ τούτων
αὐαινομένων[2] φλέβες λείπονται μόνον. τούτων δ'
αἴτιον ὅτι τὸ αἷμα καὶ τὸ ἀνάλογον τούτῳ δυνάμει
σῶμα καὶ σὰρξ ἢ τὸ ἀνάλογόν ἐστιν· καθάπερ οὖν

---

[1] ll. 10-13, quae praecedentia ll. 4-7 repetunt, secludenda.
[2] αὐαινομένων attice Bekker.

---

[a] This seems to be an unnecessary repetition of the last
sentence but one.

250

which this material is carried. As regards the manner in which animals are nourished, the source of the nourishment, and the processes by which they take it up from the stomach, it is more appropriate to consider these subjects and to discuss them in the treatise on *Generation*.

[But since the parts of the body are composed out of blood, as has been said, it is easy to see why the course of the blood-vessels passes throughout the whole body. The blood must be everywhere in the body and everywhere at hand if every one of the parts is constructed out of it.] [a]

The system of blood-vessels in the body may be compared to those water-courses which are constructed in gardens : they start from one source, or spring, and branch off into numerous channels, and then into still more, and so on progressively, so as to carry a supply to every part of the garden. And again, when a house is being built, supplies of stones are placed all alongside the lines of the foundations. These things are done because (*a*) water is the material out of which the plants in the garden grow, and (*b*) stones are the material out of which the foundations are built. In the same way, Nature has provided for the irrigation of the whole body with blood, because blood is the material out of which it is all made. This becomes evident in cases of severe emaciation, when nothing is to be seen but the blood-vessels : just as the leaves of vines and fig-trees and similar plants, when they wither, leave behind nothing but the veins. The explanation of this is that the blood (or its counterpart) is, potentially, the body (that is, flesh—or its counterpart). Thus, just as in the irrigation system the

668 a

ἐν ταῖς ὀχετείαις αἱ μέγισται τῶν τάφρων διαμένουσιν, αἱ δ' ἐλάχισται πρῶται καὶ ταχέως ὑπὸ
τῆς ἰλύος ἀφανίζονται, πάλιν δ' ἐκλειπούσης
80 φανεραὶ γίνονται, τὸν αὐτὸν τρόπον καὶ τῶν φλεβῶν
αἱ μὲν μέγισται διαμένουσιν, αἱ δ' ἐλάχισται γίνονται σάρκες ἐνεργείᾳ, δυνάμει δ' εἰσὶν οὐδὲν
ἧσσον φλέβες. διὸ καὶ σωζομένων τῶν σαρκῶν
καθ' ὁτιοῦν αἷμα ῥεῖ διαιρουμένων· καίτοι ἄνευ μὲν
35 φλεβὸς οὐκ ἔστιν αἷμα, φλέβιον[1] δ' οὐδὲν δῆλον,
ὥσπερ οὐδ' ἐν τοῖς ὀχετοῖς αἱ τάφροι πρὶν ἢ τὴν
668 b ἰλὺν ἐξαιρεθῆναι.

Ἐκ μειζόνων δ' εἰς ἐλάσσους αἱ φλέβες ἀεὶ
προέρχονται ἕως τοῦ γενέσθαι τοὺς πόρους ἐλάσσους τῆς τοῦ αἵματος παχύτητος· δι' ὧν τῷ μὲν
αἵματι δίοδος οὐκ ἔστι, τῷ δὲ περιττώματι τῆς
ὑγρᾶς ἰκμάδος, ὃν καλοῦμεν ἱδρῶτα, καὶ τοῦτο
5 διαθερμανθέντος τοῦ σώματος καὶ τῶν φλεβίων
ἀναστομωθέντων. ἤδη δέ τισιν ἱδρῶσαι συνέβη
αἱματώδει περιττώματι διὰ καχεξίαν, τοῦ μὲν
σώματος ῥυάδος καὶ μανοῦ γενομένου, τοῦ δ' αἵματος ἐξυγρανθέντος δι' ἀπεψίαν, ἀδυνατούσης τῆς
ἐν τοῖς φλεβίοις θερμότητος πέσσειν δι' ὀλιγότητα.
10 (εἴρηται γὰρ ὅτι πᾶν τὸ κοινὸν γῆς καὶ ὕδατος
παχύνεται πεσσόμενον, ἡ δὲ τροφὴ καὶ τὸ αἷμ.
μικτὸν ἐξ ἀμφοῖν.) ἀδυνατεῖ δὲ πέσσειν ἡ θερμότης
οὐ μόνον διὰ τὴν αὑτῆς ὀλιγότητα ἀλλὰ καὶ διὰ
πλῆθος καὶ ὑπερβολὴν τῆς εἰσφερομένης τροφῆς·

---

[1] φλεβίον Bekker.

---

[a] Could Aristotle have seen a case of haematoporphyria?

biggest channels persist whereas the smallest ones quickly get obliterated by the mud, though when the mud abates they reappear ; so in the body the largest blood-vessels persist, while the smallest ones become flesh in actuality, though potentially they are blood-vessels as much as ever before. Accordingly we find that, as long as the flesh is in a sound condition, wherever it is cut, blood will flow ; and although no blood-vessels are visible, they must be there (because we cannot have blood without blood-vessels)—just as the irrigating channels are there right enough, but are not visible until they are cleared of mud.

The blood-vessels get progressively smaller as they go on until their channel is too small for the blood to pass through. But, although the blood cannot get through them, the residue of the fluid moisture, which we call sweat, can do so, and this happens when the body is thoroughly heated and the blood-vessels open wider at their mouths. In some cases, the sweat consists of a blood-like residue [a] : this is due to a bad general condition, in which the body has become loose and flabby, and the blood watery owing to insufficient concoction, which in its turn is due to the weakness and scantiness of the heat in the small blood-vessels. (We have already said that all compounds of earth and water are thickened by concoction, and this category includes food and blood.) The heat may, as I say, be in itself too scanty to be able to cause concoction, or it may be that it is scanty in comparison with the amount of food that enters the body, if

See A. E. Garrod, *Inborn Errors of Metabolism*, Oxford, 1923, pp. 136 ff. Also H. Günther, *Deutsches Archiv f. klin. Medizin*, 1920, *134*, pp. 257 ff.

668 b

γίνεται δὲ πρὸς ταύτην ὀλίγη. ἡ δ' ὑπερβολὴ
15 δισσή· καὶ γὰρ τῷ ποσῷ καὶ τῷ ποιῷ· οὐ γὰρ πᾶν
ὁμοίως εὔπεπτον. (ῥεῖ δὲ μάλιστα τὸ αἷμα κατὰ
τοὺς εὐρυχωρεστάτους τῶν πόρων· διόπερ ἐκ τῶν
μυκτήρων καὶ τῶν οὔλων καὶ τῆς ἕδρας, ἐνίοτε δὲ
καὶ ἐκ τοῦ στόματος αἱμορροΐδες ἄπονοι γίνονται,
καὶ οὐχ ὥσπερ ἐκ τῆς ἀρτηρίας μετὰ βίας.)
20 Διεστῶσαι δ' ἄνωθεν ἥ τε μεγάλη φλὲψ καὶ ἡ
ἀορτή, κάτω δ' ἐναλλάσσουσαι συνέχουσι τὸ σῶμα.
προϊοῦσαι γὰρ σχίζονται κατὰ τὴν διφυΐαν τῶν
κώλων, καὶ ἡ μὲν ἐκ τοῦ ἔμπροσθεν εἰς τοὔπισθεν
προέρχεται, ἡ δ' ἐκ τοῦ ὄπισθεν εἰς τοὔμπροσθεν,
25 καὶ συμβάλλουσιν εἰς ἕν· ὥσπερ γὰρ ἐν τοῖς πλεκο-
μένοις ἐγγίνεται τὸ συνεχὲς μᾶλλον, οὕτω καὶ διὰ
τῆς τῶν φλεβῶν ἐναλλάξεως συνδεῖται τῶν σωμάτων τὰ πρόσθια τοῖς ὄπισθεν. ὁμοίως δὲ καὶ ἀπὸ
τῆς καρδίας ἐν τοῖς ἄνω τόποις συμβαίνει. τὸ δὲ
30 μετ' ἀκριβείας ὡς ἔχουσιν αἱ φλέβες πρὸς ἀλλήλας,
ἔκ τε τῶν ἀνατομῶν δεῖ θεωρεῖν καὶ ἐκ τῆς ζωικῆς
ἱστορίας.[b]

Καὶ περὶ μὲν φλεβῶν καὶ καρδίας εἰρήσθω,
περὶ δὲ τῶν ἄλλων σπλάγχνων σκεπτέον κατὰ τὴν
αὐτὴν μέθοδον.

VI. Πλεύμονα μὲν οὖν ἔχει διὰ τὸ πεζὸν εἶναί τι
γένος τῶν ζῴων. ἀναγκαῖον μὲν γὰρ γίνεσθαι τῷ
35 θερμῷ κατάψυξιν, ταύτης δὲ δεῖται θύραθεν τὰ
669 a ἔναιμα τῶν ζῴων· θερμότερα γάρ. τὰ δὲ μὴ ἔναιμα

---

[a] The posterior *vena cava.*
[b] *Hist. An.*, especially 511 b 11—515 a 26.

this is excessive ; and this excess may be due either to the quantity of it or (since some substances are less patient of concoction than others) to its quality. (Haemorrhage occurs most where the passages are widest, as from the nostrils, the gums and the fundament, and occasionally from the mouth. At these places it is not painful ; when, however, it occurs from the windpipe, it is violent.)

The Great Blood-vessel [a] and the Aorta, which in the upper part are some distance from each other, lower down change sides, and thus hold the body compact. That is to say, when they reach the place where the legs diverge, they divide into two, and the Great Blood-vessel goes over to the back from the front, and the Aorta to the front from the back ; and thus they unite the body together, for this changing over of the blood-vessels binds together the front and the back of the body just as the crossing of the strands in plaiting or twining makes the material hold together more stoutly. A similar thing occurs in the upper part of the body, where the blood-vessels that lead from the heart are interchanged. For an exact description of the relative disposition of the blood-vessels, the treatises on *Anatomy* and the *Researches upon Animals* [b] should be consulted.

We have now finished our discussion of the heart and the blood-vessels, and we must go on to consider the remaining viscera on the same lines.

VI. First the Lung. The reason why any group of animals possesses a lung is because they are land-creatures. It is necessary to have some means for cooling the heat of the body ; and blooded animals are so hot that this cooling must come from outside

Lung.

255

669 a

καὶ τῷ συμφύτῳ πνεύματι δύναται καταψύχειν.
ἀνάγκη δὲ καταψύχειν ἔξωθεν ἢ ὕδατι ἢ ἀέρι.
διόπερ τῶν μὲν ἰχθύων οὐδεὶς ἔχει πλεύμονα, ἀλλ'
ἀντὶ τούτου βράγχια, καθάπερ εἴρηται ἐν τοῖς περὶ
5 ἀναπνοῆς· ὕδατι γὰρ ποιεῖται τὴν κατάψυξιν· τὰ
δ' ἀναπνέοντα τῷ ἀέρι, διόπερ πάντα τὰ ἀναπνέοντα ἔχει πλεύμονα. ἀναπνεῖ δὲ τὰ μὲν πεζὰ
πάντα, ἔνια δὲ καὶ τῶν ἐνύδρων, οἷον φάλαινα καὶ
δελφὶς καὶ τὰ ἀναφυσῶντα κήτη πάντα· πολλὰ γὰρ
10 τῶν ζῴων ἐπαμφοτερίζει τὴν φύσιν, καὶ τῶν τε
πεζῶν καὶ τὸν ἀέρα δεχομένων διὰ τὴν τοῦ σώματος
κρᾶσιν ἐν ὑγρῷ διατελεῖ τὸν πλεῖστον χρόνον, καὶ
τῶν ἐν τῷ ὑγρῷ μετέχει τοσοῦτον ἔνια τῆς πεζῆς
φύσεως ὥστ' ἐν τῷ πνεύματι αὐτῶν εἶναι τὸ τέλος
τοῦ ζῆν.

Τοῦ δ' ἀναπνεῖν ὁ πλεύμων ὄργανόν ἐστι, τὴν μὲν
15 ἀρχὴν τῆς κινήσεως ἔχων ἀπὸ τῆς καρδίας, ποιῶν
δ' εὐρυχωρίαν τῇ εἰσόδῳ τοῦ πνεύματος διὰ τὴν
αὑτοῦ σομφότητα καὶ τὸ μέγεθος· αἰρομένου μὲν
γὰρ εἰσρεῖ τὸ πνεῦμα, συνιόντος δ' ἐξέρχεται πάλιν.
τὸ δὲ πρὸς τὴν ἅλσιν εἶναι τὸν πλεύμονα τῆς καρδίας οὐκ εἴρηται καλῶς· ἐν ἀνθρώπῳ τε γὰρ συμ-
20 βαίνει μόνον ὡς εἰπεῖν τὸ τῆς πηδήσεως διὰ τὸ
μόνον ἐν ἐλπίδι γίνεσθαι καὶ προσδοκίᾳ τοῦ μέλ-
λοντος, ἀπέχει τ' ἐν τοῖς πλείστοις πολὺν τόπον καὶ
κεῖται τὴν θέσιν ἀνωτέρω τοῦ πλεύμονος, ὥστε
μηδὲν συμβάλλεσθαι τὸν πλεύμονα πρὸς τὴν ἅλσιν
τῆς καρδίας.

Διαφέρει δ' ὁ πλεύμων πολὺ τοῖς ζῴοις. τὰ μὲν

---

[a] See above, on 659 b 17. [b] 476 a 6.
[c] See above, on 650 b 19 ff.
[d] This view is expressed by Timaeus in Plato's *Timaeus*, 70 c.

them, though the bloodless ones can do their own cooling by means of the connate *pneuma.*[a] Now external cooling must be effected either by water or by air. This explains why none of the fishes has a lung. They are water-cooled, and instead of a lung they have gills (see the treatise on *Respiration*).[b] Animals that breathe, on the other hand, are air-cooled, and so they all have a lung. All land-animals breathe ; so do some of the water-animals (*e.g.* the whale, the dolphin, and all the spouting cetacea). This is not surprising, for many animals are intermediate between the two : some that are land-animals and breathe spend most of their time in the water owing to the blend [c] in their bodies ; and some of the water-animals partake of the nature of land-animals to such an extent that the limiting condition of life for them lies in their breath.

Now the organ of breathing is the lung. It has its source of motion in the heart, and it affords a wide space for the breath to come into because it is large and spongy : when the lung rises up, the breath rushes in, and when it contracts the breath goes out again. The theory [d] that the lung is provided as a cushion for the throbbings of the heart is not correct. This leaping of the heart is practically not found except in man, and that is because man is the only animal that has hope and expectation of the future. Besides, in most animals the heart is a long way off from the lung and lies well above it, and so the lung cannot be of any assistance in absorbing the throbbings of the heart.[e]

There are many differences in the lung. Some

---

[e] In quadrupeds the lung is above the heart, but not in man, owing to the difference of posture.

669 a

25 γὰρ ἔναιμον ἔχει καὶ μέγαν, τὰ δ' ἐλάττω καὶ
σομφόν, τὰ μὲν ζῳοτόκα διὰ τὴν θερμότητα τῆς
φύσεως μείζω καὶ πολύαιμον, τὰ δ' ὠοτόκα ξηρὸν
καὶ μικρόν, δυνάμενον δὲ μεγάλα διίστασθαι ἐν τῷ
ἐμφυσᾶσθαι, ὥσπερ τὰ τετράποδα μὲν ὠοτόκα δὲ
30 τῶν πεζῶν, οἷον οἵ τε σαῦροι καὶ αἱ χελῶναι καὶ
πᾶν τὸ τοιοῦτον γένος, ἔτι δὲ πρὸς τούτοις ἡ τῶν
πτηνῶν φύσις καὶ καλουμένων ὀρνίθων. πάντων
γὰρ τούτων σομφὸς ὁ πλεύμων καὶ ὅμοιος ἀφρῷ·
καὶ γὰρ ὁ ἀφρὸς ἐκ πολλοῦ μικρὸς γίνεται συγχεό-
μενος, καὶ ὁ τούτων πλεύμων μικρὸς καὶ ὑμενώδης.
35 διὸ καὶ ἄδιψα καὶ ὀλιγόποτα ταῦτα πάντα, καὶ
δύναται πολὺν ἐν τῷ ὑγρῷ ἀνέχεσθαι χρόνον· ἅτε
γὰρ ὀλίγον ἔχοντα θερμὸν ἱκανῶς ἐπὶ πολὺν χρόνον
669 b καταψύχεται ὑπ' αὐτῆς τῆς τοῦ πλεύμονος κινήσεως, ὄντος ἀερώδους καὶ κενοῦ.[1]

(Συμβέβηκε δὲ καὶ τὰ μεγέθη τούτων ἐλάττω τῶν
ζῴων ὡς ἐπίπαν εἰπεῖν· τὸ γὰρ θερμὸν αὐξητικόν,
ἡ δὲ πολυαιμία θερμότητος σημεῖον. ἔτι δ' ὀρθοῖ
5 τὰ σώματα μᾶλλον, διόπερ ἄνθρωπος μὲν τῶν
ἄλλων ὀρθότατον, τὰ δὲ ζῳοτόκα τῶν ἄλλων τετρα-
πόδων· οὐδὲν γὰρ ὁμοίως τρωγλοδυτεῖ τῶν ζῳοτόκων, οὔτ' ἄπουν οὔτε πεζεῦον.)

Ὅλως μὲν οὖν ὁ πλεύμων ἐστὶν ἀναπνοῆς χάριν,
ἄναιμος δὲ καὶ τοιοῦτος γένους τινὸς ἕνεκεν ζῴων·
10 ἀλλ' ἀνώνυμον τὸ κοινὸν ἐπ' αὐτῶν, καὶ οὐχ ὥσπερ

---

[1] ὄντος . . . κενοῦ Thurot : οὔσης . . . κενῆς vulg.

[a] Cf. 653 a 30 ff.

animals have a large lung, which contains blood ; others a small one and spongy. In the Vivipara it is large and has much blood in it because these creatures have a hot nature : in the Ovipara it is dry and small, but it can expand to a great size when inflated : examples of these are : among land-animals, the oviparous quadrupeds like the lizards, tortoises, and all such creatures, and in addition to these the tribe of winged things, the birds. All these have a spongy lung, which, like froth, runs together and contracts from a large volume into a small one. So it counts as small ; and also it is membranous. As a result, all these creatures are not much subject to thirst, and drink but little ; and also they can bear to remain a long time under the water : this is because their heat is scanty and can therefore be sufficiently cooled over a long period by the mere motion of the lung, which is void and air-like.

(Consequently, one may add, in general these creatures are smaller in size than the majority of animals, as growth is promoted by heat, and a plentiful supply of blood is a sign that heat is present. Furthermore, heat tends to make the body upright,[a] which explains why man is the most upright among the animals and the Vivipara the most upright among the quadrupeds. And there are no viviparous creatures, either with or without feet, so fond of creeping into holes as the Ovipara are.)

The lung, then, is present for the sake of the breathing : this is its function always. Sometimes, to serve the purpose of a particular group, it is bloodless, and such as has been described above. There is no common name which is applied to all animals that have lungs. But there ought to be : because

869 b

ὁ ὄρνις ὠνόμασται ἐπί τινος γένους. διὸ ὥσπερ τὸ
ὄρνιθι εἶναι ἔκ τινός ἐστι, καὶ ἐκείνων ἐν τῇ οὐσίᾳ
ὑπάρχει τὸ πλεύμονα ἔχειν.

VII. Δοκεῖ δὲ τῶν σπλάγχνων τὰ μὲν εἶναι
μονοφυῆ, καθάπερ καρδία καὶ πλεύμων, τὰ δὲ
15 διφυῆ, καθάπερ νεφροί, τὰ δ' ἀπορεῖται ποτέρως
ἔχει. φανείη γὰρ ἂν ἐπαμφοτερίζειν τούτοις τὸ
ἧπαρ καὶ ὁ σπλήν· καὶ γὰρ ὡς μονοφυὲς ἑκάτερον,
καὶ ὡς ἀνθ' ἑνὸς δύο παραπλησίαν ἔχοντα τὴν
φύσιν. ἔστι δὲ πάντα διφυᾶ. τὸ δ' αἴτιον ἡ τοῦ
σώματος διάστασις διφυὴς μὲν οὖσα, πρὸς μίαν δὲ
20 συντελοῦσα ἀρχήν· τὸ μὲν γὰρ ἄνω καὶ κάτω, τὸ δ'
ἔμπροσθεν καὶ ὄπισθεν, τὸ δὲ δεξιὸν καὶ ἀριστερόν
ἐστιν. διόπερ καὶ ὁ ἐγκέφαλος βούλεται διμερὴς
εἶναι πᾶσι καὶ τῶν αἰσθητηρίων ἕκαστον. κατὰ
τὸν αὐτὸν δὲ λόγον ἡ καρδία ταῖς κοιλίαις. ὁ δὲ
πλεύμων ἔν γε¹ τοῖς ᾠοτόκοις τοσοῦτον διέστηκεν
25 ὥστε δοκεῖν δύ' ἔχειν αὐτὰ πλεύμονας. οἱ δὲ
νεφροὶ καὶ παντὶ δῆλοι· κατὰ δὲ τὸ ἧπαρ καὶ τὸν
σπλῆνα δικαίως ἄν τις ἀπορήσειεν. τούτου δ'
αἴτιον ὅτι ἐν μὲν τοῖς ἐξ ἀνάγκης ἔχουσι σπλῆνα
δόξειεν ἂν οἷον νόθον εἶναι ἧπαρ ὁ σπλήν, ἐν δὲ τοῖς
μὴ ἐξ ἀνάγκης ἔχουσιν, ἀλλὰ πάμμικρον ὥσπερ
30 σημείου χάριν, ἐναργῶς διμερὲς τὸ ἧπάρ ἐστι, καὶ
τὸ μὲν ⟨μεῖζον⟩² εἰς τὰ δεξιά, τὸ δ' ἔλαττον εἰς τὰ-
ριστερὰ βούλεται τὴν θέσιν ἔχειν. οὐ μὴν ἀλλὰ καὶ
ἐν τοῖς ᾠοτόκοις ἧττον μὲν ἢ ἐπὶ τούτων φανερόν,
ἐνίοις δὲ [κἀκεῖ ὥσπερ ἔν τισι]³ ζῳοτόκοις ἐπιδήλως
διέστηκεν, οἷον κατά τινας τόπους οἱ δασύποδες δύο

¹ γε Peck : τε vulg.　　² μεῖζον conieceram ; maior pars Σ.
³ seclusi : ὥσπερ ἔν τισι om. EY : κἀκείνων coni. Th.

260

the possession of a lung is one of their essential characteristics, just as there are certain characteristics which are included in the essence of a "bird," the name which is applied to another such class.

VII. Some of the viscera appear to be single (*e.g.* the heart and the lung); others double (*e.g.* the kidneys); and some it is difficult to place under either heading. The liver and the spleen apparently are intermediate; they can be considered either as each being a single organ, or else as two organs taking the place of one and having a similar character. In fact, however, all of them are double. And the reason for this is that the structure of the body is double, though its halves are combined under one source. We have upper and lower halves, front and back halves, right and left halves. Thus even the brain as well as each of the sense-organs tends in all animals to be double; so does the heart—it has cavities. In the Ovipara the lung is so much divided that they appear to have two lungs. The kidneys are obviously double; but there is fair room for hesitation about the liver and spleen. This is because in those animals which of necessity have a spleen, the spleen looks rather like a bastard liver, while in those which have a spleen though not of necessity—*i.e.* a very small one, as it were by way of a token—the liver is patently double, and the larger part of it tends to lie towards the right, the smaller towards the left. Still, there are cases even among the Ovipara where this division is less distinct than in those just described, while in some Vivipara the division is unmistakable—*e.g.* in some districts

Why the viscera are double.

Liver and spleen.

261

669 b

35 δοκοῦσιν ἧπατ᾽ ἔχειν, καθάπερ τῶν ἰχθύων ἕτεροί
τέ τινες καὶ οἱ σελαχώδεις.

670 a Διὰ δὲ τὸ τὴν θέσιν ἔχειν τὸ ἧπαρ ἐν τοῖς δεξιοῖς
μᾶλλον ἡ τοῦ σπληνὸς γέγονε φύσις, ὥστ᾽ ἀναγκαῖον μέν πως, μὴ λίαν δ᾽ εἶναι πᾶσι τοῖς ζῴοις.

Τοῦ μὲν οὖν διφυῆ τὴν φύσιν εἶναι τῶν σπλάγχνων αἴτιον, ὥσπερ εἴπομεν, τὸ δύ᾽ εἶναι τὸ δεξιὸν
5 καὶ τὸ ἀριστερόν· ἑκάτερον γὰρ ζητεῖ τὸ ὅμοιον,
ὥσπερ καὶ αὐτὰ βούλεται παραπλησίαν καὶ διδύμην
ἔχειν τὴν φύσιν, καὶ καθάπερ[1] ἐκεῖνα δίδυμα μέν,
συνήρτηται δ᾽ εἰς ἕν, καὶ τῶν σπλάγχνων ὁμοίως
ἕκαστον.

Ἔστι δὲ σπλάγχνα τὰ κάτω τοῦ ὑποζώματος
κοινῇ μὲν πάντα τῶν φλεβῶν χάριν, ὅπως οὖσαι
10 μετέωροι μένωσι τῷ τούτων συνδέσμῳ πρὸς τὸ
σῶμα. καθάπερ ἄγκυραι γὰρ βέβληνται πρὸς τὸ
σῶμα διὰ τῶν ἀποτεταμένων μορίων· ἀπὸ μὲν τῆς
μεγάλης φλεβὸς πρὸς τὸ ἧπαρ καὶ τὸν σπλῆνα,
τούτων γὰρ τῶν σπλάγχνων ἡ φύσις οἷον ἧλοι πρὸς
τὸ σῶμα προσλαμβάνουσιν αὐτήν, εἰς μὲν τὰ
15 πλάγια τοῦ σώματος τό θ᾽ ἧπαρ καὶ ὁ σπλὴν τὴν
φλέβα τὴν μεγάλην (ἀπὸ ταύτης γὰρ εἰς αὐτὰ μόνον
διατείνουσι φλέβες), εἰς δὲ τὰ ὄπισθεν οἱ νεφροί.
πρὸς δ᾽ ἐκείνους οὐ μόνον ἀπὸ τῆς μεγάλης
φλεβὸς ἀλλὰ καὶ ἀπὸ τῆς ἀορτῆς τείνει φλὲψ εἰς
ἑκάτερον.

Ταῦτα δὴ συμβαίνει διὰ τούτων τῇ συστάσει
20 τῶν ζῴων· καὶ τὸ μὲν ἧπαρ καὶ ὁ σπλὴν βοηθεῖ
πρὸς τὴν πέψιν τῆς τροφῆς (ἔναιμα γὰρ ὄντα θερ-

---

[1] καὶ καθάπερ PZ: καὶ om. vulg.

hares appear to have a couple of livers ; so do certain fishes, especially the cartilaginous ones.[a]

The spleen owes its existence to the liver being placed somewhat over to the right-hand side of the body : this makes the spleen a necessity in a way, though not an urgent one, for all animals.

Thus, the reason why the viscera are double in their formation is, as we have said, that the body is two-sided, having right and left. Each of the two aims at similarity, just as the sides themselves strive to have a similar nature, and to be as like as twins ; and just as the sides, though dual, are conjoined together into a unity, so also it is with the several viscera.

The viscera which are below the diaphragm are all of them present for the sake of the blood-vessels, in order that the latter may have freedom of carriage and at the same time be attached to the body by means of the viscera, which act as a bond. Indeed, there are, as it were, anchor-lines thrown out to the body through the extended parts : e.g. from the Great Blood-vessel to the liver and to the spleen, for these viscera act, as it were, like rivets and fasten it to the body ; that is to say, the liver and the spleen fasten the Great Blood-vessel to the sides of the body (since blood-vessels pass to them from it alone), while the kidneys fasten it to the rear parts. And to the kidneys—to each of them—there is a blood-vessel passing not only from the Great Blood-vessel but also from the Aorta.

These advantages, then, accrue to the animal organism from the lower viscera. Liver and spleen also assist in the concoction of the food, since they both

[a] Sharks, etc.

670 a

μὴν ἔχει τὴν φύσιν), οἱ δὲ νεφροὶ πρὸς τὸ περίττωμα τὸ εἰς τὴν κύστιν ἀποκρινόμενον.

Καρδία μὲν οὖν καὶ ἧπαρ πᾶσιν ἀναγκαῖα τοῖς
ζῴοις, ἡ μὲν διὰ τὴν τῆς θερμότητος ἀρχήν (δεῖ γὰρ
25 εἶναί τινα οἷον ἑστίαν, ἐν ᾗ κείσεται τῆς φύσεως τὸ
ζωπυροῦν, καὶ τοῦτο εὐφύλακτον, ὥσπερ ἀκρόπολις
οὖσα τοῦ σώματος), τὸ δ' ἧπαρ τῆς πέψεως χάριν.
πάντα δὲ δεῖται τὰ ἔναιμα δυοῖν τούτοιν, διόπερ
ἔχει πάντα τὰ ἔναιμα δύο τὰ σπλάγχνα ταῦτα[1]· ὅσα
30 δ' ἀναπνεῖ, καὶ πλεύμονα τρίτον.

Ὁ δὲ σπλὴν κατὰ συμβεβηκὸς ἐξ ἀνάγκης ὑπάρχει τοῖς ἔχουσιν, ὥσπερ καὶ τὰ περιττώματα,
τό τ' ἐν τῇ κοιλίᾳ καὶ τὸ περὶ τὴν κύστιν. διόπερ
ἔν τισιν ἐκλείπει κατὰ τὸ μέγεθος, ὥσπερ τῶν τε
πτερωτῶν ἐνίοις, ὅσα θερμὴν ἔχει τὴν κοιλίαν, οἷον
670 b περιστερά, ἱέραξ, ἰκτῖνος, καὶ ἐπὶ τῶν ᾠοτόκων
δὲ καὶ τετραπόδων ὁμοίως (μικρὸν γὰρ πάμπαν
ἔχουσιν), καὶ πολλοῖς τῶν λεπιδωτῶν· ἅπερ καὶ
κύστιν οὐκ ἔχει διὰ τὸ τρέπεσθαι τὸ περίττωμα διὰ
μανῶν τῶν σαρκῶν εἰς πτερὰ καὶ λεπίδας. ὁ γὰρ
5 σπλὴν ἀντισπᾷ ἐκ τῆς κοιλίας τὰς ἰκμάδας τὰς
περιττευούσας, καὶ δύναται συμπέττειν αἱματώδης
ὤν. ἂν δὲ τὸ περίττωμα πλεῖον ᾖ ἢ ὀλιγόθερμος ὁ
σπλήν, νοσακερὰ γίνεται πλήρης[2] ⟨οὖσα⟩[3] τροφῆς·
καὶ διὰ τὴν ἐνταῦθα παλίρροιαν τῆς ὑγρότητος πολλοῖς αἱ κοιλίαι σκληραὶ γίνονται σπληνιῶσιν, ὥσ-
10 περ τοῖς λίαν οὐρητικοῖς, διὰ τὸ ἀντιπερισπᾶσθαι

---

[1] ταῦτα P : ταῦτα μόνον vulg.
[2] πλήρης EYZ : πλήρη vulg.
[3] ⟨οὖσα⟩ Peck.

have blood in them and so are hot. The kidneys assist in connexion with the residue which is excreted into the bladder.

Now the heart and the liver are *necessary* to all animals. The heart is necessary because there must be a source of heat: there must be, as it were, a hearth, where that which kindles the whole organism shall reside; and this part must be well guarded, being as it were the citadel of the body. The liver is necessary for the sake of effecting concoction. All blooded creatures must have these two viscera, and that is why these two are always present in them. A third, the lung, is present in those animals that breathe.

But the spleen, where present, is present *of necessity* Spleen in the sense of being an incidental concomitant, as are the residues in the stomach and in the bladder. So in some animals the spleen is deficient in size, as in certain birds which have a hot stomach, *e.g.* the pigeon, the hawk, and the kite; the same applies to the oviparous quadrupeds (all of these have an extremely small spleen) and to many of the scaly creatures. These animals just mentioned also lack a bladder, because their flesh is porous enough to enable the residues formed to pass through it and produce feathers and scales. For the spleen draws off the residual humours from the stomach and in virtue of its blood-like nature can assist in the concoction of them. If, however, the residue is too bulky or the spleen has too little heat, the stomach gets full of nourishment and becomes diseased. And in many cases, when the spleen is ailing, the stomach becomes hardened owing to the fluid which runs back into it. This happens with

τὰς ὑγρότητας. οἷς δὲ ὀλίγη περίττωσις γίνεται,
καθάπερ τοῖς ὀρνέοις καὶ τοῖς ἰχθύσι, τὰ μὲν οὐ
μέγαν ἔχει, τὰ δὲ σημείου χάριν. καὶ ἐν τοῖς
τετράποσι δὲ τοῖς ᾠοτόκοις μικρὸς καὶ στιφρὸς καὶ
15 νεφρώδης ὁ σπλήν ἐστι διὰ τὸ τὸν πλεύμονα σομφὸν
εἶναι καὶ ὀλιγοποτεῖν καὶ τὸ περιγινόμενον περίττωμα τρέπεσθαι εἰς τὸ σῶμα καὶ τὰς φολίδας,
ὥσπερ εἰς τὰ πτερὰ τοῖς ὄρνισιν.

Ἐν δὲ τοῖς κύστιν ἔχουσι καὶ τὸν πλεύμονα
ἔναιμον ὑγρός ἐστι διά τε τὴν εἰρημένην αἰτίαν καὶ
διὰ τὸ τὴν φύσιν τὴν τῶν ἀριστερῶν ὅλως ὑγρο-
20 τέραν εἶναι καὶ ψυχροτέραν. διῄρηται γὰρ τῶν
ἐναντίων ἕκαστον πρὸς τὴν συγγενῆ συστοιχίαν,
οἷον δεξιὸν ἐναντίον ἀριστερῷ καὶ θερμὸν ἐναντίον
ψυχρῷ· καὶ σύστοιχα γὰρ ἀλλήλοις εἰσὶ τὸν εἰρημένον τρόπον.

Οἱ δὲ νεφροὶ τοῖς ἔχουσιν οὐκ ἐξ ἀνάγκης ἀλλὰ
τοῦ εὖ καὶ καλῶς ἕνεκεν ὑπάρχουσιν· τῆς γὰρ
25 περιττώσεως χάριν τῆς εἰς τὴν κύστιν ἀθροιζομένης
εἰσὶ κατὰ τὴν ἰδίαν φύσιν, ἐν ὅσοις πλεῖον ὑπόστημα γίνεται τὸ τοιοῦτον, ὅπως βέλτιον ἀποδιδῷ
ἡ κύστις τὸ αὑτῆς ἔργον.

Ἐπεὶ δὲ τῆς αὐτῆς ἕνεκα χρείας τούς τε νεφροὺς
συμβέβηκεν ἔχειν τὰ ζῷα καὶ τὴν κύστιν, λεκτέον
30 περὶ κύστεως νῦν, ὑπερβάντας τὸν ἐφεξῆς τῶν
μορίων ἀριθμόν· περὶ γὰρ φρενῶν οὐδέν πω δι-

---

[a] The reference to the " columns " or " double list " is
not clear. There was a Pythagorean συστοιχία; this and
other συστοιχίαι are mentioned in Ross's note on his translation of *Met.* 986 a 23.

[b] *i.e.* left and cold are both in the same column ; right and
hot are both in the other column.

those who make water excessively : the fluids are
drawn back again into the stomach. But in animals
where the amount of residue produced is small, as in
birds and fishes, the spleen is either small or present
simply by way of a token. In the oviparous quadrupeds, too, the spleen is small and compact, and
like a kidney, because the lung is spongy and the
animals drink little, and also because the residue
which is produced is applied for the benefit of the
body itself and of the scaly plates which cover it,
just as in birds it is applied for the benefit of the
feathers.

In those animals, however, which possess a bladder,
and whose lung contains blood, the spleen is watery.
The reason already given partly explains this. Another is that the left side of the body is generally
more watery and colder than the right. As we know,
the opposites are divided up into two columns,[a] so
that each is classed with those that are akin to it, *e.g.*
right is in the opposite column to left, and hot to
cold ; and thus some of them stand together in the
same column, as I have just indicated.[b]

Kidneys are present in some animals, but not
of necessity ; they are present to serve a *good purpose* ; that is to say, their particular nature enables
them to cope with the residue which collects in the
bladder, in those cases where this deposit is somewhat
abundant, and to help the bladder to perform its
function *better*.

Since the bladder is present in animals to serve
precisely the same purpose as the kidneys, we must
now say something about it. This will involve a
departure from the serial order in which the parts
actually come, for we have said nothing so far about

670 b

ὥρισται, τοῦτο δέ τι τῶν περὶ τὰ σπλάγχνα μορίων
ἐστίν.

VIII. Κύστιν δ' οὐ πάντ' ἔχει τὰ ζῷα, ἀλλ'
ἔοικεν ἡ φύσις βουλομένη ἀποδιδόναι τοῖς ἔχουσι
671 a τὸν πλεύμονα ἔναιμον μόνον, τούτοις δ' εὐλόγως.
διὰ γὰρ τὴν ὑπεροχὴν τῆς φύσεως, ἣν ἔχουσιν ἐν
τῷ μορίῳ τούτῳ, διψητικά τε ταῦτ' ἐστὶ μάλιστα
τῶν ζῴων, καὶ δεῖται τροφῆς οὐ μόνον τῆς ξηρᾶς
ἀλλὰ καὶ τῆς ὑγρᾶς πλείονος, ὥστ' ἐξ ἀνάγκης καὶ
5 περίττωμα γίνεσθαι πλεῖον καὶ μὴ τοσοῦτον μόνον
ὅσον ὑπὸ τῆς κοιλίας πέττεσθαι καὶ ἐκκρίνεσθαι
μετὰ τοῦ ταύτης περιττώματος. ἀνάγκη τοίνυν
εἶναί τι δεκτικὸν καὶ τούτου τοῦ περιττώματος.
διόπερ ὅσα πλεύμονα ἔχει τοιοῦτον, ἅπαντ' ἔχει
κύστιν· ὅσα δὲ μὴ τοιοῦτον, ἀλλ' ἢ ὀλιγόποτά ἐστι
10 διὰ τὸ πλεύμονα ἔχειν σομφόν, ἢ ὅλως τὸ ὑγρὸν
προσφέρεται οὐ ποτοῦ χάριν ἀλλὰ τροφῆς, οἷον τὰ
ἔντομα καὶ οἱ ἰχθύες, ἔτι δὲ πτερωτά ἐστιν ἢ
λεπιδωτὰ ἢ φολιδωτά, ταῦτα δι' ὀλιγότητά τε τῆς
τοῦ ὑγροῦ προσφορᾶς καὶ διὰ τὸ τρέπεσθαι εἰς
ταῦτα τὸ περιγινόμενον τοῦ περιττώματος οὐδὲν
15 ἔχει τούτων κύστιν, πλὴν αἱ χελῶναι τῶν φολιδωτῶν, καὶ ἐνταῦθ' ἡ φύσις κεκολόβωται μόνον· αἴτιον
δ' ὅτι αἱ μὲν θαλάττιαι σαρκώδη καὶ ἔναιμον ἔχουσι
τὸν πλεύμονα καὶ ὅμοιον τῷ βοείῳ, αἱ δὲ χερσαῖαι
μείζω ἢ κατὰ λόγον. ἔτι δὲ διὰ τὸ ὀστρακῶδες
20 καὶ πυκνὸν εἶναι τὸ περιέχον οὐ διαπνέοντος τοῦ
ὑγροῦ διὰ μανῶν τῶν σαρκῶν, οἷον τοῖς ὄρνισι καὶ
τοῖς ὄφεσι καὶ τοῖς ἄλλοις τοῖς φολιδωτοῖς, ὑπὸ

the diaphragm, though this is one of the parts that are near the viscera.

VIII. The bladder is not present in all animals : Bladder. Nature seems to have intended only those animals which have blood in their lung to have a bladder. And this is quite reasonable, when we remember that such animals have an excess of the natural substance which constitutes the lung, and are therefore more subject to thirst than any others ; *i.e.* they need a larger amount of fluid food as well as of the ordinary solid food, and the necessary result of this is that a larger amount of residue also is produced, too large in fact for all of it to be concocted by the stomach and excreted with its own proper residue ; hence it is necessary to have some part that will receive this additional residue. This shows us why all animals which have blood in their lung possess a bladder too. As for those whose lung is spongy and which therefore drink little, or which take fluids not as something to drink but as food (*e.g.* insects and fishes), or which are covered with feathers or scales or scaly plates, not one of these has a bladder, owing to the small amount of fluid which they take and owing to the fact that the surplus residue goes to form feathers or scales or scaly plates, as the particular case may be. Exceptions to this are the Tortoises : though scaly-plated they have a bladder. In them the natural formation has simply been stunted. The cause of this is that in the sea-varieties the lung is fleshy and contains blood, and is similar to the lung of the ox ; while in the land-varieties it is disproportionately large. And whereas in birds and snakes and the other scaly-plated creatures the moisture exhales through the porous flesh, in these it

στασις γίνεται τοσαύτη ὥστε δεῖσθαι τὴν φύσιν
αὐτῶν ἔχειν τι μόριον δεκτικὸν καὶ ἀγγειῶδες.
κύστιν μὲν οὖν ταῦτα μόνον τῶν τοιούτων ἔχει διὰ
25 ταύτην τὴν αἰτίαν, ἡ μὲν θαλαττία μεγάλην, αἱ δὲ
χερσαῖαι μικρὰν πάμπαν.

IX. Ὁμοίως δ' ἔχει καὶ περὶ νεφρῶν. οὐδὲ γὰρ
νεφροὺς οὔτε τῶν πτερωτῶν καὶ λεπιδωτῶν οὐδὲν
ἔχει οὔτε τῶν φολιδωτῶν, πλὴν αἱ θαλάττιαι
χελῶναι καὶ αἱ χερσαῖαι· ἀλλ' ὡς τῆς εἰς τοὺς
30 νεφροὺς τεταγμένης σαρκὸς οὐκ ἐχούσης χώραν
ἀλλὰ διεσπαρμένης εἰς πολλά, πλατέα νεφροειδῆ ἐν
ἐνίοις τῶν ὀρνίθων ἐστίν. ἡ δ' ἐμὺς οὔτε κύστιν
οὔτε νεφροὺς ἔχει· διὰ τὴν μαλακότητα γὰρ τοῦ
χελωνίου εὐδιάπνουν γίνεται τὸ ὑγρόν. ἡ μὲν οὖν
ἐμὺς διὰ ταύτην τὴν αἰτίαν οὐκ ἔχει τῶν μορίων
οὐδέτερον· τοῖς δ' ἄλλοις ζῴοις τοῖς ἔχουσιν ἔν-
35 αιμον, ὥσπερ εἴρηται, τὸν πλεύμονα πᾶσι συμ-
671 b βέβηκεν ἔχειν νεφρούς. καταχρῆται γὰρ ἡ φύσις
ἅμα τῶν τε φλεβῶν χάριν καὶ πρὸς τὴν τοῦ
ὑγροῦ περιττώματος ἔκκρισιν· φέρει γὰρ εἰς
αὐτοὺς πόρος ἐκ τῆς μεγάλης φλεβός.

Ἔχουσι δ' οἱ νεφροὶ πάντες κοῖλον, ἢ πλεῖον ἢ
5 ἔλαττον, πλὴν οἱ τῆς φώκης· οὗτοι δ' ὅμοιοι τοῖς
βοείοις ὄντες στερεώτατοι πάντων εἰσίν. ὅμοιοι δὲ
καὶ οἱ τοῦ ἀνθρώπου τοῖς βοείοις· εἰσὶ γὰρ ὥσπερ
συγκείμενοι ἐκ πολλῶν νεφρῶν μικρῶν καὶ οὐχ
ὁμαλεῖς, ὥσπερ οἱ τῶν προβάτων καὶ τῶν ἄλλων
τῶν τετραπόδων. διὸ καὶ τὸ ἀρρώστημα τοῖς

---

[a] Greek, " hemys." This description, which does not fit

cannot do so, because the integument which surrounds
them is dense, like a shell; and so the excretion is
produced in such quantities that the Tortoises need
some part which shall act as a vessel to receive it.
That, then, is why they are the only animals of the
kind which have a bladder. In the sea-tortoise it is
large, in the land-tortoise quite small.

IX. Much the same may be said of the kidneys as _{Kidneys.}
of the bladder. Kidneys are not present in any of
the animals that have feathers or scales or scaly plates,
except the two sorts of tortoises just mentioned. In
some birds, however, there are flat kidney-shaped
objects, as if the flesh that was allotted to form the
kidneys had found no room for its proper function
and had been scattered to form several organs. The
Emys [a] has neither bladder nor kidneys: this is because it has a soft shell which allows the moisture to
transpire freely through it. But, as I said before, all
the other animals whose lung contains blood have
kidneys, since Nature makes use of them for two purposes : (1) to subserve the blood-vessels ; and (2) to
excrete the fluid residue. (A channel leads into them
from the Great Blood-vessel.)

There is always a hollow (*lumen*), varying in size,
in the kidneys, except in the seal, whose kidneys are
more solid than any others and in shape resemble
those of the ox. Human kidneys too resemble those
of the ox : they are, as it were, made up out of a
number of small kidneys,[b] and have not an even
surface like those of the sheep and other quadrupeds.
Thus, when once an ailment attacks the human

any animal now known as Emys, seems to be that of some
freshwater tortoise.
[b] This is not true of the normal adult, but it is true of
the foetus.

671 b

10 ἀνθρώποις δυσαπάλλακτον αὐτῶν ἐστιν, ἂν ἅπαξ
νοσήσωσιν· συμβαίνει γὰρ ὥσπερ πολλοὺς νεφροὺς
νοσούντων χαλεπωτέραν εἶναι τὴν ἴασιν ἢ τῶν ἕνα
νοσούντων.

Ὁ δ᾽ ἀπὸ τῆς φλεβὸς τείνων πόρος οὐκ εἰς τὸ
κοῖλον τῶν νεφρῶν κατατελευτᾷ, ἀλλ᾽ εἰς τὸ σῶμα
καταναλίσκεται τῶν νεφρῶν· διόπερ ἐν τοῖς κοίλοις
15 αὐτῶν οὐκ ἐγγίνεται αἷμα, οὐδὲ πήγνυται τελευ-
τώντων. ἐκ δὲ τοῦ κοίλου τῶν νεφρῶν φέρουσι
πόροι ἄναιμοι εἰς τὴν κύστιν δύο νεανικοί, ἐξ
ἑκατέρου εἷς, καὶ ἄλλοι ἐκ τῆς ἀορτῆς ἰσχυροὶ καὶ
συνεχεῖς. ταῦτα δ᾽ ἔχει τὸν τρόπον τοῦτον ὅπως
ἐκ μὲν τῆς φλεβὸς τὸ περίττωμα τῆς ὑγρότητος
20 βαδίζῃ εἰς τοὺς νεφρούς, ἐκ δὲ τῶν νεφρῶν ἡ
γινομένη ὑπόστασις διηθουμένων τῶν ὑγρῶν διὰ
τοῦ σώματος τῶν νεφρῶν εἰς τὸ μέσον συρρέῃ, οὗ
τὸ κοῖλον οἱ πλεῖστοι ἔχουσιν αὐτῶν (διὸ καὶ δυσ-
ωδέστατον τοῦτο τῶν σπλάγχνων ἐστίν)· ἐκ δὲ τοῦ
μέσου διὰ τούτων τῶν πόρων εἰς τὴν κύστιν ἤδη
25 μᾶλλον ὡς περίττωμα ἀποκρίνεται. καθώρμισται
δ᾽ ἡ κύστις ἐκ τῶν νεφρῶν· τείνουσι γάρ, ὥσπερ
εἴρηται, πόροι ἰσχυροὶ πρὸς αὐτήν.

Οἱ μὲν οὖν νεφροὶ διὰ ταύτας τὰς αἰτίας εἰσί, καὶ
τὰς δυνάμεις ἔχουσι τὰς εἰρημένας.

Ἐν πᾶσι δὲ τοῖς ἔχουσι νεφροὺς ὁ δεξιὸς ἀνωτέρω
τοῦ ἀριστεροῦ ἐστιν· διὰ γὰρ τὸ τὴν κίνησιν εἶναι
30 ἐκ τῶν δεξιῶν καὶ ἰσχυροτέραν διὰ ταῦτ᾽ εἶναι τὴν

---

^a The ureters.

272

kidneys, the trouble is not easily removed, because it is as though the patient had many kidneys diseased and not one only ; and so the cure is more difficult to effect.

The channel which runs from the Great Blood-vessel to the kidneys does not debouch into the hollow part of the kidneys, but the whole of what it supplies is spent upon the body of the kidneys; thus no blood goes into the hollows, and at death none congeals there. From the hollow part of the kidneys two sturdy channels [a] lead into the bladder, one from each ; these contain no blood. Other channels come from the Aorta to the kidneys ; these are strong, continuous ones. This arrangement is on purpose to enable the residue from the moisture to pass out of the blood-vessel into the kidneys, and so that when the fluid percolates through the body of the kidneys the excretion that results may collect into the middle of the kidneys, where the hollow is in most cases. (This explains, incidentally, why the kidney is the most ill-scented of all the viscera.) From the middle of the kidney the fluid is passed off through the aforesaid channels into the bladder ; by which time it has practically taken on the character of excremental residue. The bladder is actually moored to the kidneys : as has been stated, there are strong channels extending from them to it.

We have now given the causes for which the kidneys exist, as well as their character and functions.

The right kidney is always higher up than the left. The reason for this is that as motion always begins on the right-hand side, the parts that are on that side are stronger than those on the other ; and owing to this

671 b

φύσιν τὴν τῶν δεξιῶν, δεῖ προοδοποιήσασθαι διὰ
τὴν κίνησιν πρὸς τὸ ἄνω ταῦτα[1] τὰ μόρια μᾶλλον,
ἐπεὶ καὶ τὴν ὀφρὺν τὴν δεξιὰν αἴρουσι μᾶλλον καὶ
ἐπικεκαμμένην ἔχουσι τῆς ἀριστερᾶς μᾶλλον. καὶ
35 διὰ τὸ ἀνεσπάσθαι ἀνώτερον τὸν δεξιὸν νεφρὸν τὸ
ἧπαρ ἅπτεται τοῦ δεξιοῦ νεφροῦ ἐν πᾶσιν· ἐν τοῖς
672 a δεξιοῖς γὰρ τὸ ἧπαρ.

Ἔχουσι δ' οἱ νεφροὶ μάλιστα τῶν σπλάγχνων
πιμελήν, ἐξ ἀνάγκης μὲν διὰ τὸ διηθεῖσθαι τὸ
περίττωμα διὰ τῶν νεφρῶν· τὸ γὰρ λειπόμενον
αἷμα καθαρὸν ὂν εὔπεπτόν ἐστι, τέλος δ' εὐπεψίας
5 αἱματικῆς πιμελὴ καὶ στέαρ ἐστίν. (ὥσπερ γὰρ ἐν
τοῖς πεπυρωμένοις ξηροῖς, οἷον τῇ τέφρᾳ, ἐγκατα-
λείπεταί τι πῦρ, οὕτω καὶ ἐν τοῖς πεπεμμένοις
ὑγροῖς· ἐγκαταλείπεται γάρ τι τῆς εἰργασμένης
θερμότητος μόριον. διόπερ τὸ λιπαρὸν κοῦφόν ἐστι
καὶ ἐπιπολάζει ἐν τοῖς ὑγροῖς.) ἐν αὐτοῖς μὲν οὖν
10 οὐ γίνεται τοῖς νεφροῖς διὰ τὸ πυκνὸν εἶναι τὸ
σπλάγχνον, ἔξω δὲ περιίσταται πιμελὴ μὲν ἐν τοῖς
πιμελώδεσι, στέαρ δ' ἐν τοῖς στεατώδεσιν· ἡ δὲ
διαφορὰ τούτων εἴρηται πρότερον ἐν ἑτέροις.

Ἐξ ἀνάγκης μὲν οὖν πιμελώδεις γίνονται διὰ
ταύτην τὴν αἰτίαν ἐκ τῶν συμβαινόντων ἐξ ἀνάγκης
15 τοῖς ἔχουσι νεφρούς, ἕνεκα δὲ σωτηρίας καὶ τοῦ
θερμὴν εἶναι τὴν φύσιν τὴν τῶν νεφρῶν. ἔσχατοί
τε γὰρ ὄντες ἀλέας δέονται πλείονος· τὸ μὲν γὰρ
νῶτον σαρκῶδές ἐστιν, ὅπως ᾖ προβολὴ τοῖς περὶ

---

[1] ταῦτα Peck : πάντα vulg.

---

[a] See Book II. ch. v.

motion they are bound to make their way upwards before the ones on the left. Thus people raise the right eyebrow more than the left, and it is more arched. A result of this drawing up of the right kidney is that in all animals the liver, which is on the right side of the body, is in contact with it.

The kidneys contain more fat than any other of the viscera. This is partly a necessary consequence upon the percolation of the residue through the kidneys : in other words, the blood which gets left behind there is easy of concoction because it is pure, and when blood undergoes complete concoction the final products are lard and suet. (A parallel is to be found in the case of solid substances which have undergone combustion : *e.g.* a certain amount of fire gets left behind in ash. So, in fluid substances which have undergone concoction : some portion of the heat which has been generated remains behind. That is why oily substances are light and come to the top of fluids.) This fat is not formed actually in the kidneys themselves, because they are so dense : it collects outside them. In some it has the form of lard, in others the form of suet, according to the character of the animal. (The difference between the two has been explained already in another connexion.) [a]

This formation of lard, then, about the kidneys is the necessary consequence upon the conditions which necessarily obtain in animals that possess kidneys. But there is another reason for its formation, and that is, on *purpose* to safeguard the kidneys themselves and to preserve their natural heat. The kidneys are the outermost of all the viscera, and therefore they need more warmth. Whereas the back is liberally supplied with flesh, which enables it to act as a

672 a

τὴν καρδίαν σπλάγχνοις, ἡ δ' ὀσφὺς ἄσαρκος
(ἄσαρκοι γὰρ αἱ καμπαὶ πάντων)· ἀντὶ σαρκὸς οὖν
20 ἡ πιμελὴ πρόβλημα γίνεται τοῖς νεφροῖς. ἔτι δὲ
διακρίνουσι καὶ πέττουσι τὴν ὑγρότητα μᾶλλον
πίονες ὄντες· τὸ γὰρ λιπαρὸν θερμόν, πέττει δ' ἡ
θερμότης.

Διὰ ταύτας μὲν οὖν τὰς αἰτίας οἱ νεφροὶ πιμελώδεις εἰσίν, ἐν πᾶσι δὲ τοῖς ζῴοις ὁ δεξιὸς ἀπιμελώ-
τερός ἐστιν. αἴτιον δὲ τὸ τὴν φύσιν ξηρὰν εἶναι
25 τὴν τῶν δεξιῶν καὶ κινητικωτέραν· ἡ δὲ κίνησις
ἐναντία· τήκει γὰρ τὸ πῖον μᾶλλον.

Τοῖς μὲν οὖν ἄλλοις ζῴοις συμφέρει τε τοὺς
νεφροὺς ἔχειν πίονας, καὶ πολλάκις ἔχουσιν ὅλους
περίπλεως· τὸ δὲ πρόβατον ὅταν τοῦτο πάθῃ
ἀποθνήσκει. ἀλλ' ἂν καὶ πάνυ πίονες ὦσιν, ὅμως
30 ἐλλείπει τι, ἂν μὴ κατ' ἀμφοτέρους, ἀλλὰ κατὰ τὸν
δεξιόν.[1] αἴτιον δὲ τοῦ μόνον ἢ μάλιστα τοῦτο
συμβαίνειν ἐπὶ τῶν προβάτων, ὅτι τοῖς μὲν πιμε-
λώδεσιν ὑγρὸν τὸ πῖον, ὥστ' οὐχ ὁμοίως ἐγκατα-
κλειόμενα τὰ πνεύματα ποιεῖ τὸν πόνον. τοῦ δὲ
σφακελισμοῦ τοῦτ' αἴτιόν ἐστιν· διὸ καὶ τῶν ἀν-
35 θρώπων τοῖς πονοῦσι τοὺς νεφρούς, καίπερ τοῦ πιαί-
νεσθαι συμφέροντος, ὅμως ἂν λίαν γίνωνται πίονες,
ὀδύναι θανατηφόροι συμβαίνουσιν. τῶν δ' ἄλλων
672 b τοῖς στεατώδεσιν ἧττον πυκνὸν τὸ στέαρ ἢ τοῖς
προβάτοις. καὶ τῷ πλήθει πολὺ τὰ πρόβατα ὑπερ-
βάλλει· γίνεται γὰρ περίνεφρα τάχιστα τῶν ζῴων
τὰ πρόβατα πάντων. ἐγκατακλειομένης οὖν τῆς
ὑγρότητος καὶ τῶν πνευμάτων διὰ τὸν σφακελισμὸν

[1] ἀλλ' ἂν . . . δεξιόν post εἰσίν l. 23 transponit Thurot.

protection for the viscera about the heart, the loin, in common with all parts that bend, is not so supplied ; and this fat we have been speaking of serves as a safeguard to the kidneys in place of flesh. Further, the kidneys are better able to secrete and to concoct the fluid if they are fat, because fat is hot and heat causes concoction.

These are the reasons why the kidneys are fat. In all animals, however, the right kidney has less fat than the left. This is because the right-hand side is dry and solid and more adapted for motion than the left ; and motion is an enemy to fat, because it tends to melt it.

Now it is an advantage to all animals to have fat kidneys, and often they are completely filled with fat. The sheep is an exception : if this happens to a sheep it dies. But even if the kidneys are as fat as can be, there is always some portion which is clear of fat, if not in both kidneys, at any rate in the right one. The reason why this happens solely (or more especially) to sheep is as follows. Some animals have their fat in the form of lard, which is fluid, and thus the wind cannot so easily get shut up within and cause trouble. When this happens, however, it causes rot. Thus, too, in the case of human beings who suffer from their kidneys, although it is an advantage for them to be fat, yet if they become unduly fat, pains result which prove fatal. As for the animals whose fat is in the form of suet, none has such dense suet as the sheep has ; and moreover, in the sheep the amount of it is much greater ; the fact that they get fat about the kidneys much more quickly than any other animal shows this. So when the moisture and the wind get shut up within, rot is produced, which rapidly kills

672 b

5 ἀναιροῦνται ταχέως· διὰ γὰρ τῆς ἀορτῆς καὶ τῆς
φλεβὸς εὐθὺς ἀπαντᾷ τὸ πάθος πρὸς τὴν καρδίαν·
οἱ δὲ πόροι συνεχεῖς ἀπὸ τούτων τῶν φλεβῶν εἰσι
πρὸς τοὺς νεφρούς.

Περὶ μὲν οὖν τῆς καρδίας καὶ πλεύμονος εἴρηται,
καὶ περὶ ἥπατος καὶ σπληνὸς καὶ νεφρῶν. X. τυγ-
10 χάνει δὲ ταῦτα κεχωρισμένα ἀλλήλων τῷ διαζώ-
ματι. τοῦτο δὲ τὸ διάζωμα καλοῦσί τινες φρένας·
ὃ διορίζει τόν τε πλεύμονα καὶ τὴν καρδίαν. καλεῖται
δὲ τοῦτο τὸ διάζωμα ἐν τοῖς ἐναίμοις, ὥσπερ καὶ
εἴρηται, φρένες. ἔχει δὲ πάντα τὰ ἔναιμα αὐτό,
καθάπερ καρδίαν καὶ ἧπαρ. τούτου δ᾽ αἴτιον ὅτι
15 τοῦ διορισμοῦ χάριν ἐστὶ τοῦ τε περὶ τὴν κοιλίαν
τόπου καὶ τοῦ περὶ τὴν καρδίαν, ὅπως ἡ τῆς
αἰσθητικῆς ψυχῆς ἀρχὴ ἀπαθὴς ᾖ καὶ μὴ ταχὺ
καταλαμβάνηται διὰ τὴν ἀπὸ τῆς τροφῆς γινομένην
ἀναθυμίασιν καὶ τὸ πλῆθος τῆς ἐπεισάκτου θερμότητος. ἐπὶ γὰρ τοῦτο διέλαβεν ἡ φύσις, οἷον
20 παροικοδόμημα ποιήσασα καὶ φραγμὸν τὰς φρένας,
καὶ διεῖλε τό τε τιμιώτερον καὶ τὸ ἀτιμότερον ἐν
ὅσοις ἐνδέχεται διελεῖν τὸ ἄνω καὶ κάτω· τὸ μὲν
γὰρ ἄνω ἐστὶν οὗ ἕνεκεν καὶ βέλτιον, τὸ δὲ κάτω
τὸ τούτου ἕνεκεν καὶ ἀναγκαῖον, τὸ τῆς τροφῆς
δεκτικόν.

Ἔστι δὲ τὸ διάζωμα πρὸς μὲν τὰς πλευρὰς
25 σαρκωδέστερον καὶ ἰσχυρότερον, κατὰ μέσον δ᾽
ὑμενωδέστερον· οὕτω γὰρ πρὸς τὴν ἰσχὺν καὶ τὴν
τάσιν χρησιμώτερον. διότι δὲ πρὸς τὴν θερμότητα
τὴν κάτωθεν οἷον παραφυάδες εἰσί, σημεῖον ἐκ τῶν
278

the sheep off. The disease makes its way directly to the heart through the Aorta and the Great Blood-vessel, since there are continuous passages leading from these to the kidneys.

We have now spoken of the heart and the lung; and also of the liver, the spleen and the kidneys. X. These two sets of viscera are separated from each other by the diazoma, which some call the *phrenes* (diaphragm). This divides off the heart and the lung. In blooded animals it is called *phrenes*, as I have said. All blooded creatures have one, just as they all have heart and liver. The reason for this is that the diaphragm serves to divide the part round the stomach from the part round the heart, to ensure that the source of the sensory Soul may be unaffected, and not be quickly overwhelmed by the exhalation that comes up from the food when it is eaten and by the amount of heat introduced into the system. For this purpose, then, Nature made the division, and constructed the *phrenes* to be, as it were, a partition-wall and a fence; and thus, in those creatures where it is possible to divide the upper from the lower, she divided off the nobler parts from the less noble ones; for it is the upper which is "better," that *for the sake of which* the lower exists, while the lower is "necessary," existing *for the sake of* the upper, by acting as a receptacle for the food.

Towards the ribs the diaphragm is fleshier and stronger, while in the middle it is more like a membrane: this makes it more serviceable as regards strength and extensibility. An indication to show why there are, as it were, "suckers," to keep off the heat which comes up from below, is provided by

Diaphragm.

συμβαινόντων· ὅταν γὰρ διὰ τὴν γειτνίασιν ἑλκύσωσιν ὑγρότητα θερμὴν καὶ περιττωματικήν, εὐθὺς
30 ἐπιδήλως ταράττει τὴν διάνοιαν καὶ τὴν αἴσθησιν,
διὸ καὶ καλοῦνται φρένες ὡς μετέχουσαί τι τοῦ
φρονεῖν. αἱ δὲ μετέχουσι μὲν οὐδέν, ἐγγὺς δ᾿
οὖσαι τῶν μετεχόντων ἐπίδηλον ποιοῦσι τὴν μεταβολὴν τῆς διανοίας. διὸ καὶ λεπταὶ κατὰ μέσον
εἰσίν, οὐ μόνον ἐξ ἀνάγκης, ὅτι σαρκώδεις οὔσας τὰ
35 πρὸς τὰς πλευρὰς ἀναγκαῖον εἶναι σαρκωδεστέρας,
ἀλλ᾿ ἵν᾿ ὅτι ὀλιγίστης μετέχωσιν ἰκμάδος· σαρκώ-
673 a δεις γὰρ ἂν οὖσαι καὶ εἶχον καὶ εἷλκον μᾶλλον
ἰκμάδα πολλήν. ὅτι δὲ θερμαινόμεναι ταχέως
ἐπίδηλον ποιοῦσι τὴν αἴσθησιν, σημαίνει καὶ τὸ
περὶ τοὺς γέλωτας συμβαῖνον· γαργαλιζόμενοί τε
γὰρ ταχὺ γελῶσι, διὰ τὸ τὴν κίνησιν ἀφικνεῖσθαι
5 ταχὺ πρὸς τὸν τόπον τοῦτον, θερμαινόμενον[1] δ᾿
ἠρέμα ποιεῖν ὅμως ἐπίδηλον καὶ κινεῖν τὴν διάνοιαν παρὰ τὴν προαίρεσιν. τοῦ δὲ γαργαλίζεσθαι
μόνον ἄνθρωπον αἴτιον ἥ τε λεπτότης τοῦ δέρματος
καὶ τὸ μόνον γελᾶν τῶν ζῴων ἄνθρωπον. ὁ δὲ
γαργαλισμὸς γέλως ἐστὶ διὰ κινήσεως[2] τοιαύτης
10 τοῦ μορίου τοῦ περὶ τὴν μασχάλην.

Συμβαίνειν δέ φασι καὶ περὶ τὰς ἐν τοῖς πολέμοις
πληγὰς εἰς τὸν τόπον τὸν περὶ τὰς φρένας γέλωτα
διὰ τὴν ἐκ τῆς πληγῆς γινομένην θερμότητα. τοῦτο

---

[1] θερμαινόμενον Peck : θερμαίνουσι vulg. : -ουσα SZ : -ουσαν
PUY. [2] κνήσεως Langkavel.

[a] The *Risus Sardonicus* : see Allbutt and Rolleston, *A System of Medicine*[2] (1910), viii. 642.

what actually happens: whenever, owing to their proximity, they draw up the hot residual fluid, this at once causes a recognizable disturbance of the intelligence and of sensation. And that is why they are called *phrenes* : as if they took a part in the act of thinking (*phronein*). This of course they do not do ; but their proximity to those organs which do so take part makes the change of condition in the intelligence recognizable. That, too, is why the *phrenes* are thin in the middle ; this is not due entirely to necessity (though as they are fleshy to begin with, the parts of them nearest the ribs must of necessity be more fleshy still); there is another reason, which is, to enable them to have as little moisture in them as possible, since if they had been wholly of flesh they would have tended to draw to themselves and to retain a large quantity of moisture. Another indication that it is when heated that they quickly make the sensation recognizable is afforded by what happens when we laugh. When people are tickled, they quickly burst into laughter, and this is because the motion quickly penetrates to this part, and even though it is only gently warmed, still it produces a movement (independently of the will) in the intelligence which is recognizable. The fact that human beings only are susceptible to tickling is due (1) to the fineness of their skin and (2) to their being the only creatures that laugh. Tickling means, simply, laughter produced in the way I have described by a movement applied to the part around the armpit.

It is said that when in war men are struck in the part around the diaphragm, they laugh[a] on account of the heat which arises owing to the blow.

673 a

γὰρ μᾶλλόν ἐστιν ἀξιοπίστων ἀκοῦσαι λεγόντων ἢ
τὸ περὶ τὴν κεφαλήν, ὡς ἀποκοπεῖσα φθέγγεται
15 τῶν ἀνθρώπων. λέγουσι γάρ τινες ἐπαγόμενοι καὶ
τὸν Ὅμηρον, ὡς διὰ τοῦτο ποιήσαντος

    φθεγγομένη δ' ἄρα τοῦ γε κάρη κονίῃσιν
    ἐμίχθη,

ἀλλ' οὐ φθεγγομένου. περὶ δὲ Ἀρκαδίαν[1] οὕτω
τὸ τοιοῦτον διεπίστευσαν ὥστε καὶ κρίσιν ἐποιή-
σαντο περί τινος τῶν ἐγχωρίων. τοῦ γὰρ ἱερέως
20 τοῦ ὁπλοσμίου Διὸς ἀποθανόντος, ὑφ' ὅτου δὲ ἀδήλου ὄντος,[2] ἔφασάν τινες ἀκοῦσαι τῆς κεφαλῆς
ἀποκεκομμένης λεγούσης πολλάκις

    ἐπ' ἀνδρὸς ἄνδρα Κερκιδᾶς ἀπέκτεινεν·

διὸ καὶ ζητήσαντες ᾧ ὄνομα ἦν ἐν τῷ τόπῳ
Κερκιδᾶς, ἔκριναν. ἀδύνατον δὲ φθέγγεσθαι κεχω-
ρισμένης τῆς ἀρτηρίας καὶ ἄνευ τῆς ἐκ τοῦ πλεύ-
25 μονος κινήσεως. παρά τε τοῖς βαρβάροις, παρ'
οἷς ἀποτέμνουσι ταχέως τὰς κεφαλάς, οὐδέν πω
τοιοῦτον συμβέβηκεν. ἔτι δ' ἐπὶ τῶν ἄλλων ζῴων
διὰ τίν' αἰτίαν οὐ γίνεται; [τὸ μὲν γὰρ τοῦ γέλω-
τος πληγεισῶν τῶν φρενῶν εἰκότως, οὐδὲν γὰρ γελᾷ
τῶν ἄλλων· προϊέναι δέ ποι τὸ σῶμα τῆς κεφαλῆς
30 ἀφῃρημένης οὐδὲν ἄλογον, ἐπεὶ τά γ' ἄναιμα καὶ

---

[1] ἀρκαδίαν Z, probat J. Schaefer *de Jove apud Cares culto*,
pp. 370 sq.: Καρίαν vulg.: καρ . . αν E: καρ   P.
[2] δὲ ἀδήλου ὄντος Peck: δὲ δὴ ἀδήλως vulg.: codd. varia.

---

[a] *Iliad*, x. 457 and *Od.* xxii. 329. In both places the
text of Homer has φθεγγομένου (" As he spake . . .").

    [b] The Berlin text here reads " Caria," but the Oxford ms.
Z reads " Arcadia." A cult of Zeus *hoplosmios* is attested
only for Methydrion, a town in Arcadia, and the name
Kerkidas is found in Arcadia, not in Caria. (See A. B.

This may be so ; and those who assert it are more credible than those who tell the tale of how a man's head speaks after it is cut off. Sometimes they cite Homer in support, who (so they say) was referring to this when he wrote

As it spake, his head was mingled with the dust

(not

As he spake, his head was mingled with the dust.) [a]

In Arcadia [b] this kind of thing was at one time so firmly believed that one of the inhabitants was actually brought into court on the strength of it. The priest of Zeus *hoplosmios* had been killed, but no one knew who had done it. Certain persons, however, affirmed that they had heard the man's head, after it had been cut off, repeating the following line several times

'Twas Kerkidas did slaughter man on man.

So they set to work and found someone in the district who bore this name and brought him to trial. Of course, speech is impossible once the windpipe has been severed and no motion is forthcoming from the lung. And among the barbarians, where they cut heads off with expedition, nothing of this sort has taken place so far. Besides, why does it not occur with the other animals ? [For (*a*) the story about the laughter when the diaphragm has been struck is plausible, for none of the others laughs ; and (*b*) that the body should go forward some distance after the head has been cut off, is not at all absurd, since bloodless animals at any rate actually go on

Cook, *Zeus*, ii. 290, who gives the evidence, and J. Schaefer, *De Jove apud Cares culto*, 1912, pp. 370 f.)

ζῇ πολὺν χρόνον· δεδήλωται δὲ περὶ τῆς αἰτίας
αὐτῶν ἐν ἑτέροις.]¹

Τίνος μὲν οὖν ἕνεκέν ἐστιν ἕκαστον τῶν σπλάγχνων, εἴρηται· γέγονε δ' ἐξ ἀνάγκης ἐπὶ τοῖς ἐντὸς
πέρασι τῶν φλεβῶν, ἐξιέναι τε γὰρ ἰκμάδα ἀναγ-
673 b καῖον, καὶ ταύτην αἱματικήν, ἐξ ἧς συνισταμένης
καὶ πηγνυμένης γίνεσθαι τὸ σῶμα τῶν σπλάγχνων·
διόπερ αἱματικά, καὶ αὐτοῖς μὲν ὁμοίαν ἔχουσι τὴν
τοῦ σώματος φύσιν, τοῖς δ' ἄλλοις ἀνομοίαν.

XI. Πάντα δὲ τὰ σπλάγχνα ἐν ὑμένι ἐστίν·
5 προβολῆς τε γὰρ δεῖ πρὸς τὸ ἀπαθῆ εἶναι, καὶ
ταύτης ἐλαφρᾶς, ὁ δ' ὑμὴν τὴν φύσιν τοιοῦτος·
πυκνὸς μὲν γὰρ ὥστ' ἀποστέγειν, ἄσαρκος δὲ ὥστε
μὴ ἕλκειν μηδ' ἔχειν ἰκμάδα, λεπτὸς δ' ὅπως κοῦ-
φος ᾖ καὶ μηδὲν ποιῇ βάρος. μέγιστοι δὲ καὶ
ἰσχυρότατοι τῶν ὑμένων εἰσὶν οἵ τε περὶ τὴν
10 καρδίαν καὶ περὶ τὸν ἐγκέφαλον, εὐλόγως· ταῦτα
γὰρ δεῖται πλείστης φυλακῆς· ἡ μὲν γὰρ φυλακὴ
περὶ τὰ κύρια, ταῦτα δὲ κύρια μάλιστα τῆς ζωῆς.

XII. Ἔχουσι δ' ἔνια μὲν τῶν ζώων πάντα τὸν
ἀριθμὸν αὐτῶν, ἔνια δ' οὐ πάντα· ποῖα δὲ ταῦτα καὶ
διὰ τίν' αἰτίαν, εἴρηται πρότερον. καὶ τῶν ἐχόντων
15 δὲ ταῦτα διαφέρουσιν· οὐ γὰρ ὁμοίας οὔτε τὰς
καρδίας ἔχουσι πάντα τὰ ἔχοντα καρδίαν, οὔτε τῶν
ἄλλων ὡς εἰπεῖν οὐδέν. τό τε γὰρ ἧπαρ τοῖς μὲν
πολυσχιδές ἐστι τοῖς δὲ μονοφυέστερον, πρῶτον

¹ codd. edd. varia ; corrupta et inepta seclusi.

living for a long time. The reason for these phenomena has been explained elsewhere.]

We have now said what is the *purpose* for which
each of the viscera is present; but also they have
been formed *of necessity* at the inner ends of the blood-
vessels, because moisture, *i.e.* moisture of a blood-
like nature, must of necessity make its way out there,
and, as it sets and solidifies, form the substance of
the viscera. That, too, is why they are blood-like in
character, and why the substance of all of them is
similar, though different from that of the other
parts.

XI. All the viscera are enclosed in membranes. Membranes.
Some covering is needed to ensure their safety, and
it must be a light one. These conditions are fulfilled
by a membrane, which is close-textured, thus making
a good protection; does not consist of flesh, and
therefore does not draw in moisture or retain it; is
thin, therefore light, and causes no burden. The
biggest and strongest membranes are those round
the heart and the brain, which is natural enough, as
it is always the controlling power which has to be
protected; therefore the heart and the brain, which
have the supreme controlling power over the life of
the body, need the most protection.

XII. Some animals possess a full complement of Variations
in the
viscera.
viscera, some do not. We have already stated what
animals have less than the full number, and the
reason. But also, the same viscera are different in
the various animals that have them. For instance,
the heart is not identical in all the animals which have
a heart; nor is any other of the viscera. The liver
illustrates this: in some it is split into several parts,
in some almost undivided. This variation of form is

673 b

αὐτῶν τῶν ἐναίμων καὶ ζωοτόκων· ἔτι δὲ μᾶλλον
καὶ πρὸς ταῦτα καὶ πρὸς ἄλληλα διαφέρει τά τε τῶν
20 ἰχθύων καὶ ⟨τῶν⟩[1] τετραπόδων καὶ ᾠοτόκων. τὸ δὲ
τῶν ὀρνίθων μάλιστα προσεμφερὲς τῷ τῶν ζωοτόκων ἐστὶν ἥπατι· καθαρὸν γὰρ καὶ ἔναιμον τὸ
χρῶμα αὐτῶν ἐστι καθάπερ κἀκείνων. αἴτιον δὲ
τὸ τὰ σώματα τούτων εὐπνούστατα εἶναι καὶ μὴ
πολλὴν ἔχειν φαύλην περίττωσιν. διόπερ ἔνια καὶ
25 οὐκ ἔχει χολὴν τῶν ζωοτόκων· τὸ γὰρ ἧπαρ συμ-
βάλλεται πολὺ μέρος πρὸς εὐκρασίαν τοῦ σώματος
καὶ ὑγίειαν· ἐν μὲν γὰρ τῷ αἵματι μάλιστα τὸ
τούτων τέλος, τὸ δ' ἧπαρ αἱματικώτατον μετὰ τὴν
καρδίαν τῶν σπλάγχνων. τὰ δὲ τῶν τετραπόδων καὶ
ᾠοτόκων καὶ τῶν ἰχθύων ἔνωχρα τῶν πλείστων,
30 ἐνίων δὲ καὶ φαῦλα παντελῶς, ὥσπερ καὶ τὰ σώματα φαύλης τετύχηκε κράσεως, οἷον φρύνης καὶ
χελώνης καὶ τῶν ἄλλων τῶν τοιούτων.

Σπλῆνα δ' ἔχει τὰ μὲν κερατοφόρα καὶ δίχαλα
στρογγύλον, καθάπερ αἲξ καὶ πρόβατον καὶ τῶι
ἄλλων ἕκαστον, εἰ μή τι διὰ μέγεθος εὐαυξέστεροι
674 a ἔχει κατὰ μῆκος, οἷον ὁ τοῦ βοὸς πέπονθεν· τὰ δε
πολυσχιδῆ πάντα μακρόν, οἷον ὗς καὶ ἄνθρωπος και
κύων, τὰ δὲ μώνυχα μεταξὺ τούτων καὶ μικτόν· τῇ
μὲν γὰρ πλατὺν ἔχει τῇ δὲ στενόν, οἷον ἵππος καὶ
ὀρεὺς καὶ ὄνος.

5 XIII. Οὐ μόνον δὲ διαφέρει τὰ σπλάγχνα τῆς
σαρκὸς τῷ ὄγκῳ τοῦ σώματος, ἀλλὰ καὶ τῷ τὴν[2]
μὲν ἔξω τὰ δ' ἔσω τὴν θέσιν ἔχειν. αἴτιον δ' ὅτι

---

[1] ⟨τῶν⟩ Peck.   [2] τὴν ESUYZ : τὰ vulg.

ᵃ See above, on 650 b 24.   Cf. 677 a 19 ff.

found first of all even among the viviparous blooded animals ; but it is more noticeable among the fishes and oviparous quadrupeds, whose livers differ not only from those of the Vivipara, but also from each other's. In birds, the liver very closely resembles that of the Vivipara : in both, its colour is pure and blood-like. The reason for which is, that their bodies give a very free passage to the breath, which means that they retain very little foul residue; hence, indeed, some of the Vivipara have no gall-bladder, and this is largely due to the very considerable assistance given by the liver in maintaining a good blend*a* and healthiness in the body. This is because the purpose which these viscera serve lies chiefly in the blood, and after the heart the liver contains more blood than any other of the viscera. In most of the oviparous quadrupeds and the fishes the liver is yellowish, and in some of them it is altogether bad-looking, on a par with the bad blend of the rest of their bodies. This happens in the toad, the tortoise, and the like.

As for the spleen : In horned animals that have cloven hoofs it is rounded : *e.g.* in the goat, the sheep, and similar animals ; unless greatness of size has made it grow out at some point lengthways, as in the case of the ox. In all the polydactylous animals the spleen is long, as in the pig, in man, and in the dog. In animals with solid hoofs the spleen is intermediate between the two and has the characteristics of both : in one place it is broad, in another narrow, as exemplified in the horse, the mule, and the ass.

XIII. Now the viscera differ from the flesh not only in the bulkiness of their mass, but also in their situation, for the flesh is on the outside of the body, while they are inside. The reason for this is that

874 a

τὴν φύσιν ἔχει κοινωνοῦσαν ταῖς φλεψί, καὶ τὰ μὲν τῶν φλεβῶν χάριν, τὰ δ᾽ οὐκ ἄνευ φλεβῶν ἐστιν. XIV. Ὑπὸ δὲ τὸ ὑπόζωμα κεῖται ἡ κοιλία τοῖς

10 ζῴοις, τοῖς μὲν ἔχουσιν οἰσοφάγον ᾗ τελευτᾷ τοῦτο τὸ μόριον, τοῖς δὲ μὴ ἔχουσιν εὐθὺς πρὸς τῷ στόματι· τῆς δὲ κοιλίας ἐχόμενον τὸ καλούμενον ἔντερον.

Δι᾽ ἣν δ᾽ αἰτίαν ἔχει ταῦτα τὰ μόρια τῶν ζῴων ἕκαστον, φανερὸν πᾶσιν. καὶ γὰρ δέξασθαι τὴν εἰσελθοῦσαν τροφὴν καὶ τὴν ἐξικμασμένην ἀναγ-

15 καῖον ἐκπέμψαι, καὶ μὴ τὸν αὐτὸν τόπον εἶναι τῆς τ᾽ ἀπέπτου καὶ τοῦ περιττώματος, εἶναί τέ τινα δεῖ τόπον ἐν ᾧ μεταβάλλει. τὸ μὲν γὰρ τὴν εἰσ-ελθοῦσαν ἕξει μόριον, τὸ δὲ τὸ περίττωμα τὸ ἄχρη-στον· ὥσπερ δὲ χρόνος ἕτερος ἑκατέρου τούτων, ἀναγκαῖον διειλῆφθαι καὶ τοῖς τόποις. ἀλλὰ περὶ

20 μὲν τούτων ἐν τοῖς περὶ τὴν γένεσιν καὶ τὴν τροφὴν οἰκειότερός ἐστιν ὁ διορισμός· περὶ δὲ τῆς διαφορᾶς τῆς κοιλίας καὶ τῶν συντελῶν μορίων νῦν ἐπι-σκεπτέον.

Οὔτε γὰρ τοῖς μεγέθεσιν οὔτε τοῖς εἴδεσιν ὁμοίας ἔχουσιν ἀλλήλοις τὰ ζῷα· ἀλλ᾽ ὅσα μέν ἐστιν αὐτῶν ἀμφώδοντα τῶν ἐναίμων καὶ τῶν ζωοτόκων, μίαν

25 ἔχει κοιλίαν, οἷον ἄνθρωπος καὶ κύων καὶ λέων καὶ τἆλλα ὅσα πολυδάκτυλα, καὶ ὅσα μώνυχα, οἷον ἵππος, ὀρεύς, ὄνος, καὶ ὅσα δίχαλα μὲν ἀμφώδοντα δέ, οἷον ὗς, πλὴν εἴ[1] τι διὰ μέγεθος τοῦ σώματος

---

[1] ὑσπλὴξ ἢ εἰ ESUY (ἢ om. E) : ὑσπλὴξ πλὴν εἰ P et corr. U : ὑσπληγξ in ras. et supra καὶ χοῖρος Z². tum πλὴν εἰ Z¹ : ὗς, εἰ μή Bekker : ὗς, πλὴν εἰ μή Buss.

ᵃ See De gen. an. Bk. II. chh. 6 and 7.

288

their nature shares that of the blood-vessels : some of them exist for the sake of the blood-vessels, others do not exist apart from the blood-vessels.

XIV. Below the diaphragm is the Stomach, which is placed where the oesophagus ends (if there is an oesophagus ; if not, immediately next to the mouth). Next after the stomach and continuous with it is what is called the Gut. Stomach and Intestines.

It must be obvious to everyone why all animals have these parts. It is a necessity for them to have some receptacle for the food they take in, and to expel it again when its moisture has been extracted from it ; and there must be two different places for these two things—the unconcocted food and the residue ; there must also be another place in which the change from one to the other is effected. Two receptacles, then, one for the incoming food, one for the residue which is no more use—as there is a separate time for these so there must be a separate place. However, it will be more appropriate to go into these matters in our treatise on *Generation and Nutrition.*[a] At the present we must consider the variations that are to be found in the stomach and its subsidiary parts.

The stomach differs both in size and appearance in different animals. Those of the blooded Vivipara which have front teeth in both jaws have one stomach ; *e.g.* man, the dog, the lion, and the other polydactyls ; so also those that have solid hoofs, *e.g.* the horse, the mule, the ass ; and those which although they are cloven-hoofed have front teeth in both jaws, *e.g.* the pig. These rules apply unless the size of the frame and the character of the food

674 a

καὶ τὴν τῆς τροφῆς δύναμιν, οὖσαν οὐκ εὔπεπτον
30 ἀλλ' ἀκανθώδη καὶ ξυλικήν, ἔχει πλείους, οἷον
κάμηλος, ὥσπερ καὶ τὰ κερατοφόρα· τὰ γὰρ
κερατοφόρα οὐκ ἔστιν ἀμφώδοντα. διὰ τοῦτο δὲ καὶ
ἡ κάμηλος οὐ τῶν ἀμφωδόντων ἐστίν, ἀκέρατος
οὖσα, διὰ τὸ ἀναγκαιότερον εἶναι αὐτῇ τὴν κοιλίαν
ἔχειν τοιαύτην ἢ τοὺς προσθίους ὀδόντας. ὥστ'

674 b ἐπεὶ ταύτην ὁμοίαν ἔχει τοῖς μὴ ἀμφώδουσι, καὶ τὰ
περὶ τοὺς ὀδόντας ὁμοίως ἔχει αὐτῇ, ὡς οὐδὲν ὄντας
προέργου. ἅμα δὲ καὶ ἐπεὶ ἡ τροφὴ ἀκανθώδης,
τὴν δὲ γλῶτταν ἀνάγκη σαρκώδη εἶναι, πρὸς
σκληρότητα τοῦ οὐρανοῦ κατακέχρηται τῷ ἐκ τῶν
5 ὀδόντων γεώδει ἡ φύσις. καὶ μηρυκάζει δ' ἡ
κάμηλος ὥσπερ τὰ κερατοφόρα, διὰ τὸ τὰς κοιλίας
ὁμοίας ἔχειν τοῖς κερατοφόροις. τούτων δ' ἕκαστον
πλείους ἔχει κοιλίας, οἷον πρόβατον, βοῦς, αἴξ,
ἔλαφος, καὶ τἆλλα τὰ τοιαῦτα τῶν ζῴων, ὅπως
ἐπειδὴ τῆς ἐργασίας ἐλλείπει περὶ τὴν τροφὴν ἡ
10 λειτουργία ἡ τοῦ στόματος διὰ τὴν ἔνδειαν τῶν
ὀδόντων, ἡ τῶν κοιλιῶν ἑτέρα πρὸς ἑτέρας δέχηται[1]
τὴν τροφήν, ἡ μὲν ἀκατέργαστον, ἡ δὲ κατειργασμένην μᾶλλον, ἡ δὲ πάμπαν, ἡ δὲ λείαν. διὸ τὰ
τοιαῦτα τῶν ζῴων πλείους ἔχει τόπους καὶ μόρια.
15 καλοῦνται δὲ ταῦτα κοιλία καὶ κεκρύφαλος καὶ
ἐχῖνος καὶ ἤνυστρον. ὃν δ' ἔχει τρόπον ταῦτα πρὸς

---

[1] δέχηται Peck : δεχομένη vulg.

modify them : for instance, if the food is thorny and
woody and therefore not easy to concoct, in which
case the animal has several stomachs, *e.g.* the camel ;
so also have the horned animals, as they have not front
teeth in both jaws. Thus also the camel has not the
two rows of front teeth either, although it has no
horns ; this is because it is more necessary for the
camel to have several stomachs than to have all these
front teeth. So, as it resembles the animals which
lack the upper front teeth in that it has several
stomachs, therefore the arrangement of its teeth
is that which normally accompanies the multiple
stomachs : in other words, it lacks these front teeth,
as they would be no use to it. And also, as its food
is thorny, and as the tongue has of necessity to be
of a fleshy character, Nature has made use of the
earthy matter saved from the missing teeth to make
the roof of the mouth hard. Again, the camel
ruminates as the horned animals do, because it has
stomachs that resemble theirs. Every one of the
horned animals (such as the sheep, the ox, the goat,
the deer, and the like) has several stomachs ; and the
purpose of them is this : Since the mouth is deficient
in teeth, the service which it performs upon the food
is deficient ; and so one stomach after another
receives the food, which is quite untreated when it
enters the first stomach, more treated in the next,
completely treated in the next, and a smooth pulp
in the next. And that is why these animals have
several such places or parts, the names of which are
(1) the paunch (*rumen*), (2) the net or honeycomb-bag
(*reticulum*), (3) the manyplies (*omasum*), (4) the reed [a]
(*abomasum*). For the relation of these to each other

---

[a] Or, true stomach.

674 b

ἄλληλα τῇ θέσει καὶ τοῖς εἴδεσιν, ἔκ τε τῆς ἱστορίας
τῆς περὶ τὰ ζῷα δεῖ θεωρεῖν καὶ ἐκ τῶν ἀνατομῶν.

Διὰ τὴν αὐτὴν δ' αἰτίαν καὶ τὸ τῶν ὀρνίθων
γένος ἔχει διαφορὰν περὶ τὸ τῆς τροφῆς δεκτικὸν
20 μόριον. ἐπεὶ γὰρ οὐδὲ ταῦτα ὅλως τὴν τοῦ στόματος ἀποδίδωσι λειτουργίαν (ἀνόδοντα γάρ) καὶ
οὔθ' ᾧ διαιρήσει οὔθ' ᾧ λεανεῖ τὴν τροφὴν ἔχουσι,
διὰ τοῦτο τὰ μὲν πρὸ τῆς κοιλίας ἔχουσι τὸν
καλούμενον πρόλοβον ἀντὶ τῆς τοῦ στόματος ἐργασίας, οἱ δὲ τὸν οἰσοφάγον πλατύν, ἢ πρὸ τῆς κοιλίας
25 αὐτοῦ μέρος τι ὀγκῶδες ἐν ᾧ προθησαυρίζουσι τὴν
ἀκατέργαστον τροφήν, ἢ τῆς κοιλίας αὐτῆς τι
ἐπανεστηκός, οἱ δ' αὐτὴν τὴν κοιλίαν ἰσχυρὰν καὶ
σαρκώδη πρὸς τὸ δύνασθαι πολὺν χρόνον θησαυρί-
ζειν καὶ πέττειν ἀλείαντον οὖσαν τὴν τροφήν· τῇ
δυνάμει γὰρ καὶ τῇ θερμότητι τῆς κοιλίας ἡ φύσις
30 ἀναλαμβάνει τὴν τοῦ στόματος ἔνδειαν. εἰσὶ δέ
τινες οἳ τούτων οὐδὲν ἔχουσιν, ἀλλὰ τὸν πρόλοβον[1]
μακρόν, ὅσα μακροσκελῆ καὶ ἕλεια, διὰ τὴν τῆς
τροφῆς ὑγρότητα. αἴτιον δ' ὅτι ἡ τροφὴ πᾶσι
τούτοις εὐλέαντος, ὥστε συμβαίνειν διὰ ταῦτα τῶν
τοιούτων τὰς κοιλίας εἶναι ὑγρὰς [διὰ τὴν ἀπεψίαν
καὶ τὴν τροφήν].[2]

675 a Τὸ δὲ τῶν ἰχθύων γένος ἔχει μὲν ὀδόντας, τούτους δὲ καρχαρόδοντας σχεδὸν ὡς εἰπεῖν πάντες[3]·
ὀλίγον γάρ τί ἐστι γένος τὸ μὴ τοιοῦτον, οἷον ὁ
καλούμενος σκάρος, ὃς δὴ καὶ δοκεῖ μηρυκάζειν

---

[1] πρόλοβον] στόμαχον Ogle, collato Hist. An. 509 a 9.
[2] secludenda.
[3] πάντες Ogle: πάντας vulg.

---

[a] At 507 a 36 ff. [b] The gizzard.
[c] Ogle reads " oesophagus."

as regards position and appearance, the *Researches
upon Animals* [a] and the treatises on *Anatomy* should be
consulted.

The same reason as has just been described accounts
for the difference which presents itself in birds in the
part which receives the food. Birds, like the other
animals, do not get the full service from the mouth in
dealing with the food—since they have no teeth at
all, and they have nothing with which to bite up or
grind down the food ; and so some of them have,
before the stomach, what is called the crop, to perform the work instead of the mouth. Others have
a broad oesophagus ; or their oesophagus has a
bulge in it, just before it reaches the stomach, in
which they keep a preliminary store of untreated
food ; or some part of the stomach itself sticks out.
Others have a strong and fleshy stomach [b] which
is thus able to store the food up for a long period
and to concoct it although it has not been ground
down ; thus Nature makes up for the deficiency of
the mouth by means not only of the heat of the
stomach but also by its special character. Other
birds have none of these devices, but a long crop,[c]
because their food is moist : these are the long-legged
marsh birds. The reason for this is that the food
which all of these take is easily ground down, and
the result is that the stomachs of birds of this sort
are moist [owing to the unconcocted and moist state
of the food].

The tribe of fishes have teeth : practically all have
saw-teeth. There is one small group to which this
does not apply, *e.g.* the Scarus,[d] as it is called, and
it seems reasonable to suppose that this is why

[d] The parrot-fish ; see above, 662 a 7.

5 εὐλόγως διὰ ταῦτα μόνος· καὶ γὰρ τὰ μὴ ἀμφώδοντα κερατοφόρα δὲ μηρυκάζει. ὀξεῖς δὲ πάντας[1]
ἔχουσιν, ὥστε διελεῖν μὲν δύνανται, φαύλως δὲ δι-
ελεῖν· ἐνδιατρίβειν γὰρ οὐχ οἷόν τε χρονίζοντας· διόπερ οὐδὲ πλατεῖς ἔχουσιν ὀδόντας, οὐδ᾽ ἐνδέχεται λεαίνειν· μάτην ἂν οὖν εἶχον. ἔτι δὲ στόμαχον
10 οἱ μὲν ὅλως οὐκ ἔχουσιν, οἱ δὲ βραχύν. ἀλλὰ πρὸς
τὴν βοήθειαν τῆς πέψεως οἱ μὲν ὀρνιθώδεις ἔχουσι
τὰς κοιλίας καὶ σαρκώδεις, οἷον κεστρεύς, οἱ δὲ
πολλοὶ παρὰ τὴν κοιλίαν ἀποφυάδας πυκνάς, ἵν᾽
ἐν ταύταις ὥσπερ ἐν προλακκίοις θησαυρίζοντες
συσσήπωσι καὶ πέττωσι τὴν τροφήν. ἔχουσι δ᾽
15 ἐναντίως οἱ ἰχθύες τοῖς ὄρνισι τὰς ἀποφυάδας· οἱ
μὲν γὰρ ἰχθύες ἄνω πρὸς τῇ κοιλίᾳ, τῶν δ᾽ ὀρνίθων
οἱ ἔχοντες ἀποφυάδας κάτω πρὸς τῷ τέλει τοῦ
ἐντέρου. ἔχουσι δ᾽ ἀποφυάδας ἔνια καὶ τῶν ζῳοτόκων ἐντερικὰς κάτω διὰ τὴν αὐτὴν αἰτίαν.

Τὸ δὲ τῶν ἰχθύων γένος ἅπαν, διὰ τὸ ἐνδεεστέρως
20 ἔχειν τὰ περὶ τὴν τῆς τροφῆς ἐργασίαν, ἀλλ᾽
ἄπεπτα διαχωρεῖν, λαίμαργον πρὸς τὴν τροφήν
ἐστι, καὶ τῶν ἄλλων δὲ πάντων ὅσα εὐθύεντερα·
ταχείας γὰρ γινομένης τῆς διαχωρήσεως, καὶ διὰ
ταῦτα βραχείας οὔσης τῆς ἀπολαύσεως, ταχεῖαν
ἀναγκαῖον γίνεσθαι πάλιν καὶ τὴν ἐπιθυμίαν.

25 Τὰ δ᾽ ἀμφώδοντα ὅτι μὲν μικρὰν ἔχει κοιλίαν
εἴρηται πρότερον· εἰς διαφορὰς δὲ πίπτουσι δύο
πᾶσαι σχεδόν· τὰ μὲν γὰρ τῇ τῆς κυνὸς ὁμοίαν

---

[1] πάντας S : πάντες vulg.

---

[a] Probably some kind of mullet.
[b] "Caecal appendages" (Ogle), or "alimentary sacs."
[c] The vermiform appendix.

294

it alone ruminates, for horned animals which have no teeth in the upper jaw also ruminate. All teeth in fish are sharp; this enables them to bite up their food, though somewhat unsatisfactorily; this is because they cannot spend long over mastication; hence they neither have flat teeth nor may they grind the food down; therefore it would be idle to have the teeth. Furthermore, some fishes have no gullet at all, others have a short one; but, in order to assist the process of concoction, some of them, like the Kestreus,[a] have fleshy stomachs, similar to those of birds; the majority, however, have a large number of appendages [b] by the side of the stomach, in which to store up the food as it might be in additional cellars and there putrefy it up and concoct it. The appendages of fishes are, however, quite different from those of birds. In fishes they are fairly high up beside the stomach, whereas when present in birds they are down below at the end of the gut. Some of the Vivipara also have appendages [c] of this latter kind, and their purpose is the same.

The whole race of fishes is gluttonous for food, because their equipment for reducing it is defective, as a result of which most of it passes through unconcocted. Of all, those which have a straight intestine are especially gluttonous, since the food passes through quickly, which means that their enjoyment of it is brief, and therefore in its turn the desire for food must come on again very quickly.

I have already said that in animals with front teeth in both jaws the stomach is small. These stomachs fall into two main classes. Some have a stomach resembling that of the dog, some that of

675 a

ἔχουσι κοιλίαν, τὰ δὲ τῇ τῆς ὑός· ἔστι δ' ἡ μὲν τῆς
ὑὸς μείζων καί τινας ἔχουσα μετρίας πλάκας πρὸς
τὸ χρονιωτέραν γίνεσθαι τὴν πέψιν, ἡ δὲ τῆς κυνὸς
30 μικρὰ τὸ μέγεθος καὶ οὐ πολὺ τοῦ ἐντέρου ὑπερ-
βάλλουσα καὶ λεῖα τὰ ἐντός. μετὰ γὰρ τὴν κοιλίαν
ἡ τῶν ἐντέρων ἔγκειται φύσις πᾶσι τοῖς ζῴοις. ἔχει
δὲ διαφορὰς πολλάς, καθάπερ ἡ κοιλία, καὶ τοῦτο
τὸ μόριον. τοῖς μὲν γὰρ ἁπλοῦν ἐστι καὶ ὅμοιον
ἀναλυόμενον, τοῖς δ' ἀνόμοιον· ἐνίοις μὲν γὰρ εὐρύ-
35 τερον τὸ πρὸς τῇ κοιλίᾳ, τὸ δὲ πρὸς τῷ τέλει
στενότερον[1] (διόπερ αἱ κύνες μετὰ πόνου προΐενται
675 b τὴν τοιαύτην περίττωσιν), τοῖς δὲ πλείοσιν ἄνωθεν
στενότερον,[1] πρὸς τῷ τέλει δ' εὐρύτερον.

Μείζω δὲ καὶ ἀναδιπλώσεις ἔχοντα πολλὰς τὰ
τῶν κερατοφόρων ἐστί, καὶ οἱ ὄγκοι τῆς κοιλίας
τούτοις μείζους καὶ τῶν ἐντέρων διὰ τὸ μέγεθος·
5 πάντα γὰρ ὡς εἰπεῖν μεγάλα τὰ κερατοφόρα διὰ
τὴν κατεργασίαν τὴν τῆς τροφῆς. πᾶσι δὲ τοῖς μὴ
εὐθυεντέροις προϊὸν[2] εὐρύτερον γίνεται τὸ μόριον
τοῦτο, καὶ τὸ καλούμενον κόλον ἔχουσι, καὶ τοῦ
ἐντέρου τυφλόν τι καὶ ὀγκῶδες, εἶτ' ἐκ τούτου
πάλιν στενότερον[3] καὶ εἰλιγμένον. τὸ δὲ μετὰ
10 τοῦτο εὐθὺ πρὸς τὴν ἔξοδον διατείνει τοῦ περιττώματος, καὶ τοῖς μὲν τοῦτο τὸ μόριον, ὁ καλούμενος ἀρχός, κνισώδης ἐστί, τοῖς δ' ἀπίμελος.
πάντα δὲ ταῦτα μεμηχάνηται τῇ φύσει πρὸς τὰς
ἁρμοττούσας ἐργασίας περὶ τὴν τροφὴν καὶ τοῦ
γινομένου περιττώματος. προϊόντι γὰρ καὶ κατα-
βαίνοντι τῷ περιττώματι εὐρυχωρία γίνεται, καὶ
15 πρὸς τὸ μεταβάλλειν ἱσταμένῳ τοῖς εὐχιλοτέροις

---

[1] στενώτερον bis vulg.   [2] προϊὸν Peck: προΐουσιν vulg.
[3] στενότερον SU: στενώτερον vulg.

the pig. The pig's stomach is larger than the dog's, and it has some folds of medium size, so as to prolong the time of concoction. The dog's is small in size—not much bigger indeed than the gut, and its inner surface is smooth. The gut has its place next after the stomach in all animals. Like the stomach, this part too presents many various forms. In some animals it is simple and similar throughout its length, when uncoiled; in others it is not similar throughout. Thus, in some it is wider near the stomach, and narrower towards the end (that is why dogs find difficulty in discharging their excrement); in the majority, however, it is narrower at the top, and wider at the end.

In the horned animals, the intestines are longer and have many convolutions; and their bulk (as well as the bulk of the stomach) is greater, owing to the size of the animal: horned animals being, on the whole, large in size because of the ample treatment which their food receives. Except in those animals where it is straight the intestine gets wider as it proceeds, and they have what is called the colon and the blind and swollen part of the gut [a]; and then after that point it gets narrower again and convoluted. After this, it goes on in a straight line to the place where the residue is discharged; and in some this part (which is called the anus) is supplied with fat, in others it is devoid of fat. All these parts have been devised by Nature to suit their appropriate functions in treating the food and in dealing with the residue produced. As the residue proceeds on its way and goes downwards, it finds a wider space where it remains in order to undergo transformation; this is what

[a] The caecal dilatation.

τῶν ζῴων καὶ πλείονος δεομένοις τροφῆς, διὰ τὸ
μέγεθος ἢ τὴν θερμότητα τῶν τόπων. εἶτ' ἐντεῦθεν πάλιν, ὥσπερ ἀπὸ τῆς ἄνω κοιλίας δέχεται
στενότερον[1] ἔντερον, οὕτως ἐκ τοῦ κώλου καὶ τῆς
εὐρυχωρίας ἐν τῇ κάτω κοιλίᾳ πάλιν εἰς στενό-
20 τερον[1] ἔρχεται καὶ εἰς τὴν ἕλικα τὸ περίττωμα
ἐξικμασμένον πάμπαν, ὅπως ταμιεύηται ἡ φύσις
καὶ μὴ ἀθρόος ᾖ ἡ ἔξοδος τοῦ περιττώματος.

Ὅσα μὲν οὖν εἶναι δεῖ τῶν ζῴων σωφρονέστερα
πρὸς τὴν τῆς τροφῆς ποίησιν εὐρυχωρίας μὲν οὐκ
ἔχει μεγάλας κατὰ τὴν κάτω κοιλίαν, ἕλικας δ'
25 ἔχει πλείους καὶ οὐκ εὐθυέντερά ἐστιν. ἡ μὲν γὰρ
εὐρυχωρία ποιεῖ πλήθους ἐπιθυμίαν, ἡ δ' εὐθύτης
ταχυτῆτα ἐπιθυμίας· διόπερ ὅσα τῶν ζῴων ἢ ἁπλᾶς
ἔχει ἢ εὐρυχώρους τὰς ὑποδοχάς, τὰ μὲν εἰς πλῆθος
γαστρίμαργα τὰ δ' εἰς τάχος ἐστίν.

Ἐπεὶ δ' ἐν τῇ ἄνω μὲν κοιλίᾳ κατὰ τὴν πρώτην
30 εἴσοδον τῆς τροφῆς νεαρὰν ἀναγκαῖον εἶναι τὴν
τροφήν, κάτω δὲ προϊοῦσαν κοπρώδη καὶ ἐξικμασμένην, ἀναγκαῖον εἶναί τι καὶ τὸ μεταξύ,
ἐν ᾧ μεταβάλλει καὶ οὔτ' ἔτι πρόσφατος οὔτ' ἤδη
κόπρος. διὰ τοῦτο πάντα τὰ τοιαῦτα ζῷα τὴν
καλουμένην ἔχει νῆστιν καὶ ἐν τῷ μετὰ τὴν κοιλίαν
35 ἐντέρῳ τῷ λεπτῷ· τοῦτο γὰρ μεταξὺ τῆς τ' ἄνω, ἐν
ᾗ τὸ ἄπεπτον, καὶ τῆς κάτω, ἐν ᾗ τὸ ἄχρηστον ἤδη
περίττωμα. γίνεται δ' ἐν πᾶσι μέν, δήλη δ' ἐν τοῖς

---

[1] στενώτερον bis Langkavel.

[a] i.e. the " stomach."
[b] i.e. the " large intestine."

happens in the animals which need and take more food owing either to their size or to the heat of these parts of the body. After this, just as it goes into a narrower part of the intestine after it leaves the upper gut,[a] so also it goes into a narrower channel after the colon or wide part of the lower gut,[b] and into the spiral coil ; into these the residue passes when its juices have been completely exhausted. In this way Nature is enabled to keep the material in store, and the residue is prevented from passing out all at the same moment.

In those animals, however, which have to be more controlled in their feeding, there are no great wide spaces in the lower gut, but their intestine is not straight, as it contains many convolutions. Spaciousness in the gut causes a desire for bulk of food, and straightness in the intestine makes the desire come on again quickly. Hence, animals of this sort are gluttonous : those with simple receptacles eat at very short intervals of time, those with spacious ones eat very large quantities.

Since the food in the upper gut, when it has just Jejunum. entered, must of necessity be fresh, and when it has proceeded further downwards must have lost its juices and be practically dung, the organ which lies between the two must of necessity be something definite, in which the change is effected, where food is no longer fresh and not yet dung. Therefore all animals of this sort have what is called the *jejunum*, which forms part of the small intestine, which is next to the stomach. That is to say, it has its place between the upper gut, where the unconcocted food is, and the lower gut, where the now useless residue is. All these animals have the *jejunum*, but

676 a μείζοσι καὶ νηστεύσασιν ἀλλ' οὐκ ἐδηδοκόσιν· τότε
γὰρ δὴ[1] οἷον[2] μεταίχμιον γίνεται τῶν τόπων ἀμφοτέρων, ἐδηδοκότων δὲ μικρὸς ὁ καιρὸς τῆς μεταβολῆς. τοῖς μὲν οὖν θήλεσι[3] γίνεται ὅπου ἂν τύχῃ
5 τοῦ ἄνω ἐντέρου ἡ νῆστις· οἱ δ' ἄρρενες[4] ἔχουσι πρὸ
τοῦ τυφλοῦ καὶ τῆς κάτω κοιλίας.

XV. Ἔχουσι δὲ τὴν καλουμένην πυετίαν τὰ μὲν
πολυκοίλια πάντα, τῶν δὲ μονοκοιλίων δασύπους.
ἔχει δὲ τὰ ἔχοντα τῶν πολυκοιλίων τὴν πυετίαν οὔτ'
ἐν τῇ μεγάλῃ κοιλίᾳ οὔτ' ἐν τῷ κεκρυφάλῳ οὔτ' ἐν
10 τῷ τελευταίῳ τῷ ἠνύστρῳ, ἀλλ' ἐν τῷ μεταξὺ τοῦ
τελευταίου καὶ [δύο][5] τῶν πρώτων, ἐν τῷ καλουμένῳ ἐχίνῳ. ἔχει δὲ ταῦτα πάντα πυετίαν διὰ τὴν
παχύτητα τοῦ γάλακτος· τὰ δὲ μονοκοίλια οὐκ
ἔχει, λεπτὸν γὰρ τὸ γάλα τῶν μονοκοιλίων. διὸ
15 τῶν μὲν κερατοφόρων πήγνυται, τῶν δ' ἀκεράτων
οὐ πήγνυται τὸ γάλα. τῷ δὲ δασύποδι γίνεται
πυετία διὰ τὸ νέμεσθαι ὀπώδη πόαν· ὁ γὰρ τοιοῦτος χυμὸς συνίστησιν ἐν τῇ κοιλίᾳ τὸ γάλα τοῖς
ἐμβρύοις. διότι δὲ τῶν πολυκοιλίων ἐν τῷ ἐχίνῳ
γίνεται ἡ πυετία, εἴρηται ἐν τοῖς προβλήμασιν.

[1] δὴ Z : ἤδη vulg.
[2] οἷον PZ, om. vulg.
[3] θήλεσι] τελείοις Z : πλείοσι Platt.
[4] ἄρρενες] κύνες Platt.
[5] [δύο] secludendum.

---

[a] This seems to mean that when the animal is fasting the
two receptacles do not bulge, and so the *jejunum* is visible ;
and though after the animal has fed you might expect to see
the *jejunum*, because it should be full of food which is being

it is apparent only in the larger ones, and in them only when they are fasting, not when they have recently been eating, for when they are fasting, there is an interspace between the two receptacles, whereas when they have been eating, the time taken by the change is short.[a]   In females the *jejunum* can have its place in any part of the upper intestine ; in males it is placed immediately before the caecum and the lower gut.

XV. What goes by the name of Rennet is present in all animals which have a multiple stomach ; the hare is the only animal with a single stomach which has it.   In the former class the rennet is not in the paunch[b] nor in the *reticulum*, nor in the *abomasum* (the last of the stomachs) ; but in the stomach between the last one and the first ones, *i.e.* the so-called *omasum* (manyplies).[c]   All these animals have rennet because their milk is so thick ; similarly, the single-bellied animals have no rennet, because their milk is thin.   This also explains why the milk of horned animals coagulates, while that of the hornless does not.   As for the hare, it has rennet because it feeds on herbs with fig-like juice ; and this juice can coagulate the milk in the stomach of sucklings.   I have stated in the *Problems*[d] why, in the animals that have many stomachs, the rennet is formed in the manyplies.

Rennet.

transmuted inside it (see above, 675 b 32), it is not visible, because the change is effected so rapidly.
  [b] Lit. " the great stomach."
  [c] See above, 674 b 14 ff.
  [d] No such reference can be found.

Τὸν αὐτὸν δὲ τρόπον ἔχει τὰ περὶ τὰ σπλάγχνα
καὶ τὴν κοιλίαν καὶ τῶν εἰρημένων μορίων ἕκαστον
τοῖς τετράποσι μὲν ᾠοτόκοις δὲ τῶν ζῴων καὶ τοῖς
25 ἄποσιν, οἷον τοῖς ὄφεσιν. καὶ γὰρ ἡ τῶν ὄφεων
φύσις ἐστὶ συγγενὴς τούτοις· ὁμοία γάρ ἐστι σαύρῳ
μακρῷ[1] καὶ ἄποδι. τούτοις δὲ καὶ τοῖς ἰχθύσι
πάντα παραπλήσια, πλὴν τὰ μὲν ἔχει πλεύμονα διὰ
τὸ πεζεύειν, οἱ δ' οὐκ ἔχουσιν, ἀλλὰ βράγχια ἀντὶ
τοῦ πλεύμονος. κύστιν δ' οὔθ' οἱ ἰχθύες ἔχουσιν
30 οὔτε τούτων οὐδὲν πλὴν χελώνης· τρέπεται γὰρ εἰς
τὰς φολίδας τὸ ὑγρὸν ὀλιγοπότων ὄντων διὰ τὴν
ἀναιμότητα τοῦ πλεύμονος, καθάπερ τοῖς ὄρνισιν
εἰς τὰ πτερά. καὶ ἐπιλευκαίνει δὲ τὸ περίττωμα
πᾶσι καὶ τούτοις, ὥσπερ καὶ τοῖς ὄρνισιν, διότι[2] ἐν
τοῖς ἔχουσι κύστιν ἐξελθόντος τοῦ περιττώματος
35 ὑφίσταται ἁλμυρὶς γεώδης ἐν τοῖς ἀγγείοις· τὸ γὰρ
γλυκὺ καὶ πότιμον ἀναλίσκεται διὰ κουφότητα εἰς
τὰς σάρκας.

676 b Τῶν δ' ὄφεων οἱ ἔχεις πρὸς τοὺς ἄλλους ἔχουσι
τὴν αὐτὴν διαφορὰν ἣν καὶ ἐν τοῖς ἰχθύσι τὰ
σελάχη πρὸς τοὺς ἄλλους· ζῳοτοκοῦσι γὰρ ἔξω καὶ
τὰ σελάχη καὶ οἱ ἔχεις, ἐν αὑτοῖς ᾠοτοκήσαντα
πρῶτον. μονοκοίλια δὲ πάντα τὰ τοιαῦτά ἐστι,

---

[1] μακρῷ Y : μακρῷ ἢ vulg.   [2] διότι Ogle : διόπερ vulg.

BOOK IV

WHAT has been said already on the subject of the viscera, the stomach, and each of the other parts mentioned, applies to the footless creatures (such as the Serpents) as well as to the oviparous quadrupeds. Indeed, the Serpents are akin to these : for a serpent is like a long and footless lizard. A third class in which all these parts are similar is the Fishes : the only difference is that the first two classes are land-creatures and therefore have a lung, whereas fishes have no lung but gills instead. Fishes have no bladder, nor has any of these creatures (except the tortoise) ; the reason is that they drink little (because their lung is bloodless), and the moisture in them is diverted to the horny scales, just as in birds it is diverted to the feathers. And in all these creatures, as in birds, the residue [a] is white on the surface, since in those animals that have a bladder, when the residue has been voided an earthy salt deposit settles in the vessels, the sweet and non-briny portion, owing to its lightness, being used up upon the flesh.

The Vipers have the same peculiarity among the Serpents as the Selachia have among the Fishes. Both of them are externally viviparous, though they first produce their ova internally. All these

---

[a] See Introduction, pp. 32 ff.

5 καθάπερ τἆλλα τὰ ἀμφώδοντα· καὶ σπλάγχνα δὲ
πάμπαν μικρὰ ἔχει, ὥσπερ τἆλλα τὰ μὴ ἔχοντα
κύστιν. οἱ δ' ὄφεις διὰ τὴν τοῦ σώματος μορφήν,
οὖσαν μακρὰν καὶ στενήν, καὶ τὰ σχήματα τῶν
σπλάγχνων ἔχουσι διὰ ταῦτα μακρὰ καὶ τοῖς τῶν
ἄλλων ζῴων ἀνόμοια, διὰ τὸ καθάπερ ἐν τύπῳ τὰ
10 σχήματ' αὐτῶν πλασθῆναι διὰ τὸν τόπον.

Ἐπίπλοον δὲ καὶ μεσεντέριον καὶ τὰ περὶ τὴν
τῶν ἐντέρων φύσιν, ἔτι δὲ τὸ διάζωμα καὶ τὴν
καρδίαν πάντ' ἔχει τὰ ἔναιμα τῶν ζῴων, πλεύμονα
δὲ καὶ ἀρτηρίαν πάντα πλὴν τῶν ἰχθύων. καὶ τὴν
θέσιν δὲ τῆς ἀρτηρίας καὶ τοῦ οἰσοφάγου πάντα
15 τὰ ἔχοντα ὁμοίως ἔχει διὰ τὰς εἰρημένας αἰτίας
πρότερον.

II. Ἔχει δὲ καὶ χολὴν τὰ πολλὰ τῶν ἐναίμων
ζῴων, τὰ μὲν ἐπὶ τῷ ἥπατι, τὰ δ' ἀπηρτημένην ἐπὶ
τοῖς ἐντέροις, ὡς οὖσαν οὐχ ἧττον ἐκ τῆς κάτω
κοιλίας τὴν φύσιν αὐτῆς. δῆλον δὲ μάλιστ' ἐπὶ τῶν
20 ἰχθύων· οὗτοι γὰρ ἔχουσί τε πάντες, καὶ οἱ πολλοὶ
πρὸς τοῖς[1] ἐντέροις, ἔνιοι δὲ παρ' ὅλον τὸ ἔντερον
παρυφασμένην, οἷον ἡ ἄμια· καὶ τῶν ὄφεων οἱ
πλεῖστοι τὸν αὐτὸν τρόπον. διόπερ οἱ λέγοντες τὴν
φύσιν τῆς χολῆς αἰσθήσεώς τινος εἶναι χάριν οὐ
καλῶς λέγουσιν· φασὶ γὰρ εἶναι διὰ τοῦτο, ὅπως
25 τῆς ψυχῆς τὸ περὶ τὸ ἧπαρ μόριον δάκνουσα μὲν
συνιστῇ, λυομένη δ' ἵλεων ποιῇ· τὰ μὲν γὰρ ὅλως

---

[1] τοῖς PYZ et corr. U : om. vulg.

---

[a] See 665 a 10 ff.　　　　[b] See 650 a 14.
[c] This seems to refer to the views expressed in Plato,
Timaeus, 71 D.

creatures have one stomach only, as do the other animals that have front teeth in both jaws. And their viscera are quite small, as are those of the other creatures which have no bladder. However, on account of the shape of the serpents' bodies, which is long and narrow, the shape of their viscera too is consequently long, thus differing from those of other animals. This is because the shape of them is fashioned, as though in a mould, on account of the space available for them.

All blooded animals have an omentum, a mesentery, and the whole intestinal equipment; also a diaphragm and a heart; and all but the fishes have a lung and a windpipe too. The relative positions of the windpipe and the oesophagus are the same in all of them. The reasons for this have been given already.[a]

II. The majority of the blooded animals have a gall-bladder in addition. In some it is placed up against the liver; in others it is separate from the liver and placed against the intestines, indicating that equally in these its derivation is from the lower gut.[b] This is clearest in the fishes, all of which have one, and in most of them it is placed against the intestines, though in some it runs along the whole length of the intestine, like a woven border, as in the Amia; a similar arrangement is found in most of the serpents. Hence, those who assert that the gall-bladder is present for the sake of some act of sensation are wrong. They say its purpose is as follows :— on the one hand (a) to irritate that part of the Soul which is around the liver, and so to congeal it [c]; and on the other hand (b) by running free to make that part cheerful. This cannot be true; because some

305

676 b

οὐκ ἔχει χολήν, οἷον ἵππος καὶ ὀρεὺς καὶ ὄνος καὶ
ἔλαφος καὶ πρόξ· οὐκ ἔχει δ᾽ οὐδ᾽ ἡ κάμηλος
ἀποκεκριμένην, ἀλλὰ φλέβια χολώδη μᾶλλον· οὐκ
ἔχει δ᾽ οὐδ᾽ ἡ φώκη χολήν, οὐδὲ τῶν θαλαττίων
30 δελφίς. ἐν δὲ τοῖς γένεσι τοῖς αὐτοῖς τὰ μὲν ἔχειν
φαίνεται τὰ δ᾽ οὐκ ἔχειν, οἷον ἐν τῷ τῶν μυῶν·
τούτων δ᾽ ἐστὶ καὶ ὁ ἄνθρωπος, ἔνιοι μὲν γὰρ
φαίνονται ἔχοντες χολὴν ἐπὶ τοῦ ἥπατος, ἔνιοι δ᾽
οὐκ ἔχοντες· διὸ καὶ γίνεται ἀμφισβήτησις περὶ
35 ὅλου τοῦ γένους· οἱ γὰρ ἐντυχόντες ὁποτερωσοῦν
ἔχουσι περὶ πάντων ὑπολαμβάνουσιν ὡς ἁπάντων
ἐχόντων. συμβαίνει δὲ τοιοῦτον καὶ περὶ τὰ πρόβατα καὶ τὰς αἶγας· τὰ μὲν γὰρ πλεῖστα τούτων
677 a ἔχει χολήν, ἀλλ᾽ ἐνιαχοῦ μὲν τοσαύτην ὥστε δοκεῖν
τέρας εἶναι τὴν ὑπερβολήν, οἷον ἐν Νάξῳ, ἐνιαχοῦ
δ᾽ οὐκ ἔχουσιν, οἷον ἐν Χαλκίδι τῆς Εὐβοίας κατά
τινα τόπον τῆς χώρας αὐτῶν. ἔτι δέ, ὥσπερ εἴρη-
5 ται, ἡ τῶν ἰχθύων ἀπήρτηται πολὺ τοῦ ἥπατος.
οὐκ ὀρθῶς δ᾽ ἐοίκασιν οἱ περὶ Ἀναξαγόραν ὑπο-
λαμβάνειν ὡς αἰτίαν οὖσαν τῶν ὀξέων νοσημάτων·
ὑπερβάλλουσαν γὰρ ἀπορραίνειν πρός τε τὸν πλεύμονα καὶ τὰς φλέβας καὶ τὰ πλευρά. σχεδὸν γὰρ
οἷς ταῦτα συμβαίνει τὰ πάθη τῶν νόσων, οὐκ
10 ἔχουσι χολήν, ἔν τε ταῖς ἀνατομαῖς ἂν ἐγίνετο τοῦτο
φανερόν· ἔτι δὲ τὸ πλῆθος τό τ᾽ ἐν τοῖς ἀρρωστή-
μασιν ὑπάρχον καὶ τὸ ἀπορραινόμενον ἀσύμβλητον.
ἀλλ᾽ ἔοικεν ἡ χολή, καθάπερ καὶ ἡ κατὰ τὸ ἄλλο

---

[a] This is true of quite a number of species, and as Aristotle
says, the gall-bladder is specially variable in mice. In man,
its absence is rare ; and Aristotle's statement may well be
derived from his observation of aborted embryos, in which
the gall-bladder develops somewhat late.

animals have no gall-bladder at all, such as the horse, the mule, the ass, the deer, and the roe ; and the camel has no distinct gall-bladder, but what would better be described as consisting of small biliary vessels. There is no gall-bladder in the seal, nor (among sea-animals) in the dolphin. Sometimes in the same group there are some animals which look as if they have one, and some as if they have none [a] : This is true of the Mice ; and also of the human species, as in some individuals the gall-bladder is placed against the liver and is obvious ; while in some it is missing. The result of this has been a dispute concerning the group as a whole. Whatever an observer has found to be the condition of the individuals he happens to have seen, that he holds is true of every individual throughout the group. The same has occurred with regard to sheep and goats, most of which have a gall-bladder ; but, whereas in some individuals it is so large that its excessive size is portentous (*e.g.* in Naxos), in others it is entirely absent (*e.g.* in a particular district of Chalcis, Euboea). A further point, already mentioned, is that in fishes the gall-bladder is separated from the liver by a good distance. Moreover, it is safe to say that Anaxagoras's school is wrong in holding that the gall-bladder is the cause of acute diseases : they say that when it gets too full it spurts its liquid out into the lung and blood-vessels and sides. This must be wrong, because nearly everyone who suffers from these affections actually has no gall-bladder, and this would be proved if they were dissected. Besides, there is no comparison between the amount of bile which is present in these ailments and that which is emitted from the gall-bladder. No ; it seems probable that, just as the

σῶμα γινομένη περίττωμά τί ἐστιν ἢ σύντηξις,
οὕτω καὶ ἡ ἐπὶ τῷ ἥπατι χολὴ περίττωμα εἶναι καὶ
15 οὐχ ἕνεκά τινος, ὥσπερ καὶ ἡ ἐν τῇ κοιλίᾳ καὶ
ἐν τοῖς ἐντέροις ὑπόστασις. καταχρῆται μὲν οὖν
ἐνίοτε ἡ φύσις εἰς τὸ ὠφέλιμον καὶ τοῖς περιττώ-
μασιν, οὐ μὴν διὰ τοῦτο δεῖ ζητεῖν πάντα ἕνεκα
τίνος· ἀλλὰ τινῶν ὄντων τοιούτων ἕτερα ἐξ ἀνάγκης
συμβαίνει διὰ ταῦτα πολλά.

Ὅσοις μὲν οὖν ἡ τοῦ ἥπατος σύστασις ὑγιεινή
20 ἐστι καὶ ἡ τοῦ αἵματος φύσις γλυκεῖα ἡ εἰς τοῦτ'
ἀποκρινομένη, ταῦτα μὲν ἢ πάμπαν οὐκ ἴσχει χολὴν
ἐπὶ τοῦ ἥπατος, ἢ ἔν τισι φλεβίοις, ἢ τὰ μὲν τὰ δ'
οὔ. διὸ καὶ τὰ ἥπατα τὰ τῶν ἀχόλων εὔχρω καὶ
γλυκερά ἐστιν ὡς ἐπίπαν εἰπεῖν, καὶ τῶν ἐχόντων
25 χολὴν τὸ ὑπὸ τῇ χολῇ τοῦ ἥπατος γλυκύτατόν
ἐστιν. τῶν δὲ συνισταμένων ἐξ ἧττον καθαροῦ
αἵματος τούτου[1] ἐστὶν ἡ χολὴ τὸ γινόμενον περίτ-
τωμα· ἐναντίον τε γὰρ τῇ τροφῇ τὸ περίττωμα
βούλεται εἶναι καὶ τῷ γλυκεῖ τὸ πικρόν, καὶ τὸ
αἷμα γλυκὺ τὸ ὑγιαῖνον. φανερὸν οὖν ὅτι οὔ τινος
30 ἕνεκα, ἀλλ' ἀποκάθαρμά ἐστιν ἡ χολή. διὸ καὶ
χαριέστατα λέγουσι τῶν ἀρχαίων οἱ φάσκοντες
αἴτιον εἶναι τοῦ πλείω ζῆν χρόνον τὸ μὴ ἔχειν
χολήν, βλέψαντες ἐπὶ τὰ μώνυχα καὶ τὰς ἐλάφους·
ταῦτα γὰρ ἄχολά τε καὶ ζῇ πολὺν χρόνον. ἔτι δὲ
καὶ τὰ μὴ ἑωραμένα ὑπ' ἐκείνων ὅτι οὐκ ἔχει
35 χολήν, οἷον δελφὶς καὶ κάμηλος, καὶ ταῦτα τυγχάνει μακρόβια ὄντα. εὔλογον γὰρ τὴν τοῦ ἥπατος

[1] τούτου Peck : τοῦτ' vulg.

bile elsewhere in the body is a residue or colliquescence, so this bile around the liver is a residue
and serves no *purpose*—like the sediment produced in the stomach and the intestines. I agree
that occasionally Nature turns even residues to
use and advantage, but that is no reason for trying
to discover a purpose in all of them. The truth is
that some constituents are present for a definite
purpose, and then many others are present *of
necessity* in consequence of these.

We may say, then, that in animals whose liver is
healthy in its composition, and in which the blood
that supplies the liver is sweet, there is either no
gall-bladder at all by the liver, or else the bile is in
tiny vessels, or else in some these are present and in
some not. This is why the livers of gall-bladderless
animals are, generally, of a good colour and sweet ;
and in those that have a gall-bladder the part of the
liver immediately below it is very sweet. But in those
animals which are formed out of blood which is less
pure, the bile is the residue of this ; since " residue "
means that which is the opposite of " food," and
" bitter " the opposite of " sweet "; and healthy blood
is sweet. So it is evident that bile exists for no definite purpose, but is merely an offscouring. So that
was an extremely neat remark which we find made
by some of the old authors, when they say that if you
have no gall in you your life will be longer. This
was a reference to animals with uncloven hoofs and
to deer, which have no gall-bladder, and are long-
lived. And also, certain other animals are long-
lived, such as the dolphin and camel, which, though un-
observed by them, have no gall-bladder. After all,
the liver is vital and indispensable for all blooded

677 a
φύσιν, ἐπίκαιρον οὖσαν καὶ ἀναγκαίαν πᾶσι τοῖς
677 b ἐναίμοις ζῴοις, αἰτίαν εἶναι, ποιάν τιν' οὖσαν, τοῦ
ζῆν ἐλάττω ἢ πλείω χρόνον. καὶ τὸ τούτου μὲν τοῦ
σπλάγχνου εἶναι περίττωμα τοιοῦτον, τῶν δ' ἄλλων
μηδενός, κατὰ λόγον ἐστίν. τῇ μὲν γὰρ καρδίᾳ
τοιοῦτον οὐδένα πλησιάζειν οἷόν τε χυμόν (οὐδὲν
5 γὰρ δέχεται βίαιον πάθος), τῶν δ' ἄλλων οὐδὲν
σπλάγχνων ἀναγκαῖόν ἐστι τοῖς ζῴοις, τὸ δ' ἧπαρ
μόνον· διόπερ καὶ τοῦτο συμβαίνει περὶ αὐτὸ μόνον.
ἄτοπόν τε τὸ μὴ πανταχοῦ νομίζειν, ὅπου ἄν τις ἤδη
φλέγμα ἢ τὸ ὑπόστημα τῆς κοιλίας, περίττωμα
εἶναι, ὁμοίως δὲ δῆλον ὅτι καὶ χολήν, καὶ μὴ
10 διαφέρεσθαι τοῖς τόποις.

Καὶ περὶ μὲν χολῆς, διὰ τίν' αἰτίαν τὰ μὲν ἔχει
τὰ δ' οὐκ ἔχει τῶν ζῴων, εἴρηται, III. περὶ δὲ
μεσεντερίου καὶ ἐπιπλόου λοιπὸν εἰπεῖν· ταῦτα γὰρ
ἐν τῷ τόπῳ τούτῳ καὶ μετὰ τῶν μορίων ἐστὶ
τούτων.

15 Ἔστι δὲ τὸ μὲν ἐπίπλοον ὑμὴν τοῖς μὲν στέαρ
ἔχουσι στεατώδης, τοῖς δὲ πιμελὴν πιμελώδης·
ποῖα δ' ἐστὶν ἑκάτερα τούτων, εἴρηται πρότερον.
ἤρτηται[1] δὲ τὸ ἐπίπλοον ὁμοίως τοῖς τε μονοκοιλίοις
καὶ τοῖς πολυκοιλίοις ἀπὸ μέσης τῆς κοιλίας κατὰ
τὴν ὑπογεγραμμένην οἷον ῥαφήν· ἐπέχει δὲ τό τε
20 λοιπὸν τῆς κοιλίας καὶ τὸ τῶν ἐντέρων πλῆθος
ὁμοίως τοῖς ἐναίμοις, ἔν τε τοῖς πεζοῖς καὶ τοῖς
ἐνύδροις ζῴοις.

Ἡ μὲν οὖν γένεσις ἐξ ἀνάγκης συμβαίνει τοιαύτη
τοῦ μορίου τούτου· ξηροῦ γὰρ καὶ ὑγροῦ μίγματος
θερμαινομένου τὸ ἔσχατον ἀεὶ δερματῶδες γίνεται

---

[1] ἤρκται SUYZ.

animals, and so it is quite reasonable to hold that the condition of it controls the length of its owner's life. And it is equally reasonable to hold that the liver produces a residue such as the bile although none of the other viscera does so. Take the heart : no such humour as bile could possibly come near the heart, because the heart cannot withstand any violent affection. Of the other viscera none is indispensable to an animal, except the liver only, and that is why this phenomenon occurs in connexion with the liver exclusively. And it would be absurd to say that phlegm and the sediment produced by the stomach are residues when found in some places but not in others ; and clearly the same applies to bile : its locality makes no difference.

We have now spoken of the gall-bladder, and we have shown why some animals have it and why some have not.   III. It remains to speak of the Mesentery and of the Omentum.   These are in the same region and close to the parts we have just described.

The Omentum is a membrane, formed of suet Omentum. or lard according to the animal in which it is. (We have already stated which animals contain suet and which lard.)ᵃ Whether the animal has one stomach or many, the Omentum is always fastened to the middle of the stomach, on the line marked on it like a seam ; and it covers the rest of the stomach and most of the intestines.   This is so in all blooded creatures, land- and water-animals alike.

As for the *necessary* ᵇ formation of this part, it occurs as follows.   When a mixture containing solid substance and fluid is warmed up, the surface of it always becomes skin-like and membranous ; and

ᵃ At 651 a 26 ff.   ᵇ See Introd. p. 22.

677 b

καὶ ὑμενῶδες, ὁ δὲ τόπος οὗτος τοιαύτης πλήρης
25 ἐστὶ τροφῆς. ἔτι δὲ διὰ πυκνότητα τοῦ ὑμένος τὸ
διηθούμενον τῆς αἱματώδους τροφῆς ἀναγκαῖον
λιπαρὸν εἶναι (τοῦτο γὰρ λεπτότατον) καὶ διὰ τὴν
θερμότητα τὴν περὶ τὸν τόπον συμπεττόμενον ἀντὶ
σαρκώδους καὶ αἱματώδους συστάσεως στέαρ γίνεσθαι καὶ πιμελήν. ἡ μὲν οὖν γένεσις τοῦ ἐπι-
30 πλόου συμβαίνει κατὰ τὸν λόγον τοῦτον, καταχρῆται δ' ἡ φύσις αὐτῷ πρὸς τὴν εὐπεψίαν τῆς
τροφῆς, ὅπως ῥᾷον πέττῃ καὶ θᾶττον τὰ ζῷα τὴν
τροφήν· τὸ μὲν γὰρ θερμὸν πεπτικόν, τὸ δὲ πῖον
θερμόν, τὸ δ' ἐπίπλοον πῖον. καὶ διὰ τοῦτ' ἀπὸ
35 μέσης ἤρτηται[1] τῆς κοιλίας, ὅτι τὸ ἐπέκεινα[2] μέρος
συμπέττει τὸ παρακείμενον ἧπαρ. καὶ περὶ μὲν
τοῦ ἐπιπλόου εἴρηται.

IV. Τὸ δὲ καλούμενον μεσεντέριον ἔστι μὲν ὑμήν,
διατείνει δὲ συνεχὲς ἀπὸ τῆς τῶν ἐντέρων παρα-
678 a τάσεως εἰς τὴν φλέβα τὴν μεγάλην καὶ τὴν ἀορτήν,
πλῆρες ὂν φλεβῶν πολλῶν καὶ πυκνῶν, αἳ τείνουσιν
ἀπὸ τῶν ἐντέρων εἴς τε τὴν μεγάλην φλέβα καὶ τὴν
ἀορτήν. τὴν μὲν οὖν γένεσιν ἐξ ἀνάγκης οὖσαν
5 εὑρήσομεν ὁμοίως τοῖς ἄλλοις μορίοις[3]· διὰ τίνα δ'
αἰτίαν ὑπάρχει τοῖς ἐναίμοις, φανερόν ἐστιν ἐπι-
σκοποῦσιν. ἐπεὶ γὰρ ἀναγκαῖον τὰ ζῷα τροφὴν
λαμβάνειν θύραθεν, καὶ πάλιν ἐκ ταύτης γίνεσθαι
τὴν ἐσχάτην τροφήν, ἐξ ἧς ἤδη διαδίδοται εἰς τὰ
μόρια (τοῦτο δὲ τοῖς μὲν ἀναίμοις ἀνώνυμον, τοῖς δ'

---

[1] ἤρκται EPSUYZ.
[2] ἐπέκεινα Peck : ἐπ' ἐκεῖνο vulg.
[3] ⟨τοιούτοις⟩ μορίοις Ogle : [μορίοις] ὑμέσι Platt.

the place where the Omentum is is full of nutriment of this very sort. Furthermore, owing to the thickness of the membrane, that portion of the blood-like nutriment which percolates through it must of necessity be fatty, because that is the finest in texture; and then owing to the heat in that part it will be concocted and so become suet or lard instead of some fleshy or blood-like substance. This, then, is the way in which the formation of the Omentum occurs. Nature, however, turns the Omentum to advantage in the concoction of the food, so as to enable the animal to concoct its food more easily and more quickly; for the Omentum is fat; fat things are hot, and hot things aid concoction. For this reason, too, the Omentum is fastened to the middle of the stomach; since as regards that part of the stomach which is beyond, the liver which is close by it assists it in concoction. So much for the Omentum.

IV. What is called the Mesentery is also a membrane; and it extends continuously from the line of extension of the intestines as far as the Great Blood-vessel and the Aorta. It is full of blood-vessels, which are many in number and closely packed together; and they extend from the intestines as far as the Great Blood-vessel and the Aorta. We shall find, as with the other parts, that the development and formation of the Mesentery is the result of *necessity*. As for its *purpose* in the blooded animals, that is clear enough to those who consider. Animals must of necessity take in nutriment from without; and, again, out of this the "ultimate nutriment" has to be made; and from this store the supply is distributed directly to the parts of the body. (In blooded animals this is called blood; there is no

Mesentery.

313

678 a

10 ἐναίμοις αἷμα καλεῖται), δεῖ τι εἶναι δι' οὗ εἰς τὰς
φλέβας ἐκ τῆς κοιλίας οἷον διὰ ῥιζῶν πορεύσεται ἡ
τροφή. τὰ μὲν οὖν φυτὰ τὰς ῥίζας ἔχει εἰς τὴν γῆν
(ἐκεῖθεν γὰρ λαμβάνει τὴν τροφήν), τοῖς δὲ ζῴοις ἡ
κοιλία καὶ ἡ τῶν ἐντέρων δύναμις γῆ ἐστιν, ἐξ ἧς
δεῖ λαμβάνειν τὴν τροφήν· διόπερ ἡ τοῦ μεσεν-
15 τερίου φύσις ἐστίν, οἷον ῥίζας ἔχουσα τὰς δι' αὑτῆς[1]
φλέβας. οὗ μὲν οὖν ἕνεκα τὸ μεσεντέριόν ἐστιν,
εἴρηται· τίνα δὲ τρόπον λαμβάνει τὴν τροφήν, καὶ
πῶς εἰσέρχεται διὰ τῶν φλεβῶν ἀπὸ τῆς ἐσχάτης[2]
τροφῆς εἰς τὰ μόρια πάντα[3] τὸ διαδιδόμενον εἰς τὰς
φλέβας, ἐν τοῖς περὶ τὴν γένεσιν τῶν ζῴων λεχθή-
20 σεται καὶ τὴν τροφήν.

Τὰ μὲν οὖν ἔναιμα τῶν ζῴων πῶς ἔχει μέχρι τῶν
διωρισμένων μορίων, καὶ διὰ τίνας αἰτίας, εἴρηται·
περὶ δὲ τῶν εἰς τὴν γένεσιν συντελούντων, οἷς δοκεῖ
διαφέρειν τὸ θῆλυ τοῦ ἄρρενος, ἐχόμενον μέν ἐστι
25 καὶ λοιπὸν τῶν εἰρημένων· ἀλλ' ἐπειδὴ περὶ γενέσεως λεκτέον, ἁρμόττον ἐστὶ καὶ περὶ τούτων ἐν τῇ
περὶ ἐκείνων θεωρίᾳ διελθεῖν.

V. Τὰ δὲ καλούμενα μαλάκια καὶ μαλακόστρακα
πολλὴν ἔχει πρὸς ταῦτα διαφοράν· εὐθὺς γὰρ τὴν
τῶν σπλάγχνων ἅπασαν οὐκ ἔχει φύσιν. ὁμοίως δ'
30 οὐδὲ τῶν ἄλλων ἀναίμων οὐδέν. ἔστι δὲ δύο γένη
λοιπὰ τῶν ἀναίμων, τά τ' ὀστρακόδερμα καὶ τὸ τῶν
ἐντόμων γένος. ἐξ οὗ γὰρ συνέστηκεν ἡ τῶν
σπλάγχνων φύσις, οὐδὲν τούτων ἔχει αἷμα, διὰ τὸ

---

[1] αὑτῆς Peck : αὐτῆς vulg.
[2] ἐσχάτης Peck : εἰσιούσης vulg.
[3] πάντα Ogle : ταῦτα vulg. : om. Z.

special name for it in the others.) Now there must be some passage or passages (as it might be roots) through which this nutriment shall pass from the stomach to the blood-vessels. The roots of plants are of course in the ground, because that is the source from which plants get their nutriment. For an animal, the stomach and the intestines correspond to the ground, the place from which the nutriment has to be derived. And the Mesentery exists to contain these vessels, corresponding to roots ; they pass through the inside of it. This completes my account of its Final Cause. As for the means by which the nutriment is taken up, and the way in which that portion of the ultimate nutriment which is distributed into the blood-vessels reaches all the parts of the body through them, these points will be dealt with in the treatises on the *Generation of Animals* and on *Nutrition.*

I have now described the blooded animals as far as concerns the parts that have been dealt with, and also the causes that are responsible. It remains, and would follow after this, to speak of the organs of generation, by which male and female are distinguished. But as we shall have to deal with generation itself, it is more appropriate to speak of these organs in our consideration of that subject.

V. The animals called Cephalopods and Crustacea are very different from the blooded ones. First of all, they have no visceral structure at all. This is true of all the bloodless creatures, in which are included beside Cephalopods and Crustacea two other groups, the Testacea and the Insects. This is because none of them has blood, which is the material out of which

INTERNAL PARTS OF BLOODLESS ANIMALS.

678 a
τῆς οὐσίας αὐτῶν εἶναί τι τοιοῦτον πάθος [αὐτῆς]¹·
ὅτι γάρ ἐστι τὰ μὲν ἔναιμα τὰ δ' ἄναιμα, ἐν τῷ
35 λόγῳ ἐνυπάρξει τῷ ὁρίζοντι τὴν οὐσίαν αὐτῶν. ἔτι
δ' ὧν ἕνεκεν ἔχουσι τὰ σπλάγχνα τὰ ἔναιμα τῶν
ζῴων, οὐδὲν ὑπάρξει τοῖς τοιούτοις· οὔτε γὰρ
678 b φλέβας ἔχουσιν οὔτε κύστιν οὔτ' ἀναπνέουσιν, ἀλλὰ
μόνον ἀναγκαῖον ἔχειν αὐτοῖς τὸ ἀνάλογον τῇ καρ-
δίᾳ· τὸ γὰρ αἰσθητικὸν ψυχῆς καὶ τὸ τῆς ζωῆς αἴτιον ⟨ἐν⟩² ἀρχῇ τινι τῶν μορίων καὶ τοῦ σώματος
ὑπάρχει πᾶσι τοῖς ζῴοις. τὰ δὲ πρὸς τὴν τροφὴν
5 μόρια ἔχει καὶ ταῦτα ἐξ ἀνάγκης πάντα· οἱ δὲ
τρόποι διαφέρουσι διὰ τοὺς τόπους ἐν οἷς λαμβάνουσι τὴν τροφήν.

Ἔχουσι δὲ τὰ μὲν μαλάκια περὶ τὸ καλούμενον
στόμα δύο ὀδόντας, καὶ ἐν τῷ στόματι ἀντὶ γλώττης σαρκῶδές τι, ᾧ κρίνουσι τὴν ἐν τοῖς ἐδεστοῖς
ἡδονήν. ὁμοίως δὲ καὶ τὰ μαλακόστρακα τούτοις
10 τοὺς πρώτους ὀδόντας ἔχει καὶ τὸ ἀνάλογον τῇ
γλώττῃ σαρκῶδες. ἔτι δὲ καὶ τὰ ὀστρακόδερμα
πάντα τὸ τοιοῦτον ἔχει μόριον διὰ τὴν αὐτὴν αἰτίαν
τοῖς ἐναίμοις, πρὸς τὴν τῆς τροφῆς αἴσθησιν.
ὁμοίως δὲ καὶ τὰ ἔντομα τὰ μὲν τὴν ἐξιοῦσαν ἐπι-
βοσκίδα τοῦ στόματος, οἷον τό τε τῶν μελιττῶν
15 γένος καὶ τὸ τῶν μυιῶν, ὥσπερ εἴρηται καὶ πρό-
τερον· ὅσα δὲ μή ἐστιν ἐμπροσθόκεντρα, ἐν τῷ
στόματι ἔχει τὸ τοιοῦτον μόριον, οἷον τὸ τῶν
μυρμήκων γένος καὶ εἴ τι τοιοῦτον ἕτερον. ὀδόντας
δὲ τὰ μὲν ἔχει τούτων, ἀλλοιοτέρους δέ, καθάπερ

---

¹ αὐτῆς seclusi.   ² ἐν supplevit Th.

ᵃ See Introduction, pp. 26 ff.
ᵇ These teeth are the two halves of what might be compared to a beak.

viscera are made ; and the reason for this is that a
condition of this sort is part of their being : the fact
that some animals are blooded and some bloodless
will be found to be included in the *logos* [a] which
defines their being. Further, we shall see that none
of those purposes for whose sake blooded animals
have viscera operate in these other creatures : they
have no blood-vessels and no bladder, they do not
breathe : the only organ they must necessarily have
is the counterpart of the heart, since the sensitive
part of the Soul and the original cause of life is always
situated in some place which rules the body and its
parts. Also, they all have of necessity the parts
adapted for dealing with food and nutrition; but the
manner of these varies according to the places where
they take their food.

The Cephalopods have two teeth around what is
called their mouth [b]; and inside the mouth, instead of
a tongue, they have a fleshy object, by means of
which they discriminate the savour of things to eat.
Likewise, the Crustacea have these front teeth and
the fleshy counterpart of the tongue. The Testacea
all have this latter part, too, for the same reason that
blooded animals have a tongue, viz. to perceive the
taste of the food they eat. Similarly, too, the Insects
have, some of them, a proboscis which comes out
from the mouth, as with the Bees and Flies (this has
been mentioned earlier [c]); and the ones which have no
sharp protrusion in front have a part such as this
inside the mouth, as Ants, and the like. Some of
these creatures have teeth, though somewhat different from ordinary teeth (as the Flies,[d] and Bees) ;

[c] At 661 a 21 ; *cf. Hist. An.* 528 b 28.
[d] Or " Ants " (translating Meyer's emendation).

678 b

τό τε τῶν μυιῶν¹ καὶ τὸ τῶν μελιττῶν γένος, τὰ δ'
20 οὐκ ἔχει, ὅσα ὑγρᾷ χρῆται τῇ τροφῇ· πολλὰ γὰρ
τῶν ἐντόμων οὐ τροφῆς ἔχει χάριν τοὺς ὀδόντας
ἀλλ' ἀλκῆς.

Τῶν δ' ὀστρακοδέρμων τὰ μέν, ὥσπερ ἐλέχθη καὶ
ἐν τοῖς κατ' ἀρχὰς λόγοις, τὴν καλουμένην ἔχει
γλῶτταν ἰσχυράν, οἱ δὲ κόχλοι καὶ ὀδόντας δύο,
25 καθάπερ τὰ μαλακόστρακα. μετὰ δὲ τὸ στόμα τοῖς
μαλακίοις ἐστὶ στόμαχος μακρός, τούτου δ' ἐχό-
μενος πρόλοβος οἷός περ τοῖς ὄρνισιν, εἶτα συνεχὴς
κοιλία, καὶ ταύτης ἐχόμενον ἔντερον ἁπλοῦν μέχρι
τῆς ἐξόδου. ταῖς μὲν οὖν σηπίαις καὶ τοῖς πολύποσιν ὅμοια καὶ τοῖς σχήμασι καὶ τῇ ἁφῇ τὰ περὶ
30 τὴν κοιλίαν· ταῖς δὲ καλουμέναις τευθίσι δύο μὲν
ὁμοίως αἱ κοιλιώδεις εἰσὶν ὑποδοχαί, ἧττον δὲ
προλοβώδης ἡ ἑτέρα, καὶ τοῖς σχήμασιν ἐκείνων
διαφέρουσι διὰ τὸ καὶ τὸ σῶμα πᾶν ἐκ μαλακω-
τέρας συνεστάναι σαρκός.

Ταῦτα δ' ἔχει τὰ μόρια τοῦτον τὸν τρόπον διὰ
τὴν αὐτὴν αἰτίαν ὥσπερ καὶ οἱ ὄρνιθες· οὐδὲ γὰρ
35 τούτων οὐδὲν ἐνδέχεται λεαίνειν τὴν τροφήν, διόπερ
ὁ πρόλοβός ἐστι πρὸ τῆς κοιλίας.

Πρὸς βοήθειαν δὲ καὶ σωτηρίαν ἔχει ταῦτα τὸν
679 a καλούμενον θολὸν ἐν χιτῶνι ὑμενώδει προσπεφυκότι,² τὴν ἔξοδον ἔχοντι καὶ τὸ πέρας ᾗπερ ἀφιᾶσι
τὸ περίττωμα τῆς κοιλίας κατὰ τὸν καλούμενον
αὐλόν· οὗτος δ' ἐστὶν ἐν τοῖς ὑπτίοις. ἔχει μὲν οὖν
5 πάντα τὰ μαλάκια τοῦτο τὸ μόριον ἴδιον, μάλιστα
δ' ἡ σηπία καὶ πλεῖστον· ὅταν γὰρ φοβηθῶσι καὶ

---

¹ μυιῶν] μυιῶν ζῷον EY : μυρμήκων Meyer.
² προσπεφυκότι Ogle : προσπεφυκότα vulg.

others have no teeth at all : these are the creatures whose food is fluid. Indeed, in many of the insects the purpose of the teeth is not mastication of food at all, but for use as weapons.

Of the Testacea, as we stated in the opening treatise,[a] some have a very strong tongue (so-called) ; and the Sea-snails actually have two teeth as well, like the Crustacea. In the Cephalopods there is a long gullet next after the mouth, and contiguous to that is a crop like a bird's. Continuous with this is the stomach, then immediately the intestine, which is simple and reaches to the vent. In the Sepias and Octopuses these parts round the stomach are similar both in shape and in consistency. The creatures called Calamaries, like the others, have the two gastric receptacles,[b] but the first of them is less like a crop ; and they differ in shape from the organs of the previous classes, and that is because their bodies are composed of softer flesh throughout.

These creatures have these parts arranged in this way for the same reason that birds have them: they, like birds, are unable to grind down their food ; hence the crop is placed before the stomach.

The Cephalopods, for the sake of self-defence and self-preservation, have what is called their Ink. This is contained in a membranous bag which is attached to the body, and comes to an end in an outlet where the residue from the stomach is discharged by the so-called funnel. This is on the under side of the body. All the Cephalopods have this peculiar part, but it is most remarkable in the Sepia, as well as the largest in size. When the Sepia is frightened and in terror,

---

[a] At *Hist. An.* 528 b 30 ff.
[b] Viz. the crop and the stomach.

δείσωσιν, οἷον φράγμα πρὸ τοῦ σώματος ποιοῦνται
τὴν τοῦ ὑγροῦ μελανίαν καὶ θόλωσιν. αἱ μὲν οὖν
τευθίδες καὶ πολύποδες ἔχουσιν ἄνωθεν τὸν θολὸν
ἐπὶ τῇ μύτιδι μᾶλλον, ἡ δὲ σηπία πρὸς τῇ κοιλίᾳ
10 κάτω· πλείω γὰρ ἔχει διὰ τὸ χρῆσθαι μᾶλλον.
τοῦτο δ᾽ αὐτῇ συμβαίνει διὰ τὸ πρόσγειον μὲν εἶναι
τὸν βίον αὐτῆς, μὴ ἔχειν δ᾽ ἄλλην βοήθειαν, ὥσπερ
ὁ πολύπους τὰς πλεκτάνας ἔχει χρησίμους καὶ τὴν
τοῦ χρώματος μεταβολήν, ἣ συμβαίνει αὐτῷ,
ὥσπερ καὶ ἡ τοῦ θολοῦ πρόεσις, διὰ δειλίαν. ἡ δὲ
15 τευθὶς πελάγιόν ἐστι τούτων μόνον. πλείω μὲν οὖν
ἔχει ἡ σηπία παρὰ τοῦτο τὸν θολόν, κάτωθεν δὲ διὰ
τὸ πλείω· ῥᾴδιον γὰρ προΐεσθαι καὶ πόρρωθεν ἀπὸ
τοῦ πλείονος. γίνεται δέ [ὁ θολός],[1] καθάπερ τοῖς
ὄρνισιν ὑπόστημα τὸ λευκὸν ἐπὶ τοῦ περιττώματος
γεῶδες, οὕτω καὶ τούτοις ὁ θολὸς διὰ τὸ μηδὲ ταῦτ᾽
20 ἔχειν κύστιν· ἀποκρίνεται γὰρ τὸ γεωδέστατον εἰς
αὐτόν, καὶ τῇ σηπίᾳ πλεῖστον διὰ τὸ πλεῖστον ἔχειν
γεῶδες. σημεῖον δὲ τὸ σήπιον τοιοῦτον ὄν· τοῦτο
γὰρ ὁ μὲν πολύπους οὐκ ἔχει, αἱ δὲ τευθίδες χονδρῶδες καὶ λεπτόν. (δι᾽ ἣν δ᾽ αἰτίαν τὰ μὲν οὐκ
ἔχει τὰ δ᾽ ἔχει, καὶ ποῖόν τι τούτων ἔχει ἑκάτερον,
εἴρηται.[2])
25 Ἀναίμων δ᾽ ὄντων καὶ διὰ τοῦτο κατεψυγμένων
καὶ φοβητικῶν, ὥσπερ ἐνίοις ὅταν δείσωσιν ἡ
κοιλία ταράττεται, τοῖς δ᾽ ἐκ τῆς κύστεως ῥεῖ
περίττωσις, καὶ τούτοις τοῦτο συμβαίνει μὲν ἐξ

---

[1] [ὁ θολός] seclusi : ὁ om. P.
[2] εἴρηται πρότερον P.

---

[a] The *mytis*, which is the same as the *mecon*, is an excretory
organ, and corresponds to the liver. See below, 679 b 11.
[b] *Cf.* above, 676 a 32.

it produces this blackness and muddiness in the water, as it were a shield held in front of the body. Now the Calamaries and Octopuses have this ink-bag in the upper region of the body, quite near the *mytis*[a]; whereas in the Sepia it is lower down, against the stomach, since it has a larger supply because it uses it more. This circumstance is due (1) to its living near the lánd and (2) to its having no other means of defence—nothing like the Octopus, for instance, which has its twining feet, which are useful for this purpose ; it can also change its colour, and it does so (just as the Sepia emits its ink) when put in fear. Of all these, only the Calamary lives well out at sea and gets protection thereby. Hence, compared with it, the Sepia has a larger supply of ink ; and because this is larger, it is lower in the body, as it is easy for it to be emitted even to a considerable distance when the supply is great. The ink is earthy in its nature, like the white deposit on the excrement of birds, and it is produced by these creatures for the same reason— they, like birds, have no urinary bladder[b]; so the earthiest matter is excreted into this ink, especially in the Sepia, for the Sepia contains an exceptionally large amount of earthy matter. An indication of this is its bone, which is earthy. The Octopuses do not have this bone, and in the Calamary it is cartila- ginous and slight. (We have said why some of these animals have this part and why some have not, and what in each case its character is.)

These animals, as they have no blood, are cold and liable to take fright. While in some other animals fear causes a disturbance of the stomach, and in some the discharge of residue from the bladder, in these creatures its effect is to make them discharge their

679 a

ἀνάγκης ἀφιέναι διὰ δειλίαν, ὥσπερ ἐκ κύστεως
τοῖς ἐπουροῦσιν, ἡ δὲ φύσις ἅμα τῷ τοιούτῳ περιτ-
30 τώματι καταχρῆται πρὸς βοήθειαν καὶ σωτηρίαν
αὐτῶν.

Ἔχουσι δὲ καὶ τὰ μαλακόστρακα, τά τε καρα-
βοειδῆ καὶ οἱ καρκίνοι, δύο μὲν ὀδόντας τοὺς
πρώτους, καὶ μεταξὺ τὴν σάρκα τὴν γλωσσοειδῆ,
ὥσπερ εἴρηται καὶ πρότερον, εὐθὺς δ' ἐχόμενον τοῦ
στόματος στόμαχον μικρὸν κατὰ μέγεθος τῶν
35 σωμάτων [τὰ μείζω πρὸς τὰ ἐλάττω]¹· τούτου δὲ
κοιλίαν ἐχομένην, ἐφ' ἧς οἵ τε κάραβοι καὶ ἔνιοι
τῶν καρκίνων ὀδόντας ἔχουσιν ἑτέρους διὰ τὸ τοὺς
679 b ἄνω μὴ διαιρεῖν ἱκανῶς, ἀπὸ δὲ τῆς κοιλίας ἔντερον
ἁπλοῦν κατ' εὐθὺ μέχρι πρὸς τὴν ἔξοδον τοῦ
περιττώματος.

Ἔχει δὲ καὶ τῶν ὀστρακοδέρμων ἕκαστον ταῦτα
τὰ μόρια, τὰ μὲν διηρθρωμένα μᾶλλον τὰ δ' ἧττον·
ἐν δὲ τοῖς μείζοσι διαδηλότερά ἐστιν ἕκαστα τού-
5 των. οἱ μὲν οὖν κόχλοι καὶ ὀδόντας ἔχουσι σκλη-
ροὺς καὶ ὀξεῖς, ὥσπερ εἴρηται πρότερον, καὶ τὸ
μεταξὺ σαρκῶδες ὁμοίως τοῖς μαλακίοις καὶ μαλακοστράκοις, καὶ τὴν προβοσκίδα, καθάπερ εἴρηται,
μεταξὺ κέντρου καὶ γλώττης, τοῦ δὲ στόματος
ἐχόμενον οἷον ὀρνιθώδη τινὰ πρόλοβον, τούτου δ'
10 ἐχόμενον στόμαχον· τούτου δ' ἔχεται ἡ κοιλία, ἐν ᾗ
ἡ καλουμένη μήκων, ἀφ' ἧς συνεχές ἐστιν ἔντερον
ἁπλῆν τὴν ἀρχὴν ἔχον ἀπὸ τῆς μήκωνος· ἔστι γὰρ
ἐν πᾶσι τοῖς ὀστρακηροῖς περίττωμα τοῦτο τὸ
μάλιστα δοκοῦν εἶναι ἐδώδιμον. ἔχει δ' ὁμοίως τῷ

¹ seclusit Rackham.

322

ink; and though this is an effect due to necessity, like the discharge of urine in the others, yet Nature makes good use of this residue at the same time for the animal's defence and preservation.

The Crustacea as well, that is, both the Crabs and the Caraboids, have the two front teeth, and between the teeth they have the tongue-like flesh, as has already been stated[a]; and immediately next to the mouth they have a gullet which is quite small compared with the animal's size; and immediately after that the stomach; and on this the Carabi and some of the Crabs have another set of teeth, since the upper ones do not masticate the food sufficiently. From the stomach a simply formed intestine runs straight to the vent where residues are discharged.

These parts are present in every one of the Testacea as well, more distinct in some, less in others. They are more clearly marked in the larger animals. Take the Sea-snails. These have (1) as stated already, the teeth, which are hard and sharp, (2) the fleshy object in between them, similarly to the Crustacea and Cephalopods; (3) the proboscis, as already mentioned,[b] something between a sting and a tongue; (4) immediately after the mouth is a sort of bird's crop, and (5) after that the gullet; (6) continuous with that is the stomach, and (7) in the stomach is what is known as the *mecon*[c]; and (8) attaching to this is an intestine: this intestine begins directly from the *mecon*. This residue (the *mecon*) appears to be the most tasty piece in all the Testacea. The other creatures that have spiral shells (*e.g.* the

[a] At 678 b 10.
[b] At 661 a 15 ff.
[c] The hepatopancreas or liver; see above, 679 a 9.

15 κόχλῳ καὶ τἆλλα τὰ στρομβώδη, οἷον πορφύραι
καὶ κήρυκες.

Ἔστι δὲ γένη καὶ εἴδη πολλὰ τῶν ὀστρακο-
δέρμων· τὰ μὲν γὰρ στρομβώδη ἐστίν, ὥσπερ τὰ
νῦν εἰρημένα, τὰ δὲ δίθυρα, τὰ δὲ μονόθυρα. τρόπον
δέ τινα καὶ τὰ στρομβώδη διθύροις ἔοικεν· ἔχει γὰρ
ἐπιπτύγματ᾽ ἐπὶ τῷ φανερῷ τῆς σαρκὸς πάντα τὰ
20 τοιαῦτα ἐκ γενετῆς, οἷον αἵ τε πορφύραι καὶ
κήρυκες καὶ οἱ νηρεῖται καὶ πᾶν τὸ τοιοῦτον γένος,
πρὸς βοήθειαν· ᾗ γὰρ μὴ προβέβληται τὸ ὄστρακον,
ῥᾴδιον ταύτῃ βλάπτεσθαι ὑπὸ τῶν θύραθεν προσπιπτόντων. τὰ μὲν οὖν μονόθυρα διὰ τὸ προσπεφυκέναι σώζεται τῷ πρανὲς ἔχειν τὸ ὄστρακον,
25 καὶ γίνεται ἀλλοτρίῳ φράγματι τρόπον τινὰ δί-
θυρον, οἷον αἱ καλούμεναι λεπάδες· τὰ δὲ δίθυρα,
οἷον κτένες καὶ μύες, τῷ συνάγειν, τὰ δὲ στρομβώδη
τούτῳ τῷ ἐπικαλύμματι, ὥσπερ δίθυρα γινόμενα ἐκ
μονοθύρων. ὁ δ᾽ ἐχῖνος μάλιστα πάντων ἀλεωρὰν
ἔχει· κύκλῳ γὰρ τὸ ὄστρακον συνηρεφὲς καὶ κε-
30 χαρακωμένον ταῖς ἀκάνθαις. ἴδιον δ᾽ ἔχει τῶν
ὀστρακοδέρμων τοῦτο, καθάπερ εἴρηται πρότερον.

Τῶν δὲ μαλακοστράκων καὶ τῶν ὀστρακοδέρμων
συνέστηκεν ἡ φύσις τοῖς μαλακίοις ἀντικειμένως·
τοῖς μὲν γὰρ ἔξω τὸ σαρκῶδες, τοῖς δ᾽ ἐντός, ἐκτὸς
δὲ τὸ γεῶδες. ὁ δ᾽ ἐχῖνος οὐδὲν ἔχει σαρκῶδες.
35 Πάντα μὲν οὖν ἔχει, καθάπερ εἴρηται, καὶ τἆλλα
τὰ ὀστρακόδερμα στόμα τε καὶ τὸ γλωττοειδὲς καὶ
κοιλίαν καὶ τοῦ περιττώματος τὴν ἔξοδον, διαφέρει

ᵃ The *operculum*.

Purpuras and the Whelks) are similar to the Sea-snails in structure.

There are very many genera and species of Testacea. Some have spiral shells, like the ones just mentioned ; some are bivalves, some univalves. In a way, the spiral shells resemble the bivalves, as they have, all of them, from birth, a covering *a* over the exposed part of their flesh, *e.g.* the Purpuras, the Whelks, the Nerites, and the whole tribe of them. This covering serves as a protection ; for in any place where the animal has no shell to protect it, it could quite easily be injured by the impact of external objects. The univalves' means of preservation is this : they cling to some object, and have their shell on the upper side ; so they become in a way bivalves in virtue of the borrowed protection afforded by the object to which they cling. Example, the Limpets. The bivalves proper (*e.g.* Scallops and Mussels) get their protection by closing themselves up ; the spiral-shelled creatures by the covering I mentioned, which, as it were, turns them from univalves into bivalves. The Sea-urchin has a better defence system than any of them : he has a good thick shell all round him, fortified with a palisade of spines. As I stated previously, the Sea-urchin is the only one of the Testacea which possesses this peculiarity.

The natural structure of the Crustacea and of the Testacea is the reverse of that of the Cephalopods. The latter have their fleshy part outside, the former have the earthy part outside and the fleshy inside. The Sea-urchin, however, has no fleshy part at all.

All these parts, as described—mouth, tongue-like object, stomach, vent for the residue—are present in the rest of the Testacea too, but they differ in

680 a δὲ τῇ θέσει καὶ τοῖς μεγέθεσιν. ὃν δὲ τρόπον ἔχει
τούτων ἕκαστον, ἔκ τε τῶν ἱστοριῶν τῶν περὶ τὰ
ζῷα θεωρείσθω καὶ ἐκ τῶν ἀνατομῶν· τὰ μὲν γὰρ
τῷ λόγῳ τὰ δὲ πρὸς τὴν ὄψιν αὐτῶν σαφηνίζειν δεῖ
μᾶλλον.

Ἰδίως δ᾽ ἔχουσι τῶν ὀστρακοδέρμων οἵ τ᾽ ἐχῖνοι
5 καὶ τὸ τῶν καλουμένων τηθύων γένος. ἔχουσι δ᾽ οἱ
ἐχῖνοι ὀδόντας μὲν πέντε καὶ μεταξὺ τὸ σαρκῶδες,
ὅπερ ἐπὶ πάντων ἐστὶ τῶν εἰρημένων, ἐχόμενον δὲ
τούτου στόμαχον, ἀπὸ δὲ τούτου τὴν κοιλίαν εἰς
πολλὰ διῃρημένην, ὡσπερανεὶ πολλὰς τοῦ ζῴου
κοιλίας ἔχοντος. κεχωρισμέναι μὲν γάρ εἰσι καὶ
10 πλήρεις περιττώματος, ἐξ ἑνὸς δ᾽ ἤρτηνται τοῦ
στομάχου καὶ τελευτῶσι πρὸς μίαν ἔξοδον τὴν τοῦ
περιττώματος. παρὰ δὲ τὴν κοιλίαν σαρκῶδες μὲν
οὐδὲν ἔχουσιν, ὥσπερ εἴρηται, τὰ δὲ καλούμενα ᾠὰ
πλείω τὸν ἀριθμὸν ἐν ὑμένι χωρὶς ἕκαστον, καὶ
κύκλῳ ἀπὸ τοῦ στόματος μέλαν᾽ ἄττα διεσπαρμένα
15 χύδην, ἀνώνυμα. ὄντων δὲ πλειόνων γενῶν (οὐ γὰρ
ἓν εἶδος τῶν ἐχίνων πάντων ἐστί) πάντες μὲν ἔχουσι
ταῦτα τὰ μόρια, ἀλλ᾽ οὐκ ἐδώδιμα πάντες τὰ
καλούμενα ᾠά, καὶ μικρὰ πάμπαν ἔξω τῶν ἐπι-
πολαζόντων. ὅλως δὲ τοῦτο καὶ περὶ τἆλλα συμ-
20 βέβηκε τὰ ὀστρακόδερμα· καὶ γὰρ αἱ σάρκες οὐχ
ὁμοίως ἐδώδιμοι πάντων, καὶ τὸ περίττωμα, ἡ
καλουμένη μήκων, ἐνίων μὲν ἐδώδιμος ἐνίων δ᾽ οὐκ
ἐδώδιμος. ἔστι δὲ τοῖς στρομβώδεσιν ἐν τῇ ἑλίκῃ

---

[a] *Hist. An.* 528 b 10 ff.

[b] This seems to imply that diagrams or illustrations accompanied the treatises.

[c] These form what is compared to a lantern at *Hist. An.* 531 a 5, hence the name, " lantern of Aristotle."

their position and size. For the details of these, consult the *Researches upon Animals*[a] and the *Dissections*. Some points are better explained by inspection[b] than in words.

The Sea-urchin and the genus of Ascidians are peculiar among the Testacea. The Sea-urchin has five teeth,[c] and between them it has the fleshy substance (the same as in all the above-mentioned creatures); after that, the gullet, after that, the stomach, which is divided into several compartments, so that the animal seems to have several stomachs. But although they are separated from each other and are full of residue, they all spring from the gullet and they all terminate in the residual vent. Apart from the stomach, these creatures contain no fleshy substance, as I have said. They have, however, what are called ova[d]; there are several of them and each is in a separate membrane; and scattered at random round the body, beginning from the mouth, are certain black objects,[e] which have no name. There are several kinds of Sea-urchin, and in all of them these parts are present. Not all, however, have edible[f] ova, and, except in the common[g] varieties, they are quite small. There is a similar distinction among the other Testacea: the flesh is not equally edible in all of them, and in some of them the residue (the so-called *mecon*) is edible, in others not. In the spiral shells, the *mecon* is in the spiral, in univalves

[d] These are really ovaries (or testes): gonads.
[e] These may be the ambulacral vesicles, but the identification is not certain.
[f] See the story of the Spartan in Athenaeus iii. 41.
[g] The word translated "common" may mean "living near the surface."

680 a

τοῦτο, τοῖς δὲ μονοθύροις ἐν τῷ πυθμένι, οἷον ταῖς
λεπάσι, τοῖς δὲ διθύροις πρὸς τῇ συναφῇ· τὸ δ' ᾠὸν
25 καλούμενον ἐν τοῖς δεξιοῖς, ἐν δὲ τοῖς ἐπὶ θάτερα ἡ
ἔξοδος τοῦ περιττώματος τοῖς διθύροις. καλεῖται
δ' ᾠὸν οὐκ ὀρθῶς ὑπὸ τῶν καλούντων· τοῦτο γάρ
ἐστιν οἷον τοῖς ἐναίμοις, ὅταν εὐθηνῶσιν, ἡ πιότης.
διὸ καὶ γίνεται κατὰ τούτους τοὺς καιροὺς τοῦ
ἐνιαυτοῦ ἐν οἷς εὐθηνοῦσιν, ἔν τε τῷ ἔαρι καὶ
μετοπώρῳ· ἐν γὰρ τῷ ψύχει καὶ ταῖς ἀλέαις πο-
30 νοῦσι πάντα τὰ ὀστρακόδερμα, καὶ φέρειν οὐ
δύνανται τὰς ὑπερβολάς. σημεῖον δὲ τὸ συμβαῖνον
ἐπὶ τῶν ἐχίνων· εὐθύς τε γὰρ γινόμενοι ἔχουσι καὶ
ἐν ταῖς πανσελήνοις μᾶλλον, οὐ διὰ τὸ νέμεσθαι
καθάπερ τινὲς οἴονται μᾶλλον, ἀλλὰ διὰ τὸ ἀλεεινο-
τέρας εἶναι τὰς νύκτας διὰ τὸ φῶς τῆς σελήνης.
35 δύσριγα γὰρ ὄντα διὰ τὸ ἄναιμα εἶναι δέονται ἀλέας.
διὸ καὶ ἐν τῷ θέρει μᾶλλον πανταχοῦ εὐθηνοῦσιν,
680 b πλὴν οἱ ἐν τῷ Πυρραίῳ εὐρίπῳ· ἐκεῖνοι δ' οὐχ
ἧττον τοῦ χειμῶνος· αἴτιον δὲ τὸ νομῆς εὐπορεῖν
τότε μᾶλλον, ἀπολειπόντων τῶν ἰχθύων τοὺς τόπους κατὰ ταύτην τὴν ὥραν.

Ἔχουσι δ' οἱ ἐχῖνοι πάντες ἴσα τε τῷ ἀριθμῷ τὰ
5 ᾠὰ καὶ περιττά· πέντε γὰρ ἔχουσιν, τοσούτους δὲ
καὶ τοὺς ὀδόντας καὶ τὰς κοιλίας. αἴτιον δ' ὅτι τὸ
ᾠὸν ἐστι, καθάπερ εἴρηται πρότερον, οὐκ ᾠὸν ἀλλὰ
τοῦ ζῴου εὐτροφία. γίνεται δὲ τοῦτο ἐπὶ θάτερα

---

a This is true of the sea-urchins in the Red Sea, though not
of the Mediterranean ones. The former have a cycle corresponding exactly to that of the moon. The five roes, ovaries, or
testes are large and swollen during the week preceding each
of the summer full moons, and the spawning of the eggs
takes place during the few days before and after full moon.
For a most interesting discussion of this and kindred matters

(like limpets) it is in the tip; in bivalves it is near the hinge. In the bivalves the so-called ovum is on the right-hand side, and the residual vent on the left. " Ovum " is a misnomer; actually it corresponds to fat in blooded creatures when they are in good condition; and that is why it appears only in spring and autumn, which are the seasons when they are in good condition. In great cold and great heat all the Testacea are hard put to it; they cannot endure inordinate temperatures. The behaviour of the Sea-urchins is a good illustration of this: they have ova in them as soon as they are born, and at the time of full moon these increase in size [a]; and this is not, as some think, because the creatures eat more then, but because the nights are warmer owing to the moonlight. These creatures have need of the heat because they are bloodless and therefore adversely affected by cold. That is why they are in better condition during the summer, and this is true of them in all localities except the strait of Pyrrha,[b] where they flourish equally well in winter, and the reason for this is that in winter they have a more plentiful supply of foodstuff, due to the fish leaving the district at that season.

The Sea-urchins all have the same number of ova— an odd number, five, identical with the number of teeth and stomachs which they have. This is accounted for by the " ovum " not being really an ovum (as I said before) but simply a result of good nourishment. The " ovum " is found in Oysters too, though

[a] see H. M. Fox, *Selene*, especially pp. 35 ff., and id. *Proc. Roy. Soc. B.*, 1923, 95, 523.

[b] In Lesbos, leading to the lagoon, one of Aristotle's favourite hunting-grounds: see *Hist. An.* 544 a 21 (sea-urchin), 548 a 9, 603 a 21, 621 b 12. *Cf. Gen. An.* 763 b 2.

μόνον ἐν τοῖς ὀστρέοις, τὸ καλούμενον ᾠόν. ταὐτὸ
δὲ τοῦτό ἐστι καὶ τὸ ἐν τοῖς ἐχίνοις. ἐπεὶ τοίνυν
10 ἐστὶ σφαιροειδὴς ὁ ἐχῖνος, καὶ οὐχ ὥσπερ ἐπὶ τῶν
ἄλλων ὀστρέων τοῦ σώματος κύκλος εἷς, ὁ δ' ἐχῖνος
οὐ τῇ μὲν τοιοῦτος τῇ δ' οὔ, ἀλλὰ πάντῃ ὅμοιος
(σφαιροειδὴς γάρ), ἀνάγκη καὶ τὸ ᾠὸν ὁμοίως ἔχειν·
οὐ γάρ ἐστιν, ὥσπερ τοῖς ἄλλοις, τὸ κύκλῳ ἀν-
όμοιον· ἐν μέσῳ γὰρ ἡ κεφαλὴ πᾶσιν αὐτοῖς, τῷ δ'
15 ἄνω τὸ τοιοῦτον μόριον. ἀλλὰ μὴν οὐδὲ συνεχὲς
οἷόν τ' εἶναι τὸ ᾠόν—οὐδὲ γὰρ τοῖς ἄλλοις—ἀλλ' ἐπὶ
θάτερα τοῦ κύκλου μόνον. ἀνάγκη τοίνυν, ἐπεὶ
τοῦτο μὲν ἁπάντων κοινόν, ἴδιον δ' ἐκείνου εἶναι
τὸ σῶμα σφαιροειδές, μὴ εἶναι ἄρτια τὰ ᾠά. κατὰ
διάμετρον γὰρ ἂν ἦν, διὰ τὸ ὁμοίως δεῖν ἔχειν τὸ
20 ἔνθεν καὶ ἔνθεν, εἰ ἦν ἄρτια [καὶ κατὰ διάμετρον]¹·
οὕτως δ' ἐχόντων ἐπ' ἀμφότερα ἂν τοῦ κύκλου
εἶχον τὸ ᾠόν. τοῦτο δ' οὐκ ἦν οὐδ' ἐπὶ τῶν ἄλλων
ὀστρέων· ἐπὶ θάτερα γὰρ τῆς περιφερείας ἔχουσι τὰ
ὄστρεα καὶ οἱ κτένες τὸ τοιοῦτον μόριον. ἀνάγκη
τοίνυν τρία ἢ πέντε εἶναι ἢ ἄλλον τιν' ἀριθμὸν
25 περιττόν. εἰ μὲν οὖν τρία εἶχε, πόρρω λίαν ⟨ἂν⟩²
ἦν, εἰ δὲ πλείω τῶν πέντε, συνεχὲς ἄν· τούτων δὲ
τὸ μὲν οὐ βέλτιον, τὸ δ' οὐκ ἐνδεχόμενον. ἀνάγκη
ἄρα πέντ' αὐτοὺς ἔχειν τὰ ᾠά.

Διὰ τὴν αὐτὴν δ' αἰτίαν καὶ ἡ κοιλία τοιαύτη
ἔσχισται καὶ τὸ τῶν ὀδόντων τοσοῦτόν ἐστι πλῆθος.
ἕκαστον γὰρ τῶν ᾠῶν, οἷον σῶμά τι τοῦ ζῴου ὄν,
30 πρὸς τὸν τρόπον τὸν τῆς κοιλίας³ ὅμοιον ἔχειν

---

¹ secludenda.　　　　　² ⟨ἂν⟩ Ogle.
³ κοιλίας Ogle : ζωῆς vulg.

on one side of the body only ; it is the same as that of
the Sea-urchin. Now the Sea-urchin is spherical,
and is not just one flat disk like the Oysters ; thus,
being spherical, it is not different shapes in different
directions, but equiform in all directions ; hence of
necessity its " ovum " is correspondingly arranged,
since this creature's perimeter is not, as in the others,
non-equiform [a] : they all have their head in the
centre, whereas the Sea-urchin's is at the top.
Yet even so the " ovum " cannot be continuous,
since no other of the Testacea has it thus ; it
is always on one side of the disk only. Hence,
since this is a common property of all species
of Testacea, and the Sea-urchin is peculiar in having
a spherical shape, the result follows of necessity that
the Sea-urchins cannot have an even number of ova.
If they were even, they would have to be arranged in
diametrically opposite positions, because both sides
would have to be alike, and then there would be ova
on both sides of the circumference ; but this arrangement is not found in any of the other Ostreae ; both
Oysters and Scallops have ova on one side only of
their circumference. Therefore there must be three,
or five, or some other odd number of ova in the Sea-
urchin. If there were three, they would be too far
apart ; if more than five, they would be quite continuous ; the former would not subserve a good
purpose, the latter is impossible. Therefore the
Sea-urchin must of necessity have five ova.

For the same cause the creature's stomach is cloven
into five and it has five teeth. Each of the ova, being,
as it were, a body belonging to the creature, must
conform to the general character of the stomach,

---

[a] That is, it is circular in all planes, not in one only.

680 b

ἀναγκαῖον· ἐντεῦθεν γὰρ ἡ αὔξησις. μιᾶς μὲν γὰρ οὔσης ἢ πόρρω ἂν ἦσαν, ἢ πᾶν ἂν κατεῖχε τὸ κύτος, ὥστε καὶ δυσκίνητον εἶναι τὸν ἐχῖνον καὶ μὴ πληροῦσθαι τῆς τροφῆς τὸ ἀγγεῖον· πέντε δ' ὄντων τῶν διαλειμμάτων ἀνάγκη πρὸς ἑκάστῳ οὖσαν
35 πενταχῇ διῃρῆσθαι. διὰ τὴν αὐτὴν δ' αἰτίαν καὶ τὸ τῶν ὀδόντων ἔστι τοσοῦτον πλῆθος[1]· τὸ γὰρ
681 a ὅμοιον οὕτως ἂν ἡ φύσις εἴη ἀποδεδωκυῖα τοῖς εἰρημένοις μορίοις.

Διότι μὲν οὖν περιττὰ καὶ τοσαῦτα τὸν ἀριθμὸν ἔχει ὁ ἐχῖνος τὰ ᾠά, εἴρηται· διότι δ' οἱ μὲν πάμπαν μικρὰ οἱ δὲ μεγάλα, αἴτιον τὸ θερμοτέρους εἶναι τὴν φύσιν τούτους· πέττειν γὰρ τὸ θερμὸν δύναται
5 τὴν τροφὴν μᾶλλον, διόπερ περιττώματος πλήρεις οἱ ἄβρωτοι μᾶλλον. καὶ παρασκευάζει κινητικω- τέρους ἡ τῆς φύσεως θερμότης, ὥστε νέμεσθαι καὶ μὴ μένειν ἑδραίους. σημεῖον δὲ τούτου τὸ ἔχειν τοὺς τοιούτους ἀεί τι ἐπὶ τῶν ἀκανθῶν ὡς κινου- μένους πυκνά· χρῶνται γὰρ ποσὶ ταῖς ἀκάνθαις.
10 Τὰ δὲ τήθυα μικρὸν τῶν φυτῶν διαφέρει τὴν φύσιν, ὅμως δὲ ζωτικώτερα τῶν σπόγγων· οὗτοι γὰρ πάμπαν ἔχουσι φυτοῦ δύναμιν. ἡ γὰρ φύσις μεταβαίνει συνεχῶς ἀπὸ τῶν ἀψύχων εἰς τὰ ζῷα διὰ τῶν ζώντων μὲν οὐκ ὄντων δὲ ζῴων, οὕτως ὥστε δοκεῖν πάμπαν μικρὸν διαφέρειν θατέρου
15 θάτερον τῷ σύνεγγυς ἀλλήλοις. ὁ μὲν οὖν σπόγγος,

---

[1] hinc manus recentior E (= E).

---

[a] This is true ; but motion is effected mainly by the tube-feet, not noticed by Aristotle (vide Ogle).
[b] The " sea-squirts."

because growth has its origin from the stomach. Now if there were only one stomach, either the ova would be too far away from it, or the stomach would entirely fill up the cavity, which would make it difficult for the Sea-urchin to move about and to find sufficient food to replenish itself. But, as it is, there are five ova separated by five intervals, and so there must be five departments of the stomach, one for each interval. For the same reason there are five teeth, since this enables Nature to assign one tooth alike to each ovum and each department of the stomach.

I have now stated why the Sea-urchin has an odd number of ova, and why it has five of them. Now some Sea-urchins have quite small ones, and some large : the reason for this is that the latter have a hotter constitution, and the heat enables them to concoct their food better. This explains why the uneatable ones tend to be full of residue. This natural heat also induces the creatures to move about, and so instead of remaining settled in one place they keep on the move as they feed. An indication of this is that Sea-urchins of this sort always have something sticking on to their spines (which they use as feet),[a] which suggests that they are continually moving about.

The Ascidians[b] differ very little in their nature from plants, but they are more akin to animals than the Sponges are, which are completely plants. Nature passes in a continuous gradation from lifeless things to animals, and on the way there are living things which are not actually animals, with the result that one class is so close to the next that the difference seems infinitesimal. Now a sponge, as I said just now, is in

681 a

ὥσπερ εἴρηται, καὶ τῷ ζῆν προσπεφυκὼς μόνον,
ἀπολυθεὶς δὲ μὴ ζῆν, ὁμοίως ἔχει τοῖς φυτοῖς
παντελῶς· τὰ δὲ καλούμενα ὁλοθούρια καὶ οἱ πνεύ-
μονες, ἔτι δὲ καὶ ἕτερα τοιαῦτ' ἐν τῇ θαλάττῃ
μικρὸν διαφέρει τούτων τῷ ἀπολελύσθαι· αἴσθησιν
20 μὲν γὰρ οὐδεμίαν ἔχει, ζῇ δ' ὥσπερ ὄντα φυτὰ
ἀπολελυμένα. ἔστι δὲ καὶ ἐν τοῖς ἐπιγείοις φυτοῖς
ἔνια τοιαῦτα, ἃ καὶ ζῇ καὶ γίνεται τὰ μὲν ἐν ἑτέροις
φυτοῖς, τὰ δὲ καὶ ἀπολελυμένα, οἷον καὶ τὸ ἐκ τοῦ
Παρνασσοῦ καλούμενον ὑπό τινων ἐπίπετρον· τοῦτο
γὰρ ζῇ πολὺν χρόνον κρεμάμενον ἄνω ἐπὶ τῶν
25 παττάλων. ἔστι δ' ὅτε καὶ τὰ τήθυα, καὶ εἴ τι
τοιοῦτον ἕτερον γένος, τῷ μὲν προσπεφυκὸς ζῆν
μόνον φυτῷ παραπλήσιον, τῷ δ' ἔχειν τι σαρκῶδες
δόξειεν ἂν ἔχειν τιν' αἴσθησιν· ἄδηλον δὲ τοῦτο
ποτέρως θετέον.

Ἔχει δὲ τοῦτο τὸ ζῷον δύο πόρους καὶ μίαν
30 διαίρεσιν, ᾗ τε δέχεται τὴν ὑγρότητα τὴν εἰς
τροφήν, καὶ ᾗ πάλιν διαπέμπει τὴν ὑπολειπομένην
ἰκμάδα· περίττωμα γὰρ οὐδέν ἐστι δῆλον ἔχον,
ὥσπερ τἆλλα τὰ ὀστρακόδερμα. διὸ μάλιστα καὶ
τοῦτο, κἂν εἴ τι ἄλλο τοιοῦτον τῶν ζῴων, φυτικὸν
δίκαιον καλεῖν· οὐδὲ γὰρ τῶν φυτῶν οὐδὲν ἔχει
35 περίττωμα. διὰ μέσου δὲ λεπτὸν διάζωμα, ἐν ᾧ
τὸ κύριον ὑπάρχειν εὔλογον τῆς ζωῆς. ἃς δὲ
καλοῦσιν οἱ μὲν κνίδας οἱ δ' ἀκαλήφας, ἔστι μὲν οὐκ
681 b ὀστρακόδερμα, ἀλλ' ἔξω πίπτει τῶν διῃρημένων
γενῶν, ἐπαμφοτερίζει δὲ τοῦτο καὶ φυτῷ καὶ ζῴῳ

---

all respects like a plant : it lives only while it is
growing on to something, and when it is pulled off it
dies. What are called Holothuria and the Sea-lungs [a]
and other similar sea-animals differ only slightly
from the sponges in being unattached. They have
no power of sensation, but they live just as if they
were plants unattached to the soil. Even among
land-plants such instances exist : living and growing
either on other plants or quite unattached : for
example, the plant found on Parnassus, sometimes
called the Epipetron (Rockplant). If you hang this
up on the pegs [b] it will keep alive for a considerable time. Sometimes it is doubtful whether these
Ascidians and any other such group of creatures
ought to be classed as plants or as animals : In so far
as they live only by growing on to some other object
they approach the status of a plant ; but yet they
have some fleshy substance and therefore probably
are capable of sensation of a kind.

This particular creature (the Ascidian) has two
orifices and one septum; by one orifice it takes in fluid
matter for food, by the other it discharges the surplus
moisture ; so far as can be seen it has no residue like
the other Testacea. And as no plant ever has any
residue this is a strong justification for classing it
(and any other such animal) as a plant. Through its
middle there runs a thin partition, and it is reasonable to suppose that the governing and vital part of
the creature is situated here. As for what are called
Knides or Acalephae,[c] they are not Testacea, it is
true, but fall outside the defined groups. In their
nature they incline towards the plants on one side

Those common to the Mediterranean are more virulent in
their stinging powers than those of the north.

681 b

τὴν φύσιν. τῷ μὲν γὰρ ἀπολύεσθαι καὶ προσ-
πίπτειν πρὸς τὴν τροφὴν ἐνίας αὐτῶν ζωικόν ἐστι,
5 καὶ τῷ αἰσθάνεσθαι τῶν προσπιπτόντων· ἔτι δὲ τῇ
τοῦ σώματος τραχύτητι χρῆται πρὸς τὴν σωτηρίαν·
τῷ δ' ἀτελὲς εἶναι καὶ προσφύεσθαι ταχέως ταῖς
πέτραις τῷ γένει τῶν φυτῶν παραπλήσιον, καὶ τῷ
περίττωμα μηδὲν ἔχειν φανερόν, στόμα δ' ἔχειν.
ὅμοιον δὲ τούτῳ καὶ τὸ τῶν ἀστέρων ἐστὶ γένος—
10 καὶ γὰρ τοῦτο προσπῖπτον ἐκχυμίζει πολλὰ τῶν
ὀστρέων—τοῖς τ' ἀπολελυμένοις τῶν εἰρημένων
ζῴων, οἷον τοῖς τε μαλακίοις καὶ τοῖς μαλακοστράκοις. ὁ δ' αὐτὸς λόγος καὶ περὶ τῶν ὀστρακοδέρμων.

Τὰ μὲν οὖν μόρια τὰ περὶ τὴν τροφήν, ἅπερ
ἀναγκαῖον πᾶσιν ὑπάρχειν, ἔχει τὸν προειρημένον
15 τρόπον, δεῖ δὲ δηλονότι καὶ τῶν τοῖς ἐναίμοις
ὑπαρχόντων κατὰ τὸ κύριον τῶν αἰσθήσεων ἔχειν
ἀνάλογόν τι μόριον· τοῦτο γὰρ δεῖ πᾶσιν ὑπάρχειν
τοῖς ζῴοις. ἔστι δὲ τοῦτο τοῖς μὲν μαλακίοις ἐν
ὑμένι κείμενον ὑγρόν, δι' οὗπερ ὁ στόμαχος τέταται
πρὸς τὴν κοιλίαν, προσπέφυκε δὲ πρὸς τὰ πρανῆ
20 μᾶλλον, καὶ καλεῖται μύτις ὑπό τινων. τοιοῦτον δ'
ἕτερον καὶ τοῖς μαλακοστράκοις ἐστί, καὶ καλεῖται
κἀκεῖνο μύτις. ἔστι δ' ὑγρὸν καὶ σωματῶδες ἅμα
τοῦτο τὸ μόριον, τείνει δὲ δι' αὐτοῦ, καθάπερ
εἴρηται, διὰ μέσου μὲν ὁ στόμαχος· εἰ γὰρ ἦν
μεταξὺ τούτου καὶ τοῦ πρανοῦς, οὐκ ἂν ἠδύνατο
25 λαμβάνειν ὁμοίως διάστασιν εἰσιούσης τῆς τροφῆς
διὰ τὴν τοῦ νώτου σκληρότητα. ἐπὶ δὲ τῆς μύτιδος
τὸ ἔντερον ἔξωθεν, καὶ ὁ θολὸς πρὸς τῷ ἐντέρῳ,

---

^a That is, dorsal.

and the animals on the other. Towards the animals, because some of them detach themselves and fasten upon their food, and are sensible of objects that come up against them ; and also because they make use of the roughness of their body for self-preservation. Towards the plants, because they are incomplete, and quickly attach themselves to rocks ; and further, because they have no residue that can be seen, though they have a mouth. The group of Starfish resembles these creatures ; Starfish too fasten on to their food, and by doing this to oysters suck large numbers of them dry. But Starfish also resemble those unattached creatures of which we spoke, the Cephalopods and the Crustacea. The same may be said of the Testacea.

The parts connected with nutrition are such as I have now described. These must of necessity be present in all animals. But there is yet another part which every animal must have. These creatures must have some part which is analogous to the parts which in blooded animals are connected with the control of sensation. In the Cephalopods this consists of a fluid contained in a membrane, through which the gullet extends towards the stomach. It is attached to the body rather towards the upper ^a side. Some call it the *mytis*. An organ just like this, also called the *mytis*, is present in the Crustacea. This part is fluid and corporeal at the same time. The gullet, as I said, extends through the middle of it. If the gullet had been placed between the *mytis* and the dorsal side, the gullet would not have been able to distend sufficiently when the food enters, owing to the hardness of the back. The intestine is placed up against the outer surface of the *mytis*, and the ink-bag

681 b

ὅπως ὅτι πλεῖστον ἀπέχῃ τῆς εἰσόδου καὶ τὸ δυσχερὲς ἄποθεν ᾖ τοῦ βελτίονος καὶ τῆς ἀρχῆς. ὅτι δ' ἐστὶ τὸ ἀνάλογον τῇ καρδίᾳ τοῦτο τὸ μόριον,
30 δηλοῖ ὁ τόπος (οὗτος γάρ ἐστιν ὁ αὐτός) καὶ ἡ γλυκύτης τῆς ὑγρότητος ὡς οὖσα πεπεμμένη καὶ αἱματώδης.

Ἐν δὲ τοῖς ὀστρακοδέρμοις ἔχει μὲν τὸν αὐτὸν τόπον[1] τὸ κύριον τῆς αἰσθήσεως, ἧττον δ' ἐπίδηλον. πλὴν δεῖ ζητεῖν ἀεὶ περὶ μεσότητα ταύτην τὴν ἀρχήν, ὅσα μὲν μόνιμα, τοῦ δεχομένου μορίου τὴν
35 τροφήν, καὶ δι' οὗ ποιεῖται τὴν ἀπόκρισιν ἢ τὴν σπερματικὴν ἢ τὴν περιττωματικήν, ὅσα δὲ
682 a καὶ πορευτικὰ τῶν ζῴων, ἀεὶ ἐν[2] τῷ μέσῳ τῶν δεξιῶν καὶ τῶν ἀριστερῶν.

Τοῖς δ' ἐντόμοις τὸ μὲν τῆς τοιαύτης ἀρχῆς μόριον, ὥσπερ ἐν τοῖς πρώτοις ἐλέχθη λόγοις, μεταξὺ κεφαλῆς καὶ τοῦ περὶ τὴν κοιλίαν ἐστὶ κύτους. τοῦτο δὲ τοῖς μὲν πολλοῖς ἐστιν ἕν, τοῖς
5 δὲ πλείω, καθάπερ τοῖς ἰουλώδεσι καὶ μακροῖς· διόπερ διατεμνόμενα ζῇ. βούλεται μὲν γὰρ ἡ φύσις ἐν πᾶσι μόνον ἓν ποιεῖν τὸ τοιοῦτον, καὶ δυναμένη μὲν ποιεῖ μόνον ἕν, οὐ δυναμένη δὲ πλείω.[3] δῆλον δ' ἐν ἑτέροις ἑτέρων μᾶλλον.

Τὰ δὲ πρὸς τὴν τροφὴν μόρια οὐ πᾶσιν ὁμοίως,
10 ἀλλὰ διαφορὰν ἔχει πολλήν. ἐντὸς γὰρ τοῦ στόματος ἐνίοις μέν ἐστι τὸ καλούμενον κέντρον, ὡσπερανεὶ σύνθετον καὶ ἔχον γλώττης καὶ χειλῶν

---

[1] τόπον Rackham : τρόπον vulg.   [2] ἐν P : om. vulg.
[3] sic SUY (δυνάμενα bis S) : καὶ δυναμένην μέν, ἓν ποιεῖ μόνον· οὐ δυναμένη δὲ πλείω Z : οὐ δυναμένη δ' ἐνεργείᾳ ποιεῖ μόνον ἕν, δυνάμει δὲ πλείω· vulg. (cf. 667 b 25).

up against the intestine ; this is to ensure that it and
its unpleasantness are kept as far as possible from the
body's entrance and from the sovereign and most
noble part.   The *mytis* occupies a place which corresponds exactly with that of the heart in blooded
creatures : which shows that it is the counterpart of
it.[a]   Another proof of this is that the fluid in it is
sweet—that is, it has undergone concoction and is of
the nature of blood.

In the Testacea the part which rules sensation
occupies the same place but is not so easy to pick out.
But this source of control should always be looked for
around some middle position in these creatures : in
stationary ones, in the midst between the part which
receives the food and the part where the seed or the
residue is emitted ; and in those which move about,
always midway between the right side and the left.

In insects the part where this control is placed, as
was said in the first treatise,[b] is situated between the
head and the cavity where the stomach is.   In the
majority there is one such part, but in creatures like
the Centipede, that is, which are long in the body,
there are more than one : so if the creatures are cut
up they go on living.   Now Nature's desire is to make
this part a unity in all creatures, and when she can,
she makes it a unity, when she cannot, a plurality.[c]
This is clearer in some cases than in others.

The parts connected with nutrition are by no means
alike in all insects ; indeed they exhibit great
differences.   For instance : Some have what is
known as a sting inside the mouth—a sort of combination of tongue and lips,—which possesses the

---

[a] The heart of invertebrates escaped the notice of Aristotle.
[b] At *Hist. An.* 531 b 34.   [c] *Cf.* 667 b 22 ff.

682 a

ἅμα δύναμιν· τοῖς δὲ μὴ ἔχουσιν ἔμπροσθεν τὸ
κέντρον ἐστὶν ἐντὸς τῶν ὀδόντων τοιοῦτον αἰσθητήριον. τούτου δ' ἐχόμενον πᾶσιν ἔντερον εὐθὺ καὶ
15 ἁπλοῦν μέχρι τῆς ἐξόδου τοῦ περιττώματος· ἐνίοις
δὲ τοῦτο ἑλίκην ἔχει. τὰ δὲ κοιλίαν μετὰ τὸ στόμα,
ἀπὸ δὲ τῆς κοιλίας τὸ ἔντερον εἰλιγμένον, ὅπως
ὅσα βρωτικώτερα καὶ μείζω τὴν φύσιν ὑποδοχὴν
ἔχῃ πλείονος τροφῆς. τὸ δὲ τῶν τεττίγων γένος
20 ἰδίαν ἔχει μάλιστα τούτων φύσιν· τὸ γὰρ αὐτὸ
μόριον ἔχει στόμα καὶ γλῶτταν συμπεφυκός, δι'
οὗ καθαπερεὶ διὰ ῥίζης δέχεται τὴν τροφὴν ἀπὸ
τῶν ὑγρῶν. πάντα μὲν οὖν ἐστιν ὀλιγότροφα τὰ
ἔντομα τῶν ζῴων, οὐχ οὕτω διὰ μικρότητα ὡς
διὰ ψυχρότητα (τὸ γὰρ θερμὸν καὶ δεῖται τροφῆς
καὶ πέττει τὴν τροφὴν ταχέως, τὸ δὲ ψυχρὸν ἄ-
25 τροφον), μάλιστα δὲ τὸ τῶν τεττίγων γένος· ἱκανὴ
γὰρ τροφὴ τῷ σώματι ἡ ἐκ τοῦ πνεύματος ὑπο-
μένουσα ὑγρότης, καθάπερ τοῖς ἐφημέροις ζῴοις
(γίνεται δὲ ταῦτα περὶ τὸν Πόντον), πλὴν ἐκεῖνα
μὲν ζῇ μιᾶς ἡμέρας χρόνον, ταῦτα δὲ πλειόνων
μὲν ἡμερῶν, ὀλίγων δὲ τούτων.

30 Ἐπεὶ δὲ περὶ τῶν ἐντὸς ὑπαρχόντων μορίων τοῖς
ζῴοις εἴρηται, πάλιν περὶ τῶν λοιπῶν τῶν ἐκτὸς
ἐπανιτέον. ἀρκτέον δ' ἀπὸ τῶν νῦν εἰρημένων,
ἀλλ' οὐκ ἀφ' ὧν ἀπελίπομεν, ὅπως ἀπὸ τούτων
διατριβὴν ἐλάττω ἐχόντων ἐπὶ τῶν τελείων καὶ
ἐναίμων ζῴων ὁ λόγος σχολάζῃ μᾶλλον.

35 VI. Τὰ μὲν οὖν ἔντομα τῶν ζῴων οὐ πολυμερῆ
μὲν τὸν ἀριθμόν ἐστιν, ὅμως δ' ἔχει πρὸς ἄλληλα
340

character of both. Those that have no sting in front
have a sense-organ of that sort behind the teeth.
After the mouth, in all insects comes the intestine,
which is straight and simple right up to the residual
vent. (Sometimes, however, it has a spiral in it.)
And some there are which have the stomach next
after the mouth, while from the stomach runs a
twisted intestine ; this gives the bigger and more
gluttonous insects room for a larger amount of food.
Of all these creatures the grasshoppers are the most
peculiar. In them the mouth and tongue are united
so as to make one single part, and through this they
draw up their nourishment from fluid substances as
through a root. All insects take but little nourishment ; and this is not so much because they are
small as because they are cold. (Heat needs nourishment and quickly concocts it ; cold needs none.)
This is most marked in the grasshoppers. They find
sufficient nourishment in the moisture which the air
deposits ; so do the one-day creatures which occur
around the Black Sea. Still, they live only for the
space of a day ; whereas the grasshoppers live for
several, though not many, days.

Now that we have spoken of the internal parts of
animals, we must go back and deal with the remainder
of the external parts. We had better begin with the
creatures of which we have just been speaking, and
not go back to the point where we left the external
parts. This will mean that we take first those which
need less discussion, and that will give more time for
speaking of the " perfect " animals, *i.e.* the blooded
ones.

VI. Insects first, then. Though their parts are not EXTERNAL
numerous, insects differ from one another. They all PARTS OF
BLOODLESS
ANIMALS.

341

682 a

διαφοράς. πολύποδα μὲν γάρ ἐστι πάντα διὰ τὸ
682 b πρὸς τὴν βραδυτῆτα καὶ κατάψυξιν τῆς φύσεως τὴν
πολυποδίαν ἀνυτικωτέραν αὐτοῖς ποιεῖν τὴν κίνησιν·
καὶ μάλιστα πολύποδα τὰ μάλιστα κατεψυγμένα διὰ
τὸ μῆκος οἷον τὸ τῶν ἰούλων γένος. ἔτι δὲ διὰ τὸ
5 ἀρχὰς ἔχειν πλείονας αἵ τ᾽ ἐντομαί εἰσι καὶ πολύποδα κατὰ ταῦτα¹ ἐστιν.

Ὅσα δ᾽ ἐλάττονας ἔχει πόδας, πτηνὰ ταῦτ᾽ ἐστὶ
πρὸς τὴν ἔλλειψιν τὴν τῶν ποδῶν. αὐτῶν δὲ τῶν
πτηνῶν ὧν μέν ἐστιν ὁ βίος νομαδικὸς καὶ διὰ τὴν
τροφὴν ἀναγκαῖον ἐκτοπίζειν, τετράπτερά τέ ἐστι
καὶ τὸν τοῦ σώματος ἔχει κοῦφον ὄγκον, οἷον αἵ τε
10 μέλιτται καὶ τὰ σύμφυλα ζῷα ταύταις· δύο γὰρ ἐφ᾽
ἑκάτερα πτερὰ² ἔχουσιν. ὅσα δὲ μικρὰ τῶν τοιούτων, δίπτερα, καθάπερ τὸ τῶν μυιῶν γένος. τὰ δὲ
βαρέα³ καὶ τοῖς βίοις ἑδραῖα πολύπτερα μὲν ὁμοίως
ταῖς μελίτταις⁴ ἐστίν, ἔχει δ᾽ ἔλυτρα τοῖς πτεροῖς,
15 οἷον αἵ τε μηλολόνθαι καὶ τὰ τοιαῦτα τῶν ἐντόμων,
ὅπως σῴζῃ τὴν τῶν πτερῶν δύναμιν· ἑδραίων γὰρ
ὄντων εὐδιάφθορα μᾶλλόν ἐστι τῶν εὐκινήτων,
διόπερ ἔχει φραγμὸν πρὸ αὐτῶν. καὶ ἄσχιστον δὲ
τούτων ἐστὶ τὸ πτερὸν καὶ ἄκαυλον· οὐ γάρ ἐστι
πτερὸν ἀλλ᾽ ὑμὴν δερματικός, ὃς διὰ ξηρότητα ἐξ
20 ἀνάγκης ἀφίσταται τοῦ σώματος αὐτῶν ψυχομένου
τοῦ σαρκώδους.

Ἔντομα δ᾽ ἐστὶ διά τε τὰς εἰρημένας αἰτίας, καὶ
ὅπως σῴζηται δι᾽ ἀπάθειαν συγκαμπτόμενα· συν-
ελίττεται γὰρ τὰ μῆκος ἔχοντ᾽ αὐτῶν, τοῦτο δ᾽ οὐκ
ἂν ἐγίνετ᾽ αὐτοῖς μὴ οὖσιν ἐντόμοις. τὰ δὲ μὴ

---

¹ ταῦτά Peck : ταυτὰ Y : ταῦτ᾽ vulg. : ταύτας Ogle.
² πτερὰ τοῦ σώματος vulg. : τοῦ σ. delevi.
³ βαρέα Ogle : βραχέα vulg.   ⁴ μελίτταις ⟨οὐκ⟩ Platt.

342

have numerous feet ; this is in order to make their (a) Insects.
motion quicker, and to counteract their natural slow-
ness and coldness. Those which are most subject to
coldness owing to their length (*e.g.* the Centipedes)
have the greatest number of feet. Furthermore,
these creatures have several sources of control ; and
on that account they have the " insections " in their
bodies, and the numerous feet which are placed in
precise correspondence.

Those that have fewer feet are winged by way of
compensation. Some of these flying insects live a
wandering life and have to go abroad in search of
food ; so they have a light body and four wings, two
on either side ; such are the bees and the kindred
tribes. The small ones have only two wings all told
—like the flies. Those that are heavy and sedentary
in their habits have the larger number of wings like
the bees, but they have shards round their wings
(*e.g.* the Melolonthae [a] and similar insects) to preserve
them in their proper condition ; for, as these creatures
are sedentary, their wings are more liable to be
destroyed than those of the nimbler insects ; and
that is why there is this protection round them.
An insect's wing is not divided, and it has no shaft.
In fact, it is not a wing at all, but a membrane of skin,
which being dry detaches itself of necessity from the
creature's body as the fleshy part cools off.

I have already stated some reasons why these
creatures have " insected " bodies : there is another,
viz. it is so that they may curl up and thus escape
injury and remain safe. It is the long ones that roll
themselves up, and this would be impossible for them
if they were not insected. Those that do not roll up

---

[a] Perhaps cockchafers (Ogle).

682 b

ἑλικτὰ αὐτῶν σκληρύνεται μᾶλλον συνιόντα εἰς τὰς
25 τομάς. δῆλον δὲ τοῦτο γίνεται θιγγανόντων, οἷον
ἐπὶ τῶν καλουμένων κανθάρων· φοβηθέντα γὰρ
ἀκινητίζει, καὶ τὸ σῶμα γίνεται σκληρὸν αὐτῶν.
ἀναγκαῖον δ' ἐντόμοις αὐτοῖς εἶναι· τοῦτο γὰρ ἐν
τῇ οὐσίᾳ αὐτῶν ὑπάρχει τὸ πολλὰς ἔχειν ἀρχάς, καὶ
30 ταύτῃ προσέοικε τοῖς φυτοῖς. ὥσπερ γὰρ τὰ φυτά,
καὶ ταῦτα διαιρούμενα δύναται ζῆν, πλὴν ταῦτα μὲν
μέχρι τινός, ἐκεῖνα δὲ καὶ τέλεια γίνεται τὴν φύσιν
καὶ δύο ἐξ ἑνὸς καὶ πλείω τὸν ἀριθμόν.

Ἔχει δ' ἔνια τῶν ἐντόμων καὶ κέντρα πρὸς
βοήθειαν τῶν βλαπτόντων. τὸ μὲν οὖν κέντρον
35 τοῖς μὲν ἔμπροσθέν ἐστι τοῖς δ' ὄπισθεν, τοῖς μὲν
ἔμπροσθεν κατὰ τὴν γλῶτταν, τοῖς δ' ὄπισθεν κατὰ
τὸ οὐραῖον. ὥσπερ γὰρ τοῖς ἐλέφασι τὸ τῶν
ὀσμῶν αἰσθητήριον γεγένηται χρήσιμον πρός τε
683 a τὴν ἀλκὴν καὶ τὴν τῆς τροφῆς χρῆσιν, οὕτως τῶν
ἐντόμων ἐνίοις τὸ κατὰ τὴν γλῶτταν τεταγμένον·
αἰσθάνονταί τε γὰρ τούτῳ τῆς τροφῆς καὶ ἀναλαμ-
βάνουσι καὶ προσάγονται αὐτήν. ὅσα δὲ μή ἐστιν
αὐτῶν ἐμπροσθόκεντρα, ὀδόντας ἔχει τὰ μὲν
5 ἐδωδῆς χάριν τὰ δὲ τοῦ λαμβάνειν καὶ προσάγεσθαι
τὴν τροφήν, οἷον οἵ τε μύρμηκες καὶ τὸ τῶν μελιττῶν πασῶν γένος. ὅσα δ' ὀπισθόκεντρά ἐστι, διὰ
τὸ θυμὸν ἔχειν ὅπλον ἔχει τὸ κέντρον. ἔχουσι δὲ
τὰ μὲν ἐν ἑαυτοῖς τὰ κέντρα, καθάπερ αἱ μέλιτται
καὶ οἱ σφῆκες, διὰ τὸ πτηνὰ εἶναι· λεπτὰ μὲν γὰρ
10 ὄντα καὶ ἔξω εὔφθαρτα ⟨ἂν⟩[1] ἦν· εἰ δὲ παχέα ἦν[2]
ὥσπερ τοῖς σκορπίοις, βάρος ἂν παρεῖχεν. τοῖς δὲ

---

[1] ⟨ἂν⟩ Ogle.   [2] δὲ παχέα ἦν Platt: δ' ἀπεῖχεν vulg.

increase their hardness by closing up the insections. This is obvious if you touch them—*e.g.* the insects called Canthari (dung-beetles) are frightened when touched and become motionless, and their bodies become hard. But also it is *necessary* for them to be insected, for it is of their essential being to have numerous sources of control; and herein they resemble plants. Plants can live when they are cut up; so can insects. There is a difference, however, for whereas the period of survival of a divided insect is limited, a plant can attain the perfection of its nature when divided, and so two plants or more come out of one.

Some of the insects have a sting as well, for defence against attackers. In some the sting is in front, by the tongue; in others it is behind at the tail-end. Consider the elephant's trunk: this is its organ of smell; but the elephant uses it as a means of exerting force as well as for the purposes of nutrition. Compare with this the sting of insects: when, as in some of them, it is ranged alongside the tongue, not only do they get their sensation of the food by means of it, but they also pick up the food with it and convey it to the mouth. Those which have no sting in front have teeth; which some of them use for eating, others for picking up the food and conveying it to the mouth, as do the ants and the whole tribe of bees. Those that have a sting at the back are fierce creatures and the sting serves them as a weapon. Sometimes the sting is well inside the body, as in bees and wasps. This is because they are winged, and a delicate sting on the outside of the body would be easily destroyed; on the other hand, a thick one such as scorpions have would weigh them down. Scorpions

σκορπίοις πεζοῖς οὖσι καὶ κέρκον[1] ἔχουσιν ἀναγκαῖον ἐπὶ ταύτῃ[2] ἔχειν τὸ κέντρον, ἢ μηθὲν χρήσιμον
εἶναι πρὸς τὴν ἀλκήν. δίπτερον δ' οὐθέν ἐστιν
ὀπισθόκεντρον· διὰ τὸ ἀσθενῆ γὰρ καὶ μικρὰ εἶναι
15 δίπτερά ἐστιν· ἱκανὰ γὰρ τὰ μικρὰ αἴρεσθαι ὑπὸ
τῶν ἐλαττόνων τὸν ἀριθμόν. διὰ ταὐτὸ δὲ τοῦτο
καὶ ἔμπροσθεν ἔχει τὸ κέντρον· ἀσθενῆ γὰρ ὄντα
μόλις δύναται τοῖς ὄπισθεν[3] τύπτειν. τὰ δὲ
πολύπτερα, διὰ τὸ μείζω τὴν φύσιν εἶναι, πλειόνων
τετύχηκε πτερῶν καὶ ἰσχύει τοῖς ὄπισθεν μορίοις.
20 βέλτιον δ' ἐνδεχομένου μὴ ταὐτὸ ὄργανον ἐπὶ
ἀνομοίας ἔχειν χρήσεις, ἀλλὰ τὸ μὲν ἀμυντικὸν
ὀξύτατον, τὸ δὲ γλωττικὸν σομφὸν καὶ σπαστικὸν
τῆς τροφῆς. ὅπου γὰρ ἐνδέχεται χρῆσθαι δυσὶν
ἐπὶ δύ' ἔργα καὶ μὴ ἐμποδίζειν πρὸς ἕτερον,
οὐδὲν ἡ φύσις εἴωθε ποιεῖν ὥσπερ ἡ χαλκευτικὴ
25 πρὸς εὐτέλειαν ὀβελισκολύχνιον· ἀλλ' ὅπου μὴ
ἐνδέχεται, καταχρῆται τῷ αὐτῷ ἐπὶ πλείω ἔργα.[a]

Τοὺς δὲ πόδας τοὺς προσθίους μείζους ἔνια τούτων ἔχει, ὅπως ἐπειδὴ διὰ τὸ σκληρόφθαλμα εἶναι
οὐκ ἀκριβῆ τὴν ὄψιν ἔχουσι, τὰ προσπίπτοντα τοῖς
προσθίοις ἀποκαθαίρωσι σκέλεσιν· ὅπερ καὶ φαί-
30 νονται ποιοῦσαι αἵ τε μυῖαι καὶ τὰ μελιττώδη τῶν
ζῴων· ἀεὶ γὰρ χαρακίζουσι τοῖς προσθίοις σκέλεσιν.
τὰ δ' ὀπίσθια μείζω τῶν μέσων διά τε τὴν βάδισιν
καὶ πρὸς τὸ αἴρεσθαι ῥᾷον ἀπὸ τῆς γῆς ἀναπετ-

---

[1] κέρκον Z (coniecerat Ogle) : κέντρον vulg.
[2] ταύτῃ Ogle : ταῦτ' vulg.
[3] ὄπισθεν Ogle, Thurot : ἔμπροσθεν vulg.

[a] The principle of "division of labour" in a living organism,
not stated again until 1827 (by Milne Edwards). See Ogle's
note.

themselves, being land-creatures and having a tail,
are bound to have their sting on their tail ; otherwise
it would be no use for exerting force. No two-
winged insect has a sting at the rear ; these are small
weak creatures, and can be supported by a smaller
number of wings : that is why they have only two.
The same reason explains why they have their sting
in the front : owing to their weakness they cannot
well deliver a blow with their hind parts. Many-
winged creatures, on the other hand, owe their
greater number of wings to their own greater size,
and so too their hind parts are stronger and bear the
sting. It is better, when it is possible, that one and the
same organ should not be put to dissimilar uses ; that
is, there should be an organ of defence which is very
sharp, and another organ to act as a tongue, which
should be spongy and able to draw up nourishment.
And thus, whenever it is possible to employ two
organs for two pieces of work without their getting in
each other's way, Nature provides and employs two.[a]
Her habits are not those of the coppersmith who for
cheapness' sake makes you a spit-and-lampstand
combination. Still, where two are impossible, Nature
employs the same organ to perform several pieces
of work.

Some insects, whose eyesight is not distinct owing
to their eyes being made of some hard substance, have
specially long forefeet, which enable them to clear
away anything that comes down on to the eyes.
Flies and bees and the like are obvious examples :
they are always crossing their front legs. These
creatures' hind legs are longer than their middle ones
for two reasons : (1) to assist them in walking, and
(2) to lift them more easily off the ground when they

683 a

όμενα. ὅσα δὲ πηδητικὰ αὐτῶν ἔτι μᾶλλον τοῦτο
φανερόν, οἷον αἵ τ᾽ ἀκρίδες καὶ τὸ τῶν ψυλλῶν
35 γένος· ὅταν γὰρ κάμψαντ᾽ ἐκτείνῃ πάλιν, ἀναγκαῖον
ἀπὸ τῆς γῆς ἦρθαι. οὐκ ἔμπροσθεν δ᾽ ἀλλ᾽
683 b ὄπισθεν μόνον ἔχουσι τὰ πηδαλιώδη αἱ ἀκρίδες·
τὴν γὰρ καμπὴν ἀναγκαῖον εἴσω κεκλάσθαι, τῶν
δὲ προσθίων κώλων οὐδέν ἐστι τοιοῦτον. ἐξάποδα
δὲ τὰ τοιαῦτα πάντ᾽ ἐστὶ σὺν τοῖς ἁλτικοῖς μορίοις.

VII. Τῶν δ᾽ ὀστρακοδέρμων οὐκ ἔστι τὸ σῶμα
5 πολυμερές. τούτου δ᾽ αἴτιον τὸ μόνιμον αὐτῶν
εἶναι τὴν φύσιν· πολυμερέστερα γὰρ ἀναγκαῖον
εἶναι τῶν ζώων τὰ κινητικὰ διὰ τὸ ⟨πλείους⟩¹ εἶναι
αὐτῶν πράξεις· ὀργάνων γὰρ δεῖται πλειόνων τὰ
πλειόνων μετέχοντα κινήσεων. τούτων δὲ τὰ μὲν
ἀκίνητα πάμπαν ἐστί, τὰ δὲ μικρᾶς μετέχει κι-
10 νήσεως· ἀλλ᾽ ἡ φύσις πρὸς σωτηρίαν αὐτοῖς τὴν
τῶν ὀστράκων σκληρότητα περιέθηκεν. ἔστι δὲ
τὰ μὲν μονόθυρα τὰ δὲ δίθυρα αὐτῶν, τὰ δὲ στρομβώδη, καθάπερ εἴρηται πρότερον· καὶ τούτων τὰ
μὲν ἑλίκην ἔχοντα, οἷον κήρυκες, τὰ δὲ σφαιροειδῆ
μόνον, καθάπερ τὸ τῶν ἐχίνων γένος. καὶ τῶν
15 διθύρων τὰ μέν ἐστιν ἀναπτυκτά, οἷον κτένες καὶ
μύες (ἐπὶ θάτερα γὰρ συγκέκλεισται, ὥστε ἀν-
οίγεσθαι ἐπὶ θάτερα καὶ συγκλείεσθαι), τὰ δ᾽ ἐπ᾽
ἄμφω συμπέφυκεν, οἷον τὸ τῶν σωλήνων γένος.
ἅπαντα δὲ τὰ ὀστρακόδερμα, καθάπερ τὰ φυτά,
20 κάτω τὴν κεφαλὴν ἔχει. τούτου δ᾽ αἴτιον ὅτι
κάτωθεν λαμβάνει τὴν τροφήν, ὥσπερ τὰ φυτὰ
ταῖς ῥίζαις. συμβαίνει οὖν αὐτοῖς τὰ μὲν κάτω
ἄνω ἔχειν, τὰ δ᾽ ἄνω κάτω. ἐν ὑμένι δ᾽ ἐστί, δι᾽

---

¹ ⟨πλείους⟩ Peck : ⟨πολλὰς⟩ Platt.

rise in flight. This peculiarity is even more noticeable in the leaping insects, such as locusts and the
various sorts of fleas, which first bend their hind legs
and then stretch them out again, and this forces them
to rise up from the ground. The rudder-shaped legs
which locusts have are at the rear only and not in
front ; this is because the joint must bend inwards,[a]
and no front limb satisfies this condition. All
these creatures have six feet, inclusive of the parts
used for leaping.

VII. In Testacea the body is not divided into
several parts, owing to their being of stationary
habits, as opposed to creatures which move about :
the latter are bound to have more parts to their body
because their activities are more numerous, and the
more motions of which a species is capable, the more
organs it requires. Now some of the Testacea are
altogether stationary : others move about but little ;
and so, to keep them safe, Nature has compassed
them about with hard shells. Some of them are (as I
said earlier [b]) one-valved, some two-valved ; and some
conical, either spiral like the Whelks, or spherical
like the Sea-urchins. The two-valved shells are
divided into (a) those which open—i.e. which have a
joint on one side and can open and shut on the other ;
e.g. the scallops and mussels ; (b) those which are
joined together on both sides, e.g. the group of razor-
fishes. In all Testacea, just as in plants, the head is
down below. The reason for this is that they take up
their food from below, as plants take it up by their
roots ; so they have their nether parts above and their
upper parts below. These creatures are enveloped
in a membrane, and through this they strain fresh-

---

[a] See note on 693 b 3, p. 433.   [b] At 679 b 16.

οὗ διηθεῖ τὸ πότιμον καὶ λαμβάνει τὴν τροφήν.
ἔχει δὲ κεφαλὴν μὲν πάντα, τὰ δὲ τοῦ σώματος
μόρια παρὰ τὸ τῆς τροφῆς δεκτικὸν ἀνώνυμα
τἆλλα.

25 VIII. Τὰ δὲ μαλακόστρακα πάντα καὶ πορευτικά,
διὸ ποδῶν ἔχει πλῆθος. ἔστι δὲ γένη μὲν τέτταρα
τὰ μέγιστ᾽ αὐτῶν οἵ τε καλούμενοι κάραβοι καὶ
ἀστακοὶ καὶ καρίδες καὶ καρκίνοι· τούτων δ᾽
ἑκάστου πλείω εἴδη ἐστὶ διαφέροντα οὐ μόνον κατὰ
30 τὴν μορφὴν ἀλλὰ καὶ κατὰ τὸ μέγεθος πολύ· τὰ
μὲν γὰρ μεγάλα τὰ δὲ μικρὰ πάμπαν αὐτῶν ἐστιν.
τὰ μὲν οὖν καρκινώδη καὶ καραβώδη παρόμοι᾽
ἐστὶ τῷ χηλὰς ἔχειν ἀμφότερα. ταύτας δ᾽ οὐ
πορείας ἔχουσι χάριν, ἀλλὰ πρὸς τὸ λαβεῖν καὶ
κατασχεῖν ἀντὶ χειρῶν. διὸ καὶ κάμπτουσιν ἐναν-
35 τίως ταύτας τοῖς ποσίν· τοὺς μὲν γὰρ ἐπὶ τὸ κοῖλον
τὰς δ᾽ ἐπὶ τὸ περιφερὲς κάμπτουσι καὶ ἑλίσσουσιν·
οὕτω γὰρ χρήσιμαι πρὸς τὸ λαβοῦσαι προσφέρεσθαι
684 a τὴν τροφήν.

Διαφέρουσι δ᾽ ᾗ οἱ μὲν κάραβοι ἔχουσιν οὐράν,
οἱ δὲ καρκίνοι οὐκ ἔχουσιν οὐράν· τοῖς μὲν γὰρ διὰ
τὸ νευστικοῖς εἶναι χρήσιμος ἡ οὐρά (νέουσι γὰρ
ἀπερειδόμενοι οἷον πλάταις αὐταῖς), τοῖς δὲ καρ-
5 κίνοις οὐδὲν χρήσιμον διὰ τὸ πρόσγειον εἶναι τὸν
βίον[1] αὐτῶν καὶ εἶναι τρωγλοδύτας. ὅσοι δ᾽ αὐτῶν
πελάγιοί εἰσι, διὰ τοῦτο πολὺ ἀργοτέρους ἔχουσι
τοὺς πόδας[2] πρὸς τὴν πορείαν, οἷον αἵ τε μαῖαι
καὶ οἱ Ἡρακλεωτικοὶ καλούμενοι καρκίνοι, ὅτι
ὀλίγῃ κινήσει χρῶνται, ἀλλ᾽ ἡ σωτηρία αὐτοῖς
10 τῷ ὀστρειώδεις εἶναι γίνεται· διὸ αἱ μὲν μαῖαι

---

[1] τὸ βίον Bekker per typothetae errorem.
[2] αὐτῶν post πόδας vulg.: om. Y.

water to drink, which is their way of taking nourishment. All of them possess a head, but except for the part which takes in the food none of the other parts has a special name.

VIII. All the Crustacea can walk on land as well as swim; and hence they all have numerous feet. There are four main groups of Crustacea, called (1) Carabi; (2) Astaci; (3) Carides; and (4) Carcini.[a] Each of these contains several species which differ not only in shape, but also in size, and that considerably, for some species are large, others extremely small. The Carcinoid and the Caraboid crustacea resemble each other, in both having claws. These claws are not for the sake of locomotion, but serve instead of hands, for catching and holding; and that is why they bend in an opposite direction to the feet, which bend and twist toward the concave side, while the claws bend toward the convex side. This makes the claws serviceable for catching hold of the food and conveying it to the mouth.

The two groups, Carabi and Carcini, differ in that the former have a tail and the latter have not. The Carabi find a tail useful because they are swimmers: they propel themselves with it as though with oars. A tail would be useless to the Carcini, which spend their lives near the land and creep into holes and crannies. Those that live out at sea and move about but little, and owe their safety to their shelly exterior, have for these reasons feet which are considerably less effective for locomotion: examples of this are the

(c) Crustacea.

---

[a] Roughly, these four divisions may be represented by our own groups, thus: (1) lobsters; (2) crayfish; (3) prawns and shrimps; (4) crabs.

684 a

λεπτοσκελεῖς, οἱ δ' Ἡρακλεωτικοὶ μικροσκελεῖς
εἰσίν.

Οἱ δὲ πάμπαν μικροὶ καρκίνοι, οἳ ἁλίσκονται ἐν
τοῖς μικροῖς ἰχθυδίοις, ἔχουσι τοὺς τελευταίους
πλατεῖς πόδας, ἵνα πρὸς τὸ νεῖν αὐτοῖς χρήσιμοι
ὦσιν, ὥσπερ πτερύγια ἢ πλάτας ἔχοντες τοὺς πόδας.

Αἱ δὲ καρίδες τῶν μὲν καρκινοειδῶν διαφέρουσι
15 τῷ ἔχειν κέρκον, τῶν δὲ καραβοειδῶν διὰ τὸ μὴ
ἔχειν χηλάς· ἃς οὐκ ἔχουσι διὰ τὸ πλείους ἔχειν
πόδας, ἐνταῦθα γὰρ ἡ ἐκεῖθεν ἀνήλωται αὔξησις.
πλείους δ' ἔχουσι πόδας, ὅτι νευστικώτερά ἐστιν
ἢ πορευτικώτερα.

Τὰ δ' ἐν τοῖς ὑπτίοις μόρια καὶ περὶ τὴν κεφαλὴν τὰ μὲν εἰς τὸ δέξασθαι τὸ ὕδωρ καὶ ἀφεῖναι
20 ἔχουσι βραγχοειδῆ· πλακωδέστερα δὲ τὰ κάτω αἱ
θήλειαι τῶν ἀρρένων καράβων ἔχουσι, καὶ τὰ ἐν
τῷ ἐπιπτύγματι δασύτερα αἱ θήλειαι καρκίνοι
τῶν ἀρρένων, διὰ τὸ ἐκτείνειν τὰ ᾠὰ πρὸς αὐτά,
ἀλλὰ μὴ ἄποθεν, ὥσπερ οἱ ἰχθύες καὶ τἆλλα τὰ
⟨ᾠὰ⟩[1] τίκτοντα· εὐρυχωρέστερα γὰρ ὄντα καὶ μείζω
25 χώραν ἔχει τοῖς ᾠοῖς μᾶλλον. οἱ μὲν οὖν κάραβοι
καὶ οἱ καρκίνοι πάντες τὴν δεξιὰν ἔχουσι χηλὴν
μείζω καὶ ἰσχυροτέραν· τοῖς γὰρ δεξιοῖς πάντα
πέφυκε τὰ ζῷα δρᾶν μᾶλλον, ἡ δὲ φύσις ἀποδίδωσιν ἀεὶ τοῖς χρῆσθαι δυναμένοις ἕκαστον ἢ μόνως
ἢ μᾶλλον, οἷον χαυλιόδοντας καὶ ὀδόντας καὶ
30 κέρατα καὶ πλῆκτρα καὶ πάντα τὰ τοιαῦτα μόρια,
ὅσα πρὸς βοήθειαν καὶ ἀλκήν ἐστιν.[2]

Οἱ δ' ἀστακοὶ μόνοι, ὁποτέραν ἂν τύχωσιν
ἔχουσι μείζω τῶν χηλῶν, καὶ αἱ θήλειαι καὶ οἱ

---

[1] ⟨ᾠὰ⟩ Peck : τήκοντα S : κυΐσκοντα PY : ᾠοτοκοῦντα Ogle.
[2] ἐστιν Peck : εἰσιν vulg.

Maiae [a] (whose legs are thin) and the crabs called Heracleotic (whose legs are short).

The little tiny crabs, which are found among the catch with small fishes, have their hindmost feet flat, like fins or oars, to make them useful for swimming.

The Carides differ from the Carcinoids in having a tail, and from the Caraboids just mentioned in not having claws. Claws are absent because they have more feet : the material for their growth has gone into the feet. And they have more feet because they swim about more or move about more.

As for the parts on the under [b] surface around the head, in some animals these are formed like gills so as to let in the water and to discharge it ; the lower parts, however, of female crabs are flatter in formation than those of male ones, and also the appendages on the flap are hairier. This is because they deposit their eggs there instead of getting rid of them, as the fishes and the other oviparous animals do. These appendages are wider and larger and so can provide more space for the eggs. In all the Carabi and in all the Carcini the right claw is bigger and stronger than the left. This is because all animals in their activities naturally use the right side more ; and Nature always assigns an instrument, either exclusively or in a better form, to those that can use it. This holds good for tusks, teeth, horns, spurs and all such parts which serve animals for assistance and offence.

In Lobsters only, whether male or female, it is a matter of chance which claw is the bigger. The

---

[a] Probably the spiny spider-crab.
[b] That is, ventral.

ἄρρενες. αἴτιον δὲ τοῦ μὲν ἔχειν χηλὰς ὅτι ἐν τῷ
35 γένει εἰσὶ τῷ ἔχοντι χηλάς· τοῦτο δ᾽ ἀτάκτως
684 b ἔχουσιν ὅτι πεπήρωνται καὶ οὐ χρῶνται ἐφ᾽ ὃ
πεφύκασιν, ἀλλὰ πορείας χάριν.

Καθ᾽ ἕκαστον δὲ τῶν μορίων, τίς ἡ θέσις αὐτῶν
καὶ τίνες διαφοραὶ πρὸς ἄλληλα, τῶν τ᾽ ἄλλων καὶ
τίνι διαφέρει τὰ ἄρρενα τῶν θηλειῶν, ἔκ τε τῶν
5 ἀνατομῶν θεωρείσθω καὶ ἐκ τῶν ἱστοριῶν τῶν
περὶ τὰ ζῷα.

IX. Τῶν δὲ μαλακίων περὶ μὲν τῶν ἐντὸς
εἴρηται πρότερον, ὥσπερ καὶ περὶ τῶν ἄλλων
ζῴων· ἐκτὸς δ᾽ ἔχει τό τε τοῦ σώματος κύτος,
ἀδιόριστον ὄν, καὶ τούτου πόδας ἔμπροσθεν περὶ
τὴν κεφαλήν, ἐντὸς μὲν τῶν ὀφθαλμῶν, περὶ δὲ
10 τὸ στόμα καὶ τοὺς ὀδόντας. τὰ μὲν οὖν ἄλλα ζῷα
τὰ ἔχοντα πόδας τὰ μὲν ἔμπροσθεν ἔχει καὶ
ὄπισθεν, τὰ δ᾽ ἐκ τοῦ πλαγίου, ὥσπερ τὰ πολύποδα
καὶ ἄναιμα τῶν ζῴων· τοῦτο δὲ τὸ γένος ἰδίως
τούτων· πάντας γὰρ ἔχουσι τοὺς πόδας ἐπὶ τὸ
καλούμενον ἔμπροσθεν. τούτου δ᾽ αἴτιον ὅτι
15 συνῆκται αὐτῶν τὸ ὄπισθεν πρὸς τὸ ἔμπροσθεν,
ὥσπερ τῶν ὀστρακοδέρμων τοῖς στρομβώδεσιν.
ὅλως γὰρ τὰ ὀστρακόδερμα ἔχει τῇ μὲν ὁμοίως
τοῖς μαλακοστράκοις, τῇ δὲ τοῖς μαλακίοις. ᾗ
μὲν γὰρ ἔξωθεν τὸ γεῶδες ἐντὸς δὲ τὸ σαρκῶδες,
τοῖς μαλακοστράκοις, τὸ δὲ σχῆμα τοῦ σώματος
20 ὃν τρόπον συνέστηκε, τοῖς μαλακίοις, τρόπον μέν

---

[a] See *Hist. An.* 525 a 30—527 b 34, 541 b 19 ff.

[b] At 678 b 24 ff.

[c] The theory that the cuttle-fish is comparable to a verte-
brate bent double was put forward in a paper read before the
Academy of Sciences in 1830, and was the origin of the famous

reason why they have claws is because they belong
to a group which has claws ; and they have them
in this irregular way because they themselves are
deformed and use the claws not for their natural
purpose but for locomotion.

For an account of every one of the parts, of
their position, and of the differences between them,
including the differences between the male and the
female, consult the Anatomical treatises and the
Inquiries upon Animals.[a]

IX. With regard to the Cephalopods, their internal (*d*) Cepha-
parts have already been described, as have those lopods.
of the other animals.[b] The external parts include
(1) the trunk of the body, which is undefined, and
(2) in front of this, the head, with the feet round it :
the feet are not beyond the eyes, but are outside the
mouth and the teeth. Other footed animals either
have some of their feet in front and some at the
back ; or else arranged along the sides—as with the
bloodless animals that have numerous feet. The
Cephalopods, however, have an arrangement of
their own. All their feet are on what may be
called the front. The reason for this is that their
back half is drawn up on to the front half,[c] just as
in the conical-shelled Testacea.   And generally,
though in some respects the Testacea resemble the
Crustacea, in others they resemble the Cephalopods.
In having their earthy material outside and their
fleshy material inside, they resemble the Crustacea ;
but as regarding the formation and construction of
their body they resemble the Cephalopods — all of

controversy between G. St-Hilaire and Cuvier about unity of
type. This controversy excited Goethe more than the revolu-
tion of the same year. (Ogle.)

684 b

τινα πάντα, μάλιστα δὲ τῶν στρομβωδῶν τὰ
ἔχοντα τὴν ἑλίκην· ἀμφοτέρων γὰρ τοῦτον ἔχει
τὸν τρόπον ἡ φύσις[1]· et propter hoc ambulant uniformiter ⟨ἀλλ' οὐ⟩[2] καθάπερ συμβέβηκεν ἐπὶ τῶν
τετραπόδων ζῴων καὶ τῶν ἀνθρώπων. homo vero
25 habet os in capite, scilicet in parte superiori corporis,
ἔπειτα τὸν στόμαχον, ἔπειτα δὲ τὴν κοιλίαν, ἀπὸ δὲ
ταύτης τὸ ἔντερον μέχρι τῆς διεξόδου τοῦ περιττώματος. τοῦτον μὲν οὖν τὸν τρόπον ἔχει τοῖς
ἐναίμοις ζῴοις, καὶ μετὰ τὴν κεφαλήν ἐστιν ὁ καλούμενος θώραξ, καὶ τὰ περὶ τοῦτον· τὰ δὲ λοιπὰ μόρια
30 τούτων τε χάριν καὶ ἕνεκα τῆς κινήσεως προσέθηκεν
ἡ φύσις, οἷον τά τε πρόσθια κῶλα καὶ τὰ ὄπισθεν.
βούλεται δὲ καὶ τοῖς μαλακοστράκοις καὶ τοῖς
ἐντόμοις ἥ γ' εὐθυορία τῶν ἐντοσθιδίων τὸν αὐτὸν
ἔχειν τρόπον, κατὰ δὲ τὰς ὑπηρεσίας τὰς ἔξωθεν
κινητικὰς διαφέρει τῶν ἐναίμων. τὰ δὲ μαλάκιά
τε καὶ ⟨τὰ⟩[3] στρομβώδη τῶν ὀστρακοδέρμων ἔχει

---

[1] sequitur locus corruptus. quae corrigi possunt sec. vers.
arabicam correxi, supposititia eieci, amissa e versione latina
Mich. Scot supplevi. text. vulg. habet ἡ φύσις ὥσπερ εἴ τις
νοήσειεν ἐπ' εὐθείας, καθάπερ συμβέβηκεν ἐπὶ τῶν τετραπόδων
ζῴων καὶ τῶν ἀνθρώπων, πρῶτον μὲν ἐπὶ ἄκρῳ τῷ ἄνω στόματι
τῆς εὐθείας κατὰ τὸ Α, ἔπειτα ⟨κατὰ adduut PY⟩ τὸ Β τὸν
στόμαχον, [τὸ δὲ om. PY] Γ τὴν κοιλίαν· ἀπὸ δὲ τοῦ ἐντέρου
μέχρι τῆς διεξόδου τοῦ περιττώματος, ᾗ τὸ Δ. τοῦτον μὲν οὖν
τὸν τρόπον ἔχει τοῖς ἐναίμοις ζῴοις, καὶ περὶ τοῦτό ἐστιν ἡ
κεφαλὴ καὶ ὁ θώραξ καλούμενος (καλ. θώραξ SU)· τὰ δὲ λοιπά,
etc. vide et quae p. 432 scripsi.
[2] ⟨ἀλλ' οὐ⟩ Peck.   [3] ⟨τὰ⟩ Peck.

them do so to some extent, but most markedly those conical Testacea which have a spiral shell, since both these classes have this natural structure [a]; *and therefore they walk with an even gait,* and not as is the case with quadrupeds and man.[b] *Now man has his mouth placed in his head, viz. in the upper part of the body,* and after that the gullet, then the stomach, and after that the intestine which reaches as far as the vent where the residue is discharged. This is the arrangement in the blooded animals, *i.e.,* after the head comes what is known as the trunk, and the parts adjoining. The remaining parts (*e.g.* the limbs at front and back) have been added by Nature for the sake of those which I have just mentioned and also to make movement possible. Now in the Crustacea too and in the Insects the internal parts tend to be in a straight alignment of this kind; though with regard to the external parts which subserve locomotion their arrangement differs from that of the blooded animals. The Cephalopods and the conical-shelled Testacea have the same

[a] The passage which follows has been badly corrupted by references to a diagram which have ousted the text. The words in italics have been translated from the Arabic version, of which Michael Scot's Latin translation is given opposite, in default of the original Greek. See supplementary note on p. 432.

[b] This refers to their uneven progression by moving first one side of the body and then the other. The Testacea, however, " have no right and left" (*De incessu an.* 714 b 9), and their movement was evidently an awkward problem for Aristotle. He reserves them until the very end of the *De incessu,* and he has to admit that they move, although they ought not to do so ! They move παρὰ φύσιν. The mechanism of their motion can be detected by the microscope, and is known as ciliary. See also *De incessu,* 706 a 13, 33, *Hist. An.* 528 b 9.

685 a αὐτοῖς μὲν παραπλησίως, τούτοις δ' ἀντεστραμ-
μένως· κέκαμπται γὰρ ἡ τελευτὴ πρὸς τὴν ἀρχήν,
ὥσπερ ἂν εἴ τις τὴν εὐθεῖαν [ἐφ' ἧς τὸ Ε]¹ κάμψας
προσαγάγοι τὸ Δ πρὸς τὸ Α. οὕτως γὰρ κειμένων
νῦν τῶν ἐντοσθίων περίκειται τοῖς μὲν μαλακίοις τὸ
5 κύτος, ὃ καλεῖται μόνον ἐπὶ τῶν πολυπόδων κεφαλή·
τοῖς δ' ὀστρακοδέρμοις τὸ τοιοῦτόν ἐστιν ὁ στρόμ-
βος. διαφέρει δ' οὐδὲν ἄλλο πλὴν ὅτι τοῖς μὲν
μαλακὸν τὸ πέριξ, τοῖς δὲ σκληρὸν περὶ τὸ σαρκῶδες
περιέθηκεν ἡ φύσις, ὅπως σώζηται διὰ τὴν δυσκινη-
σίαν· καὶ διὰ τοῦτο τὸ περίττωμα τοῖς τε μαλακίοις
10 ἐξέρχεται περὶ τὸ στόμα καὶ τοῖς στρομβώδεσι,
πλὴν τοῖς μὲν μαλακίοις κάτωθεν, τοῖς δὲ στρομβώδεσιν ἐκ τοῦ πλαγίου.

Διὰ ταύτην μὲν οὖν τὴν αἰτίαν τοῖς μαλακίοις οἱ
πόδες τοῦτον ἔχουσι τὸν τρόπον, καὶ ὑπεναντίως
ἢ τοῖς ἄλλοις. ἔχουσι δ' ἀνομοίως αἱ σηπίαι καὶ
15 αἱ τευθίδες τοῖς πολύποσι διὰ τὸ νευστικαὶ μόνον
εἶναι, τοὺς δὲ καὶ πορευτικούς. αἱ μὲν γὰρ τοὺς
ἄνωθεν τῶν ὀδόντων ⟨ἓξ μικροὺς⟩² ἔχουσι, καὶ
τούτων τοὺς ἐσχάτους δύο μείζους, τοὺς δὲ λοιποὺς
τῶν ὀκτὼ δύο κάτωθεν μεγίστους πάντων.³ ὥσπερ
γὰρ τοῖς τετράποσι τὰ ὀπίσθια ἰσχυρότερα κῶλα,
καὶ ταύταις μέγιστοι οἱ κάτωθεν ⟨πόδες⟩⁴· τὸ γὰρ
20 φορτίον οὗτοι ἔχουσι καὶ κινοῦσι μάλιστα. καὶ οἱ
ἔσχατοι δύο μείζους τῶν μέσων, ὅτι τούτοις συν-

¹ seclusi ; post ἧς add. Ζ τὸ ὅλον φησί. vid. p. 432.
² Schneider ex Gazae vers. (senos exiguos) ; sex Σ ; μικροὺς
Ζ (sed ποδῶν pro ὀδόντων), idem F teste Buss.
³ πάντων Ogle : τούτων vulg.   ⁴ ⟨πόδες⟩ Rackham.

arrangement as one another, but it differs completely
from that of the others, as the tail-end of these
creatures is bent right over to meet the front,
just as if I were to bend the straight line over
until the point D met the point A. Such

$$\overline{\text{A} \quad \text{B} \quad \text{C} \quad \text{D}}$$

then, is the disposition of their internal parts.
Round them, in Cephalopods, is situated the sac (in
the Octopuses and in them only it is called the head):
in the Testacea the corresponding thing is the conical
shell. The only difference is that in the one case
the surrounding substance is soft, and in the other
Nature has surrounded the flesh with something
hard, to give them the preservation they need owing
to their bad locomotion. As a result of the above-
mentioned arrangement, in both sets the residue
leaves at a point near the mouth : in the Cephalopods
under the mouth, in the conical Testacea at the side
of it.

So what we have said explains why the feet of
Cephalopods are where they are, quite differently
placed from all other animals' feet. Sepias and
Calamaries, however, being swimmers merely, differ
from the Octopuses, which are walkers as well ; they
have six small feet above the teeth, and of these the
ones at each end are larger ; the remaining two out
of the total eight are down below and largest of
all. These creatures have their strongest feet down
below, just as quadrupeds have their strongest limbs
at the back ; and the reason is that they carry the
weight of the body and they chiefly are responsible
for locomotion. The two outer feet are larger than
the inner ones because they have to help the others

ὑπηρετοῦσιν. ὁ δὲ πολύπους τοὺς ἐν μέσῳ τέτταρας
μεγίστους.

Πόδας μὲν οὖν πάντα ἔχουσι ταῦτα ὀκτώ, ἀλλ'
αἱ μὲν σηπίαι καὶ αἱ τευθίδες βραχεῖς, τὰ δὲ
πολυποδώδη μεγάλους. τὸ γὰρ κύτος τοῦ σώματος
25 αἱ μὲν μέγα ἔχουσιν τὰ[1] δὲ μικρόν, ὥστε τοῖς μὲν
ἀφεῖλεν ἀπὸ τοῦ σώματος, πρὸς δὲ τὸ μῆκος τῶν
ποδῶν προσέθηκεν ἡ φύσις, ταῖς δ' ἀπὸ τῶν
ποδῶν λαβοῦσα τὸ σῶμα ηὔξησεν. διόπερ τοῖς
μὲν οὐ μόνον πρὸς τὸ νεῖν χρήσιμοι οἱ πόδες ἀλλὰ
καὶ πρὸς τὸ βαδίζειν, ταῖς δ' ἄχρηστοι· μικροὶ γάρ,
30 τὸ δὲ κύτος μέγα ἔχουσιν. ἐπεὶ δὲ βραχεῖς ἔχουσι
τοὺς πόδας καὶ ἀχρήστους πρὸς τὸ ἀντιλαμβάνεσθαι
καὶ μὴ ἀποσπᾶσθαι[2] ἀπὸ τῶν πετρῶν, ὅταν κλύδων
ᾖ καὶ χειμών, καὶ πρὸς τὸ τὰ ἄποθεν προσάγεσθαι,
διὰ ταῦτα προβοσκίδας ἔχουσι δύο μακράς, αἷς
35 ὁρμοῦσί τε καὶ ἀποσαλεύουσιν ὥσπερ πλοῖον ὅταν
685 b χειμὼν ᾖ, καὶ τὰ ἄποθεν θηρεύουσι καὶ προσάγονται
ταύταις αἵ τε σηπίαι καὶ αἱ τευθίδες. οἱ δὲ πολύποδες οὐκ ἔχουσι τὰς προβοσκίδας διὰ τὸ τοὺς
πόδας αὐτοῖς εἶναι πρὸς ταῦτα χρησίμους. ἐνίοις[3]
δὲ κοτυληδόνες πρὸς τοῖς ποσὶ καὶ πλεκτάναι
5 πρόσεισι, δύναμιν ἔχουσαι[4] καὶ σύνθεσιν τοιαύτην
οἵανπερ τὰ πλέγματα οἷς οἱ ἰατροὶ οἱ ἀρχαῖοι τοὺς
δακτύλους ἐνέβαλλον· οὕτω καὶ ἐκ τῶν ἰνῶν

---

[1] τὰ Peck: οἱ vulg.
[2] ἀποσπᾶσθαι Bekker: ἀντισπᾶσθαι codd.
[3] ἐνίοις Peck: ὅσοις vulg.
[4] ἔχουσαι P: ἔχουσι vulg.

---

[a] The use of these σαῦραι or σειραί is described by Hippocrates, Περὶ ἄρθρων (Littré iv. 318-320 ; L.C.L. iii. 390 : "The
tubes woven out of palm-tissue are satisfactory means of

in performing their duty. In the Octopuses, however, the four middle feet are the biggest.

And although all these creatures have eight feet, the Sepia's and the Calamary's are short ones, since their bodies are large in the trunk, and the Octopus's feet are long, because his body is small. Thus in one case the substance which she took from the body Nature has given towards lengthening the feet, and in the other she has taken away from the feet and made the body itself bigger. Hence it results that the Octopuses have feet which will serve them for walking as well as for swimming, whereas the other creatures' feet will not do so, being small, while the body itself is big. And inasmuch as these creatures' feet are short, and useless for holding on tightly to the rock in a storm when there is a strong sea running, or for bringing to the mouth objects that are at a distance, by way of compensation they have two long proboscves, with which during a storm they moor themselves up and ride at anchor like a ship; therewith also they hunt distant prey and bring it to their mouths. These things the Sepias and Calamaries do. The Octopuses have no proboscves because their feet serve these purposes. Some creatures have suckers and twining tentacles as well as feet: these have the same character and function as well as the same structure as those plaited tubes which the early physicians used for reducing dislocated fingers.[a] They are similarly made out of plaited fibres, and their

reduction, if you make extension of the finger both ways, grasping the tube at one end and the wrist at the other." The σαύρα was thus a tube open at both ends. A similar passage in Diocles *ap.* Apollonius of Kitium, no doubt taken from Hippocrates, refers to " the σειραί which children plait " (L.C.L. iii. 453).

πεπλεγμέναι εἰσίν, καὶ¹ ἕλκουσι τὰ σαρκία καὶ
τὰ ἐνδιδόντα. περιλαμβάνει μὲν γὰρ χαλαρὰ ὄντα·
ὅταν δὲ συντείνῃ, πιέζει καὶ ἔχεται τοῦ ἐντὸς
θιγγάνοντος παντός.

10 Ὥστ' ἐπεὶ ἄλλο οὐκ ἔστιν ᾧ προσάξονται, ἀλλ'
ἢ τὰ μὲν τοῖς ποσὶ τὰ δὲ ταῖς προβοσκίσι, ταύτας
ἔχουσι πρὸς ἀλκὴν καὶ τὴν ἄλλην βοήθειαν² ἀντὶ
χειρῶν.

Τὰ μὲν οὖν ἄλλα δικότυλά ἐστι, γένος δέ τι πολυπόδων μονοκότυλον. αἴτιον δὲ τὸ μῆκος καὶ ἡ λεπτότης τῆς φύσεως αὐτῶν· μονοκότυλον γὰρ ἀναγ-
15 καῖον εἶναι τὸ στενόν. οὐκ οὖν ὡς βέλτιστον ἔχουσιν,
ἀλλ' ὡς ἀναγκαῖον διὰ τὸν ἴδιον λόγον τῆς οὐσίας.

Πτερύγιον δ' ἔχουσι ταῦτα πάντα κύκλῳ περὶ τὸ
κύτος. τοῦτο δ' ἐπὶ μὲν τῶν ἄλλων συναπτόμενον
καὶ συνεχές ἐστι, καὶ ἐπὶ τῶν μεγάλων τευθῶν·
αἱ δ' ἐλάττους καὶ καλούμεναι τευθίδες πλατύτερόν
20 τε τοῦτο ἔχουσι καὶ οὐ στενόν, ὥσπερ αἱ σηπίαι
καὶ οἱ πολύποδες, καὶ τοῦτ' ἀπὸ μέσου ἠργμένον,
καὶ οὐ κύκλῳ διὰ παντός. τοῦτο δ' ἔχουσιν
ὅπως νέωσι καὶ πρὸς τὸ διορθοῦν, ὥσπερ τοῖς μὲν
πτηνοῖς τὸ ὀρροπύγιον, τοῖς δ' ἰχθύσι τὸ οὐραῖον.
ἐλάχιστον δὲ τοῦτο καὶ ἥκιστα ἐπίδηλον τοῖς
25 πολύποσίν ἐστι διὰ τὸ μικρὸν ἔχειν τὸ κύτος καὶ
διορθοῦσθαι τοῖς ποσὶν ἱκανῶς.

Περὶ μὲν οὖν τῶν ἐντόμων καὶ μαλακοστράκων
καὶ ὀστρακοδέρμων καὶ μαλακίων εἴρηται, καὶ
περὶ τῶν ἐντὸς μορίων καὶ τῶν ἐκτός.

30 Χ. Πάλιν δ' ἐξ ὑπαρχῆς περὶ τῶν ἐναίμων καὶ

---

¹ καὶ Ogle: αἷς vulg.
² ἄλλην χρείαν καὶ βοήθειαν Υ, Ogle.

action is to draw flesh and yielding substances, as follows. First they encircle the object while they are still relaxed; then they contract, and by so doing compress and hold fast the whole of whatever is in contact with their inner surface.

So, as these creatures have nothing else with which to convey objects to the mouth except the feet (in some species) and the probosces (in others), they possess these organs in lieu of hands to serve them as weapons and generally to assist them otherwise.

All these creatures have two rows of suckers, except a certain kind of Octopus, and these have only one, because owing to their length and slimness they are so narrow that they cannot possibly have another. Thus they have the one row only, not because this arrangement is the *best*, but because it is *necessitated* by the particular and specific character of their being.

All these animals have a fin which forms a circle round the sac. In most of them it is a closed and continuous circle, as it is in the large Calamaries (*teuthi*), while in the smaller ones called *teuthides* it is quite wide (not narrow as in the Sepias and Octopuses), and furthermore it begins at the middle and does not go round the whole way. They have this fin to enable them to swim and to steer their course, and it answers to a bird's tail-feathers and a fish's tail-fin. In the Octopuses this fin is extremely small and insignificant because their body is small and can be steered well enough by means of the feet.

This brings to an end our description of the internal and external parts of the Insects, the Crustacea, the Testacea, and the Cephalopods.

X. Now we must go back and begin again with

685 b

ζωοτόκων ἐπισκεπτέον, ἀρξαμένοις ἀπὸ τῶν ὑπο-
λοίπων καὶ πρότερον εἰρημένων μορίων· τούτων δὲ
διορισθέντων περὶ τῶν ἐναίμων καὶ ᾠοτόκων τὸν
αὐτὸν τρόπον ἐροῦμεν.

35 Τὰ μὲν οὖν μόρια τὰ περὶ τὴν κεφαλὴν τῶν ζῴων
εἴρηται πρότερον, καὶ τὰ περὶ τὸν καλούμενον
αὐχένα καὶ τράχηλον. ἔχει δὲ κεφαλὴν πάντα τὰ
686 a ἔναιμα ζῷα· τῶν δ' ἀναίμων ἐνίοις ἀδιόριστον
τοῦτο τὸ μόριον, οἷον τοῖς καρκίνοις. αὐχένα οὖν
τὰ μὲν ζωοτόκα πάντ' ἔχει, τῶν δ' ᾠοτόκων τὰ
μὲν ἔχει τὰ δ' οὐκ ἔχει· ὅσα μὲν γὰρ πνεύμονα
5 ἔχει, καὶ αὐχένα ἔχει, τὰ δὲ μὴ ἀναπνέοντα θύραθεν
οὐκ ἔχει τοῦτο τὸ μόριον.
Ἔστι δ' ἡ μὲν κεφαλὴ μάλιστα τοῦ ἐγκεφάλου
χάριν· ἀνάγκη γὰρ τοῦτο τὸ μόριον ἔχειν τοῖς ἐναίμοις, καὶ ἐν ἀντικειμένῳ τόπῳ τῆς καρδίας,
διὰ τὰς εἰρημένας πρότερον αἰτίας. ἐξέθετο δ' ἡ
10 φύσις ἐν αὐτῇ καὶ τῶν αἰσθήσεων ἐνίας διὰ τὸ
σύμμετρον εἶναι τὴν τοῦ αἵματος κρᾶσιν καὶ ἐπι-
τηδείαν πρός τε τὴν τοῦ ἐγκεφάλου ἀλέαν καὶ
πρὸς τὴν τῶν αἰσθήσεων ἡσυχίαν καὶ ἀκρίβειαν.
ἔτι δὲ τρίτον μόριον ὑπέθηκε τὸ τὴν τῆς τροφῆς
εἴσοδον δημιουργοῦν· ἐνταῦθα γὰρ ὑπέκειτο συμ-
μέτρως μάλιστα· οὔτε γὰρ ἄνωθεν κεῖσθαι τῆς
15 καρδίας καὶ τῆς ἀρχῆς ἐνεδέχετο τὴν κοιλίαν, οὔτε
κάτωθεν οὔσης ὃν τρόπον ἔχει νῦν ἐνεδέχετο τὴν
εἴσοδον ἔτι κάτω εἶναι τῆς καρδίας· πολὺ γὰρ ἂν[1]
τὸ μῆκος ἦν τοῦ σώματος, καὶ πόρρω λίαν τῆς
κινούσης ἀρχῆς καὶ πεττούσης. ἡ μὲν οὖν κεφαλὴ
τούτων χάριν ἐστίν, ὁ δ' αὐχὴν τῆς ἀρτηρίας χάριν·

[1] ἂν P, om. vulg.

the blooded viviparous animals. Some of the parts which we have already enumerated still remain to be described, and we will take these first. This done, we will describe similarly the blooded Ovipara.

We have already [a] spoken of the parts around the head, and what is called the neck, and the throat. All blooded animals have a head, but in some of the bloodless ones the head is indistinct (*e.g.* in crabs). All Vivipara have a neck, but not all Ovipara : to be precise, only those which breathe in air from without and have a lung.

EXTERNAL PARTS OF BLOODED ANIMALS. (a) Vivipara:

The presence of the head is mainly for the sake of the brain. Blooded creatures must have a brain, which (for reasons aforeshown) [b] must be set in some place opposite to the heart. But in addition, Nature has put some of the senses up in the head, apart from the rest, because the blend of its blood is well proportioned and suitable for securing not only warmth for the brain but also quiet and accuracy for the senses. There is yet a third part which Nature has disposed of in the head, viz. the part which manages the intake of food ; it was put here because this gave the best-ordered arrangement. It would have been impossible to put the stomach above the source and sovereign part, the heart ; and it would have been impossible to make the entrance for the food below the heart, even with the stomach below the heart as it actually is, because then the length of the body would be very great, and the stomach would be too far away from the source which provides motion and concoction. These then are the three parts for whose sake the head exists. The neck exists for the sake of the

Head and neck.

---

[a] At 655 b 27—665 a 25.   [b] At 652 b 17 ff.

686 a

20 πρόβλημα γάρ ἐστι, καὶ σώζει ταύτην καὶ τὸν οἰσοφάγον κύκλῳ περιέχων. τοῖς μὲν οὖν ἄλλοις ἐστὶ
καμπτὸς καὶ σφονδύλους ἔχων, οἱ δὲ λύκοι καὶ
λέοντες μονόστουν τὸν αὐχένα ἔχουσιν. ἔβλεψε
γὰρ ἡ φύσις ὅπως πρὸς τὴν ἰσχὺν χρήσιμον αὐτὸν
ἔχωσι μᾶλλον ἢ πρὸς τὰς ἄλλας βοηθείας.

Ἐχόμενα δὲ τοῦ αὐχένος καὶ τῆς κεφαλῆς τά τε
25 πρόσθια κῶλα τοῖς ζῴοις ἐστὶ καὶ θώραξ. ὁ μὲν
οὖν ἄνθρωπος ἀντὶ σκελῶν καὶ ποδῶν τῶν προσθίων
βραχίονας καὶ τὰς καλουμένας ἔχει χεῖρας. ὀρθὸν
μὲν γάρ ἐστι μόνον τῶν ζῴων διὰ τὸ τὴν φύσιν
αὐτοῦ καὶ τὴν οὐσίαν εἶναι θείαν· ἔργον δὲ τοῦ
θειοτάτου τὸ νοεῖν καὶ φρονεῖν· τοῦτο δ' οὐ ῥᾴδιον
30 πολλοῦ τοῦ ἄνωθεν ἐπικειμένου σώματος· τὸ γὰρ
βάρος δυσκίνητον ποιεῖ τὴν διάνοιαν καὶ τὴν
κοινὴν αἴσθησιν. διὸ πλείονος γινομένου τοῦ
βάρους καὶ τοῦ σωματώδους ἀνάγκη ῥέπειν τὰ
σώματα πρὸς τὴν γῆν, ὥστε πρὸς τὴν ἀσφάλειαν
ἀντὶ βραχιόνων καὶ χειρῶν τοὺς προσθίους πόδας
35 ὑπέθηκεν ἡ φύσις τοῖς τετράποσιν. τοὺς μὲν
686 b γὰρ ὀπισθίους δύο πᾶσιν ἀναγκαῖον τοῖς πορευτικοῖς ἔχειν, τὰ δὲ τοιαῦτα τετράποδα ἐγένετο οὐ
δυναμένης φέρειν τὸ βάρος τῆς ψυχῆς. πάντα γάρ
ἐστι τὰ ζῷα νανώδη τἆλλα παρὰ τὸν ἄνθρωπον·
νανῶδες γάρ ἐστιν οὗ τὸ μὲν ἄνω μέγα, τὸ δὲ
5 φέρον τὸ βάρος καὶ πεζεῦον μικρόν· ἄνω δ' ἐστὶν
ὁ καλούμενος θώραξ, ἀπὸ τῆς κεφαλῆς μέχρι τῆς

---

a For the " general " or " common " sense see *De mem.*
450 a 10, etc. ; and *cf. De part. an.* 656 a 28, 665 a 12. The
" general " sense is not another sense over and above the
ordinary five, but rather the common nature inherent in
366

windpipe : it acts as a shield and keeps the windpipe and the oesophagus safe by completely encircling them. The neck is flexible and has a number of vertebrae in all animals except the wolf and the lion whose neck consists of one bone only, for Nature's object was to provide these with a neck that should be useful for its strength rather than for other purposes.

The anterior limbs and the trunk are continuous with the head and neck. Man, instead of forelegs and forefeet, has arms and hands. Man is the only animal that stands upright, and this is because his nature and essence is divine. Now the business of that which is most divine is to think and to be intelligent ; and this would not be easy if there were a great deal of the body at the top weighing it down, for weight hampers the motion of the intellect and of the general sense.[a] Thus, when the bodily part and the weight of it become excessive, the body itself must lurch forward towards the ground ; and then, for safety's sake, Nature provided forefeet instead of arms and hands—as has happened in quadrupeds. All animals which walk must have two hind feet, and those I have just mentioned became quadrupeds because their soul could not sustain the weight bearing it down. Compared with man, all the other animals are dwarf-like. By " dwarf-like " I mean to denote that which is big at the top (*i.e.* big in the " trunk," or the portion from the head to the residual vent), and small where the weight is supported and where

Marginal note: Limbs, and their relative sizes.

them all ; thus Aristotle (*De somno*) argues that their simultaneous inactivity during sleep is not a mere coincidence but is due to the inactivity of the central perceptive faculty of which they are differentiations. Among the functions of the " general " sense are : discrimination between the objects of two senses, and the perceiving that we perceive.

367

ἐξόδου τοῦ περιττώματος. τοῖς μὲν οὖν ἀνθρώποις
τοῦτο πρὸς τὸ κάτω σύμμετρον, καὶ πολλῷ
ἔλαττόν ἐστι τελειουμένοις· νέοις δ' οὖσι τοὐναντίον τὰ μὲν ἄνω μεγάλα, τὸ δὲ κάτω μικρόν
10 (διὸ καὶ ἕρπουσι, βαδίζειν δ' οὐ δύνανται, τὸ δὲ
πρῶτον οὐδ' ἕρπουσιν, ἀλλ' ἀκινητίζουσιν)· νάνοι
γάρ εἰσι τὰ παιδία πάντα. προϊοῦσι δὲ τοῖς μὲν
ἀνθρώποις αὔξεται τὰ κάτωθεν· τοῖς δὲ τετράποσι
τοὐναντίον τὰ κάτω μέγιστα τὸ πρῶτον, προϊόντα
δ' αὔξεται ἐπὶ τὸ ἄνω, τοῦτο δ' ἐστὶ τὸ ἀπὸ τῆς
ἕδρας ἐπὶ τὴν κεφαλὴν κύτος. διὸ καὶ τῷ ὕψει οἱ
15 πῶλοι τῶν ἵππων οὐδὲν ἢ μικρὸν ἐλάττους εἰσί,
καὶ νέοι μὲν ὄντες θιγγάνουσι τῷ ὄπισθεν σκέλει
τῆς κεφαλῆς, πρεσβύτεροι δ' ὄντες οὐ δύνανται.
τὰ μὲν οὖν μώνυχα καὶ δίχηλα τοῦτον ἔχει τὸν
τρόπον, τὰ δὲ πολυδάκτυλα καὶ ἀκέρατα νανώδη
μέν ἐστιν, ἧττον δὲ τούτων· διὸ καὶ τὴν αὔξησιν
20 πρὸς τὰ ἄνω τὰ κάτω κατὰ λόγον ποιεῖται τῆς
ἐλλείψεως.

Ἔστι δὲ καὶ τὸ τῶν ὀρνίθων καὶ τὸ τῶν ἰχθύων
γένος καὶ πᾶν τὸ ἔναιμον, ὥσπερ εἴρηται, νανῶδες.
διὸ καὶ ἀφρονέστερα πάντα τὰ ζῷα τῶν ἀνθρώπων
ἐστίν. καὶ γὰρ τῶν ἀνθρώπων, οἷον τά τε παιδία
πρὸς τοὺς ἄνδρας καὶ αὐτῶν τῶν ἐν ἡλικίᾳ οἱ
25 νανώδεις τὴν φύσιν, ἐὰν καί τιν' ἄλλην δύναμιν
ἔχωσι περιττήν, ἀλλὰ τῷ τὸν νοῦν ἔχειν ἐλ-
λείπουσιν. αἴτιον δ', ὥσπερ εἴρηται πρότερον, ὅτι
ἡ τῆς ψυχῆς ἀρχὴ πολλοῖς δὴ[1] δυσκίνητός ἐστι
καὶ σωματώδης. ἔτι δ' ἐλάττονος γινομένης τῆς

---

[1] πολλοῖς δὴ Peck: πολλῷ δὴ vulg.: add. καὶ Y, Platt, qui
et insuper addit ⟨βαρεῖ σώματι καταφερομένη⟩.

locomotion is effected. In man, the size of the trunk
is proportionate to the lower portions, and as a man
grows up it becomes much smaller in proportion.
In infancy the reverse is found : the upper portion is
large and the lower is small (and that is why infants
cannot walk but crawl about, and at the very beginning cannot even crawl, but remain where they
are). In other words, all children are dwarfs. Now,
in man, as time proceeds, the lower portion grows :
Not so with the quadruped animals : their lower
portion is biggest at the beginning, and as time
proceeds the top portion grows (*i.e.* the trunk, the
portion between the head and the seat). Thus foals
are quite or almost as high as horses, and at that age a
foal can touch its head with its hind leg, but not when
it is older.[a] What has been said holds good of the
animals that have solid hoofs or cloven. The polydactylous, hornless animals are indeed dwarf-like
too, but not so markedly, and so the growth of their
lower portions compared with the upper is proportionate to the smaller deficiency.

The whole groups of birds and fishes are dwarf-like ;
indeed, so is every animal with blood in it, as I have
said. This is why all animals are less intelligent than
man. Even among human beings, children, when
compared with adults, and dwarf adults when compared with others, may have some characteristics in
which they are superior, but in intelligence, at any
rate, they are inferior. And the reason, as aforesaid, is that in very many of them the principle of the
soul is sluggish and corporeal. And if the heat which

---

[a] These observations are entirely correct. *Cf.* Ogle's
quotation *ad loc.* from T. H. Huxley. See also *Hist. an.*
500 b 26 ff.

686 b

αἱρούσης θερμότητος καὶ τοῦ γεώδους πλείονος, τά
30 τε σώματα ἐλάττονα τῶν ζῴων ἐστὶ καὶ πολύποδα,
τέλος δ' ἄποδα γίνεται καὶ τεταμένα πρὸς τὴν γῆν.
μικρὸν δ' οὕτω προβαίνοντα καὶ τὴν ἀρχὴν ἔχουσι
κάτω, καὶ τὸ κατὰ τὴν κεφαλὴν μόριον τέλος
ἀκίνητόν ἐστι καὶ ἀναίσθητον, καὶ γίνεται φυτόν,
35 ἔχον τὰ μὲν ἄνω κάτω, τὰ δὲ κάτω ἄνω· αἱ γὰρ
ῥίζαι τοῖς φυτοῖς στόματος καὶ κεφαλῆς ἔχουσι
687 a δύναμιν, τὸ δὲ σπέρμα τοὐναντίον· ἄνω γὰρ καὶ
ἐπ' ἄκροις γίνεται τοῖς πτόρθοις.

Δι' ἣν μὲν οὖν αἰτίαν τὰ μὲν δίποδα τὰ δὲ πολύποδα τὰ δ' ἄποδα τῶν ζῴων ἐστί, καὶ διὰ τίν'
αἰτίαν τὰ μὲν φυτὰ τὰ δὲ ζῷα γέγονεν, εἴρηται,
5 καὶ διότι μόνον ὀρθόν ἐστι τῶν ζῴων ὁ ἄνθρωπος·
ὀρθῷ δ' ὄντι τὴν φύσιν οὐδεμία χρεία σκελῶν τῶν
ἐμπροσθίων, ἀλλ' ἀντὶ τούτων βραχίονας καὶ χεῖρας
ἀποδέδωκεν ἡ φύσις. Ἀναξαγόρας μὲν οὖν φησι
διὰ τὸ χεῖρας ἔχειν φρονιμώτατον εἶναι τῶν ζῴων
ἄνθρωπον· εὔλογον δὲ διὰ τὸ φρονιμώτατον εἶναι
10 χεῖρας λαμβάνειν. αἱ μὲν γὰρ χεῖρες ὄργανόν
εἰσιν, ἡ δὲ φύσις ἀεὶ διανέμει, καθάπερ ἄνθρωπος
φρόνιμος, ἕκαστον τῷ δυναμένῳ χρῆσθαι (προσήκει γὰρ τῷ ὄντι αὐλητῇ δοῦναι μᾶλλον αὐλοὺς
ἢ τῷ αὐλοὺς ἔχοντι προσθεῖναι αὐλητικήν)· τῷ γὰρ
μείζονι καὶ κυριωτέρῳ προσέθηκε τοὔλαττον, ἀλλ'
15 οὐ τῷ ἐλάττονι τὸ τιμιώτερον καὶ μεῖζον. εἰ οὖν
οὕτως βέλτιον, ἡ δὲ φύσις ἐκ τῶν ἐνδεχομένων

---

[a] With the terminology used in ll. 28-29 cf. Hippocrates,
Περὶ διαίτης, i. 35.
[b] That is, it answers to residue in animals ; cf. 655 b 35.

raises the organism up wanes still further while the
earthy matter waxes,[a] then the animals' bodies wane,
and they will be many-footed ; and finally they lose
their feet altogether and lie full length on the ground.
Proceeding a little further in this way, they actually
have their principal part down below, and finally the
part which answers to a head comes to have neither
motion nor sensation ; at this stage the creature
becomes a plant, and has its upper parts below and its
nether parts aloft ; for in plants the roots have the
character and value of mouth and head, whereas the
seed counts as the opposite,[b] being produced in the
upper part of the plant on the ends of the twigs.

We have now stated why it is that some animals
have two feet, some many, some none at all ; why
some creatures are plants and some animals ; and
why man is the only one of the animals that stands
upright. And since man stands upright, he has no
need of legs in front ; instead of them Nature has
given him arms and hands. Anaxagoras indeed
asserts that it is his possession of hands that makes
man the most intelligent of the animals ; but surely
the reasonable point of view is that it is because he
is the most intelligent animal that he has got hands.
Hands are an instrument ; and Nature, like a sensible human being, always assigns an organ to the
animal that can use it (as it is more in keeping to
give flutes to a man who is already a flute-player
than to provide a man who possesses flutes with the
skill to play them) ; thus Nature has provided that
which is less as an addition to that which is greater
and superior ; not *vice versa*. We may conclude, then,
that, if this is the *better* way, and if Nature always does
the *best* she can in the circumstances, it is not true

687 a

ποιεῖ τὸ βέλτιστον, οὐ διὰ τὰς χεῖράς ἐστιν ὁ
ἄνθρωπος φρονιμώτατος, ἀλλὰ διὰ τὸ φρονιμώτατον εἶναι τῶν ζῴων ἔχει χεῖρας. ὁ γὰρ φρονιμώτατος πλείστοις ἂν ὀργάνοις ἐχρήσατο καλῶς,
20 ἡ δὲ χεὶρ ἔοικεν εἶναι οὐχ ἓν ὄργανον ἀλλὰ πολλά·
ἔστι γὰρ ὡσπερεὶ ὄργανον πρὸ ὀργάνων. τῷ οὖν
πλείστας δυναμένῳ δέξασθαι τέχνας τὸ ἐπὶ
πλεῖστον τῶν ὀργάνων χρήσιμον τὴν χεῖρα ἀποδέδωκεν ἡ φύσις.

'Αλλ' οἱ λέγοντες ὡς συνέστηκεν οὐ καλῶς ὁ
ἄνθρωπος ἀλλὰ χείριστα τῶν ζῴων (ἀνυπόδητόν
25 τε γὰρ αὐτὸν εἶναί φασι καὶ γυμνὸν καὶ οὐκ
ἔχοντα ὅπλον πρὸς τὴν ἀλκήν) οὐκ ὀρθῶς λέγουσιν.
τὰ μὲν γὰρ ἄλλα μίαν ἔχει βοήθειαν, καὶ μεταβάλλεσθαι ἀντὶ ταύτης ἑτέραν οὐκ ἔστιν, ἀλλ'
ἀναγκαῖον ὥσπερ ὑποδεδεμένον ἀεὶ καθεύδειν καὶ
πάντα πράττειν, καὶ τὴν περὶ τὸ σῶμα ἀλεώραν
μηδέποτε καταθέσθαι, μηδὲ μεταβάλλεσθαι ὃ δὴ
30 ἐτύγχανεν[1] ὅπλον ἔχον[2]· τῷ δὲ ἀνθρώπῳ τάς τε
687 b βοηθείας πολλὰς ἔχειν καὶ ταύτας ἀεὶ ἔξεστι
μεταβάλλειν, ἔτι δ' ὅπλον οἷον ἂν βούληται καὶ
ὅπου ἂν[3] βούληται ἔχειν. ἡ γὰρ χεὶρ καὶ ὄνυξ καὶ
χηλὴ καὶ κέρας γίνεται καὶ δόρυ καὶ ξίφος
καὶ ἄλλο ὁποιονοῦν ὅπλον καὶ ὄργανον· πάντα γὰρ
5 ἔσται ταῦτα διὰ τὸ πάντα δύνασθαι λαμβάνειν καὶ
ἔχειν αὐτήν· εὖ[4] δὲ συμμεμηχάνηται[5] καὶ τὸ εἶδος[6]
τῇ φύσει τῆς χειρός, διαιρετὴ γὰρ καὶ πολυσχιδής·

---

[1] ἐτύγχανεν ἐν U[1] : τυγχάνει ἐν Th. ; hic alia omnino Σ
[2] ἔχον Z, et corr. U : ἔχων vulg.
[3] ὅπου ἂν] ὁπόταν Ogle.
[4] ἔχειν αὐτήν· εὖ P : ἔχειν ταύτῃ vulg.
[5] συμμεμηχάνηται Ogle : συμμεμηχανῆσθαι vulg.
[6] εἶδος καὶ vulg. : εἶδος PSUYZ.

to say that man is the most intelligent animal because he possesses hands, but he has hands because he is the most intelligent animal. We should expect the most intelligent to be able to employ the greatest number of organs or instruments to good purpose ; now the hand would appear to be not one single instrument but many, as it were an instrument that represents many instruments. Thus it is to that animal (viz. man) which has the capability for acquiring the greatest number of crafts that Nature has given that instrument (viz. the hand) whose range of uses is the most extensive.

Now it must be wrong to say, as some do, that the structure of man is not good, in fact, that it is worse than that of any other animal. Their grounds are : that man is barefoot, unclothed, and void of any weapon of force. Against this we may say that all the other animals have just one method of defence and cannot change it for another : they are forced to sleep and perform all their actions with their shoes on the whole time, as one might say ; they can never take off this defensive equipment of theirs, nor can they change their weapon, whatever it may be. For man, on the other hand, many means of defence are available, and he can change them at any time, and above all he can choose what weapon he will have and where. Take the hand : this is as good as a talon, or a claw, or a horn, or again, a spear or a sword, or any other weapon or tool : it can be all of these, because it can seize and hold them all. And Nature has admirably contrived the actual shape of the hand so as to fit in with this arrangement. It is not all of one piece, but it branches into several pieces ; which gives the possi-

687 b

ἔνι γὰρ ἐν τῷ διαιρετὴν εἶναι καὶ συνθετὴν εἶναι,
ἐν τούτῳ δ' ἐκεῖνο οὐκ ἔστιν. καὶ χρῆσθαι ἑνὶ[1]
10 καὶ δυοῖν καὶ πολλαχῶς ἔστιν. καὶ αἱ καμπαὶ τῶν
δακτύλων καλῶς ἔχουσι πρὸς τὰς λήψεις καὶ
πιέσεις. καὶ ἐκ πλαγίου εἷς, καὶ οὗτος βραχὺς
καὶ παχὺς ἀλλ' οὐ μακρός· ὥσπερ γὰρ εἰ μὴ ἦν
χεὶρ ὅλως, οὐκ ἂν ἦν λῆψις, οὕτω κἂν εἰ μὴ ἐκ
πλαγίου οὗτος ἦν. οὗτος γὰρ κάτωθεν ἄνω πιέζει,
15 ὅπερ οἱ ἕτεροι ἄνωθεν κάτω· δεῖ δὲ τοῦτο συμβαίνειν, εἰ μέλλει ἰσχυρὸς ὥσπερ σύναμμα ἰσχυρὸν
συνδεῖν, ἵνα ἰσάζῃ εἷς ὢν πολλοῖς. καὶ βραχὺς
διά τε τὴν ἰσχὺν καὶ διότι οὐδὲν ὄφελος εἰ μακρός.
(καὶ ὁ ἔσχατος δὲ μικρὸς ὀρθῶς, καὶ ὁ μέσος
μακρός, ὥσπερ κώπη μεσόνεως[2]· μάλιστα γὰρ τὸ
20 λαμβανόμενον ἀνάγκη περιλαμβάνεσθαι κύκλῳ
κατὰ τὸ μέσον πρὸς τὰς ἐργασίας.) καὶ διὰ τοῦτο
καλεῖται μέγας μικρὸς ὤν, ὅτι ἄχρηστοι ὡς
εἰπεῖν οἱ ἄλλοι ἄνευ τούτου. εὖ δὲ καὶ τὸ τῶν
ὀνύχων μεμηχάνηται· τὰ μὲν γὰρ ἄλλα ζῷα ἔχει
καὶ πρὸς χρῆσιν αὐτούς, τοῖς δ' ἀνθρώποις ἐπι-
25 καλυπτήρια· σκέπασμα γὰρ τῶν ἀκρωτηρίων εἰσίν.

Αἱ δὲ καμπαὶ τῶν βραχιόνων ἔχουσι πρός τε
τὴν τῆς τροφῆς προσαγωγὴν καὶ πρὸς τὰς ἄλλας
χρήσεις ἐναντίως τοῖς τετράποσιν. ἐκείνοις μὲν
γὰρ ἀναγκαῖον εἴσω κάμπτειν τὰ ἐμπρόσθια κῶλα
(χρῶνται γὰρ ὡς[3] ποσίν) ἵν' ᾖ χρήσιμα πρὸς τὴν

---

[1] ἑνὶ] μιᾷ Ogle.
[2] μεσόνεως Schneider: μέσον νέως vulg.
[3] ὡς P, om. vulg.

---

[a] That is, the pieces. Ogle's suggested emendation
would be translated "use the hands singly." The two
transpositions suggested for this passage by Ogle seem unnecessary.

bility of its coming together into one solid piece, whereas the reverse order of events would be impossible. Also, it is possible to use them [a] singly, or two at a time, or in various ways. Again, the joints of the fingers are well constructed for taking hold of things and for exerting pressure. One finger is placed sideways : this is short and thick, not long like the others. It would be as impossible to get a hold if this were not placed sideways as if no hand were there at all. It exerts its pressure upwards from below, whereas the others act downwards from above ; and this is essential for a strong tight grip (like that of a strong clamp), so that it may exert a pressure equivalent to that of the other four. It is short, then, first, for strength, but also because it would be no good if it were long. (The end finger also is small—this is as it should be—and the middle one is long like an oar amidships, because any object which is being grasped for active use has to be grasped right around the middle.) And on this account it is called " big " although it is small, because the other fingers are practically useless without it. The nails, too, are a good piece of planning. In man they serve as coverings : a guard, in fact, for the tip of the fingers. In animals they serve for practical use as well.[b]

The joints of the arms in man bend in the opposite direction to those of quadrupeds : this is to facilitate the bringing of food to the mouth, and other uses to which they are put. Quadrupeds must be able to bend their fore limbs inwards [c] so that they may be serviceable in locomotion, since they use them as

[b] That is, as tools.
[c] See note on 693 b 3, p. 433.

30 πορείαν, ἐπεὶ θέλει γε κἀκείνων τοῖς πολυδακτύλοις
οὐ μόνον πρὸς τὴν πορείαν χρήσιμ' εἶναι τὰ ἔμπροσθεν σκέλη, ἀλλὰ καὶ ἀντὶ χειρῶν, ὥσπερ καὶ
φαίνεται χρώμενα· καὶ γὰρ λαμβάνουσι καὶ ἀμύ-
688 a νονται τοῖς προσθίοις. τὰ δὲ μώνυχα τοῖς ὀπισθίοις·
οὐ γὰρ ἔχει αὐτοῖς τὰ πρόσθια σκέλη ἀνάλογον τοῖς
ἀγκῶσι καὶ ταῖς χερσίν. τῶν δὲ πολυδακτύλων
ἔνια καὶ διὰ τοῦτο καὶ πενταδακτύλους ἔχει τοὺς
5 προσθίους πόδας, τοὺς δ' ὄπισθεν τετραδακτύλους,
οἷον λέοντες καὶ λύκοι, ἔτι δὲ κύνες καὶ παρδάλεις·
ὁ γὰρ πέμπτος ὥσπερ ὁ τῆς χειρὸς γίνεται μέγας
[πέμπτος].[1] τὰ δὲ μικρὰ τῶν πολυδακτύλων καὶ
τοὺς ὀπισθίους ἔχει πενταδακτύλους διὰ τὸ
ἑρπυστικὰ εἶναι, ὅπως τοῖς ὄνυξι πλείοσιν οὖσιν
10 ἀντιλαμβανόμενα ῥᾷον ἀνέρπῃ πρὸς τὸ μετεωρό-
τερον καὶ ὑπὲρ κεφαλῆς.

Μεταξὺ δὲ τῶν ἀγκώνων τοῖς ἀνθρώποις, τοῖς
δ' ἄλλοις τῶν ἐμπροσθίων σκελῶν, τὸ καλούμενον
στῆθός ἐστι, τοῖς μὲν ἀνθρώποις ἔχον πλάτος εὐλόγως (οὐ γὰρ κωλύουσιν οἱ ἀγκῶνες ἐκ πλαγίου
προσκείμενοι τοῦτον εἶναι τὸν τόπον πλατύν), τοῖς
15 δὲ τετράποσι διὰ τὴν ἐπὶ τὸ πρόσθιον τῶν κώλων
ἔκτασιν ἐν τῷ πορεύεσθαι καὶ μεταβάλλειν τὸν
τόπον στενὸν τοῦτ' ἐστὶ τὸ μόριον. καὶ διὰ τοῦτο
τὰ μὲν τετράποδα τῶν ζῴων οὐκ ἔχει μαστοὺς ἐν
τῷ τόπῳ τούτῳ· τοῖς δ' ἀνθρώποις διὰ τὴν εὐρυχωρίαν καὶ τὸ σκεπάζεσθαι δεῖν τὰ περὶ τὴν
20 καρδίαν, διὰ τοῦτο ὑπάρχοντος τοῦ τόπου σαρκώδους οἱ μαστοὶ διήρθρωνται, σαρκώδεις ὄντες
τοῖς μὲν ἄρρεσι διὰ τὴν εἰρημένην αἰτίαν, ἐπὶ δὲ

---

[1] πέμπτος seclusi.

feet; though even among quadrupeds the polydactylous ones tend to use the fore limbs not only for
locomotion but also instead of hands; and this can
actually be seen happening: they take hold of things
and defend themselves with their fore limbs. (Solid-
hoofed animals, on the other hand, do this with their
hind limbs, as their forelegs have nothing that corresponds to elbows and hands.) This explains why
some polydactylous quadrupeds actually have five
toes on their forefeet (lions, wolves, dogs and leo-
pards, for instance), although there are only four on
their hind feet: the fifth one, like the fifth [a] digit
on the hand, is a "big" one.[b] However, the small
polydactylous quadrupeds have five toes on their
hind feet too, because they are creepers; and this
gives them more nails, and so enables them to get a
better hold and creep up more easily to greater
heights and above your head.

Between the arms in man (in other animals between the forelegs) is what is known as the breast. In
man the breast is broad, and reasonably so, for the
arms are placed at the side and so do not in any way
prevent this part from being wide. In the quadrupeds, however, it is narrow, because as they walk
about and change their position the limbs have to be
extended forwards. And on this account, in quadrupeds, the mammae are not on the breast. In man,
on the other hand, as the space here is wide, and the
parts around the heart need some covering, the
breast is fleshy in substance and the mammae
are placed on it and are distinct. In the male they
are themselves fleshy for the reason just given. In

Breast

[a] Now generally called the "first."
[b] And needed when the foot is used as a hand.

τῶν θηλειῶν παρακέχρηται καὶ πρὸς ἕτερον ἔργον
ἡ φύσις, ὅπερ φαμὲν αὐτὴν πολλάκις ποιεῖν· ἀπο-
25 τίθεται γὰρ ἐνταῦθα τοῖς γεννωμένοις τροφήν. δύο
δ᾽ εἰσὶν οἱ μαστοὶ διὰ τὸ δύο τὰ μόρια εἶναι, τό τ᾽
ἀριστερὸν καὶ τὸ δεξιόν. καὶ σκληρότεροι μέν,
διωρισμένοι δὲ διὰ τὸ καὶ τὰς πλευρὰς συνάπτεσθαι
μὲν ἀλλήλαις[1] κατὰ τὸν τόπον τοῦτον, μὴ ἐπίπονον
δ᾽ εἶναι τὴν φύσιν αὐτῶν. τοῖς δ᾽ ἄλλοις ζῴοις ἐν
30 μὲν τῷ στήθει μεταξὺ τῶν σκελῶν ἀδύνατόν ἐστιν
ἔχειν ἢ χαλεπὸν[2] τοὺς μαστούς (ἐμποδίζοιεν μὲν
γὰρ ἂν πρὸς τὴν πορείαν), ἔχουσι δ᾽ ἤδη πολλοὺς
τρόπους.[3] τὰ μὲν γὰρ ὀλιγοτόκα καὶ μώνυχα καὶ
κερατοφόρα ἐν τοῖς μηροῖς ἔχουσι τοὺς μαστούς,
καὶ τούτους δύο, τὰ δὲ πολυτόκα ἢ πολυσχιδῆ τὰ
35 μὲν περὶ τὴν γαστέρα πλαγίους καὶ πολλούς, οἷον
688 b ὗς καὶ κύων, τὰ δὲ δύο μόνους, περὶ μέσην μέντοι
γαστέρα, οἷον λέων. τούτου δ᾽ αἴτιον οὐχ ὅτι
ὀλιγοτόκον, ἐπεὶ τίκτει ποτὲ πλείω δυοῖν, ἀλλ᾽ ὅτι
οὐ πολυγάλακτον· ἀναλίσκει γὰρ εἰς τὸ σῶμα τὴν
λαμβανομένην τροφήν, λαμβάνει δὲ σπάνιον διὰ τὸ
5 σαρκοφάγον εἶναι.

Ὁ δ᾽ ἐλέφας δύο μόνον ἔχει, τούτους δ᾽ ὑπὸ ταῖς
μασχάλαις τῶν ἐμπροσθίων σκελῶν. αἴτιον δὲ τοῦ
μὲν δύο ἔχειν ὅτι μονοτόκον ἐστί, τοῦ δὲ μὴ ἐν τοῖς
μηροῖς ὅτι πολυσχιδές (οὐδὲν γὰρ ἔχει πολυσχιδὲς
ἐν τοῖς μηροῖς), ἄνω δὲ πρὸς ταῖς μασχάλαις,

---

[1] ἀλλήλας Bekker per typothetae errorem.
[2] ἢ χαλεπὸν P: vulg. non habet.
[3] fort. τόπους Rackham (sic etiam E teste Buss. et Z)

the female, Nature employs them for an additional
function (a regular practice of hers, as I maintain),
by storing away in them nourishment for the offspring. There are two mammae because the body
has two parts, the right and the left. The fact that
they are somewhat hard and at the same time two in
number is accounted for by the ribs being joined together at this place and by the nature of the mammae
not being at all burdensome. In other animals it is
either impossible or difficult for the mammae to be
situated upon the breast, *i.e.* in between the legs,
since they would be a hindrance to walking ; but, excluding that particular position, there are numerous
ways in which they are placed. Animals which have
small litters, both those that have solid hoofs and those
that carry horns, have their mammae by the thighs ;
and there are two of them. Animals that have large
litters or are polydactylous, either have numerous
mammae placed at the sides upon the abdomen—
e.g. swine and dogs; or have only two, set in the middle
of the abdomen—*e.g.* the lion.[a] The reason for this is
not that the lion has few cubs at a birth, because
sometimes the number exceeds two, but that it is
deficient in milk. It uses up all the food it gets upon
the upkeep of the body, and as it is a flesh-eater it
gets food but rarely.

The elephant has only two mammae (this is because
it has its young one at a time), and they are under the
axillae of the forelegs and not by the thighs because
the elephant is polydactylous and no polydactylous
animal has them there. They are high up, near the
axillae, because that is the place of the foremost

[a] This, like many of Aristotle's statements about the lion,
is incorrect.

688 b

10 ὅτι πρῶτοι οὗτοι τῶν μαστῶν τοῖς πολλοὺς ἔχουσι
μαστούς, καὶ ἱμῶνται γάλα πλεῖστον. σημεῖον
δὲ τὸ ἐπὶ τῶν ὑῶν συμβαῖνον· τοῖς γὰρ πρώτοις
γενομένοις τῶν χοίρων τοὺς πρώτους παρέχουσι
μαστούς· ᾧ οὖν τὸ πρῶτον γινόμενον ἓν μόνον
ἐστί, τούτῳ τοὺς μαστοὺς ἀναγκαῖον ἔχειν τοὺς
πρώτους· πρῶτοι δ᾽ εἰσὶν οἱ ὑπὸ ταῖς μασχάλαις.
15 ὁ μὲν οὖν ἐλέφας διὰ ταύτην τὴν αἰτίαν δύο ἔχει
καὶ ἐν τούτῳ τῷ τόπῳ, τὰ δὲ πολυτόκα περὶ τὴν
γαστέρα. τούτου δ᾽ αἴτιον ὅτι πλειόνων δεῖ μαστῶν τοῖς πλείω μέλλουσιν ἐκτρέφειν· ἐπεὶ οὖν ἐπὶ
πλάτος οὐχ οἷόν τε ἀλλ᾽ ἢ δύο μόνους ἔχειν διὰ τὸ
δύο εἶναι τό τ᾽ ἀριστερὸν καὶ τὸ δεξιόν, ἐπὶ μῆκος
20 ἀναγκαῖον ἔχειν· ὁ δὲ μεταξὺ τόπος τῶν ἔμπροσθεν
σκελῶν καὶ τῶν ὄπισθεν ἔχει μῆκος μόνον. τὰ
δὲ μὴ πολυσχιδῆ ἀλλ᾽ ὀλιγοτόκα ἢ κερατοφόρα ἐν[1]
τοῖς μηροῖς ἔχει τοὺς μαστούς, οἷον ἵππος, ὄνος,
κάμηλος (ταῦτα γὰρ μονοτόκα, καὶ τὰ μὲν μώνυχα,
25 τὸ δὲ δίχηλον), ἔτι δ᾽ ἔλαφος καὶ βοῦς καὶ αἲξ καὶ
τἆλλα πάντα τὰ τοιαῦτα. αἴτιον δ᾽ ὅτι τούτοις
ἡ αὔξησις ἐπὶ τὸ ἄνω τοῦ σώματός ἐστιν. ὥσθ᾽
ὅπου συλλογὴ καὶ περιουσία γίνεται τοῦ περιττώματος καὶ αἵματος (οὗτος δ᾽ ὁ τόπος ἐστὶν ὁ
κάτω καὶ περὶ τὰς ἐκροάς), ἐνταῦθα ἐποίησεν ἡ
φύσις τοὺς μαστούς· ὅπου γὰρ κίνησις γίνεται τῆς
30 τροφῆς, ἐντεῦθεν καὶ λαβεῖν ἐστιν αὐτοῖς δυνατόν.
ἄνθρωπος μὲν οὖν καὶ ὁ θῆλυς καὶ ὁ ἄρρην ἔχει
μαστούς, ἐν δὲ τοῖς ἄλλοις ἔνια τῶν ἀρρένων οὐκ
ἔχει, οἷον ἵπποι οἱ μὲν οὐκ ἔχουσιν οἱ δ᾽ ἔχουσιν,
ὅσοι ἐοίκασι τῇ μητρί.

[1] καὶ ἐν vulg.: καὶ del. Ogle.

mammae in those that have many, and these are the ones that yield the most milk. An illustration of this is the case of the sow : a sow will offer the first of its mammae to the first ones of the litter. Thus, where the first of an animal's litter amounts to one and no more, such an animal must possess these first mammae, and " the first mammae " means those under the axillae. This explains, then, the number and position of the elephant's mammae. The animals that have large litters have their mammae upon the abdomen. Why is this ? They have numerous young to feed, and so they need numerous mammae. Now as the body has two sides, right and left, the mammae cannot be more than two deep across the body, and so they have to be disposed lengthwise, and the only place where there is sufficient length for this is between the front and hind legs. Non-polydactylous animals which yet produce few at a birth, or carry horns, have their mammae by the thighs, as the horse and the ass (both solid-hoofed) and the camel (cloven-hoofed), all of which bear their young singly ; also the deer, the ox, the goat, and all such animals. The reason for which is, that in them the growth of the body proceeds in an upward direction ; so the place where the superfluous residue and blood collects is down below, near the places of efflux, and there Nature has made the mammae ; for where the food is set in motion, there is the very place where they can get it. In man, both male and female have mammae, but some males of other animals have none, as *e.g.* stallions, some of which have none, while others, which resemble their dams, have them.

Καὶ περὶ μὲν μαστῶν εἴρηται, μετὰ δὲ τὸ στῆθος
35 ὁ περὶ τὴν κοιλίαν ἐστὶ τόπος, ἀσύγκλειστος ταῖς
689 a πλευραῖς διὰ τὴν εἰρημένην ἔμπροσθεν αἰτίαν,
ὅπως μὴ ἐμποδίζωσι μήτε τὴν ἀνοίδησιν τῆς
τροφῆς, ἣν ἀναγκαῖον συμβαίνειν θερμαινομένης
αὐτῆς, μήτε τὰς ὑστέρας τὰς περὶ τὴν κύησιν.

Τέλος δὲ τοῦ καλουμένου θώρακός ἐστι τὰ μόρια
5 τὰ περὶ τὴν τῆς περιττώσεως ἔξοδον, τῆς τε ξηρᾶς
καὶ τῆς ὑγρᾶς. καταχρῆται δ' ἡ φύσις τῷ αὐτῷ
μορίῳ ἐπί τε τὴν τῆς ὑγρᾶς ἔξοδον περιττώσεως
καὶ περὶ τὴν ὀχείαν, ὁμοίως ἔν τε τοῖς θήλεσι καὶ
τοῖς ἄρρεσιν,[1] ἔξω τινῶν ὀλίγων πᾶσι τοῖς ἐναίμοις,
ἐν δὲ τοῖς ζῳοτόκοις πᾶσιν. αἴτιον δ' ὅτι ἡ γονὴ
10 ὑγρόν ἐστί τι καὶ περίττωμα. (τοῦτο δὲ νῦν μὲν
ὑποκείσθω, ὕστερον δὲ δειχθήσεται περὶ αὐτοῦ.)
τὸν αὐτὸν δὲ τρόπον καὶ ἐν τοῖς θήλεσι τά τε
καταμήνια, καὶ ᾗ προίενται τὴν γονήν[2]· διορισθή-
σεται δὲ καὶ περὶ τούτων ὕστερον, νῦν δ' ὑποκείσθω μόνον ὅτι περίττωμα καὶ τὰ καταμήνια τοῖς
15 θήλεσιν· ὑγρὰ δὲ τὴν φύσιν τὰ καταμήνια καὶ ἡ
γονή, ὥστε[3] τῶν ὁμοίων εἰς τὰ αὐτὰ[4] μόρια τὴν
ἔκκρισιν εἶναι κατὰ λόγον ἐστίν. ἐντὸς δὲ πῶς
ἔχει, καὶ πῇ διαφέρουσι τά τε περὶ τὸ σπέρμα καὶ
τὰ περὶ τὴν κύησιν, ἔκ τε τῆς ἱστορίας τῆς περὶ
τὰ ζῷα φανερὸν καὶ τῶν ἀνατομῶν, καὶ ὕστερον
20 λεχθήσεται ἐν τοῖς περὶ γενέσεως. ὅτι δ' ἔχει καὶ

---

[1] τοῖς ἄρρεσιν Ogle : τῶν ἀρρένων vulg.
[2] καὶ εἰ προίενταί τινα γονήν Platt.
[3] post ὥστε vulg. habet τῶν αὐτῶν καί : Ogle del.
[4] τὰ αὐτὰ Peck : ταῦτα τὰ vulg.

This concludes our remarks on the mammae.

After the breast comes the region around the stomach, which is not enclosed by the ribs for the reason stated earlier,[a] viz. to avoid interference (*a*) with the food when it swells, as it must do when it is heated, and (*b*) with the womb during pregnancy.

At the end of what is called the trunk are the parts that have to do with the discharge of the residue, both solid and fluid. Nature employs one and the same part for the discharge of the fluid residue and for copulation in all blooded animals (with a few exceptions), male and female alike, and in all Vivipara without exception. The reason is that the semen is a fluid, and a residue. (This statement may stand for the present : the proof of it will be given later on.[b]) The same applies to the catamenia in females, and the part where they emit the semen.[c] This also will be dealt with particularly later on. For the present, let the statement stand simply that the catamenia in females (like the semen in males) are a residue. Now both semen and catamenia are fluids, so it is reasonable that things which are alike should be discharged through the same parts. A clear account of the internal structure of these parts, showing the differences between the parts connected with semen and those connected with conception, is given in the *Researches upon Animals*[d] and the *Dissections*, and there will be a discussion of them in the book on

Excretory organs.

---

[a] At 655 a 2.

[b] In *De gen. an.* 724 b 21 ff.

[c] This seems to agree with what Aristotle says on the subject in the *Hist. An.*, but contradicts what he says in *De gen. an.* Platt's suggested emendation would make the translation read : " and to the semen, if so be they emit any."

[d] At 493 a 24–b 6, 497 a 24 ff., book iii, ch. 1.

689 a

τὰ σχήματα τῶν μορίων τούτων πρὸς τὴν ἐργασίαν
ἀναγκαίως, οὐκ ἄδηλον. ἔχει δὲ διαφορὰς τὸ τῶν
ἀρρένων ὄργανον κατὰ τὰς τοῦ σώματος διαφοράς.
οὐ γὰρ ὁμοίως ἅπαντα νευρώδη τὴν φύσιν ἐστίν.
ἔτι δὲ μόνον τοῦτο τῶν μορίων ἄνευ νοσερᾶς μετα-
25 βολῆς αὔξησιν ἔχει καὶ ταπείνωσιν· τούτων γὰρ τὸ
μὲν χρήσιμον πρὸς τὸν συνδυασμόν, τὸ δὲ πρὸς τὴν
τοῦ ἄλλου σώματος χρείαν· ἀεὶ γὰρ ὁμοίως ἔχον
τἆλλα¹ ἐνεπόδιζεν ἄν. συνέστηκε δὲ τὴν φύσιν
ἐκ τοιούτων τὸ μόριον τοῦτο ὥστε δύνασθαι ταῦτ'
ἀμφότερα συμβαίνειν· τὸ μὲν γὰρ ἔχει νευρῶδες
30 τὸ δὲ χονδρῶδες, διόπερ συνιέναι τε δύναται καὶ
ἔκτασιν ἔχειν καὶ πνεύματός ἐστι δεκτικόν. τὰ
μὲν οὖν θήλεα τῶν τετραπόδων πάντ' ἐστὶν ὀπι-
σθουρητικὰ διὰ τὸ πρὸς τὴν ὀχείαν οὕτως εἶναι
αὐτοῖς χρησίμην τὴν θέσιν, τῶν δ' ἀρρένων ὀλίγα
ἐστὶν ὀπισθουρητικά, οἷον λύγξ, λέων, κάμηλος,
689 b δασύπους· μώνυχον δ' οὐδέν ἐστιν ὀπισθουρητικόν.

Τὰ δ' ὄπισθεν καὶ τὰ περὶ τὰ σκέλη τοῖς ἀνθρώποις ἰδίως ἔχει πρὸς τὰ τετράποδα. κέρκον δ' ἔχει
πάντα σχεδόν, οὐ μόνον τὰ ζῳοτόκα ἀλλὰ καὶ τὰ
ᾠοτόκα· καὶ γὰρ ἂν μὴ μέγεθος αὐτοῖς ἔχον τύχῃ²
5 τοῦτο τὸ μόριον, ἀλλὰ σημείου³ γ' ἕνεκεν ἔχουσί
τινα στόλον. ὁ δ' ἄνθρωπος ἄκερκον μέν ἐστιν,
ἰσχία δ' ἔχει, τῶν δὲ τετραπόδων οὐδέν. ἔτι δὲ καὶ
τὰ σκέλη ὁ μὲν ἄνθρωπος σαρκώδη καὶ μηροὺς καὶ
κνήμας,⁴ τὰ δ' ἄλλα πάντ' ἄσαρκα ἔχει, οὐ μόνον τὰ
ζῳοτόκα ἀλλ' ὅλως ὅσα σκέλη ἔχει τῶν ζῴων·
10 νευρώδη γὰρ ἔχει καὶ ὀστώδη καὶ ἀκανθώδη.
τούτων δ' αἰτία μία τίς ἐστιν ὡς εἰπεῖν ἁπάντων,

---

¹ ἔχον τἆλλα Peck : ἔχοντα vulg.
² τύχῃ Rackham : ᾖ vulg.

Generation.[a]   Still, it is clear that the actual forms of
these parts is determined of necessity by the function
they have to perform.   The male organ, however,
exhibits differences corresponding to those of the
body as a whole, for some animals are more sinewy,
some less.   Further, this organ is the only one which
increases and subsides apart from any change due
to disease.   Its increasing in size is useful for copulation, its contraction for the employment of the rest
of the body, since it would be a nuisance to the
other parts if it were always extended.   And so it
is composed of substances which make both conditions possible : it contains both sinew and cartilage ;
and so it can contract and expand and admits air
into itself.   All female quadrupeds discharge the
urine backwards, as this arrangement is useful to
them for copulation.   A few males do this (among
them are the lynx, the lion, the camel, and the
hare), but no solid-hoofed animal does so.

The rear parts and the parts around the legs are Rear parts.
peculiar in man compared with the quadrupeds, nearly
all of which (Ovipara as well as Vivipara) have a tail,
which even if it is not of any great size, still is present
for a token as a sort of stump.   Man has no tail, but
he has buttocks, which no quadruped possesses.[b]   In
man, the legs, both in thighs and calves, are fleshy :
in all other animals that have them (not only Vivipara) the legs are fleshless, being sinewy, bony and
spinous.   One might say that there is a single explanation which covers them all, which is, that man is

[a] At 716 a 2—721 a 29.
[b] There seems to be something wrong with this statement,
but perhaps when taken in conjunction with the whole of the
argument which follows, it may appear less unjustifiable.

[3] σημείου Buss. : σμικροῦ vulg.   [4] κνήμας] πόδας Y.

689 b

διότι μόνον ἐστὶν ὀρθὸν τῶν ζῴων ἄνθρωπος. ἵν'
οὖν φέρῃ ῥᾳδίως τἄνω κοῦφα ὄντα, ἀφελοῦσα τὸ
σωματῶδες ἀπὸ τῶν ἄνω πρὸς τὰ κάτω τὸ βάρος
ἡ φύσις προσέθηκεν· διόπερ τὰ ἰσχία σαρκώδη
15 ἐποίησε καὶ μηροὺς καὶ γαστροκνημίας. ἅμα δὲ
τήν τε τῶν ἰσχίων φύσιν καὶ πρὸς τὰς ἀναπαύσεις
ἀπέδωκε χρήσιμον· τοῖς μὲν γὰρ τετράποσιν ἄκοπον
τὸ ἑστάναι, καὶ οὐ κάμνουσι τοῦτο ποιοῦντα συνεχῶς (ὥσπερ γὰρ κατακείμενα διατελεῖ ὑπο-
κειμένων τεττάρων ἐρεισμάτων), τοῖς δ' ἀνθρώποις
20 οὐ ῥᾴδιον ὀρθῶς ἑστῶσι διαμένειν, ἀλλὰ δεῖται τὸ
σῶμα ἀναπαύσεως καὶ καθέδρας. ὁ μὲν οὖν ἄνθρωπος ἰσχία τ' ἔχει καὶ τὰ σκέλη σαρκώδη διὰ
τὴν εἰρημένην αἰτίαν, καὶ διὰ ταῦτα ἄκερκον (ἥ
τε γὰρ ἐκεῖσε[1] τροφὴ πορευομένη εἰς ταῦτα ἀνα-
25 λίσκεται, καὶ διὰ τὸ ἔχειν ἰσχία ἀφῄρηται ἡ τῆς
οὐρᾶς ἀναγκαία χρῆσις), τὰ δὲ τετράποδα καὶ
τἆλλα ζῷα ἐξ ἐναντίας· νανώδεσι γὰρ οὖσι πρὸς τὸ
ἄνω τὸ βάρος καὶ τὸ σωματῶδες ἐπίκειται πᾶν,
ἀφῃρημένον ἀπὸ τῶν κάτωθεν· διόπερ ἀνίσχια καὶ
σκληρὰ τὰ σκέλη ἔχουσιν. ὅπως δ' ἐν φυλακῇ καὶ
30 σκέπῃ ᾖ τὸ λειτουργοῦν μόριον τὴν ἔξοδον τοῦ
περιττώματος, τὴν καλουμένην οὐρὰν καὶ κέρκον
αὐτοῖς ἀπέδωκεν ἡ φύσις, ἀφελομένη τῆς εἰς τὰ
σκέλη γιγνομένης τροφῆς.

(Ὁ δὲ πίθηκος διὰ τὸ τὴν μορφὴν ἐπαμφοτερίζειν
καὶ μηδετέρων τ' εἶναι καὶ ἀμφοτέρων, διὰ τοῦτ'
οὔτ' οὐρὰν ἔχει οὔτ' ἰσχία, ὡς μὲν δίπους ὢν οὐράν,
ὡς δὲ τετράπους ἰσχία.)

690 a Τῶν δὲ καλουμένων κέρκων διαφοραί τ' εἰσὶ

---

[1] ἐκεῖσε Peck : ἐκεῖ vulg.

the only animal that stands upright. Hence, Nature, so as to make the upper parts light and easy to carry, took off the corporeal matter from the top and transferred the weight down below ; and that is how she came to make the buttocks and the thighs and the calves of the legs fleshy. At the same time, in making the buttocks fleshy, Nature made them useful for resting the body. Quadrupeds find it no trouble to remain standing, and do not get tired if they remain continually on their feet—the time is as good as spent lying down, because they have four supports underneath them. But human beings cannot remain standing upright continually with ease ; the body needs rest ; it must be seated. That, then, is why man has buttocks and fleshy legs, and for the same reason he has no tail : the nourishment gets used up for the benefit of the buttocks and legs before it can get as far as the place for the tail. Besides, the possession of buttocks takes away the need and necessity of a tail. But in quadrupeds and other animals it is the opposite : they are dwarf-like, which means that their heavy corporeal substance is in the upper part of them and does not come into the lower parts ; and as a result they have no buttocks and their legs are hard. Yet to ensure that the part which serves them for the discharge of the residue shall be guarded and covered over, Nature has assigned to them tails or scuts by taking off somewhat of the nourishment which would otherwise go into the legs.

(The Ape is, in form, intermediate between the two, man and quadruped, and belongs to neither, or to both, and consequently he has no tail, *qua* biped, and no buttocks, *qua* quadruped.)

There are numerous differences in the various tails,

690 a

πλείους καὶ ἡ φύσις παρακαταχρῆται καὶ ἐπὶ τούτων, οὐ μόνον πρὸς φυλακὴν καὶ σκέπην τῆς ἕδρας,
ἀλλὰ καὶ πρὸς ὠφέλειαν καὶ χρῆσιν τοῖς ἔχουσιν.

5 Οἱ δὲ πόδες τοῖς μὲν τετράποσι διαφέρουσιν· τὰ
μὲν γὰρ μώνυχα αὐτῶν ἐστι τὰ δὲ δίχηλα τὰ δὲ
πολυσχιδῆ, μώνυχα μὲν ὅσοις διὰ μέγεθος καὶ τὸ
πολὺ γεῶδες ἔχειν ἀντὶ κεράτων καὶ ὀδόντων εἰς
τὴν τοῦ ὄνυχος φύσιν τὸ τοιοῦτον μόριον ἔλαβεν
ἀπόκρισιν, καὶ διὰ πλῆθος ἀντὶ πλειόνων ὀνύχων
10 εἷς ὄνυξ ἡ ὁπλή ἐστιν. καὶ ἀστράγαλον δὲ διὰ
τοῦτο οὐκ ἔχουσιν ὡς ἐπὶ τὸ πολὺ εἰπεῖν, καὶ διὰ¹
τὸ δυσκινητοτέραν εἶναι τὴν καμπὴν τοῦ ὄπισθεν
σκέλους ἀστραγάλου ἐνόντος· θᾶττον γὰρ ἀνοίγεται
καὶ κλείεται τὰ μίαν ἔχοντα γωνίαν ἢ πλείους, ὁ
δ' ἀστράγαλος γόμφος ὢν ὥσπερ ἀλλότριον κῶλον
15 ἐμβέβληται τοῖς δυσί, βάρος μὲν παρέχων, ποιοῦν
δ' ἀσφαλεστέραν τὴν βάσιν. διὰ γὰρ τοῦτο καὶ ἐν
τοῖς ἐμπροσθίοις οὐκ ἔχουσιν ἀστράγαλον τὰ ἔχοντα
ἀστράγαλον, ἀλλ' ἐν τοῖς ὄπισθεν, ὅτι δεῖ ἐλαφρὰ
εἶναι τὰ ἡγούμενα καὶ εὔκαμπτα, τὸ δ' ἀσφαλὲς καὶ
τὴν τάσιν ἐν τοῖς ὄπισθεν. ἔτι δὲ πρὸς τὸ ἀμύνε-
20 σθαι ἐμβριθεστέραν ποιεῖ τὴν πληγήν· τὰ δὲ τοιαῦτα
τοῖς ὄπισθεν χρῆται κώλοις, λακτίζοντα τὸ λυποῦν.

Τὰ δὲ δίχηλα ἔχει ἀστράγαλον (κουφότερα γὰρ
τὰ ὄπισθεν), καὶ διὰ τὸ ἔχειν ἀστράγαλον καὶ οὐ
μώνυχά ἐστιν, ὡς τὸ ἐκλεῖπον ὀστῶδες ἐκ τοῦ

---

¹ καὶ διὰ SUZ Ogle : διὰ vulg.

ᵃ The word used in the Greek is " part." See Introd. p. 28.
ᵇ See Introduction, pp. 38-39.

which provide another example of Nature's habit of using an organ for secondary purposes, for she employs the tail not only as a guard and covering for the fundament but also in other serviceable ways.

There are differences too in the feet of quadrupeds. Hoofs, etc. Some have a solid hoof, some a cloven hoof; others have a foot that is divided into several parts. Solid hoofs are present in those animals which are large and contain much earthy substance,[a] which instead of making horns and teeth forms an abscession[b] so as to produce nail, and owing to the abundance of it, it produces not several separate nails but a single one, in other words, a hoof. Because of this, these animals in general have no hucklebone; and also because the presence of a hucklebone makes it rather difficult to bend the hind leg freely, since a limb that has one angle can be bent to and fro more quickly than one that has several. It is a sort of connecting-rod, and therefore practically interpolates another bit of a limb between the two, thereby increasing the weight; but it makes the animal's footing more reliable. This explains why, when hucklebones are present, they are present in the hind limbs only, never in the front: the front limbs have to be light and flexible because they go first, while the hind limbs must be reliable and able to stretch. Further, a hucklebone puts more force into a blow—a useful point in self-defence—and animals which have one use their hind limbs in this way: if anything hurts them they kick out at it.

Cloven-hoofed animals have a hucklebone, as their hind limbs are on the light side; and that is the very reason why they are cloven-hoofed: the bony substance stays in the joint and therefore is deficient in

690 a

ποδὸς ἐν τῇ κάμψει μένον. τὰ δὲ πολυδάκτυλα
25 οὐκ ἔχει ἀστράγαλον· οὐ γὰρ ἂν ἦν πολυδάκτυλα,
ἀλλὰ τοσοῦτον ἐσχίζετο τὸ πλάτος ὅσον ἐπέχει
ὁ ἀστράγαλος. διὸ καὶ τῶν ἐχόντων αὐτὸν τὰ
πλείω δίχηλα.

Ὁ δ᾽ ἄνθρωπος πόδας μεγίστους ἔχει τῶν ζῴων
ὡς κατὰ μέγεθος, εὐλόγως· μόνον γὰρ ἕστηκεν
ὀρθόν, ὥστε τοὺς μέλλοντας δύ᾽ ὄντας ἕξειν πᾶν τὸ
30 τοῦ σώματος βάρος δεῖ μῆκος ἔχειν καὶ πλάτος.
καὶ τὸ τῶν δακτύλων δὴ μέγεθος ἐναντίως ἔχει ἐπί
τε τῶν ποδῶν καὶ τῶν χειρῶν κατὰ λόγον· τῶν
μὲν γὰρ τὸ λαμβάνειν ἔργον καὶ πιέζειν, ὥστε δεῖ
690 b μακροὺς ἔχειν (τῷ γὰρ καμπτομένῳ μέρει περιλαμβάνει ἡ χείρ), τῶν δὲ τὸ βεβηκέναι ἀσφαλῶς,
πρὸς δὲ[1] τοῦτο δεῖ τὸ μόριον εἶναι μεῖζον[2] τὸ
ἄσχιστον τοῦ ποδὸς τῶν δακτύλων. ἐσχίσθαι δὲ
βέλτιον ἢ ἄσχιστον εἶναι τὸ ἔσχατον· ἅπαν γὰρ ἂν
5 συμπαθὲς ἦν ἑνὸς μορίου πονήσαντος, ἐσχισμένῳ[3]
δ᾽ εἰς δακτύλους τοῦτ᾽ οὐ συμβαίνει ὁμοίως. ἔτι
δὲ καὶ βραχεῖς ὄντες ἧττον ⟨ἂν⟩ βλάπτοιντο.[4] διὸ
πολυσχιδεῖς οἱ πόδες τῶν ἀνθρώπων, οὐ μακρο-
δάκτυλοι δ᾽ εἰσίν. τὸ δὲ τῶν ὀνύχων γένος διὰ
τὴν αὐτὴν αἰτίαν καὶ ἐπὶ τῶν χειρῶν ἔχουσιν· δεῖ
10 γὰρ σκέπεσθαι τὰ ἀκρωτήρια μάλιστα διὰ τὴν
ἀσθένειαν.

Περὶ μὲν οὖν τῶν ἐναίμων ζῴων καὶ ζῳοτόκων
καὶ πεζῶν εἴρηται σχεδὸν περὶ πάντων· XI. τῶν
δ᾽ ἐναίμων ζῴων ᾠοτόκων δὲ τὰ μέν ἐστι τετρά-

---

[1] πρὸς δὲ Ogle : ὥστε vulg.
[2] μεῖζον Platt, Th. : νομίζειν vulg.
[3] ἐσχισμένῳ Peck : -ον PY : -ων vulg. : -ου Ogle.
[4] ⟨ἂν⟩ Platt, Th. : βλάπτοιντο Υ : συμβλάπτοιντο vulg.

890

the foot. The polydactylous animals have no huckle-bone, otherwise they would not be polydactylous, and the divisions of the foot would cover only so much width as the hucklebone itself. So most of the animals which have a hucklebone are cloven-hoofed.

Man of all the animals has the largest feet for his size, and reasonably so, since he is the only one of them that stands upright, and as the feet have to bear the whole weight of the body and there are only two of them, they must be both long and broad. Also the toes are short compared with the fingers, and this too is reasonable. The business of the hands is to take hold and to keep hold of things, and this is done by means of that part of the hands which bends ; therefore the fingers must be long. The business of the feet is to get a firm and reliable footing ; and to secure this the undivided part of the foot must be greater than the toes. And it is better to have the tip of the foot divided than not, for otherwise, if one part were affected the whole foot would suffer as well, whereas this is to some degree avoided by the division of the tip of the foot into toes. Again, short toes are less liable to injury than long ones would be. All this indicates why the human foot has toes and why they are short. There are nails on the toes for the same reason that there are nails on the fingers : the extremities have but little strength and there-fore specially need to be protected.

We have now dealt with practically all the blooded animals that are viviparous and live on the land.

XI. We now pass on to another class of blooded animals, the oviparous, some of which have four feet, *(b)* Ovipara : (i.) Serpents and quadrupeds.

690 b

ποδα τὰ δ' ἄποδα. τοιοῦτον δ' ἐν μόνον γένος
15 ἐστὶν ἄπουν, τὸ τῶν ὄφεων· ἡ δ' αἰτία τῆς ἀποδίας
αὐτῶν εἴρηται ἐν τοῖς περὶ τῆς πορείας τῶν ζῴων
διωρισμένοις. τὰ δ' ἄλλα παραπλησίαν ἔχει τὴν
μορφὴν τοῖς τετράποσι καὶ ᾠοτόκοις.[1]

Ἔχει δὲ τὰ ζῷα ταῦτα κεφαλὴν μὲν καὶ τὰ ἐν
αὐτῇ μόρια διὰ τὰς αὐτὰς αἰτίας τοῖς ἄλλοις τοῖς
20 ἐναίμοις ζῴοις, καὶ γλῶτταν ἐν τῷ στόματι πλὴν
τοῦ ποταμίου κροκοδείλου· οὗτος δ' οὐκ ἂν δόξειεν
ἔχειν, ἀλλὰ τὴν χώραν μόνον. αἴτιον δ' ὅτι τρόπον
μέν τινα ἅμα χερσαῖος καὶ ἔνυδρός ἐστιν· διὰ μὲν
οὖν τὸ χερσαῖος εἶναι ἔχει χώραν γλώττης, διὰ δὲ
τὸ ἔνυδρος ἄγλωττος. οἱ γὰρ ἰχθύες, καθάπερ εἴρη-
25 ται πρότερον, οἱ μὲν οὐ δοκοῦσιν ἔχειν, ἂν μὴ σφόδρα ἀνακλίνῃ τις, οἱ δ' ἀδιάρθρωτον ἔχουσιν. αἴτιον
δ' ὅτι ὀλίγη τούτοις χρεία[2] τῆς γλώττης διὰ τὸ μὴ
ἐνδέχεσθαι μασᾶσθαι μηδὲ προγεύεσθαι, ἀλλ' ἐν τῇ
καταπόσει γίνεσθαι τὴν αἴσθησιν καὶ τὴν ἡδονὴν
πᾶσι τούτοις τῆς τροφῆς. ἡ μὲν γὰρ γλῶττα τῶν
30 χυμῶν ποιεῖ τὴν αἴσθησιν, τῶν δὲ ἐδεστῶν ἐν τῇ
καθόδῳ ἡ ἡδονή· καταπινομένων γὰρ αἰσθάνονται
τῶν λιπαρῶν καὶ θερμῶν καὶ τῶν ἄλλων τῶν
τοιούτων. ἔχει μὲν οὖν καὶ τὰ ζῳοτόκα ταύτην
τὴν αἴσθησιν (καὶ σχεδὸν τῶν πλείστων ὄψων καὶ
691 a ἐδεστῶν ἐν τῇ καταπόσει τῇ τάσει τοῦ οἰσοφάγου
γίνεται ἡ χάρις· διὸ οὐχ οἱ αὐτοὶ περὶ τὰ πόματα
καὶ τοὺς χυμοὺς ἀκρατεῖς εἰσι καὶ τὰ ὄψα καὶ τὴν

---

[1] ᾠοτόκοις PUYZ : ζῳοτόκοις vulg.
[2] ἦν τούτοις χρεία S : ἦν χρεία τούτοις vulg. : ἦν dedevi.

---

* At De inc. an. 708 a 9 ff ; see also infra, 696 a 10.
b At 660 b 13-25.

and some no feet at all. Actually there is only one group that has no feet, the Serpents ; and the reason why they have none has been stated in my treatise on the *Locomotion of Animals*.[a] In other respects their conformation is similar to that of the oviparous quadrupeds.

These animals have a head, and the parts that compose it, for the same reasons that other blooded creatures have one, and they have a tongue inside the mouth—all except the river crocodile, which apparently has none, but only a space for it ; and the reason is that in a way he is both a land-animal and a water-animal. In virtue of being a land-animal, he has a space for a tongue ; as a water-animal, he is tongueless. This agrees with our previous statement,[b] that some fishes appear to have no tongue unless you pull the mouth very well open, others have one which is not distinctly articulated. The reason for this is that these creatures have not much need for a tongue because they cannot chew their food or even taste it before they eat it : they can perceive the pleasantness of it only while they are swallowing it. This is because the perception of juices is effected by the tongue ; whereas the pleasantness of solid food is perceived while it is passing down the gullet, and thus oily food and hot food and the like are perceived while they are being swallowed. Of course the Vivipara as well as these creatures have this power of perception (indeed, the enjoyment derived from practically all edible dainties takes place while they are being swallowed and is due to the distension of the oesophagus—which is why intemperate appetite for edible dainties is not found in the same animals as intemperate appetite for drink and juices) ;

393

691 a

ἐδωδήν), ἀλλὰ τοῖς μὲν ἄλλοις ζῴοις καὶ ἡ κατὰ
5 τὴν γεῦσιν ὑπάρχει αἴσθησις, ἐκείνοις δ' ἄνευ
ταύτης μόνη¹ ἡ ἑτέρα. τῶν δὲ τετραπόδων καὶ
ᾠοτόκων οἱ σαῦροι, ὥσπερ καὶ οἱ² ὄφεις, δικρόαν
ἔχουσι τὴν γλῶτταν καὶ ἐπ' ἄκρου τριχώδη πάμπαν,
καθάπερ εἴρηται πρότερον. ἔχουσι δὲ καὶ αἱ φῶκαι
δικρόαν τὴν γλῶτταν· διὸ καὶ λίχνα³ πάντα τὰ ζῷά
ἐστι ταῦτα.

10 Ἔστι δὲ καὶ καρχαρόδοντα τὰ τετράποδα τῶν
ᾠοτόκων, ὥσπερ οἱ ἰχθύες. τὰ δ' αἰσθητήρια
πάντα ὁμοίως ἔχουσι τοῖς ἄλλοις ζῴοις, οἷον τῆς
ὀσφρήσεως μυκτῆρας καὶ ὄψεως ὀφθαλμοὺς καὶ
ἀκοῆς ὦτα, πλὴν οὐκ ἐπανεστηκότα, καθάπερ οὐδ'
οἱ ὄρνιθες, ἀλλὰ τὸν πόρον μόνον· αἴτιον δ' ἀμφο-
15 τέροις ἡ τοῦ δέρματος σκληρότης· τὰ μὲν γὰρ
πτερωτὰ αὐτῶν ἐστι, ταῦτα δὲ πάντα φολιδωτά,
ἔστι δ' ἡ φολὶς ὅμοιον χώρᾳ λεπίδος, φύσει δὲ
σκληρότερον. δηλοῖ δ' ἐπὶ τῶν χελωνῶν τοῦτο
καὶ ἐπὶ τῶν μεγάλων ὄφεων καὶ τῶν ποταμίων
κροκοδείλων· ἰσχυρότεραι γὰρ γίνονται τῶν ὀστῶν
ὡς οὖσαι τοιαῦται τὴν φύσιν.

20 Οὐκ ἔχουσι δὲ τὰ ζῷα ταῦτα τὴν ἄνω βλεφαρίδα,
ὥσπερ οὐδ' οἱ ὄρνιθες, ἀλλὰ τῇ κάτω μύουσι διὰ
τὴν αἰτίαν τὴν εἰρημένην ἐπ' ἐκείνων. τῶν μὲν οὖν
ὀρνίθων ἔνιοι καὶ σκαρδαμύττουσιν ὑμένι ἐκ τῶν
κανθῶν, ταῦτα δὲ τὰ ζῷα οὐ σκαρδαμύττει· σκληρ-
25 οφθαλμότερα γάρ ἐστι τῶν ὀρνίθων. αἴτιον δ' ὅτι
ἐκείνοις χρησιμωτέρα ἡ ὀξυωπία⁴ πτηνοῖς οὖσι πρὸς

---

¹ δ' ἄνευ ταύτης μόνη Peck: δ' ἂν ἦ ὥσπερ μόνη Υ: δ'
ὡσπερανεὶ vulg.; plurima hic transposuit Ogle.
² καὶ οἱ Υ: οἱ vulg.

394

but whereas the rest of the animals have the power of perception by taste as well, these are without it and possess the other one only. Among oviparous quadrupeds, lizards (and serpents too) have a two-forked tongue, the tips of which are as fine as hairs. (This has been stated earlier.[a]) Seals also have a forked tongue. This forked tongue explains why all these animals are so dainty in their food.

The four-footed Ovipara also have sharp interfitting teeth, as Fishes have. Their sense-organs are all similar to those of other animals : nostrils for smell, eyes for sight, and ears for hearing—though their ears do not stand out : they are merely a duct, as in birds ; and in both groups the cause is the same, viz. the hardness of their integument. Birds are covered with feathers, and these creatures are all covered with horny scales which correspond in position to the scales of fishes, but are harder in substance. This is clearly illustrated by the tortoises, the great snakes, and the river crocodiles, where the scales are made of the same material as the bones and actually grow stronger than the bones.

These animals, like birds, have no upper eyelid ; they close their eyes with the lower lid. The reason which was given[b] for birds applies to them too. Some birds can also blink by means of a membrane which comes out of the corner of the eye ; but these animals do not do this, since their eyes are harder than birds' eyes. The reason for this is that keen sight is of considerable use to birds in their daily

---

[a] At 660 b 9. 　　　　　　　[b] At 657 b 6 ff.

---

[3] λίχνα Karsch: ἰσχνὰ vulg.
[4] ὀξυωπία καὶ τὸ πόρρω προϊδεῖν UY.

691 a

τὸν βίον, τούτοις δ' ἧττον· τρωγλόδυτα γὰρ πάντα
τὰ τοιαῦτά ἐστιν.

Εἰς δύο δὲ διῃρημένης τῆς κεφαλῆς, τοῦ τε ἄνω
μορίου καὶ τῆς σιαγόνος τῆς κάτω, ἄνθρωπος μὲν[1]
καὶ τὰ ζῳοτόκα τῶν τετραπόδων καὶ ἄνω καὶ κάτω
30 κινοῦσι τὰς σιαγόνας καὶ εἰς τὸ πλάγιον, οἱ δ'
ἰχθύες καὶ ὄρνιθες καὶ τὰ ῳοτόκα τῶν τετραπόδων
εἰς τὸ ἄνω καὶ κάτω μόνον. αἴτιον δ' ὅτι ἡ μὲν
691 b τοιαύτη κίνησις χρήσιμος εἰς τὸ δακεῖν καὶ διελεῖν,
ἡ δ' εἰς τὸ πλάγιον ἐπὶ τὸ λεαίνειν. τοῖς μὲν οὖν
ἔχουσι γομφίους χρήσιμος ἡ εἰς τὸ πλάγιον κίνησις,
τοῖς δὲ μὴ ἔχουσιν οὐδὲν χρήσιμος, διόπερ ἀφῄρηται
πάντων τῶν τοιούτων· οὐδὲν γὰρ ποιεῖ περίεργον ἡ
5 φύσις. τὰ μὲν οὖν ἄλλα πάντα κινεῖ τὴν σιαγόνα
τὴν κάτω, ὁ δὲ ποτάμιος κροκόδειλος μόνος τὴν ἄνω.
τούτου δ' αἴτιον ὅτι πρὸς τὸ λαβεῖν καὶ κατασχεῖν
ἀχρήστους ἔχει τοὺς πόδας· μικροὶ γάρ εἰσι πάμπαν.
πρὸς οὖν ταύτας τὰς χρείας ἀντὶ ποδῶν τὸ στόμα
ἡ φύσις χρήσιμον αὐτῷ ἐποίησεν. πρὸς δὲ τὸ
10 κατασχεῖν ἢ λαβεῖν, ὁποτέρωθεν ἂν ᾖ ἡ πληγὴ
ἰσχυροτέρα, ταύτῃ χρησιμωτέρα κινουμένη ἐστίν·
ἡ δὲ πληγὴ ἰσχυροτέρα ἀεὶ ἄνωθεν ἢ κάτωθεν· ἐπεὶ
οὖν ἀμφοτέρων μὲν διὰ τοῦ στόματος ἡ χρῆσις, καὶ
τοῦ λαβεῖν καὶ τοῦ δακεῖν, ἀναγκαιοτέρα δ' ἡ τοῦ
15 κατασχεῖν μήτε χεῖρας ἔχοντι μήτε πόδας εὐφυεῖς,
χρησιμώτερον τὴν ἄνωθεν κινεῖν σιαγόνα ἢ τὴν
κάτωθεν αὐτοῖς. διὰ τὸ αὐτὸ δὲ καὶ οἱ καρκίνοι
τὸ ἄνωθεν τῆς χηλῆς κινοῦσι μόριον, ἀλλ' οὐ τὸ
κάτωθεν· ἀντὶ χειρὸς γὰρ ἔχουσι τὰς χηλάς, ὥστε
πρὸς τὸ λαβεῖν ἀλλ' οὐ πρὸς τὸ διελεῖν χρήσιμον

[1] μὲν οὖν vulg. : μὲν YZ.

life, because they fly about ; but it would be very little good to these creatures, because they all spend their time in holes and corners.

Their head has two divisions : the upper part, and the lower jaw. In man and in the viviparous quadrupeds the lower jaw moves from side to side as well as up and down ; in fishes, however, and birds and these oviparous quadrupeds it moves up and down only. The reason is that this vertical motion is useful for biting and cutting up food, while the sideways motion is useful for grinding the food down. Of course this sideways motion is useful to animals which possess grinder-teeth ; but it is of no use to those which lack grinders, and so not one of them has it. Nature never makes or does anything that is superfluous. All these animals, then, move the lower jaw—with one exception, the river crocodile, which moves the upper jaw, and the reason for this is that his feet are no use for seizing and holding things : they are too small altogether. So Nature has given him a mouth which he can use for these purposes instead of his feet. And when it comes to seizing things and holding them, the most useful direction for a blow to take is that which gives it the greatest strength. Now a blow from above is always stronger than one from below. And to an animal who has no hands and no proper feet, who has to use his mouth for seizing his food as well as for biting it, the power to seize it is the more necessary ; and therefore it is more useful to him to be able to move his upper jaw than his lower one. For the same reason crabs move the upper part of their claws and not the lower : claws are their substitute for hands, so the claws have to be useful for seizing things (not for cutting them

691 b

20 δεῖ εἶναι τὴν χηλήν· τὸ δὲ διελεῖν καὶ δακεῖν ὀδόντων ἔργον ἐστίν. τοῖς μὲν οὖν καρκίνοις καὶ τοῖς
ἄλλοις ὅσοις ἐνδέχεται σχολαίως ποιεῖσθαι τὴν
λῆψιν διὰ τὸ μὴ ἐν ὑγρῷ εἶναι τὴν χρῆσιν τοῦ
στόματος, διῄρηται, καὶ λαμβάνουσι μὲν χερσὶν ἢ
ποσί, διαιροῦσι δὲ τῷ στόματι καὶ δάκνουσιν· τοῖς
25 δὲ κροκοδείλοις ἐπ' ἀμφότερα χρήσιμον τὸ στόμα
πεποίηκεν ἡ φύσις, κινουμένων οὕτω τῶν σιαγόνων.

Ἔχουσι δὲ καὶ αὐχένα πάντα τὰ τοιαῦτα διὰ τὸ
πλεύμονα ἔχειν· δέχονται γὰρ τὸ πνεῦμα διὰ τῆς
ἀρτηρίας μῆκος ἐχούσης.

[1]Ἐπεὶ δὲ τὸ μεταξὺ κεφαλῆς καὶ ὤμων κέκληται
αὐχήν, ἥκιστα τῶν τοιούτων ὁ ὄφις δόξειεν ἂν
30 ἔχειν αὐχένα, ἀλλὰ τὸ ἀνάλογον τῷ αὐχένι, εἴ γε
δεῖ τοῖς εἰρημένοις ἐσχάτοις διορίζειν τὸ μόριον
τοῦτο. ἴδιον δὲ πρὸς τὰ συγγενῆ τῶν ζῴων
692 a ὑπάρχει τοῖς ὄφεσι τὸ στρέφειν τὴν κεφαλὴν εἰς
τοὔπισθεν ἠρεμοῦντος τοῦ λοιποῦ σώματος. αἴτιον
δ' ὅτι καθάπερ τὰ ἔντομα ἑλικτόν ἐστιν, ὥστε
εὐκάμπτους ἔχειν καὶ χονδρώδεις τοὺς σπονδύλους.
5 ἐξ ἀνάγκης μὲν οὖν διὰ ταύτην τὴν αἰτίαν τοῦτο
συμβέβηκεν αὐτοῖς, τοῦ δὲ βελτίονος ἕνεκεν πρὸς
φυλακὴν τῶν ὄπισθεν βλαπτόντων· μακρὸν γὰρ ὂν
καὶ ἄπουν ἀφυές ἐστι πρός τε τὴν στροφὴν καὶ πρὸς
τὴν τῶν ὄπισθεν τήρησιν· οὐδὲν γὰρ ὄφελος αἴρειν
μέν, στρέφειν δὲ μὴ δύνασθαι τὴν κεφαλήν. ἔχουσι
δὲ τὰ τοιαῦτα καὶ τῷ στήθει ἀνάλογον μόριον.
10 μαστοὺς δ' οὐκ ἔχουσιν οὔτ' ἐνταῦθα οὔτ' ἐν τῷ
ἄλλῳ σώματι, ὁμοίως δ' οὐδ' ὄρνις, οὐδ' ἰχθὺς
οὐδείς. αἴτιον δὲ τὸ μηδὲ γάλα ἔχειν τούτων

---

[1] hinc usque ad 695 a 22 varia codd. ; text. vulg. exhibui.

up : this, and biting, is the business of the teeth).
In crabs, then, and in other creatures which, because
their mouth does not come into action while under
water, can take their time about seizing their food,
the labour is divided : they seize their food with
their hands or feet, and cut it up and bite it with
the mouth. For the crocodile, however, by making
the jaws move as I have described, Nature has
constructed a mouth which can be used for both
these purposes.

All these animals have also a neck ; this is because
they have a lung and there is a long windpipe through
which they admit the breath to it.

Since the neck is the name given to the part of
the body between the head and the shoulders, the
serpent would appear to be the very last of these
creatures to possess one : at any rate, if the neck is
to be defined by the limits mentioned above, he has
merely something analogous to a neck. Compared
with kindred animals, serpents have this peculiarity :
they can turn their heads backwards while the rest of
the body remains still. The reason is that their body
(like an insect's) can roll up; the vertebrae are cartilaginous and flexible. This, then, is the *necessary* cause
why they have this ability; but it serves a *good* purpose
too, for it enables them to guard against attacks from
the rear, and with their long bodies devoid of feet
they are ill adapted for turning themselves round to
keep watch over the rear. To be able to raise the
head and yet unable to turn it round would be useless.
These animals have also a part which is a counterpart to the breast. But they have no mammae either
here or elsewhere ; nor have any of the birds or fishes.
This is because the mammae are receptacles, vessels,

692 a

μηθέν· ὁ δὲ μαστὸς ὑποδοχὴ καὶ ὥσπερ ἀγγεῖόν
ἐστι γάλακτος. γάλα δ' οὐκ ἔχει οὔτε ταῦτα οὔτ'
ἄλλο οὐδὲν τῶν μὴ ζωοτοκούντων ἐν αὑτοῖς, διότι
ᾠοτοκοῦσιν, ἐν δὲ τῷ ᾠῷ ἡ τροφὴ ἐγγίνεται ἐν
τοῖς ζωοτόκοις γαλακτώδης ὑπάρχουσα. σαφέ-
15 στερον δὲ περὶ αὐτῶν λεχθήσεται ἐν τοῖς περὶ
γενέσεως. περὶ δὲ τῆς τῶν σκελῶν[1] κάμψεως ἐν
τοῖς περὶ πορείας πρότερον ἐπέσκεπται κοινῇ περὶ
πάντων.[2]

Ἔχουσι δὲ καὶ κέρκον τὰ τοιαῦτα, τὰ μὲν μείζω
τὰ δ' ἐλάττω, ὑπὲρ οὗ τὴν αἰτίαν καθόλου πρότερον
εἰρήκαμεν.

20 Ἰσχνότατος δ' ὁ χαμαιλέων τῶν ᾠοτόκων καὶ
πεζῶν ἐστίν· ὀλιγαιμότατον γάρ ἐστι πάντων. ταὐτὸ
δ' αἴτιον τοῦ τῆς ψυχῆς ἤθους ἐστὶ τοῦ ζῴου·[3] πολύ-
μορφον γὰρ γίνεται διὰ τὸν φόβον, ὁ δὲ φόβος
κατάψυξις δι' ὀλιγαιμότητά ἐστι καὶ ἔνδειαν θερμότητος.

692 b Περὶ μὲν οὖν τῶν ἐναίμων ζῴων τῶν τε ἀπόδων
καὶ τετραπόδων, ὅσα μόρια τὰ ἐκτὸς ἔχει καὶ διὰ
τίνας αἰτίας, εἴρηται σχεδόν.

XII. Ἐν δὲ τοῖς ὄρνισιν ἡ πρὸς ἄλληλα διαφορὰ
ἐν τῇ τῶν μορίων ἐστὶν ὑπεροχῇ καὶ ἐλλείψει καὶ
5 κατὰ τὸ μᾶλλον καὶ ἧττον. εἰσὶ γὰρ αὐτῶν οἱ μὲν
μακροσκελεῖς οἱ δὲ βραχυσκελεῖς, καὶ τὴν γλῶτταν
οἱ μὲν πλατεῖαν ἔχουσιν οἱ δὲ στενήν· ὁμοίως δὲ
καὶ ἐπὶ τῶν ἄλλων μορίων. ἰδίᾳ δὲ μόρια ὀλίγα

---

[1] σκελῶν PZ, Ogle : καμπύλων σκελῶν Y : καμπύλων vulg.
[2] περὶ δὲ . . . πάντων fortasse secludenda.
[3] correxit Peck, cf. 667 a 11 seqq.; τούτου δ' αἴτιον τὸ ἦθος
τοῦ ζῴου τὸ τῆς ψυχῆς vulg. : αἴτιον δὲ τὸ τῆς ψυχῆς ἦθός ἐστιν
αὐτοῦ PSUZ : sed fortasse haec verba secludenda.

as it were, for the milk, and none of these creatures has any milk. Neither has any of the other animals that are not internally viviparous ; the reason is that as they produce eggs the milky nutriment which they contain goes into these eggs. A more detailed account of these matters will be given in the treatise on *Generation.*[a] With regard to the way in which they bend their legs, a general account, including all animals, has already been given in the treatise on the *Locomotion of Animals.*[b]

These creatures have a tail, some a large one, some a small one. We have already given the reason for this as generally applicable.[c]

Among the oviparous land-animals, the chameleon has the least flesh on him ; this is because he has least blood, and the same reason is at the root of the animal's habit of soul—he is subject to fear (to which his many changes in appearance are due), and fear is a process of cooling produced through scantiness of blood and insufficiency of heat.[d]

This fairly concludes our account of the external parts of the blooded animals both footless and four-footed, and of the reasons thereof.

XII. We now pass on to Birds. As among them- (ii.) Birds. selves, they differ in their parts in respect of the more and less, and excess and defect[e]—*e.g.*, some of them have long legs, some short ones : some have a broad tongue, some a narrow one ; and similarly with the other parts. Thus, as among themselves

---

[a] At 752 b 16 ff.
[b] At 712 a 1 ff. See also below, 693 b 3, and additional note on that passage, p. 433.
[c] At 689 b 1 ff.
[d] Compare the passages at 650 b 27 and 667 a 11 ff.
[e] See 644 a 19, and introductory note on p. 19.

692 b

διαφέροντα ἔχουσιν ἀλλήλων· πρὸς δὲ τὰ ἄλλα ζῷα
καὶ τῇ μορφῇ τῶν μορίων διαφέρουσιν. πτερωτοὶ
10 μὲν οὖν ἅπαντές εἰσιν, καὶ τοῦτ' ἴδιον ἔχουσι τῶν
ἄλλων. τὰ γὰρ μόρια τῶν ζῴων τὰ μὲν τριχωτά
ἐστι τὰ δὲ φολιδωτὰ τὰ δὲ λεπιδωτά, οἱ δ' ὄρνιθες
πτερωτοί. καὶ τὸ πτερὸν σχιστὸν καὶ οὐχ ὅμοιον
τῷ εἴδει τοῖς ὁλοπτέροις· τῶν μὲν γὰρ ἄσχιστον
τῶν δὲ σχιστόν ἐστι, καὶ τὸ μὲν ἄκαυλον, τὸ δ'
15 ἔχει καυλόν. ἔχουσι δὲ καὶ ἐν τῇ κεφαλῇ περιττὴν
καὶ ἴδιον τὴν τοῦ ῥύγχους φύσιν πρὸς τἆλλα· τοῖς
μὲν γὰρ ἐλέφασιν ὁ μυκτὴρ ἀντὶ χειρῶν, τῶν δ'
ἐντόμων ἐνίοις ἡ γλῶττα ἀντὶ στόματος, τούτοις
δ' ἀντὶ ὀδόντων καὶ χειλῶν τὸ ῥύγχος ὄστινον ὄν.[1]
περὶ δὲ τῶν αἰσθητηρίων εἴρηται πρότερον.

20 Αὐχένα δ' ἔχει τεταμένον τῇ φύσει, καὶ διὰ τὴν
αὐτὴν αἰτίαν ἥνπερ καὶ τἆλλα· καὶ τοῦτον τὰ μὲν
βραχὺν τὰ δὲ μακρόν, καὶ σχεδὸν ἀκόλουθον τοῖς
σκέλεσι τὰ πλεῖστα. τὰ μὲν γὰρ μακροσκελῆ
μακρὸν τὰ δὲ βραχυσκελῆ βραχὺν ἔχει τὸν αὐχένα,
χωρὶς τῶν στεγανοπόδων· τὰ μὲν γὰρ εἰ εἶχε βρα-
693 a χὺν ἐπὶ σκέλεσι μακροῖς, οὐκ ἂν ὑπηρέτει αὐτοῖς ὁ
αὐχὴν πρὸς τὴν ἀπὸ τῆς γῆς νομήν, τοῖς δ' εἰ
μακρὸς ἦν ἐπὶ βραχέσιν. ἔτι δὲ[2] τοῖς κρεωφάγοις
αὐτῶν ὑπεναντίον ἂν ἦν[3] τὸ μῆκος πρὸς τὸν βίον·
5 ὁ γὰρ μακρὸς αὐχὴν ἀσθενής, τοῖς δ' ὁ βίος ἐκ
τοῦ κρατεῖν ἐστιν. διόπερ οὐδὲν τῶν γαμψωνύχων
μακρὸν ἔχει τὸν αὐχένα. τὰ δὲ στεγανόποδα καὶ
⟨τὰ⟩[4] διῃρημένους μὲν ἔχοντα τοὺς πόδας σεσιμω-

---

[1] ὄν Y, Ogle : om. vulg.

402

they have few parts which differ from one to another. But as compared with other animals, they differ in respect of the form of their parts. One peculiarity of the birds is that they all have feathers, whereas in other animals the parts are covered with hair, or scales, or horny plates. A bird's feather is split, and therefore different in form from the wing of certain insects, which is undivided ; as well as having a shaft, whereas the insects have none. Another peculiarity of birds is the beak, an extraordinary appendage to the head. It is made of bone, and serves them instead of teeth and lips, just as the elephant's trunk takes the place of hands, and the tongue of certain insects replaces a mouth. We have spoken already of the sense-organs.[a]

Birds have a neck which sticks up, and for the same reason that other creatures have one. Some have a long neck, some a short one : in most of them it corresponds in length fairly closely to the legs, so that the long-legged birds have a long neck and the short-legged birds a short neck (web-footed birds excepted.) What assistance in getting food out of the ground would a short neck be to a bird on long legs, or a long neck to a bird on short legs ? Furthermore, the carnivorous birds would find a long neck a real disadvantage in their daily life. These birds depend for their livelihood on superior strength, and length of neck means lack of strength ; so no crook-taloned bird has a long neck. Web-footed birds, however, together with others in the same class whose

[a] In Book II. chh. 12 ff.

---

[2] δὲ Langkavel : γε Yb : om. vulg.
[3] ἂν ἦν PYb, Ogle : om. vulg.
[4] ⟨τὰ⟩ Ogle.

693 a

μένους δὲ καὶ[1] ἐν τῷ αὐτῷ γένει ὄντα τοῖς στεγανό-
ποσι, τὸν μὲν αὐχένα μακρὸν ἔχουσιν (χρήσιμος
γὰρ τοιοῦτος ὢν πρὸς τὴν τροφὴν τὴν ἐκ τοῦ
10 ὑγροῦ), τὰ δὲ σκέλη πρὸς τὴν νεῦσιν βραχέα.

Διαφορὰν δ᾽ ἔχει καὶ τὰ ῥύγχη κατὰ τοὺς βίους.
τὰ μὲν γὰρ εὐθὺ ἔχει τὰ δὲ γαμψόν, εὐθὺ μὲν ὅσα
τροφῆς ἕνεκεν, γαμψὸν δὲ τὰ ὠμοφάγα· χρήσιμον
γὰρ πρὸς τὸ κρατεῖν τὸ τοιοῦτον, τὴν δὲ τροφὴν
ἀναγκαῖον ἀπὸ ζώων πορίζεσθαι, καὶ τὰ πολλὰ
15 βιαζομένοις. ὅσων δ᾽ ἕλειος ὁ βίος καὶ ποοφάγος,
πλατὺ τὸ ῥύγχος ἔχουσιν· πρός τε γὰρ τὴν ὄρυξιν
χρήσιμον τὸ τοιοῦτον καὶ πρὸς τὴν τῆς τροφῆς
σπάσιν καὶ κουράν. ἔνια δὲ καὶ μακρὸν ἔχει τὸ
ῥύγχος τῶν τοιούτων, ὥσπερ καὶ τὸν αὐχένα, διὰ
τὸ λαμβάνειν τὴν τροφὴν ἐκ τοῦ βάθους. καὶ τὰ
πολλὰ τῶν τοιούτων καὶ τῶν στεγανοπόδων ἢ
20 ἁπλῶς ἢ κατὰ[2] μόριον[3] θηρεύοντα ζῇ τῶν ἐν τῷ
ὑγρῷ ἔνια ζῳδαρίων· καὶ γίνεται τοῖς τοιούτοις ὁ
μὲν αὐχὴν καθάπερ ἁλιευταῖς ὁ[4] κάλαμος, τὸ δὲ
ῥύγχος οἷον ἡ[5] ὁρμιὰ καὶ τὸ ἄγκιστρον.

Τὰ δὲ πρανῆ τοῦ σώματος καὶ τὰ ὕπτια, καὶ τὰ
τοῦ καλουμένου θώρακος ἐπὶ τῶν τετραπόδων,
25 ὁλοφυὴς ὁ τόπος ἐπὶ τῶν ὀρνίθων ἐστίν· καὶ ἔχουσιν
ἀπηρτημένας ἀντὶ τῶν βραχιόνων καὶ τῶν σκελῶν
693 b τῶν προσθίων[6] τὰς πτέρυγας, ἴδιόν τι μόριον,
διόπερ ἀντὶ ὠμοπλάτης τὰ τελευταῖα ἐπὶ τοῦ νώτου
τῶν πτερύγων ἔχουσιν.

Σκέλη δὲ καθάπερ ἄνθρωπος δύο, κεκαμμένα

[1] καὶ Υ*b*, Ogle : ὡς vulg.
[2] κατὰ Υ, Ogle : κατὰ τὸ vulg.
[3] post μόριον habet ταὐτὸ vulg. : ταυτὰ S : ταῦτα P : τούτοις
coni. Ogle.

feet though divided into toes yet are fashioned like a snub-nose [a]—these have long necks, because a long neck is useful to them for getting food out of the water. Their feet, on the contrary, are short so that they can swim.

Birds' beaks also differ according to their different habits of life. Some beaks are straight, some curved ; straight if they are used simply for feeding, curved if the bird eats raw meat, because a curved beak is useful for overpowering their prey, and such birds have to get their food from animals, most often by force. Those whose life is spent in swamps and are herbivorous have broad beaks, which are useful for digging and pulling up their food and for cropping plants. Some of them, however, have a long beak and a long neck as well, because they get their food from some depth. Practically all these birds and the completely or partially web-footed ones live by preying upon certain of the tiny water-animals, and their neck is to these birds what his fishing-rod is to an angler, while their beak is like a line and hook.

The under and the upper sides of the body (*i.e.* of what is called the trunk in quadrupeds) are in birds one uninterrupted whole. Instead of arms and fore-legs they have wings attached to this part (wings are another peculiarity), and hence, instead of having the shoulder-blade on their back they have the ends of the wings there.

Birds, like men, have two legs, which are bent in-

---

[a] According to Ogle, this means that the main stem of the toe corresponds to the ridge of the nose, and the lobes on either side of it to the flattened nostrils.

---

[4] ἁλιευταῖς ὁ PQSU : ἁλιευτικὸς ὁ Yb : ἁλιευτικὸς Z, vulg.

[5] ἡ Yb : om. vulg.

[6] sic Yb, Ogle : ἄπηρτ. γὰρ ἀντὶ et mox ἔχουσι post προσθίων vulg.

καθάπερ τὰ τετράποδα εἴσω, καὶ οὐχ ὥσπερ ἄνθρω-
5 πος ἔξω· τὰς δὲ πτέρυγας, ὡς τὰ πρόσθια σκέλη
τῶν τετραπόδων, ἐπὶ τὸ περιφερές. δίπουν δ' ἐξ
ἀνάγκης ἐστίν· τῶν γὰρ ἐναίμων ἡ τοῦ ὄρνιθος
οὐσία, ἅμα δὲ καὶ πτερυγωτός, τὰ δ' ἔναιμα οὐ
κινεῖται πλείοσιν ἢ τέτταρσι σημείοις. τὰ μὲν οὖν
ἀπηρτημένα μόρια τέτταρα, ὥσπερ τοῖς ἄλλοις
τοῖς πεζοῖς καὶ τοῖς πορευτικοῖς, ἔστι καὶ τοῖς
10 ὄρνισιν· ἀλλὰ τοῖς μὲν βραχίονες καὶ σκέλη, τοῖς δὲ
τετράποσι[1] σκέλη τέτταρα ὑπάρχει, τοῖς δ' ὄρνισιν
ἀντὶ τῶν προσθίων σκελῶν ἢ βραχιόνων πτέρυγες
τὸ ἴδιόν ἐστιν· κατὰ ταύτας γὰρ τονικοί[2] εἰσι, τῷ
δ' ὄρνιθι ἐν τῇ οὐσίᾳ τὸ πτητικόν ἐστιν. ὥστε
λείπεται αὐτοῖς ἐξ ἀνάγκης δίποσιν εἶναι· οὕτω γὰρ
15 τέτταρσι σημείοις κινήσονται μετὰ τῶν πτερύγων.

Στῆθος δ' ἔχουσιν ἅπαντες ὀξὺ καὶ σαρκῶδες,
ὀξὺ μὲν πρὸς τὴν πτῆσιν (τὰ γὰρ πλατέα πολὺν
ἀέρα ὠθοῦντα δυσκίνητά ἐστι), σαρκῶδες δέ, διότι
τὸ ὀξὺ ἀσθενὲς μὴ πολλὴν ἔχον σκέπην.

Ὑπὸ δὲ τὸ στῆθος κοιλία μέχρι πρὸς τὴν ἔξοδον
20 τοῦ περιττώματος καὶ τὴν τῶν σκελῶν καμπήν,
καθάπερ τοῖς τετράποσι καὶ τοῖς ἀνθρώποις. μεταξὺ μὲν οὖν τῶν πτερύγων καὶ τῶν σκελῶν ταῦτα
τὰ μόριά ἐστιν.

Ὀμφαλὸν δ' ἐν μὲν τῇ γενέσει ἅπαντα ἔχει

---

[1] sic PY*b*, Ogle : σκέλη, τοῖς δὲ τετρ. om. vulg.
[2] πτητικοί conieci ; idem Th. (*volatiles* Gaza).

[a] For an explanation of Aristotle's terminology on this
subject see additional note on p. 433.
[b] The chief difficulty in translating this passage is due to
the word τονικοί, a jargon-adjective in -ικός, which seems to
have been suggested to Aristotle's mind by the similar adjec-

wards as in the quadrupeds, not outwards as in man.[a]
The wings are bent with the convex side outwards,
like the forelegs of quadrupeds. It is inevitable that
a bird should have two feet, for (a) it belongs essentially to the blooded creatures and (b) it is winged,
and (c) four is the greatest number of motion-
points which a blooded creature can have. So there
are four parts (or limbs) attached to a bird's body,
and this corresponds exactly with the other blooded
creatures, viz. those that live and move upon the
ground. The only difference is that whereas the
latter have two arms and two legs (or, if they are
quadrupeds, four legs), the peculiarity of birds is
that they have wings instead of arms (or forelegs).
As its very essence includes the power to fly, a
bird must have something which it can stretch out,
and wings provide this.[b] So it remains that of necessity a bird shall have two feet: these with the two
wings bring up the number of its motion-points
to four.

All birds have a sharp-edged, fleshy breast:
sharp-edged, for flying (a wide surface displaces so
much air that it impedes its own motion); fleshy,
because a sharp-edged thing is weak unless it has
a good covering.

Below the breast is the stomach, which extends (as
in the quadrupeds and in man) as far as the residual
vent and the point where the legs join the body.

Those are the parts, then, which have their situation
between the wings and the legs.

Birds, in common with all animals which are pro-

tive πτητικόν in the next line. Literally, the passage reads:
" for it is at these [viz. the wings] that birds are stretchable ;
and flight-ability is included in the essence of a bird."

693 b

ὅσαπερ ζῳοτοκεῖται ἢ ᾠοτοκεῖται, τῶν δ' ὀρνίθων
αὐξηθέντων ἄδηλος. ἡ δ' αἰτία δήλη ἐν τοῖς περὶ
25 γένεσιν· εἰς γὰρ τὸ ἔντερον ἡ σύμφυσις γίνεται, καὶ
οὐχ ὥσπερ τοῖς ζῳοτόκοις τῶν φλεβῶν τι μόριόν
ἐστιν.

Ἔτι τῶν ὀρνίθων οἱ μὲν πτητικοὶ καὶ τὰς πτέρυγας
694 a μεγάλας ἔχουσι καὶ ἰσχυράς, οἷον οἱ γαμψώνυχες
καὶ ὠμοφάγοι· ἀνάγκη γὰρ πτητικοὺς[1] εἶναι διὰ τὸν
βίον, ὥσθ' ἕνεκα τούτου καὶ πλῆθος ἔχουσι πτερῶν
καὶ τὰς πτέρυγας μεγάλας. ἔστι δ' οὐ μόνον τὰ
5 γαμψώνυχα ἀλλὰ καὶ ἄλλα γένη ὀρνίθων πτητικά,
ὅσοις ἡ σωτηρία ἐν τῇ ταχυτῆτι τῆς πτήσεως ἢ
ἐκτοπιστικὸς ὁ βίος. ἔνια δ' οὐ πτητικὰ τῶν
ὀρνίθων ἐστὶν ἀλλὰ βαρέα, οἷς ὁ βίος ἐπίγειος καὶ
ἔστι καρποφάγα ἢ πλωτὰ καὶ περὶ ὕδωρ βιοτεύου-
σιν. ἔστι δὲ τὰ μὲν τῶν γαμψωνύχων σώματα
μικρὰ ἄνευ[2] τῶν πτερύγων διὰ τὸ εἰς ταύτας[3] ἀνα-
λίσκεσθαι τὴν τροφὴν ⟨καὶ⟩[4] εἰς τὰ ὅπλα καὶ τὴν
10 βοήθειαν· τοῖς δὲ μὴ πτητικοῖς τοὐναντίον τὰ σώματα ὀγκώδη, διὸ βαρέα ἐστίν. ἔχουσι δ' ἔνιοι
τῶν βαρέων βοήθειαν ἀντὶ τῶν πτερύγων τὰ καλούμενα[5] πλῆκτρα ἐπὶ τοῖς σκέλεσιν. ἅμα δ' οἱ αὐτοὶ
οὐ γίνονται πλῆκτρα ἔχοντες καὶ γαμψώνυχες·
15 αἴτιον δ' ὅτι οὐδὲν ἡ φύσις ποιεῖ περίεργον. ἔστι
δὲ τοῖς μὲν γαμψώνυχοις καὶ πτητικοῖς ἄχρηστα τὰ

---

[1] πτητικοὺς P, Rackham : πτητικὰ Yb : πτητικοῖς Z, vulg.
[2] post ἄνευ habent τῶν πτερῶν καὶ Yb.
[3] εἰς ταύτας QSUZ : ἐνταῦθα vulg.
[4] ⟨καὶ⟩ Ogle.　　　　　[5] desinit Z.

---

[a] This passage must be supplemented by reference to others
(such as *De gen. an.* 753 b 20 ff., and *Hist. An.* 561 b), in which
Aristotle speaks of *two* umbilici or umbilical cords—*i.e.* he
recognized the allantois as well as the umbilical vesicle. He

duced alive or out of eggs, have an umbilicus while they are developing, but when they are more fully grown it ceases to be visible. The reason for this is clear from what happens during their development : the umbilical cord grows on to the intestine and unites with it, and does not form a part of the system of blood-vessels, as it does in the Vivipara.[a]

The good fliers have big strong wings, *e.g.* the birds which have crooked talons and feed on raw meat : these must be good fliers owing to their habits of life, and so they have an abundance of feathers and big wings. But there are other sorts of birds which are good fliers beside these : birds whose safety lies in their speed of flight ; and migrants. Some birds are poor fliers : heavy birds, which spend their time on the ground and feed on fruits ; or birds that live on and around the water. The crook-taloned birds, leaving out of account their wings, have small bodies, because the nutriment is used up to produce the wings and weapons of offence and defensive armour. The poor fliers, on the contrary, have bulky, and therefore heavy, bodies. Some of these instead of wings have as a means of defence " spurs " on their legs. The same bird never possesses both spurs and talons, and the reason is that Nature never makes anything that is superfluous or needless. Spurs are of no use to a

states that in the bird's egg, as the embryo grows, the allantois (the " second umbilicus ") collapses first and then the " first umbilicus " (*De gen. an.* 754 a 9). Actually the reverse order is the correct one, but the interval is comparatively short. The umbilical vesicle in mammals, which shrivels very early in the process of development, escaped the notice of Aristotle, who supposed their allantois to be comparable to the umbilical vesicle of reptiles and birds. The umbilical vesicle of mammals was discovered by Needham in 1667. (See Ogle's note *ad loc.*)

694 a

πλῆκτρα· χρήσιμα γάρ ἐστιν ἐν ταῖς πεζαῖς μάχαις,
διὸ ὑπάρχει ἐνίοις τῶν βαρέων· τούτοις δ' οὐ
μόνον ἄχρηστοι ἀλλὰ καὶ βλαβεροὶ οἱ γαμψοὶ ὄνυχες
τῷ ἐμπήγνυσθαι ὑπεναντίοι πρὸς τὴν πορείαν ὄντες.
20 διὸ καὶ τὰ γαμψώνυχα πάντα φαύλως πορεύεται
καὶ ἐπὶ πέτραις οὐ καθιζάνουσιν· ὑπεναντία γὰρ
αὐτοῖς πρὸς ἀμφότερα ἡ τῶν ὀνύχων φύσις.

Ἐξ ἀνάγκης δὲ τοῦτο περὶ τὴν γένεσιν συμβέβηκεν. τὸ γὰρ γεῶδες ἐν τῷ σώματι ἐξορμώμενον[1] χρήσιμα μόρια γίνεται πρὸς τὴν ἀλκήν· ἄνω
μὲν ῥυὲν ῥύγχους ἐποίησε σκληρότητα ἢ μέγεθος,
25 ἂν δὲ κάτω ῥυῇ, πλῆκτρα ἐν τοῖς σκέλεσιν ἢ ἐπὶ
τῶν ποδῶν ὀνύχων μέγεθος καὶ ἰσχύν. ἅμα δ'
ἄλλοθι καὶ ἄλλοθι ἔκαστα τούτων οὐ ποιεῖ· δια-
σπωμένη γὰρ ἀσθενὴς γίνεται ἡ φύσις τούτου τοῦ
περιττώματος. τοῖς δὲ σκελῶν κατασκευάζει μῆ-
694 b κος. ἐνίοις δ' ἀντὶ τούτων συμπληροῖ τὸ μεταξὺ
τῶν ποδῶν· καὶ διὰ τοῦτο ἀναγκαίως οἱ πλωτοὶ
τῶν ὀρνίθων οἱ μὲν ἁπλῶς εἰσὶ στεγανόποδες, οἱ δὲ
διῃρημένην μὲν ἔχουσι τὴν καθ' ἔκαστα τῶν δακτύ-
5 λων φύσιν, πρὸς ἑκάστῳ δ' αὐτῶν προσπέφυκεν
οἷον πλάτη καθ' ὅλον συνεχής.

Ἐξ ἀνάγκης μὲν οὖν ταῦτα συμβαίνει διὰ ταύτας
τὰς αἰτίας· ὡς δὲ διὰ τὸ βέλτιον ἔχουσι τοιούτους
τοὺς πόδας τοῦ βίου χάριν, ἵνα ζῶντες ἐν ὑγρῷ καὶ
τῶν πτερύγων[2] ἀχρείων ὄντων τοὺς πόδας χρησίμους ἔχωσι πρὸς τὴν νεῦσιν. γίνονται γὰρ ὥσπερ

---

[1] ἐξορμώμενον Peck : καὶ ἔξορμον ἐκ τούτου τὰ Yb : ἔξω ῥυὲν
Langkavel ; fortasse ἐξορμᾶται καὶ ἐκ τούτου τὰ.
[2] πτερύγων Yb, Ogle : πτερῶν vulg.

410

bird that has talons and can fly well : spurs are useful for fights on the ground, and that is why certain of the heavy birds possess them, while talons would not be merely useless to them but a real disadvantage[a] : they would stick in the ground and impede the birds when walking. And in fact all crook-taloned birds do walk badly, and they never perch upon rocks ; in both instances the nature of their claws is the impediment.[a]

This state of affairs is the *necessary* result of the process of their development. There is earthy substance in the bird's body which courses along and issues out and turns into parts that are useful for weapons of offence. When it courses upwards it produces a good hard beak, or a large one ; if it courses downwards it produces spurs on the legs or makes the claws on the feet large and strong. But it does not produce spurs and large claws simultaneously, for this residual substance would be weakened if it were scattered about. Again, sometimes this substance makes the legs long ; and in some birds, instead of that, it fills in the spaces between the toes. Thus it is *of necessity* that waterbirds either are web-footed, simply, or (if they have separate toes) they have a continuous fan or blade, as it were, running the whole length of each toe and of a piece with it.

From the reasons just stated it is clear that feet of this sort are the result of *necessity*, it is true ; but they conduce to a *good* end and are meant to assist the birds in their daily life, for these birds live in the water, and while their wings are useless to them, these feet are useful and help them to swim. They

[a] See above, note on 648 a 16.

411

694 b

10 αἱ κῶπαι τοῖς πλέουσι καὶ¹ τὰ πτερύγια τοῖς ἰχθύ-
σιν· διὸ καὶ ἐὰν τῶν μὲν τὰ πτερύγια σφαλῇ, τῶν
δὲ τὸ μεταξὺ τῶν ποδῶν, οὐκέτι νέουσιν.

Ἔνιοι δὲ μακροσκελεῖς τῶν ὀρνίθων εἰσίν. αἴτιον
δ' ὅτι ὁ βίος τῶν τοιούτων ἕλειος· τὰ δ' ὄργανα
πρὸς τὸ ἔργον ἡ φύσις ποιεῖ, ἀλλ' οὐ τὸ ἔργον πρὸς
15 τὰ ὄργανα. διὰ μὲν οὖν τὸ μὴ πλωτὰ εἶναι οὐ
στεγανόποδά ἐστι, διὰ δὲ τὸ ἐν ὑπείκοντι εἶναι τὸν
βίον μακροσκελῆ καὶ μακροδάκτυλα, καὶ τὰς καμπὰς ἔχουσι πλείους ἐν τοῖς δακτύλοις οἱ πολλοὶ
αὐτῶν. ἐπεὶ δ' οὐ πτητικὰ μέν, ἐκ τῆς δ' αὐτῆς
ὕλης ἐστὶ πάντα, ἡ εἰς τὸ οὐροπύγιον αὐτοῖς τροφὴ
20 εἰς τὰ σκέλη καταναλισκομένη ταῦτα ηὔξησεν. διὸ
καὶ ἐν τῇ πτήσει ἀντ' οὐροπυγίου χρῶνται αὐτοῖς·
πέτονται γὰρ ἀποτείνοντες εἰς τὸ ὄπισθεν· οὕτω γὰρ
αὐτοῖς χρήσιμα τὰ σκέλη, ἄλλως δ' ἐμποδίζοιεν ἄν.

Τὰ δὲ βραχυσκελῆ ⟨τὰ⟩ σκέλη² πρὸς τῇ γαστρὶ
ἔχοντα πέτονται· τοῖς μὲν γὰρ αὐτῶν οὐκ ἐμποδί-
25 ζουσιν οἱ πόδες οὕτω, τοῖς δὲ γαμψώνυξι καὶ πρὸ
ἔργου εἰσὶ πρὸς τὴν ἁρπαγήν.

Τῶν δ' ἐχόντων ὀρνίθων τὸν αὐχένα μακρὸν οἱ
μὲν παχύτερον ἔχοντες πέτονται ἐκτεταμένῳ τῷ
αὐχένι, οἱ δὲ λεπτότερον³ συγκεκαμμένῳ· ἐπιπετο-
μένοις γὰρ διὰ τὴν σκέπην ἧττον εὔθρυπτόν ἐστιν.
695 a ἰσχίον δ' ἔχουσι μὲν οἱ ὄρνιθες πάντες ᾗ οὐκ ἂν
δόξαιεν ἔχειν, ἀλλὰ δύο μηροὺς διὰ τὸ τοῦ ἰσχίου
μῆκος· ὑποτέταται γὰρ μέχρι μέσης τῆς γαστρός.
αἴτιον δ' ὅτι δίπουν ἐστὶ τοῦτο τὸ ζῷον οὐκ ὀρθὸν

¹ καὶ Yb, Ogle : om. vulg.
² τὰ δὲ βραχυσκελῆ PYb ; correxi : ἔνια δὲ βραχέα ⟨τὰ Lang-
kavel⟩ σκέλη vulg.
³ λεπτότερον Peck : λεπτὸν καὶ μακρὸν vulg. : [καὶ μακρὸν]
secl. Rackham.

are like oars to a sailor or fins to a fish. A fish that
has lost its fins can no longer swim ; nor can a bird
whose webs have been destroyed.

Some birds have long legs, owing to their living in
marshes ; for Nature makes the organs to suit the
work they have to do, not the work to suit the organ.
And these birds have no webs in their feet because
they are not water birds, but because they live on
ground that gives under them they have long legs
and long toes, and most of them have additional joints
in their toes. Furthermore, though these birds are
not great fliers, they are composed of the same materials as the rest, and thus the nutriment which in the
others goes to produce the tail feathers, in these is
used up on the legs and makes them grow longer, and
when in flight these birds stretch them out behind
and use them in place of the missing tail feathers :
placed thus, the legs are useful to them ; otherwise
they would get in the way.

Short-legged birds keep their legs up against the
belly while they are flying, because if the feet are
there they are out of the way ; the crook-taloned
birds do it for an additional reason : the feet are
convenient for seizing prey.

When a bird has a long neck, this is either thick and
is held stretched out during flight ; or it is slender
and is bent up during flight, because being protected
in this way it is less easily broken if the bird flies into
anything. All birds have an ischium, but in such
a way that they would not appear to have one ; it is
so long that it reaches to the middle of the belly and
looks more like a second thigh-bone. The reason for
this is that a bird, although a biped, does not stand

695 a

⟨ὄν⟩,[1] ὡς εἴ γε εἶχε, καθάπερ ἐν τοῖς ἀνθρώποις ἢ
τοῖς τετράποσιν, ἀπὸ τῆς ἕδρας βραχὺ τὸ ἰσχίον
καὶ τὸ σκέλος εὐθὺς ἐχόμενον, ἠδυνάτει ἂν ὅλως[2]
ἑστάναι. ὁ μὲν γὰρ ἄνθρωπος ὀρθόν, τοῖς δὲ τετράποσι πρὸς τὸ βάρος σκέλη ἐμπρόσθια ὑπερήρει-
σται. οἱ δ' ὄρνιθες οὐκ ὀρθοὶ μὲν διὰ τὸ νανώδεις
εἶναι τὴν φύσιν, σκέλη δ' ἐμπρόσθια οὐκ ἔχουσιν διὰ τὸ πτέρυγας ἔχειν[3] ἀντ' αὐτῶν. ἀντὶ δὲ
τούτου μακρὸν ἡ φύσις τὸ ἰσχίον ποιήσασα εἰς
μέσον προσήρεισεν· ἐντεῦθεν δ' ὑπέθηκε τὰ σκέλη,
ὅπως ἰσορρόπου ὄντος τοῦ βάρους ἔνθεν καὶ ἔνθεν
πορεύεσθαι δύνηται καὶ μένειν.[4] δι' ἣν μὲν οὖν
αἰτίαν δίπουν ἐστὶν οὐκ ὀρθὸν ὄν, εἴρηται· τοῦ δ'
ἄσαρκα τὰ σκέλη εἶναι ἡ αὐτὴ αἰτία καὶ ἐπὶ τῶν
τετραπόδων, ὑπὲρ ἧς καὶ πρόσθεν εἴρηται.

Τετραδάκτυλοι δ' εἰσὶ πάντες οἱ ὄρνιθες ὁμοίως οἱ
στεγανόποδες τοῖς σχιζόποσιν (περὶ γὰρ τοῦ στρου-
θοῦ τοῦ Λιβυκοῦ ὕστερον διοριοῦμεν, ὅτι διχηλός,
ἅμα τοῖς λοιποῖς ἐναντιώμασιν οἷς ἔχει πρὸς τὸ τῶν
ὀρνίθων γένος). τούτων δ' οἱ μὲν τρεῖς ἔμπροσθεν, ὁ δ' εἷς ὄπισθεν πρὸς ἀσφάλειαν ἀντὶ πτέρνης·
καὶ τῶν μακροσκελῶν λείπει τοῦτο κατὰ μέγεθος,
οἷον συμβέβηκεν ἐπὶ τῆς κρεκός· πλείους δ' οὐκ
ἔχουσι δακτύλους.[5] ἐπὶ μὲν οὖν τῶν ἄλλων οὕτως
ἢ τῶν δακτύλων ἔχει θέσις, ἡ δ' ἴυγξ δύο μόνον
ἔχει τοὺς ἔμπροσθεν καὶ δύο τοὺς ὄπισθεν[6]· αἴτιον

[1] ⟨ὄν⟩ Rackham, cf. l. 14 infra.
[2] ὅλως PQU, Ogle : ὀρθὸν vulg.
[3] correxi ; ἔχουσιν· διὰ τοῦτο πτέρυγας ἔχουσιν vulg. (πτέρυγας, δὲ altero ἔχουσιν omisso, Y, Ogle, qui post διὰ τοῦτο interpungit).

414

upright; and if it had an ischium which extended only a short way from the fundament and was followed immediately by the leg (as in man and the quadrupeds), it would be unable to stand up at all. Man can stand upright, and quadrupeds have forelegs to support their forward weight; birds, however, neither stand upright (because they are dwarf-like), nor have forelegs (because they have wings instead).[a] By way of compensation, Nature has made the ischium long, reaching to the middle of the body, and has fixed it fast, while beneath it she has placed the legs, so that the weight may be equally distributed on either side and the bird enabled to walk and to stand still. This shows why birds are bipeds although they are unable to stand upright. The reason why their legs are lacking in flesh is the same as for all quadrupeds and has been stated already.[b]

All birds, web-footed or not, have four toes on each foot. (The Libyan ostrich will be dealt with later,[c] and its cloven hoof and other inconsistencies with the tribe of birds will be discussed.) Of these four toes, three are in front, and the fourth is at the back instead of a heel, for stability. In the long-legged birds this toe is deficient in length, as for instance in the Crex. Still, the number of toes does not exceed four. This arrangement of the toes holds good generally, but the wryneck is an exception, for it has only two toes in front and two at the back. This is because

[a] See above, 693 b 3 ff.
[b] See 689 b 10 ff.
[c] At the end of the book.

---

[4] μένειν Yb : μένῃ vulg.
[5] διὰ τὴν στενότητα τοῦ σκέλους add. PYb.
[6] ἔμπροσθεν . . . ὄπισθεν Karsch: ὄπισθεν . . . ἔμπροσθεν vulg.

695 a

25 δ' ὅτι ἧττόν ἐστιν αὐτῆς τὸ σῶμα προπετὲς ἐπὶ τὸ
πρόσθεν ἢ τὸ τῶν ἄλλων.

Ὄρχεις δ' ἔχουσι μὲν πάντες οἱ ὄρνιθες, ἐντὸς
δ' ἔχουσιν· ἡ δ' αἰτία ἐν τοῖς περὶ τὰς γενέσεις
λεχθήσεται τῶν ζῴων.

695 b Τὰ μὲν οὖν τῶν ὀρνίθων μόρια τὸν τρόπον ἔχει
τοῦτον.

XIII. Τὸ δὲ τῶν ἰχθύων γένος ἔτι μᾶλλον κεκολόβωται τῶν ἐκτὸς μορίων. οὔτε γὰρ σκέλη οὔτε
χεῖρας οὔτε πτέρυγας ἔχουσι (εἴρηται δὲ περὶ τού-
5 των ἡ αἰτία πρότερον), ἀλλ' ὅλον ἀπὸ τῆς κεφαλῆς
τὸ κύτος συνεχές ἐστι μέχρι τῆς οὐρᾶς. ταύτην δ'
οὐχ ὁμοίαν ἔχουσι πάντες, ἀλλὰ τὰ μὲν παραπλησίαν,[1] τῶν δὲ πλατέων ἔνια ἀκανθώδη καὶ μακράν·
ἡ ἐκεῖθεν γὰρ αὔξησις γίνεται εἰς τὸ πλάτος, οἷόν
ἐστι νάρκαις καὶ τρυγόσι καὶ εἴ τι τοιοῦτον ἄλλο
10 σέλαχός ἐστιν. τῶν μὲν οὖν τοιούτων ἀκανθῶδες
καὶ μακρὸν τὸ οὐραῖόν ἐστιν, ἐνίων δὲ σαρκῶδες μὲν
βραχὺ δὲ διὰ τὴν αὐτὴν αἰτίαν δι' ἥνπερ ταῖς
νάρκαις· διαφέρει γὰρ οὐδέν, ἢ βραχὺ μὲν σαρκωδέστερον δέ, ἢ μακρὸν μὲν ἀσαρκότερον δ' εἶναι.

Ἐπὶ δὲ τῶν βατράχων τὸ ἐναντίον συμβέβηκεν·
15 διὰ γὰρ τὸ μὴ σαρκῶδες εἶναι τὸ πλάτος αὐτῶν
τὸ ἐμπρόσθιον, ὅσον ἀφήρηται σαρκῶδες, πρὸς τὸ
ὄπισθεν αὐτῶν[2] ἔθηκεν ἡ φύσις καὶ τὴν οὐράν.

Οὐκ ἔχουσι δ' ἀπηρτημένα κῶλα οἱ ἰχθύες διὰ τὸ
νευστικὴν εἶναι τὴν φύσιν αὐτῶν κατὰ τὸν τῆς
οὐσίας λόγον, ἐπεὶ οὔτε περίεργον οὐδὲν οὔτε μάτην

---

[1] μὲν ἄλλα π. P: μὲν ἄμη π. Platt: μὲν παραπλήσια ⟨τοῖς
πτερυγίοις⟩ Ogle, similia voluit Thurot.
[2] αὐτῶν U : αὐτὸ vulg.

---

[a] See *De gen. an.* 714 b 4 ff., 719 b 11.

the weight of its body tends forward less than that of other birds.

All birds have testicles, but they are inside the body. The reason for this will be stated in the treatise on the different methods of generation among animals.[a]

This concludes our description of the parts of Birds. (iii.) Fishes

XIII. In the tribe of Fishes the external parts are still further stunted. Fishes have neither legs, hands, nor wings (the reason has been stated earlier), but the whole trunk has an uninterrupted line from head to tail. Not all fishes' tails are alike ; but the Tail. general run of them have similar tails, though some of the flat-fish have a long, spiny one, because the material for the tail's growth goes into the width of the flat body : this happens in the torpedo-fishes, in the Trygons, and any other Selachians of the same sort. These have long, spiny tails. Others have short, fleshy ones, and for the selfsame reason : it comes to the same thing whether the tail is short and has a good deal of flesh or long with little flesh.

In the fishing-frog[b] the opposite has taken place. Here, the wide, flat part of the body in front is not fleshy ; Nature has taken the fleshy material away from the front and added an equivalent amount at the back—in the tail.

Fishes have no separate limbs attached to the body. (a) This is because Nature never makes anything that is superfluous or needless, and by their essence and constitution[c] fishes are naturally swimmers and so

[b] *Lophius piscatorius*, known as the " goosefish " in U.S.A., erroneously included by Aristotle (*De gen. an.* 754 a 25) with the Selachia, though he observed that it differed in many important points.

[c] *Logos* : see Introduction, pp. 26 f.

417

695 b
20 ἡ φύσις ποιεῖ. ἐπεὶ δ' ἔναιμά ἐστι κατὰ τὴν
οὐσίαν, διὰ μὲν τὸ νευστικὰ εἶναι πτερύγια ἔχει, διὰ
δὲ τὸ μὴ πεζεύειν οὐκ ἔχει πόδας· ἡ γὰρ τῶν ποδῶν
πρόσθεσις πρὸς τὴν ἐπὶ τῷ πεδίῳ κίνησιν χρήσιμός
ἐστιν. ἅμα δὲ πτερύγια τέτταρα καὶ πόδας οὐχ
οἷόν τ' ἔχειν, οὐδ' ἄλλο κῶλον τοιοῦτον οὐδέν·
25 ἔναιμα γάρ. οἱ δὲ κορδύλοι βράγχια ἔχοντες πόδας
ἔχουσιν· πτερύγια γὰρ οὐκ ἔχουσιν, ἀλλὰ τὴν οὐρὰν
μανώδη καὶ πλατεῖαν.

Ἔχουσι δὲ τῶν ἰχθύων ὅσοι μὴ πλατεῖς, καθάπερ
βάτος καὶ τρυγών, τέτταρα πτερύγια, δύο μὲν ἐν
696 a τοῖς πρανέσι, δύο δ' ἐν τοῖς ὑπτίοις· πλείω δὲ
τούτων οὐδείς, ἄναιμοι γὰρ ἂν ἦσαν. τούτων δὲ τὰ
μὲν ἐν τῷ πρανεῖ σχεδὸν πάντες ἔχουσι, τὰ δ' ἐν
τοῖς ὑπτίοις ἔνιοι τῶν μακρῶν καὶ πάχος ἐχόντων
5 οὐκ ἔχουσιν, οἷον ἐγχέλυς καὶ γόγγρος καὶ κεστρέων
τι γένος τὸ ἐν τῇ λίμνῃ τῇ ἐν Σιφαῖς. ὅσα δ' ἐστὶ
μακροφυέστερα καὶ ὀφιώδη μᾶλλον, οἷον σμύραινα,
οὐδὲν ἔχουσι πτερύγιον ἁπλῶς, ἀλλὰ ταῖς καμπαῖς
κινοῦνται, χρώμεναι τῷ ὑγρῷ ὥσπερ οἱ ὄφεις τῇ
γῇ· τὸν αὐτὸν[1] γὰρ οἱ ὄφεις τρόπον[2] νέουσιν ὅνπερ
10 ἐπὶ τῆς γῆς ἕρπουσιν. αἰτία δὲ τοῦ μὴ ἔχειν τοὺς
ὀφιώδεις τῶν ἰχθύων πτερύγια, ἥπερ καὶ τῶν
ὄφεων τοῦ ἄποδας εἶναι. τὸ δ' αἴτιον ἐν τοῖς περὶ
πορείας καὶ κινήσεως τῶν ζῴων εἴρηται. ἢ γὰρ
κακῶς ἂν ἐκινοῦντο, τέτταρσι σημείοις κινούμενα

---

[1] τὸν αὐτὸν Peck: τοῦτον vulg.
[2] οἱ ὄφεις τὸν τρόπον Yb: τὸν deleui: τὸν τρόπον οἱ ὄφεις vulg.

---

[a] The Cordylus was probably the larval form of some triton or newt, such as *Triton alpestris* or *Salamandra atra*, which retains its gills till it is well grown (D'Arcy Thompson).
[b] *i.e.* pectoral.   [c] *i.e.* ventral.

418

need no such limbs. But also (*b*) they are essentially blooded creatures, which means that if they have four fins they cannot have any legs or any other limbs of the sort; so they have the fins because they are swimmers and do not have the feet because they are not walkers (when an animal has feet it has them because they are useful for moving about on land). The Cordylus,[a] however, has feet in addition to its gills, since it has no fins, but only a scraggy flattened-out tail.

Excluding flat-fish (like the Batos and Trygon), fish Fins. have four fins: two on their under and two on their upper surface, never more, for then they would be bloodless animals. Almost all fishes have the two upper [b] fins, but some of the large, thick-bodied fishes lack the under [c] two—as for instance the eel and the conger, and a sort of Cestreus that is found in the lake at Siphae.[d] Fishes that have even longer bodies than these, and are really more like serpents (as the Smyraena[e]), have no fins at all, and move along by bending themselves about: that is, they use the water just as serpents use the ground. And in fact serpents swim in exactly the same way as they creep on the ground. The reason why these serpent-like fishes have no fins and the reason why serpents have no feet are the same, and this has been stated in the treatises on the *Locomotion and Movement of Animals.*[f] (*a*) If they had four motion-points, their movement would be poor, because the fins would

d In Boeotia, on the south coast near Thespiae; now Tipha. Aristotle refers to this Cestreus of Siphae again, *De incessu an.* 708 a 5. *Cf.* also *Hist. An.* 504 b 33.

e Probably *Muraena Helena.*

f See *De incessu an.* 709 b 7; perhaps the other passage which Aristotle has in mind is 690 b 16, in this book.

696 a

(εἴτε γὰρ σύνεγγυς εἶχον τὰ πτερύγια, μόγις ἂν
15 ἐκινοῦντο, εἴτε πόρρω, διὰ τὸ πολὺ μεταξύ)· εἰ
δὲ πλείω τὰ κινητικὰ σημεῖα εἶχον, ἄναιμα ἂν ἦν.
ἡ δ᾽ αὐτὴ αἰτία καὶ ἐπὶ τῶν δύο μόνον ἐχόντων πτερύγια ἰχθύων· ὀφιώδη γάρ ἐστι καὶ εὐ-
μηκέστερα, καὶ χρῆται τῇ κάμψει ἀντὶ τῶν δύο
πτερυγίων. διὸ καὶ ἐν τῷ ξηρῷ ἕρπουσι καὶ ζῶσι
20 πολὺν χρόνον, καὶ τὰ μὲν οὐκ εὐθύ, τὰ δ᾽ οἰκεῖα
τῆς πεζῆς ὄντα φύσεως ἧττον ἀσπαρίζει.

Αὐτῶν δὲ τῶν πτερυγίων τὰ ἐν τοῖς πρανέσιν ἔχει
τὰ δύο ἔχοντα πτερύγια μόνον, ὅσα μὴ κωλύεται
διὰ τὸ πλάτος· τὰ δ᾽ ἔχοντα πρὸς τῇ κεφαλῇ ἔχει
διὰ τὸ μὴ ἔχειν μῆκος ἐν τῷ τόπῳ, ᾧ ἀντὶ τούτων
25 κινήσεται· ἐπὶ γὰρ τὴν οὐρὰν πρόμηκες τὸ τῶν
τοιούτων ἐστὶν ἰχθύων σῶμα. οἱ δὲ βάτοι καὶ τὰ
τοιαῦτα ἀντὶ τῶν πτερυγίων τῷ ἐσχάτῳ πλάτει
νέουσιν. τὰ δ᾽ ἧττον ἔχοντα πλάτος πτερύγια
ἔχουσιν, οἷον ἥ[1] νάρκη καὶ ὁ βάτραχος, τὰ ⟨μὲν⟩[2] ἐν
τῷ πρανεῖ κάτω διὰ τὸ πλάτος τῶν ἄνω, τὰ δ᾽ ἐν
τοῖς ὑπτίοις πρὸς τῇ κεφαλῇ (οὐ γὰρ κωλύει κινεῖ-
30 σθαι τὸ πλάτος)· ἀλλ᾽ ἀντὶ τοῦ ἄνω ἐλάττω ταῦτα
τῶν ἐν τῷ πρανεῖ ἔχει. ἡ δὲ νάρκη πρὸς τῇ οὐρᾷ
ἔχει τὰ δύο πτερύγια· ἀντὶ δὲ τῶν δύο τῷ πλάτει
χρῆται ὡς δυσὶ πτερυγίοις ἑκατέρῳ τῷ ἡμικυκλίῳ.

Περὶ δὲ τῶν ἐν τῇ κεφαλῇ μορίων καὶ αἰσθητηρίων εἴρηται πρότερον.

---

[1] τὰ δ᾽ ἧττον . . . οἷον ἡ P : ἡ δὲ tantum vulg.
[2] ⟨μὲν⟩ Langkavel.

either be very close together, or else a long way apart, and in either case would not move easily. (*b*) On the other hand, if they had more than four motion-points they would be bloodless creatures. The same reason holds good for those fishes that have only two fins. These also are serpent-like and fairly long, and they use their power of bending instead of the two missing fins. And this enables them besides to crawl about and to live a good length of time on dry land ; and it is some while before they begin to gasp ; indeed, those which are akin to the land-animals are affected even less than the others.

Except for those whose width and flatness prevents it, all fishes that have only two fins have the upper[a] ones ; and these fins are by the head, because there is no length of body just there which they could use instead of fins for propulsion—length such as fish of this sort have towards their tail-end. The Batoi and such fishes swim by means of the edge of their flat surface which they use instead of fins. Fish which are not so flat, such as the torpedo-fish and the fishing-frog, possess fins, but they have their upper fins toward their tail-end owing to the flatness of the forepart, and their under fins near the head (since the flatness of the fish does not prevent its motion) ; but the under ones are smaller than the upper ones, to make up for being placed forward. The torpedo-fish has two of his fins by his tail ; and instead of these two he uses the wide piece on each of his semi-circles[b] as though it were a fin.

We have already spoken of the parts in the head and of the sense-organs.

---

[a] *i.e.* pectoral.
[b] *Cf. De incessu an.* 709 b 17.

Ἴδιον δ' ἔχει τὸ τῶν ἰχθύων γένος πρὸς τἆλλα τὰ
696 b ἔναιμα ζῷα τὴν τῶν βραγχίων φύσιν· δι' ἣν δ'
αἰτίαν, εἴρηται ἐν τοῖς περὶ ἀναπνοῆς. καὶ ἔχει δὲ
τὰ ἔχοντα βράγχια τὰ μὲν ἐπικαλύμματα τοῖς
βραγχίοις, τὰ δὲ σελάχη πάντα[1] ἀκάλυπτα. αἴτιον
5 δ' ὅτι οἱ μὲν ἀκανθώδεις εἰσί, τὸ δ' ἐπικάλυμμα
ἀκανθῶδες, τὰ δὲ σελάχη πάντα χονδράκανθα. ἔτι
δ' ἡ κίνησις τῶν μὲν νωθρὰ[2] διὰ τὸ μὴ ἀκανθώδη
εἶναι μηδὲ νευρώδη, τῶν δ' ἀκανθωδῶν ταχεῖα· τοῦ
δ' ἐπικαλύμματος ταχεῖαν δεῖ γίνεσθαι τὴν κίνησιν·
10 ὥσπερ γὰρ πρὸς ἐκπνοὴν ἡ τῶν βραγχίων ἐστὶ
φύσις. διὰ τοῦτο τοῖς σελαχώδεσι καὶ αὐτῶν τῶν
πόρων ἡ συναγωγὴ γίνεται τῶν βραγχίων, καὶ οὐ
δεῖ ἐπικαλύμματος, ὅπως γίνηται ταχεῖα.

Οἱ μὲν οὖν αὐτῶν ἔχουσι πολλὰ βράγχια οἱ δ'
ὀλίγα, καὶ οἱ μὲν διπλᾶ οἱ δ' ἁπλᾶ· τὸ δ' ἔσχατον
ἁπλοῦν οἱ πλεῖστοι. (τὴν δ' ἀκρίβειαν ἐκ τῶν
15 ἀνατομῶν περὶ τούτων καὶ ἐν ταῖς ἱστορίαις ταῖς
περὶ τὰ ζῷα δεῖ θεωρεῖν.) αἴτιον δὲ τοῦ πλήθους
καὶ τῆς ὀλιγότητος τὸ τοῦ ἐν τῇ καρδίᾳ θερμοῦ
πλῆθος καὶ ὀλιγότης· θάττω γὰρ καὶ ἰσχυροτέραν
τὴν κίνησιν δεῖ εἶναι τοῖς πλείω ἔχουσι θερμότητα.
20 τὰ δὲ πλείω καὶ διπλᾶ βράγχια τοιαύτην ἔχει τὴν
φύσιν μᾶλλον τῶν ἁπλῶν καὶ ἐλαττόνων. διὸ καὶ
ἔνια αὐτῶν ἔξω ζῆν δύναται πολὺν χρόνον, τῶν
ἐχόντων ἐλάττω καὶ ἧττον ἐγκρατῆ τὰ βράγχια,
οἷον ἐγχέλυς καὶ ὅσα ὀφιώδη· οὐ γὰρ πολλῆς
δέονται καταψύξεως.

Ἔχει δὲ καὶ περὶ τὸ στόμα διαφοράς. τὰ μὲν
25 γὰρ κατ' ἀντικρὺ ἔχει τὸ στόμα καὶ εἰς τὸ πρόσθεν,

---

[1] (χονδράκανθα γὰρ) post πάντα vulg., om. P.
[2] ἡ κίνησις . . . νωθρὰ Y: αἱ κινήσεις . . . νωθραὶ vulg.

The peculiarity which marks off fishes from the ^{Gills.} other blooded animals is the possession of gills. It has been explained in the treatise on *Respiration* [a] why they have them. All fishes have coverings over their gills, except the Selachia, none of which have them. This is because their bones are cartilaginous, whereas other fishes' bones are of fish-spine, and this is the substance out of which the coverings are made. And again, the Selachia move sluggishly owing to their lack of fish-spine—and of sinews—while the spinous fishes move quickly, and the movement of the covering must be a quick one, for gills are a medium for expiration of a sort. On this account in the selachian group of fishes the passages of the gills can close up by themselves, and no covering is needed to make sure they close quickly.

Now some fish have many gills, some have few; some have double ones, some single. The last one is nearly always a single one. (For precise details consult the Anatomical treatises and the *Researches upon Animals.*[b]) The number of gills depends upon the amount of heat in the heart. The more heat an animal has, the quicker and stronger must be the movement of its gills; and if the gills are numerous and double they are better adapted for this than if they are few and single. And on this account, some fishes (*e.g.* the eels and the serpentine fishes) which need but little cooling, as is shown by their having only a few weakish gills, can live a long time out of water.

Fish differ also with regard to the mouth. Some ^{Mouth.} have their mouth right at the tip, straight in front;

---

[a] At 476 a 1 ff., 480 b 13 ff.
[b] At 504 b 28 ff.

696 b

τὰ δ᾽ ἐν τοῖς ὑπτίοις, οἶον οἵ τε δελφῖνες¹ καὶ τὰ
σελαχώδη· διὸ καὶ ὕπτια στρεφόμενα λαμβάνει τὴν
τροφήν. φαίνεται δ᾽ ἡ φύσις οὐ μόνον σωτηρίας
ἕνεκεν ποιῆσαι τοῦτο τῶν ἄλλων ζῴων (ἐν γὰρ τῇ
στρέψει σῴζεται τἆλλα βραδυνόντων· πάντα γὰρ
30 τὰ τοιαῦτα ζωοφάγα ἐστίν), ἀλλὰ καὶ πρὸς τὸ μὴ
ἀκολουθεῖν τῇ λαιμαργίᾳ τῇ περὶ τὴν τροφήν· ῥᾷον
γὰρ λαμβάνοντα διεφθείρετ᾽ ἂν διὰ τὴν πλήρωσιν
ταχέως. πρὸς δὲ τούτοις περιφερῆ καὶ λεπτὴν
ἔχοντα τὴν τοῦ ῥύγχους φύσιν οὐχ οἷόν τ᾽ εὐ-
διαίρετον ἔχειν.

Ἔτι δὲ καὶ τῶν ἄνω τὸ στόμα ἐχόντων τὰ μὲν
697 a ἀνερρωγὸς ἔχει τὸ στόμα τὰ δὲ μύουρον, ὅσα μὲν
σαρκοφάγα, ἀνερρωγός, ὥσπερ τὰ καρχαρόδοντα,
διὰ τὸ ἐν τῷ στόματι εἶναι τοῖς τοιούτοις τὴν ἰσχύν,
ὅσα δὲ μὴ σαρκοφάγα, μύουρον.

Τὸ δὲ δέρμα οἱ μὲν λεπιδωτὸν ἔχουσιν αὐτῶν (ἡ
5 δὲ λεπὶς διὰ λαμπρότητα καὶ λεπτότητα τοῦ σώματος ἀφίσταται), οἱ δὲ τραχύ, οἷον ῥίνη καὶ βάτος
καὶ τὰ τοιαῦτα· ἐλάχιστα δὲ τὰ λεῖα. τὰ δὲ σελάχη
ἀλεπίδωτα μὲν τραχέα δ᾽ ἐστὶ διὰ τὸ χονδράκανθα
εἶναι· τὸ γὰρ γεῶδες ἐκεῖθεν ἡ φύσις εἰς τὸ δέρμα
κατανήλωκεν.

10 Ὄρχεις δ᾽ οὐδεὶς ἔχει ἰχθὺς οὔτ᾽ ἐκτὸς οὔτ᾽ ἐντός

---

¹ δελφῖνες non probant Frantzius, Ogle; similia Hist. An.
591 b 26 secludunt Aubert et Wimmer.

ᵃ This statement about dolphins, though repeated at Hist.
an. 591 b 26, is incorrect, and as Aristotle was familiar with

others have it underneath (*e.g.* the dolphin [a] and the selachians) and that is why they turn on to their backs to get their food. It looks as if Nature made them do this partly to preserve other animals from them, for they all prey on living things, and while they are losing time turning on to their backs the other things get away safely ; but she did it also to prevent them from giving way too much to their gluttonous craving for food, since if they could get it more easily they would presently be destroyed through repletion. Another reason is that their snout is round and small and therefore cannot have much of an opening in it.

There are differences too among those that have their mouth above. With some it is a great wide opening (these are the flesh-eaters, as *e.g.* those with sharp interfitting teeth, whose strength is in their mouth) ; with others (the non-flesh-eaters) it is on a tapering snout.

As for the skin : some have a scaly skin (these Skin. scales are shiny and thin and therefore easily come loose from the body) ; others have a rough skin, *e.g.* the Rhinē and the Batos and such. Those with smooth skins are the fewest. Selachia have skins which are scaleless but rough, owing to their bones being cartilaginous : instead of using the earthy matter on the bones Nature has used it for the skin.

No fish has testicles [b] either without or within. Nor Testicles.

the creature, some editors consider this reference to be an interpolation.

[b] By this Aristotle does not mean that fish have no organ for the secretion of sperm, but that they have no organ similar in shape and consistency to those of mammalia, etc. He calls the corresponding organs in fish not testes, but tubes, or roe. Aristotle's statement does not, of course, include the Selachia, which have compact, oval testes.

697 a

(οὐδ' ἄλλο τι τῶν ἀπόδων οὐδέν, διὸ οὐδ' οἱ ὄφεις),
πόρον δὲ τοῦ περιττώματος καὶ τῶν περὶ τὴν
γένεσιν τὸν αὐτόν, καθάπερ καὶ τἆλλα ᾠοτόκα[1]
πάντα καὶ[2] τετράποδα, διὰ τὸ μὴ ἔχειν κύστιν
μηδὲ γίνεσθαι περίττωμ' αὐτοῖς ὑγρόν.

15 Τὸ μὲν οὖν τῶν ἰχθύων γένος πρὸς τἆλλα ζῷα
ταύτας ἔχει τὰς διαφοράς, οἱ δὲ δελφῖνες καὶ αἱ
φάλαιναι καὶ πάντα τὰ τοιαῦτα τῶν κητῶν βράγχια
μὲν οὐκ ἔχουσιν, αὐλὸν δὲ διὰ τὸ πνεύμονα ἔχειν·
δεχόμενα γὰρ κατὰ τὸ στόμα τὴν θάλατταν ἀφιᾶσι
κατὰ τὸν αὐλόν. ἀνάγκη μὲν γὰρ δέξασθαι τὸ
20 ὑγρὸν διὰ τὸ λαμβάνειν τὴν τροφὴν ἐν τῷ ὑγρῷ·
δεξάμενα δ' ἀφιέναι ἀναγκαῖον. τὰ μὲν οὖν βράγ-
χιά ἐστι χρήσιμα τοῖς μὴ ἀναπνέουσιν· δι' ἣν δ'
αἰτίαν, εἴρηται ἐν τοῖς περὶ ἀναπνοῆς· ἀδύνατον γὰρ
ἅμα τὸ αὐτὸ ἀναπνεῖν καὶ βράγχια ἔχειν· ἀλλὰ πρὸς
τὴν ἄφεσιν τοῦ ὕδατος ἔχουσι τὸν αὐλόν. κεῖται δ'
25 αὐτοῖς οὗτος πρὸ τοῦ ἐγκεφάλου· διελάμβανε γὰρ
ἂν ἀπὸ τῆς ῥάχεως αὐτόν. αἴτιον δὲ τοῦ πνεύμονα
ταῦτ' ἔχειν καὶ ἀναπνεῖν, ὅτι τὰ μεγάλα τῶν ζῴων
πλείονος δεῖται θερμότητος ἵνα κινῆται· διὸ ὁ
πνεύμων ἔγκειται αὐτοῖς θερμότητος ὢν πλήρης
30 αἱματικῆς. ἔστι δὲ ταῦτα τρόπον τινὰ ⟨καὶ⟩[3] πεζὰ
καὶ ἔνυδρα· τὸν μὲν γὰρ ἀέρα δέχεται ὡς πεζά,
ἄποδα δ' ἐστὶ καὶ λαμβάνει ἐκ τοῦ ὑγροῦ τὴν
697 b τροφὴν ὥσπερ τὰ ἔνυδρα. καὶ αἱ φῶκαι δὲ καὶ
αἱ νυκτερίδες διὰ τὸ ἐπαμφοτερίζειν αἱ μὲν τοῖς
ἐνύδροις καὶ πεζοῖς, αἱ δὲ τοῖς πτηνοῖς καὶ πεζοῖς,
διὰ τοῦτο ἀμφοτέρων τε μετέχουσι καὶ οὐδετέρων.

---

[1] ζωοτόκα PSUY.
[2] καὶ ⟨δίποδα καὶ⟩ Ogle.
[3] ⟨καὶ⟩ Rackham.

have any other footless animals, and this includes the serpents. In fish the passage for the residue and for the generative secretion is one and the same; and this is so in all other oviparous animals, four-footed ones included. This is because they have no bladder and produce no liquid residue.

Thus we have seen what are the differences to be noticed in fish as a group as compared with other animals. Dolphins and whales and all such Cetacea, however, have no gills, but they have a blowhole because they have a lung. They cannot help letting the sea-water enter the mouth because they feed in the water, and once it has got in they must get it out again, and they do so through the blowhole. Gills, of course, are of service herein to those creatures that do not breathe. The reason for this has been given in my book on *Respiration*[a]: no creature can breathe and at the same time have gills; instead, these Cetacea have a blowhole for getting rid of the water. It is placed in front of the brain, otherwise it would separate the brain from the spine. The reason why these creatures have a lung and breathe is that large animals need more heat than others to enable them to move; consequently they have a lung inside them full of heat derived from the blood. They are, in a way, land-animals as well as water-animals: they inhale the air, like land-animals, but they have no feet and they get their food from the water as water-animals do. Similarly, seals and bats are in an intermediate position. Seals are between land-animals and water-animals, bats between land-animals and fliers: thus they belong to both classes or to neither.

Intermediate creatures:
(i.) Cetacea.

(ii.) Seals and bats.

* References given above, see on 696 b 2.

697 b

αἵ τε γὰρ φῶκαι ὡς μὲν ἔνυδροι πόδας ἔχουσιν, ὡς
δὲ πεζαὶ πτερύγια¹ (τοὺς γὰρ ὄπισθεν πόδας ἰχθυ-
ώδεις ἔχουσι πάμπαν, ἔτι δὲ τοὺς ὀδόντας πάντας
καρχαρόδοντας καὶ ὀξεῖς)· καὶ αἱ νυκτερίδες ὡς μὲν
πτηνὰ ἔχουσι πόδας, ὡς δὲ τετράποδα οὐκ ἔχουσι,
καὶ οὔτε κέρκον ἔχουσιν οὔτ᾽ οὐροπύγιον, διὰ μὲν
10 τὸ πτηνὰ εἶναι κέρκον, διὰ δὲ τὸ πεζὰ οὐροπύγιον.
συμβέβηκε δ᾽ αὐταῖς τοῦτ᾽ ἐξ ἀνάγκης· εἰσὶ γὰρ
δερμόπτεροι, οὐδὲν δ᾽ ἔχει οὐροπύγιον μὴ σχιζό-
πτερον· ἐκ τοιούτου γὰρ πτεροῦ γίνεται τὸ οὐροπύγιον. ἡ δὲ κέρκος καὶ ἐμπόδιος ἂν ἦν ὑπάρχουσα
ἐν τοῖς πτεροῖς.

Τὸν αὐτὸν δὲ τρόπον καὶ ὁ στρουθὸς ὁ Λιβυκός·
15 τὰ μὲν γὰρ ὄρνιθος ἔχει, τὰ δὲ ζῴου τετράποδος.
ὡς μὲν γὰρ οὐκ ὢν τετράπους πτερὰ ἔχει, ὡς δ᾽
οὐκ ὢν ὄρνις οὔτε πέταται μετεωριζόμενος, καὶ τὰ
πτερὰ οὐ χρήσιμα πρὸς πτῆσιν ἀλλὰ τριχώδη· ἔτι
δὲ ὡς μὲν τετράπους ὢν βλεφαρίδας ἔχει τὰς
ἄνωθεν καὶ ψιλός ἐστι τὰ περὶ τὴν κεφαλὴν καὶ τὰ
20 ἄνω τοῦ αὐχένος, ὥστε τριχωδεστέρας ἔχειν τὰς
βλεφαρίδας, ὡς δ᾽ ὄρνις ὢν τὰ κάτωθεν ἐπτέρωται·
καὶ δίπους μέν ἐστιν ὡς ὄρνις, δίχαλος δ᾽ ὡς
τετράπους· οὐ γὰρ δακτύλους ἔχει ἀλλὰ χηλάς.
τούτου δ᾽ αἴτιον ὅτι τὸ μέγεθος οὐκ ὄρνιθος ἔχει
ἀλλὰ τετράποδος· ἐλάχιστον γὰρ ἀναγκαῖον εἶναι τὸ
25 μέγεθος ὡς καθόλου εἰπεῖν τὸ τῶν ὀρνίθων· οὐ γὰρ
ῥᾴδιον πολὺν ὄγκον κινεῖσθαι σώματος μετέωρον.

¹ πτερύγια Ogle: πτέρυγας vulg.

Seals, if regarded as water-animals, are anomalous in having feet ; if regarded as land-animals, in having fins (their hind feet are altogether like those of fishes—*i.e.* fins ; and all their teeth too are sharp and interlocking). Bats, too, if regarded as birds, are anomalous in having feet[a] ; if regarded as quadrupeds, in not having feet[b] ; furthermore, they have neither a quadruped's tail (because they are fliers) nor a bird's tail (because they are land-animals). This their lack of a tail like a bird's is a necessary consequence, since they have membranous wings, and no creature has a tail of this sort unless it has barbed feathers : such tails are always made out of barbed feathers And a tail of the other sort growing among feathers would be a definite impediment.

After the same style is the Libyan ostrich : in some points it resembles a bird, in others a quadruped. (iii.) The Ostrich. As not being a quadruped, it has feathers ; as not being a bird, it cannot rise up and fly, and it has feathers that are like hairs and useless for flight. Again, as being a quadruped, it has upper eye-lashes, and it is bald in the head and the upper part of the neck, as a result of which its eyelashes are hairier than they would otherwise be ; as being a bird, it is feathered on its lower parts. Also, as a bird, it has two feet ; but, as a quadruped, it has cloven hoofs (it has hoofs and not toes). The reason is that it has the size not of a bird but of a quadruped. Speaking generally, a bird has to be very small in size, because it is difficult for a body of large bulk to move off the ground.

---

[a] That is, of the sort that birds ought not to have, viz. on their wings.

[b] That is, of the sort that quadrupeds ought to have.

Περὶ μὲν οὖν τῶν μορίων, διὰ τίν' αἰτίαν ἕκαστόν
ἐστιν ἐν τοῖς ζῴοις, εἴρηται περὶ πάντων τῶν ζῴων
καθ' ἕκαστον· τούτων δὲ διωρισμένων ἐφεξῆς ἐστι
30 τὰ περὶ τὰς γενέσεις αὐτῶν διελθεῖν.[1]

---

[1] τούτων . . . διελθεῖν om. Yb, et statim incipiunt librum
de incessu.

We have now spoken severally of all the animals : Conclusion. we have described their parts, and stated the reason why each is present in them. Now that this is concluded, the next thing is to describe the various ways in which animals are generated.

ARISTOTLE

Commentators agree that no satisfactory sense can be obtained from the first three lines of this passage as it stands in Bekker's edition. None has so far produced a remedy; but an examination of the Arabic translation (or of Michael Scot's Latin translation made from the Arabic) shows plainly what has happened. *In neither of these two translations is there any reference whatever to a diagram until 685 a 2.* Thus the ms. from which our present Greek text is derived had been corrupted through the efforts of someone who tried to improve the text of 684 b 22-27 by inserting references to a diagram here also; and the result is that these references have caused the complete loss of one important phrase (b 22) and serious corruption of another (b 24-25). Some dislocation has also been caused in the lines following, up to line 29.

The two diagrams given in the ms. Z are obviously constructed to suit the interpolated text. One of the mss. (Merton 278) of Michael Scot's version has an entirely different diagram; the three mss. of Scot at Cambridge have no diagram at all, nor has the Arabic ms. B.M. Add. 7511.

I give below the passage as it appears in Michael Scot's version.

Natura ergo istorum duorum modorum est sicut diximus; et propter hoc ambulant uniformiter[1] *sicut accidit animalibus quadrupedibus et hominibus etiam. homo vero habet os in capite, scilicet in parte superiori corporis; deinde habet stomachum, deinde ventrem, et post ventrem intestinum perveniens ad locum exitus superfluitatis. iste ergo res in animalibus habentibus sanguinem sunt secundum hanc dispositionem, et post caput est clibanus, scilicet pectus, et quod vicinatur ei. alia vero membra sunt propter ista, etc.*

I am much indebted to Dr. R. Levy for his kindness in reading this passage for me in the Arabic in Brit. Mus. ms. Add. 7511.

---

[1] *inuniformiter* Caius 109 & Camb. U.L. Ii. 3. 16; fortasse igitur scribendum *uniformiter et non inuniformiter.*

PARTS OF ANIMALS

ADDITIONAL NOTE ON 693 b 3

Explanation of Aristotle's terminology for describing the bending of limbs.

When Aristotle is speaking about the bending of limbs,

backwards and *forwards* are relative to the direction in which the whole animal moves;

inwards and *outwards* are relative to the bulk of the body itself.

Thus, *backwards* means that the angle of the bent joint points backwards; *inwards* means that the extremity of the *limb* is brought inwards towards the body, that is, the angle of the bent *joint* points away from the main bulk of the body. ("Inward" and "outward" bending thus have no connotation of "bandy-legs" and "knock-knees.")

Example (1) 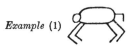 All four legs bend *inwards*;
The forelegs bend *forwards*:
The hindlegs bend *backwards*.

Example (2) 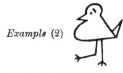 The leg bends *inwards*, and *backwards*.

(See *De incess. an.* 711 a 8 ff., *Hist. An.* 498 a 3 ff.)

ARISTOTLE

ADDITIONAL NOTE ON THE MS. Z

The following portions of the text of *De partibus* are contained in the Oxford MS. Z (see p. 50) :

fol. 60ʳ, 60ᵛ. I. 639 b 29 to 640 b 24. μέχρι to μᾶλλον ἂν
                  inclusive.
fol. 61ʳ, 61ᵛ. I. 644 a 25 to 645 a 17. καθόλου to τοῖς φυ in-
                  clusive.

Between these two folios it has apparently lost four folios, as well as one at the beginning of Book I and another at the end.

fol. 1ʳ–19ʳ.   Book II.
fol. 19ᵛ–36ʳ.   Book III, but the words οὐ πολὺ to εὐρυχώρους
                  inclusive (675 a 30–b 27) are omitted,
                  with no indication by the original scribe
                  that anything has been omitted : this
                  passage has been supplied by a later hand
                  in the margins of fol. 35ᵛ and 36ʳ and
                  on 36ᵛ.

Book IV is written by yet another (later) hand, and this Book occupies fol. 37ʳ–59ᵛ, at the end of which folio it breaks off at the words τὰ καλούμενα (694 a 13). The rest of Book IV is lost.

In the apparatus I have used the following abbreviations in quoting this MS. :

Z   Books I, II and most of III (first hand, *c.* A.D. 1000).
Z¹   indicates the reading of the first hand where this has
        been altered by another.
Z²   indicates later correctors of Z¹.
Z   indicates the readings of the MS. in Book IV.

I have collated from photostats the whole of the portion written by the first hand, and the readings of Z quoted have been confirmed by reference to the photostats.

I have used the symbol *E* when quoting the readings of E from 680 b 36 onwards, as this part of the MS. is written in a later hand.

MOVEMENT OF ANIMALS

INTRODUCTION

THAT the *De incessu animalium* is a genuine work of Aristotle himself has never been disputed. The *De motu animalium* has been regarded by many critics as a spurious work, though recent opinion has favoured its genuineness. Brandis, Rose and Zeller all condemn it, but its Aristotelian authorship has been upheld by Werner Jaeger (*Hermes*, xlviii. pp. 31 ff.), who makes out a very strong case in its favour, and by the Oxford translator, Mr. A. S. L. Farquharson. Those who deny its authenticity rely mainly on the supposition that there is a reference in 703 a 10-11 to the *De spiritu*. This treatise is generally admitted to be un-Aristotelian, but the reference, as Mr. Farquharson has pointed out, might relate equally well to numerous other passages in the Aristotelian corpus ; Michael Ephesius refers it to a treatise Περὶ τροφῆς, not otherwise known. In style, vocabulary and syntax the *De motu animalium* is entirely Aristotelian, and its doctrine corresponds with that set forth in Aristotle's genuine works.

Each treatise has its proper place in the scheme of Aristotle's biological works. Both are theoretical, the *De incessu animalium*, like the *De partibus animalium*, dealing with the material side of living things, and the *De motu animalium*, like the *De generatione animalium*, dealing with their consequential properties.

436

The chief MSS. of the *De motu animalium* are E, Y, P and S.[a] Of these E, one of the most famous of Aristotelian MSS., is the oldest ; Y is closely related to E. P and S are similarly related and form a second group.

Of the *De incessu animalium* the principal MSS. are Z, Y, U, S and P.[a] Of these Z is the oldest, and Y is closely related to it, while the other three MSS. form another group.

A full account of these MSS. and their relations to one another will be found in the Introduction (pp. iv. ff.) of W. W. Jaeger's text (Teubner, 1913).

The text used for the present translation is based on that of I. Bekker, all divergences from which are noted and the authority given for the reading adopted. Jaeger's text and *apparatus criticus* have been consulted throughout.

The Commentary of Michael Ephesius (*Commentaria in Aristotelem Graeca*, xxii. 2, Hayduck, 1904) has been of some assistance both for the text and for the interpretation, and the Latin version of Nicholaus Leonicus (died 1599), printed in the Berlin Aristotle, Vol. III, has been constantly consulted.

The two treatises have been translated into French by J. Barthélemy-Saint-Hilaire, and into English by Mr. A. S. L. Farquharson in the Oxford translation (1912). This translation with its ample explanatory notes constitutes much the most serious attempt that has been made to interpret these two treatises, and anyone who follows in Mr. Farquharson's footsteps must necessarily be heavily indebted to him.

E. S. F.

---

[a] For the meanings of these symbols see pp. 439 and 483.

ANALYSIS OF CONTENTS

ABBREVIATIONS USED IN THE APPARATUS CRITICUS

E = Codex Parisinus Regius 1853.
Y = Codex Vaticanus 261.
P = Codex Vaticanus 1339.
S = Codex Laurentianus 81. 1.
Leon. = Latin translation of Nicolaus Leonicus.
Mich. = Greek commentary of Michael Ephesius.

ΠΕΡΙ ΖΩΙΩΝ ΚΙΝΗΣΕΩΣ

698 a I. Περὶ δὲ κινήσεως τῆς τῶν ζῴων, ὅσα μὲν
αὐτῶν περὶ ἕκαστον ὑπάρχει γένος, καὶ τίνες
διαφοραί, καὶ τίνες αἰτίαι τῶν καθ᾽ ἕκαστον συμβεβηκότων αὐτοῖς, ἐπέσκεπται περὶ ἁπάντων ἐν
5 ἑτέροις· ὅλως δὲ περὶ τῆς κοινῆς αἰτίας τοῦ κινεῖσθαι κίνησιν ὁποιανοῦν (τὰ μὲν γὰρ πτήσει κινεῖται τὰ δὲ νεύσει τὰ δὲ πορείᾳ τῶν ζῴων, τὰ δὲ
κατ᾽ ἄλλους τρόπους τοιούτους) ἐπισκεπτέον νῦν.

Ὅτι μὲν οὖν ἀρχὴ τῶν ἄλλων κινήσεων τὸ
αὐτὸ ἑαυτὸ κινοῦν, τούτου[1] δὲ τὸ ἀκίνητον,
καὶ ὅτι τὸ πρῶτον κινοῦν ἀναγκαῖον ἀκίνητον
10 εἶναι, διώρισται πρότερον, ὅτεπερ καὶ περὶ κινήσεως ἀϊδίου, πότερον ἔστιν ἢ οὐκ ἔστι, καὶ εἰ
ἔστι, τίς ἐστιν. δεῖ δὲ τοῦτο μὴ μόνον τῷ λόγῳ
καθόλου λαβεῖν, ἀλλὰ καὶ ἐπὶ τῶν καθ᾽ ἕκαστα
καὶ τῶν αἰσθητῶν, δι᾽ ἅπερ καὶ τοὺς καθόλου
ζητοῦμεν λόγους, καὶ ἐφ᾽ ὧν ἐφαρμόττειν οἰόμεθα
15 δεῖν αὐτούς. φανερὸν γὰρ καὶ ἐπὶ τούτων ὅτι
ἀδύνατον κινεῖσθαι μηδενὸς ἠρεμοῦντος, πρῶτον
μὲν ἐν αὐτοῖς τοῖς ζῴοις. δεῖ γάρ, ἂν κινῆταί τι
τῶν μορίων, ἠρεμεῖν τι· καὶ διὰ τοῦτο αἱ καμπαὶ

---

[1] τούτου ΕΡΥ : τοῦτο S.

440

ON THE MOVEMENT OF ANIMALS

I. We have inquired elsewhere [a] into the details of the movement of the various kinds of animals, the differences between these movements, and the causes of the characteristics which each exhibit ; we must now inquire generally into the common cause of animal movement of whatever kind—for some animals move by flight, some by swimming, some by walking, and others by other such methods.

Now that the origin of all the other movements is that which moves itself, and that the origin of this is the immovable, and that the prime mover must necessarily be immovable, has already been determined when we were investigating [b] whether or not eternal movement exists, and if it does exist what it is. And this we must apprehend not merely in theory as a general principle but also in its individual manifestations and in the objects of sense-perception, on the basis of which we search for general theories and with which we hold that these theories ought to agree. For it is clear also in the objects of sense-perception that movement is impossible if there is nothing in a state of rest, and above all in the animals themselves. For if any one of their parts moves, another part must necessarily be at rest ; and

[a] In the *De partibus animalium.*
[b] *Physics* viii. 258 b 4-9.

τοῖς ζῴοις εἰσίν. ὥσπερ γὰρ κέντρῳ χρῶνται
ταῖς καμπαῖς, καὶ γίνεται τὸ ὅλον μέρος, ἐν ᾧ ἡ
20 καμπή, καὶ ἓν καὶ δύο, καὶ εὐθὺ καὶ κεκαμμένον,
μεταβάλλον δυνάμει καὶ ἐνεργείᾳ διὰ τὴν καμπήν.
καμπτομένου δὲ καὶ κινουμένου τὸ μὲν κινεῖται
σημεῖον τὸ δὲ μένει τῶν ἐν ταῖς καμπαῖς, ὥσπερ
ἂν εἰ τῆς διαμέτρου ἡ μὲν Α καὶ ἡ Δ μένοι, ἡ δὲ
Β κινοῖτο, καὶ γίνοιτο ἡ ΑΓ. ἀλλ᾽ ἐνταῦθα μὲν
25 δοκεῖ πάντα τρόπον ἀδιαίρετον εἶναι τὸ κέντρον
(καὶ γὰρ τὸ κινεῖσθαι, ὡς φασί, πλάττουσιν ἐπ᾽
αὐτῶν· οὐ γὰρ κινεῖσθαι[1] τῶν μαθηματικῶν
οὐδέν), τὰ δ᾽ ἐν ταῖς καμπαῖς δυνάμει καὶ ἐνεργείᾳ
698 b γίνεται ὁτὲ μὲν ἓν ὁτὲ δὲ διαιρετά. ἀλλ᾽ οὖν
ἀεὶ ἡ ἀρχὴ ἡ πρὸς ὅ, ᾗ[2] ἀρχή, ἠρεμεῖ κινουμένου
τοῦ μορίου τοῦ κάτωθεν, οἷον τοῦ μὲν βραχίονος
κινουμένου τὸ ὠλέκρανον, ὅλου δὲ τοῦ κώλου ὁ
ὦμος, καὶ τῆς μὲν κνήμης τὸ γόνυ, ὅλου δὲ τοῦ
5 σκέλους τὸ ἰσχίον. ὅτι μὲν οὖν καὶ ἐν αὐτῷ
ἕκαστόν τι δεῖ ἔχειν ἠρεμοῦν, ὅθεν ἡ ἀρχὴ
τοῦ κινουμένου ἔσται, καὶ πρὸς ὃ ἀπερειδόμενον

---

[1] κινεῖσθαι ESY : κινεῖται P.
[2] ἡ πρὸς ὅ, ᾗ Jaeger : ἡ πρὸς ὅ ἡ EY : ἡ πρώτη ᾗ S : ἡ πρόσω
(om. altero ἀρχή) P.

---

[a] e.g. the arm as an arm is one, but is divided into two at
the elbow.
[b] The term ἀρχή, which occurs frequently in this treatise,
is difficult to render in English by a single word.   It is sometimes used generally of the " origin " of movement (e.g.
701 b 33), but more often of a localized "origin" of movement,

it is on this account that animals have joints. For they use their joints as a centre, and the whole part in which the joint is situated is both one and two,[a] both straight and bent, changing potentially and actually because of the joint. And when the part is being bent and moved, one of the points in the joint moves and one remains at rest, just as would happen if A and D in the diameter of a circle were to remain still while B moved, and the radius AC were formed. (In geometrical figures, however, the centre is considered to be in every respect indivisible—for movement, too, in such figures is a figment, so they say, since in mathematics nothing actually moves,—whereas the centres in the joints are, potentially and actually, sometimes one and sometimes divided.) Be that as it may, the origin [b] to which the movement can be traced, *qua* origin, is always at rest while the part below it is in motion —the elbow-joint, for instance, when the forearm is in motion, the shoulder when the whole arm is moved, the knee when the shin is moved, and the hip when the whole leg is moved. It is obvious, then, that every animal too must have in itself something that is at rest, in order to provide that which is moved with the origin of its movement, supported

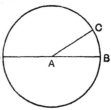

whether, as here, in a single member, or at the centre of the body, viz. the heart (701 b 25, 29), where a further idea of "ruling" seems to be implied (*e.g.* 702 a 37). It is also used sometimes in the literal sense of "beginning," and this and the meaning of "origin" of motion may occur in the same passage and cause confusion (*e.g.* 702 a 36–b 2).

καὶ ὅλον ἀθρόον κινηθήσεται καὶ κατὰ μέρος,
φανερόν.

II. Ἀλλὰ πᾶσα ἡ ἐν αὐτῷ ἠρεμία ὅμως ἄκυρος,
ἂν μή τι ἔξωθεν ᾖ ἁπλῶς ἠρεμοῦν καὶ ἀκίνητον.
10 ἄξιον δ' ἐπιστήσαντας ἐπισκέψασθαι περὶ τοῦ
λεχθέντος· ἔχει γὰρ τὴν θεωρίαν οὐ μόνον ὅσον
ἐπὶ τὰ ζῷα συντείνουσαν, ἀλλὰ καὶ πρὸς τὴν τοῦ
παντὸς κίνησιν καὶ φοράν. ὥσπερ γὰρ καὶ ἐν
αὐτῷ δεῖ τι ἀκίνητον εἶναι, εἰ μέλλει κινεῖσθαι,
οὕτως ἔτι μᾶλλον ἔξω δεῖ τι εἶναι τοῦ ζῴου
15 ἀκίνητον, πρὸς ὃ ἀπερειδόμενον κινεῖται τὸ κινούμενον. εἰ γὰρ ὑποδώσει ἀεί, οἷον τοῖς μυσὶ[1] τοῖς
ἐν τῇ γῇ[2] ἢ τοῖς ἐν τῇ ἄμμῳ πορευομένοις, οὐ
πρόεισιν, οὐδ' ἔσται οὔτε πορεία, εἰ μὴ ἡ γῆ μένοι,
οὔτε πτῆσις ἢ νεῦσις, εἰ μὴ ὁ ἀὴρ ἢ ἡ θάλαττα
ἀντερείδοι. ἀνάγκη δὲ τοῦτο ἕτερον εἶναι τοῦ
κινουμένου, καὶ ὅλον ὅλου, καὶ μόριον μηδὲν εἶναι
20 τοῦ κινουμένου τὸ οὕτως ἀκίνητον· εἰ δὲ μή, οὐ
κινηθήσεται. μαρτύριον δὲ τούτου τὸ ἀπορού-
μενον, διὰ τί ποτε τὸ πλοῖον ἔξωθεν μέν, ἄν τις
ὠθῇ τῷ κοντῷ τὸν ἱστὸν ἤ τι ἄλλο προσβάλλων
μόριον, κινεῖ ῥᾳδίως, ἐὰν δ' ἐν αὐτῷ τις ὢν τῷ
πλοίῳ τοῦτο πειρᾶται πράττειν, οὐκ ἂν κινήσειεν
25 οὔτ' ἂν ὁ Τιτυὸς οὔθ' ὁ Βορέας πνέων ἔσωθεν ἐκ
τοῦ πλοίου, εἰ τύχοι πνέων τὸν τρόπον τοῦτον ὄν-

---

[1] μυσὶν libri: ἐμύσι coni. Diels.
[2] γῇ libri: ζειᾷ coni. Farquharson.

---

[a] It is doubtful whether the MS. reading will bear this
interpretation, and ἐν τῇ γῇ is probably corrupt. It is more

444

upon which it will move both as an integral whole
and in its several parts.

II. Any quality of rest, however, in an animal is of
no effect unless there is something outside it which is
absolutely at rest and immovable. And it is worth
while to stop and consider this dictum ; for the re-
flection which it involves applies not merely to animals,
but also to the motion and progression of the universe.
For just as in the animal there must be something
which is immovable if it is to have any motion, so
a fortiori there must be something which is immovable outside the animal, supported upon which that
which is moved moves. For if that which supports
the animal is to be always giving way (as it does when
mice walk upon loose soil [a] and when persons walk on
sand), there will be no progress, that is, no walking,
unless the ground were to remain still, and no flying
or swimming unless the air or sea were to offer resistance. And that which offers resistance must be other
than that which is moved, the whole other than the
whole, and that which is thus immovable must form
no part of that which is moved ; otherwise the latter
will not move. This contention is supported by the
problem : Why can a man easily move a boat from
outside if he thrusts it along with a pole by pushing
against the mast or some other part of the boat,
but if he tries to do this when he is in the boat
itself, Tityus could not move it nor Boreas by blow-
ing from inside it, if he really blew as the artists

than likely that the comparison is with a mouse trying to
walk upon a heap of corn. Farquharson emends ἐν τῇ γῇ to
ἐν τῇ ζειᾷ, which would bear this meaning. (The form ζέη,
cp. *Petrie Pap.* ii. p. 69 (3rd cent. B.C.), would be nearer to the
MS. reading.) Diels' suggestion of ἐμύσιν for μύσιν is in-
genious, but does not give the required sense.

698 b
699 a

περ οἱ γραφεῖς ποιοῦσιν· ἐξ αὐτοῦ γὰρ τὸ πνεῦμα
ἀφιέντα γράφουσιν. ἐάν τε γὰρ ἠρέμα ῥιπτῇ τὸ
πνεῦμά τις ἐάν τ' ἰσχυρῶς οὕτως ὥστ' ἄνεμον
ποιεῖν τὸν μέγιστον, ἐάν τε ἄλλο τι ᾖ τὸ ῥιπτού-
μενον ἢ ὠθούμενον, ἀνάγκη πρῶτον μὲν πρὸς
ἠρεμοῦν τι τῶν αὐτοῦ μορίων ἀπερειδόμενον ὠθεῖν,
5 εἶτα πάλιν τοῦτο τὸ μόριον, ἢ αὐτὸ ἢ οὗ τυγχάνει
μόριον ὄν, πρὸς τῶν ἔξωθέν τι ἀποστηριζόμενον
μένειν. ὁ δὲ τὸ πλοῖον ὠθῶν ἐν τῷ πλοίῳ αὐτὸς
ὢν καὶ ἀποστηριζόμενος πρὸς τὸ πλοῖον εὐλόγως
οὐ κινεῖ τὸ πλοῖον διὰ τὸ ἀναγκαῖον εἶναι πρὸς ὃ
ἀποστηρίζεται μένειν· συμβαίνει δ' αὐτῷ τὸ αὐτὸ
10 ὅ τε κινεῖ καὶ πρὸς ὃ ἀποστηρίζεται. ἔξωθεν δ'
ὠθῶν ἢ ἕλκων κινεῖ· οὐθὲν γὰρ μέρος ἡ γῆ τοῦ
πλοίου.

III. Ἀπορήσειε δ' ἄν τις, ἆρ' εἴ τι κινεῖ τὸν
ὅλον οὐρανόν, εἶναί τε δεῖ ἀκίνητον τοῦτο καὶ[1]
μηθὲν εἶναι τοῦ οὐρανοῦ μόριον μηδ' ἐν τῷ
οὐρανῷ. εἴτε γὰρ αὐτὸ κινούμενον κινεῖ αὐτόν,
15 ἀνάγκη τινὸς ἀκινήτου θιγγάνον κινεῖν, καὶ τοῦτο
μηδὲν εἶναι μόριον τοῦ κινοῦντος· εἴτ' εὐθὺς ἀκίνητόν
ἐστι τὸ κινοῦν, ὁμοίως οὐδὲν ἔσται[2] τοῦ κινουμένου μόριον. καὶ τοῦτό γ' ὀρθῶς λέγουσιν οἱ
λέγοντες ὅτι κύκλῳ φερομένης τῆς σφαίρας οὐδ'
ὁτιοῦν μένει μόριον· ἢ γὰρ ἂν ὅλην ἀναγκαῖον ἦν
20 μένειν, ἢ διασπᾶσθαι τὸ συνεχὲς αὐτῆς. ἀλλ'
ὅτι τοὺς πόλους οἴονταί τινα δύναμιν ἔχειν, οὐθὲν

[1] τοῦτο καὶ scripsi: καὶ τοῦτο libri.
[2] ἔσται Jaeger (cum Leon.): ἔσεσθαι libri.

---

[a] Just as Odysseus' companions while seated in the ship
open the bags containing the winds, and the ship is blown
out of its course (Homer, *Od.* x. 46 ff.).

paint him *a*; for they make him emit the breath from his own lips. For whether one emits the breath gently or so strongly as to create the greatest gale (and the same is true if that which is thrown or pushed is something other than breath), it is necessary, first, that one should be supported upon one of one's own members, which is at rest, when one pushes, and secondly, that either this member itself or that of which it forms part, should remain still, resting upon something which is external to it. Now the man who tries to push the boat while he himself is in it and leaning upon it, naturally does not move the boat, because it is essential that that against which he is leaning should remain still; but in this case that which he is trying to move and that against which he is leaning, is identical. If, on the other hand, he pushes or drags the boat from outside, he can move it; for the ground is no part of the boat.

III. The difficulty may be raised, whether, if something moves the whole heaven, this motive power must be unmoved and be no part of the heaven nor in the heaven. For if it is moved itself and moves the heaven, it can only move it by being itself in contact with something that is immovable, and this can be no part of that which causes the movement; or else, if that which causes the movement is from the first immovable, it will be equally no part of that which is moved. And on this point at any rate they are quite right who say that, when the sphere is moved in a circle, no part of it whatsoever remains still; for either the whole of it must remain still, or its continuity must be rent asunder. They are not right, however, in holding that the poles possess a kind of force,

699 a

ἔχοντας μέγεθος ἀλλ' ὄντας ἔσχατα καὶ στιγμάς,
οὐ καλῶς. πρὸς γὰρ τῷ μηδεμίαν οὐσίαν εἶναι
τῶν τοιούτων μηδενός, καὶ κινεῖσθαι τὴν μίαν
κίνησιν ὑπὸ δυοῖν ἀδύνατον· τοὺς δὲ πόλους δύο
25 ποιοῦσιν. ὅτι μὲν οὖν ἔχει τι καὶ πρὸς τὴν ὅλην
φύσιν οὕτως ὥσπερ ἡ γῆ πρὸς τὰ ζῷα καὶ τὰ
κινούμενα δι' αὐτῶν,[a] ἐκ τῶν τοιούτων ἄν τις
διαπορήσειεν. οἱ δὲ μυθικῶς τὸν Ἄτλαντα
ποιοῦντες ἐπὶ τῆς γῆς ἔχοντα τοὺς πόδας δόξαιεν
ἂν ἀπὸ διανοίας εἰρηκέναι τὸν μῦθον, ὡς τοῦτον
ὥσπερ διάμετρον ὄντα καὶ στρέφοντα τὸν οὐρανὸν
30 περὶ τοὺς πόλους· τοῦτο δ' ἂν συμβαίνοι κατὰ
λόγον διὰ τὸ τὴν γῆν μένειν. ἀλλὰ τοῖς ταῦτα
λέγουσιν ἀναγκαῖον φάναι μηδὲν εἶναι μόριον
αὐτὴν τοῦ παντός. πρὸς δὲ τούτοις δεῖ τὴν ἰσχὺν
ἰσάζειν τοῦ κινοῦντος καὶ τὴν τοῦ μένοντος. ἔστι
γάρ τι πλῆθος ἰσχύος καὶ δυνάμεως καθ' ἣν μένει
35 τὸ μένον, ὥσπερ καὶ καθ' ἣν κινεῖ τὸ κινοῦν· καὶ
ἔστι τις ἀναλογία ἐξ ἀνάγκης, ὥσπερ τῶν ἐναντίων
κινήσεων, οὕτω καὶ τῶν ἠρεμιῶν. καὶ αἱ μὲν
699 b ἴσαι ἀπαθεῖς ὑπ' ἀλλήλων, κρατοῦνται δὲ κατὰ
τὴν ὑπεροχήν. διόπερ εἴτ' Ἄτλας εἴτε τι τοιοῦτόν
ἐστιν ἕτερον τὸ κινοῦν τῶν ἐντός, οὐδὲν μᾶλλον
ἀντερείδειν δεῖ τῆς μονῆς ἣν ἡ γῆ τυγχάνει μένουσα·
ἢ κινηθήσεται ἡ γῆ ἀπὸ τοῦ μέσου καὶ ἐκ τοῦ
5 αὐτῆς τόπου. ὡς γὰρ τὸ ὠθοῦν ὠθεῖ, οὕτω τὸ
ὠθούμενον ὠθεῖται, καὶ ὁμοίως κατ' ἰσχύν. κινεῖ

---

[a] *i.e.* their limbs. We should, however, perhaps read δι'
αὐτῶν "the things which move of themselves": Leon.
renders "ea quae per se moventur."

since they have no magnitude and are only extremities and points. For besides the fact that nothing of this kind has any substance, it is also impossible for a single movement to be started by a dual agency ; and they represent the poles as two. From these considerations one may hazard the suggestion that there is something which stands in the same relation to Nature as a whole as the earth stands to the animals and the things which are moved through them.[a]

Now those who in the fable represent Atlas as having his feet planted upon the earth would seem to have shown sense in the story which they tell, since they make him as it were a radius, twisting the heaven about the poles ; it would be a logical account, since the earth remains still. But those who hold this view must declare that the earth is no part of the universe ; and, further, the force of that which causes the motion and the force of that which remains still must be equal. For there must be a certain amount of force and strength in virtue of which that which remains still remains still, just as there is a force in virtue of which that which causes motion causes motion ; and there is of necessity a similar proportion between absences of motion as there is between opposite motions, and equal forces are unaffected by one another, but are overmastered by a superiority. Therefore Atlas, or whatever else it is of like kind within that causes motion, must not exert any pressure which is too strong for the equilibrium of the earth ; or else the earth will be moved away from the centre and her proper place. For as that which pushes pushes, so that which is pushed is pushed, and in exact proportion to the force exerted ; but it creates

699 b

δὲ τὸ ἠρεμοῦν πρῶτον, ὥστε μᾶλλον καὶ πλείων
ἡ ἰσχὺς ἢ ὁμοία καὶ ἴση τῆς ἠρεμίας. ὡσαύτως
δὲ καὶ τῆς[1] τοῦ κινουμένου μέν, μὴ κινοῦντος δέ.
τοσαύτην οὖν δεήσει τὴν δύναμιν εἶναι τῆς γῆς
ἐν τῷ ἠρεμεῖν ὅσην ὅ τε πᾶς οὐρανὸς ἔχει καὶ
10 τὸ κινοῦν αὐτόν. εἰ δὲ τοῦτο ἀδύνατον, ἀδύνατον
καὶ τὸ κινεῖσθαι τὸν οὐρανὸν ὑπό τινος τοιούτου
τῶν ἐντός.

IV. Ἔστι δέ τις ἀπορία περὶ τὰς κινήσεις τῶν
τοῦ οὐρανοῦ μορίων, ἣν ὡς οὖσαν οἰκείαν τοῖς
εἰρημένοις ἐπισκέψαιτ᾽ ἄν τις. ἐὰν γάρ τις ὑπερ-
βάλλῃ τῇ δυνάμει τῆς κινήσεως τὴν τῆς γῆς
15 ἠρεμίαν, δῆλον ὅτι κινήσει αὐτὴν ἀπὸ τοῦ μέσου.
καὶ ἡ ἰσχὺς δ᾽ ἀφ᾽ ἧς αὕτη ἡ δύναμις, ὅτι οὐκ
ἄπειρος, φανερόν· οὐδὲ γὰρ ἡ γῆ ἄπειρος, ὥστ᾽
οὐδὲ τὸ βάρος αὐτῆς. ἐπεὶ δὲ τὸ ἀδύνατον λέγεται
πλεοναχῶς (οὐ γὰρ ὡσαύτως τήν τε φωνὴν ἀδύνατόν
φαμεν εἶναι ὁραθῆναι καὶ τοὺς ἐπὶ τῆς σελήνης
20 ὑφ᾽ ἡμῶν· τὸ μὲν γὰρ ἐξ ἀνάγκης, τὸ δὲ πεφυκὸς
ὁρᾶσθαι οὐκ ὀφθήσεται), τὸν δ᾽ οὐρανὸν ἄφθαρτον
εἶναι καὶ ἀδιάλυτον οἰόμεθα μὲν ἐξ ἀνάγκης εἶναι,
συμβαίνει δὲ κατὰ τοῦτον τὸν λόγον οὐκ ἐξ ἀνάγκης·
πέφυκε γὰρ καὶ ἐνδέχεται εἶναι κίνησιν μείζω
25 καὶ ἀφ᾽ ἧς ἠρεμεῖ ἡ γῆ καὶ ἀφ᾽ ἧς κινοῦνται τὸ
πῦρ καὶ τὸ ἄνω σῶμα. εἰ μὲν οὖν εἰσὶν αἱ ὑπερ-
έχουσαι κινήσεις, διαλυθήσεται ταῦτα ὑπ᾽ ἀλλήλων.

---

[1] τῆς PS: ἡ Y: αἱ E.

---

[a] i.e. its central position in the universe.
[b] i.e. the region between the air and the moon (*Meteor.*
340 b 6 ff.).

motion in that which is first at rest, so that the force exerted is greater than the immobility rather than similar and equal to it, and likewise greater than the force of that which is moved but does not create movement. Therefore the power of the earth in its immobility will necessarily be as great as that possessed by the whole heaven and that which sets it in motion. If, however, this is impossible, the movement of the heaven by any such force within it is also impossible.

IV. A problem also arises about the movements of the parts of the heaven, which might well be discussed, since it is closely connected with what has been said above. If one were to overmaster the immobility of the earth by the power of motion, one will obviously move it away from the centre.[a] Moreover it is clear that the force from which this power is derived is not infinite ; for the earth is not infinite, and so its weight is not infinite either. Now the word " impossible " is used in several senses (we are using it in different senses when we say that it is impossible to see a sound, and when we say that it is impossible for us to see the men in the moon ; for the former is of necessity invisible, the latter are of such a nature as to be seen but will never be seen by us), but we hold that the heaven is of necessity impossible to destroy and dissolve, whereas according to our present argument it is not necessarily so ; for it is within the nature of things and the bounds of possibility that a motive force should exist greater both than that which causes the earth to be at rest and than that which causes the fire and upper body [b] to move. If, therefore, the overpowering motive forces exist, these will be dissolved by one another; but if they

451

εἰ δὲ μὴ εἰσὶ μέν, ἐνδέχεται δ᾽ εἶναι (ἄπειρον γὰρ
οὐκ ἐνδέχεται διὰ τὸ μηδὲν σῶμα ἐνδέχεσθαι
ἄπειρον εἶναι), ἐνδέχοιτ᾽ ἂν διαλυθῆναι τὸν οὐρανόν.
τί γὰρ κωλύει τοῦτο συμβῆναι, εἴπερ μὴ ἀδύνατον;
30 οὐκ ἀδύνατον δέ, εἰ μὴ τἀντικείμενον ἀναγκαῖον.
ἀλλὰ περὶ μὲν τῆς ἀπορίας ταύτης ἕτερος ἔστω
λόγος.

Ἆρα δὲ δεῖ ἀκίνητόν τι εἶναι καὶ ἠρεμοῦν ἔξω
τοῦ κινουμένου, μηδὲν ὂν ἐκείνου μόριον, ἢ οὔ;
καὶ τοῦτο πότερον καὶ ἐπὶ τοῦ παντὸς οὕτως
ὑπάρχειν ἀναγκαῖον; ἴσως γὰρ ἂν δόξειεν ἄτοπον
35 εἶναι, εἰ ἡ ἀρχὴ τῆς κινήσεως ἐντός. διὸ δόξειεν
ἂν τοῖς οὕτως ὑπολαμβάνουσιν εὖ εἰρῆσθαι Ὁμήρῳ·

ἀλλ᾽ οὐκ ἂν ἐρύσαιτ᾽ ἐξ οὐρανόθεν πεδίονδε
700 a Ζῆν᾽ ὕπατον πάντων, οὐδ᾽ εἰ μάλα πολλὰ κάμοιτε·
πάντες δ᾽ ἐξάπτεσθε θεοὶ πᾶσαί τε θέαιναι.

τὸ γὰρ ὅλως ἀκίνητον ὑπ᾽ οὐδενὸς ἐνδέχεται
κινηθῆναι. ὅθεν λύεται καὶ ἡ πάλαι λεχθεῖσα
ἀπορία, πότερον ἐνδέχεται ἢ οὐκ ἐνδέχεται δια-
5 λυθῆναι τὴν τοῦ οὐρανοῦ σύστασιν, εἰ ἐξ ἀκινήτου
ἤρτηται ἀρχῆς.

Ἐπὶ δὲ τῶν ζῴων οὐ μόνον τὸ οὕτως ἀκίνητον
δεῖ ὑπάρχειν, ἀλλὰ καὶ ἐν αὐτοῖς τοῖς κινουμένοις

ᵃ ἄπειρον] sc. κίνησιν. The argument is as follows: these
overpowering motive forces might exist and be dissolved by
one another, because if they can be dissolved, they are not
infinite, and the reason why they are not infinite is that they
act upon what is finite, and the infinite cannot act on the
finite (De caelo, 274 b 23 ff.).
ᵇ It is discussed in the Physics and De caelo.

do not really exist, but there is a possibility of their existing (for an infinite motive force *a* is impossible because an infinite body is also impossible), it would be possible for the heaven to be dissolved. For what is there to prevent this happening if it is not impossible ? And it is not impossible, unless the opposite proposition is inevitable. But let us leave the discussion of this question for another occasion.*b*

Must there, then, or must there not, be something immovable and at rest outside that which is moved and forming no part of it ? And must this be true also of the universe ? For it would perhaps seem strange if the origin of motion were inside. And so to those who hold this view Homer's words would seem appropriate :

> Nay, ye could never pull down to the earth from the summit of heaven,
> Zeus, the highest of all, no, not if ye toiled to the utmost.
> Come, ye gods and ye goddesses all, set your hands to the hawsers.*c*

For that which is entirely immovable cannot be moved by anything. And it is here that we must look for the solution of the problem stated some time ago, namely, whether it is possible or impossible for the composition of the heaven to be dissolved, seeing that it depends upon an origin which is immovable.

Now in the animals there must exist not only that which is immovable in this sense,*d* but there must also be something immovable in the actual things which move from place to place and which themselves

c *Iliad* viii. 20-22. The lines are quoted in the wrong order and the *textus receptus* reads μήστωρ' for πάντων.

d *i.e.* something immovable and at rest which is outside that which is moved and forms no part of it (*cf.* 699 b 32).

700 a

κατὰ τόπον ὅσα κινεῖ αὐτὰ αὐτά. δεῖ γὰρ αὐτοῦ
τὸ μὲν ἠρεμεῖν τὸ δὲ κινεῖσθαι, πρὸς ὃ ἀπερειδό-
10 μενον τὸ κινούμενον κινήσεται, οἷον ἄν τι κινῇ
τῶν μορίων· ἀπερείδεται γὰρ θάτερον ὡς πρὸς
μένον θάτερον. περὶ δὲ τῶν ἀψύχων ὅσα κινεῖται
ἀπορήσειεν ἄν τις, πότερον ἅπαντ' ἔχει ἐν ἑαυτοῖς
καὶ τὸ ἠρεμοῦν καὶ τὸ κινοῦν, καὶ πρὸς τῶν
ἔξω τι ἠρεμούντων ἀπερείδεσθαι ἀνάγκη καὶ
ταῦτα, ἢ ἀδύνατον, οἷον πῦρ ἢ γῆν ἢ τῶν ἀψύχων
15 τι, ἀλλ'¹ ὑφ' ὧν ταῦτα κινεῖται πρώτων. πάντα
γὰρ ὑπ' ἄλλου κινεῖται τὰ ἄψυχα, ἀρχὴ δὲ πάντων
τῶν οὕτως κινουμένων τὰ αὐτὰ αὐτὰ κινοῦντα.
τῶν δὲ τοιούτων περὶ μὲν τῶν ζῴων εἴρηται· τὰ
γὰρ τοιαῦτα πάντα ἀνάγκη καὶ ἐν αὐτοῖς ἔχειν
τὸ ἠρεμοῦν, καὶ ἔξω πρὸς ὃ ἀπερείσεται. εἰ δέ
20 τι ἐστὶν ἀνωτέρω καὶ πρώτως κινοῦν, ἄδηλον,
καὶ ἄλλος λόγος περὶ τῆς τοιαύτης ἀρχῆς. τὰ
δὲ ζῷα ὅσα κινεῖται, πάντα πρὸς τὰ ἔξω ἀπερ-
ειδόμενα κινεῖται, καὶ ἀναπνέοντα καὶ ἐκπνέοντα.
οὐδὲν γὰρ διαφέρει μέγα ῥῖψαι βάρος ἢ μικρόν,
ὅπερ ποιοῦσιν οἱ πτύοντες καὶ βήττοντες καὶ οἱ
25 εἰσπνέοντες καὶ ἐκπνέοντες.

V. Πότερον δ' ἐν τῷ αὐτὸ κινοῦντι κατὰ τόπον
μόνῳ δεῖ τι μένειν, ἢ καὶ ἐν τῷ ἀλλοιουμένῳ αὐτῷ
ὑφ' αὑτοῦ καὶ αὐξανομένῳ; περὶ δὲ γενέσεως
τῆς ἐξ ἀρχῆς καὶ φθορᾶς ἄλλος λόγος· εἰ γάρ ἐστιν

¹ ἀλλ' Jaeger: ἀλλὰ P: ἀλλ' ESY.

move themselves. For while one part of the animal must be in motion, another part must be at rest, supported upon which that will be moved which is moved, if, for example, it moves one of its parts ; for one part rests on another part in virtue of the fact that the latter is at rest.

But regarding inanimate things which are moved, one might raise the question whether they all possess in themselves both that which is at rest and that which creates movement, and whether they too must be supported by something external which is at rest. Or is this impossible—for example, in the case of fire or earth or any inanimate thing—but motion is due to the primary causes by which these are moved ? For all inanimate things are moved by something else, and the origin of all the things that are thus moved is the things that move themselves. Among things of this class we have already dealt with animals ; for all such things must necessarily have within themselves that which is at rest and something outside them on which they are to support themselves. But whether there is something higher and primary which moves them is uncertain, and the question of such an origin of movement is a matter for separate discussion. But animals which move all do so supported upon things outside themselves, as also when they draw their breath in and out. For it makes no difference whether they propel a great or a small weight, as those do who spit and cough, and breathe in and out.

V. But is it only in that which moves itself in respect of place that something must remain at rest, or is this also true of that in which alteration is caused by its own agency and in that which grows ? The question of original coming into being and

700 a

ἥνπερ φαμὲν πρώτη κίνησις, γενέσεως καὶ φθορᾶς
30 αὕτη αἰτία ἂν εἴη, καὶ τῶν ἄλλων δὲ κινήσεων ἴσως
πασῶν. ὥσπερ δ' ἐν τῷ ὅλῳ, καὶ ἐν τῷ ζῴῳ
κίνησις πρώτη αὕτη, ὅταν τελεωθῇ· ὥστε καὶ
αὐξήσεως, εἴ ποτε γίνεται, αὐτὸ αὑτῷ αἴτιον καὶ
ἀλλοιώσεως, εἰ δὲ μή, οὐκ ἀνάγκη. αἱ δὲ πρῶται
35 αὐξήσεις καὶ ἀλλοιώσεις ὑπ' ἄλλου γίνονται καὶ
δι' ἑτέρων· γενέσεως δὲ καὶ φθορᾶς οὐδαμῶς οἷόν
700 b τε αὐτὸ αἴτιον εἶναι αὑτῷ οὐδέν. προϋπάρχειν
γὰρ δεῖ τὸ κινοῦν τοῦ κινουμένου καὶ τὸ γεννῶν
τοῦ γεννωμένου· αὐτὸ δ' αὑτοῦ πρότερον οὐδέν
ἐστιν.

VI. Περὶ μὲν οὖν ψυχῆς, εἴτε κινεῖται ἢ μή,
5 καὶ εἰ κινεῖται, πῶς κινεῖται, πρότερον εἴρηται ἐν
τοῖς διωρισμένοις περὶ αὐτῆς.   ἐπεὶ δὲ τὰ ἄψυχα
πάντα κινεῖται ὑφ' ἑτέρου, περὶ δὲ[1] τοῦ πρώτου
κινουμένου καὶ ἀεὶ κινουμένου, τίνα τρόπον κινεῖται,
καὶ πῶς κινεῖ τὸ πρῶτον κινοῦν, διώρισται πρότερον
ἐν τοῖς περὶ τῆς πρώτης φιλοσοφίας, λοιπὸν δ'
10 ἐστὶ θεωρῆσαι πῶς ἡ ψυχὴ κινεῖ τὸ σῶμα, καὶ
τίς ἀρχὴ τῆς τοῦ ζῴου κινήσεως. τῶν γὰρ ἄλλων
παρὰ τὴν τοῦ ὅλου κίνησιν τὰ ἔμψυχα αἴτια τῆς
κινήσεως, ὅσα μὴ κινεῖται ὑπ' ἀλλήλων διὰ τὸ
προσκόπτειν ἀλλήλοις.   διὸ καὶ πέρας ἔχουσιν
αὐτῶν πᾶσαι αἱ κινήσεις· καὶ γὰρ καὶ αἱ τῶν
15 ἐμψύχων. πάντα γὰρ τὰ ζῷα καὶ κινεῖ καὶ
κινεῖται ἕνεκά τινος, ὥστε τοῦτ' ἔστιν αὐτοῖς
πάσης τῆς κινήσεως πέρας, τὸ οὗ ἕνεκα.   ὁρῶμεν

---

[1] δὲ ES : μὲν Y.

[a] τουτέστιν . . . οὐκ ἀνάγκη εἶναί τι τῶν ἀλλοιουμένων καὶ
αὐξανομένων ὑφ' αὑτῶν ἠρεμοῦν (Mich.).
[b] i.e. the Metaphysics.

corruption is a different one ; for if there is, as we assert, a primary movement, this would be the cause of coming into being and wasting away, and perhaps of all the other movements as well. And as in the universe, so in the animal, this is primary motion, when the animal comes to perfection ; so that it is itself the cause of its own growth, if this ever takes place, and of any alteration which occurs ; otherwise it is not necessary that something should remain at rest.[a] But the first growth and alteration occur through another's agency and by other means, and nothing can in any way be itself the cause of its own coming into being and wasting away ; for that which moves must be prior to that which is moved, and that which begets to that which is begotten, and nothing is prior to itself.

VI. Now whether soul is moved or not, and if it is moved, how it is moved, has already been discussed in our treatise *On Soul*. But since all inanimate things are moved by something else—and how that which is primarily and eternally moved is moved, and how the prime mover moves it, has been already set forth in our work on *First Philosophy* [b]— it remains to inquire how the soul moves the body and what is the origin of movement in an animal. For, if we exclude the movement of the universe, animate things are the cause of movement in everything else, except in things which are moved by one another through coming into collision with one another. Therefore all their movements have a limit ; for the movements of animate things have a limit. For all animals move and are moved with some object, and so this, namely their object, is the limit of all their movement. Now we see that the

457

700 b

δὲ τὰ κινοῦντα τὸ ζῷον διάνοιαν καὶ φαντασίαν
καὶ προαίρεσιν καὶ βούλησιν καὶ ἐπιθυμίαν. ταῦτα
δὲ πάντα ἀνάγεται εἰς νοῦν καὶ ὄρεξιν. καὶ γὰρ
20 ἡ φαντασία καὶ ἡ αἴσθησις τὴν αὐτὴν τῷ νῷ χώραν
ἔχουσιν· κριτικὰ γὰρ πάντα, διαφέρουσι δὲ κατὰ
τὰς εἰρημένας ἐν ἄλλοις διαφοράς. βούλησις δὲ
καὶ θυμὸς καὶ ἐπιθυμία πάντα ὄρεξις, ἡ δὲ προ-
αίρεσις κοινὸν διανοίας καὶ ὀρέξεως· ὥστε κινεῖ
πρῶτον τὸ ὀρεκτὸν καὶ τὸ διανοητόν. οὐ πᾶν
25 δὲ τὸ διανοητόν, ἀλλὰ τὸ τῶν πρακτῶν τέλος.
διὸ τὸ τοιοῦτόν ἐστι τῶν ἀγαθῶν τὸ κινοῦν, ἀλλ'
οὐ πᾶν τὸ καλόν· ᾗ γὰρ ἕνεκα τούτου ἄλλο, καὶ
ᾗ τέλος ἐστὶ τῶν ἄλλου τινὸς ἕνεκα ὄντων, ταύτῃ
κινεῖ. δεῖ δὲ τιθέναι καὶ τὸ φαινόμενον ἀγαθὸν
ἀγαθοῦ χώραν ἔχειν, καὶ τὸ ἡδύ· φαινόμενον γάρ
30 ἐστιν ἀγαθόν. ὥστε δῆλον ὅτι ἔστι μὲν ᾗ ὁμοίως
κινεῖται τὸ ἀεὶ κινούμενον ὑπὸ τοῦ ἀεὶ κινοῦντος
καὶ τῶν ζῴων ἕκαστον, ἔστι δ' ᾗ ἄλλως, διὸ καὶ
τὰ μὲν ἀεὶ κινεῖται, ἡ δὲ τῶν ζῴων κίνησις ἔχει
πέρας. τὸ δὲ ἀΐδιον καλόν, καὶ τὸ ἀληθῶς καὶ
πρώτως ἀγαθὸν καὶ μὴ ποτὲ μὲν ποτὲ δὲ μή,
35 θειότερον καὶ τιμιώτερον ἢ ὥστ' εἶναι πρότερον[1] τι[2].

Τὸ μὲν οὖν πρῶτον οὐ κινούμενον κινεῖ, ἡ δ'
701 a ὄρεξις καὶ τὸ ὀρεκτικὸν κινούμενον κινεῖ. τὸ δὲ
τελευταῖον τῶν κινουμένων οὐκ ἀνάγκη κινεῖν
οὐδέν. φανερὸν δ' ἐκ τούτων καὶ ὅτι εὐλόγως

[1] πρότερον ESY : πρὸς ἕτερον P.
[2] τι add. Jaeger.

---

[a] *De anima*, iii. 427 b 14 ff.

things which move the animal are intellect, imagination, purpose, wish and appetite. Now all these can be referred to mind and desire. For imagination and sensation cover the same ground as the mind (since they all exercise judgement) though they differ in certain aspects as has been defined elsewhere.[a] But will, temper, and appetite are all forms of desire, while purpose partakes both of intellect and of desire. So the objects of desire and intellect first set up movement—not, however, every object of intellect, but only the end in the sphere of action. So amongst good things it is the good in the sphere of action that sets up movement, and not any and every good; for it sets up movement only in so far as it is the motive of something else or the end of something which has something else as its object. And we must lay down the principle that the apparent good can take the place of a real good, and so can the pleasant, for it is an apparent good. So that it is clear that in one respect that which is eternally moved by the eternal mover, and the individual animal, are moved in a similar manner, but that in another respect they are moved differently ; and so, while other things move eternally, animal movement has a limit. Now the eternally beautiful and that which is truly and primarily good, and not at one moment good and at another not good, is too divine and precious to have anything prior to it.

The prime mover, then, moves without itself being moved, but desire and the desiderative faculty set up movement while being themselves moved. But it is not necessary that the last of a series of things which are moved should move anything ; and from this it is clear that it is only reasonable that pro-

701 a

ἡ φορὰ τελευταία τῶν γινομένων ἐν τοῖς κινουμένοις[1]· κινεῖται γὰρ καὶ πορεύεται τὸ ζῷον ὀρέξει
5 ἢ προαιρέσει, ἀλλοιωθέντος τινὸς κατὰ τὴν αἴσθησιν ἢ τὴν φαντασίαν.

VII. Πῶς δὲ νοῶν ὁτὲ μὲν πράττει ὁτὲ δ' οὐ
πράττει, καὶ κινεῖται, ὁτὲ δ' οὐ κινεῖται; ἔοικε
παραπλησίως συμβαίνειν καὶ περὶ τῶν ἀκινήτων
διανοουμένοις καὶ συλλογιζομένοις. ἀλλ' ἐκεῖ μὲν
10 θεώρημα τὸ τέλος (ὅταν γὰρ τὰς δύο προτάσεις
νοήσῃ, τὸ συμπέρασμα ἐνόησε καὶ συνέθηκεν),
ἐνταῦθα δ' ἐκ τῶν δύο προτάσεων τὸ συμπέρασμα
γίνεται ἡ πρᾶξις, οἷον ὅταν νοήσῃ ὅτι παντὶ βαδιστέον ἀνθρώπῳ, αὐτὸς δ' ἄνθρωπος, βαδίζει
εὐθέως, ἂν δ' ὅτι οὐδενὶ βαδιστέον νῦν ἀνθρώπῳ,
15 αὐτὸς δ' ἄνθρωπος, εὐθὺς ἠρεμεῖ· καὶ ταῦτα ἄμφω
πράττει, ἂν μή τι κωλύῃ ἢ ἀναγκάζῃ. ποιητέον
μοι ἀγαθόν, οἰκία δ' ἀγαθόν· ποιεῖ οἰκίαν εὐθύς.
σκεπάσματος δέομαι, ἱμάτιον δὲ σκέπασμα· ἱματίου
δέομαι. οὗ δέομαι, ποιητέον· ἱματίου δέομαι·
20 ἱμάτιον ποιητέον. καὶ τὸ συμπέρασμα, τὸ ἱμάτιον
ποιητέον, πρᾶξίς ἐστιν. πράττει δ' ἀπ' ἀρχῆς.
εἰ ἱμάτιον ἔσται, ἀνάγκη τόδε πρῶτον, εἰ δὲ τόδε,
τόδε· καὶ τοῦτο πράττει εὐθύς. ὅτι μὲν οὖν ἡ
πρᾶξις τὸ συμπέρασμα, φανερόν· αἱ δὲ προτάσεις
αἱ ποιητικαὶ διὰ δύο εἰδῶν γίνονται, διά τε τοῦ
25 ἀγαθοῦ καὶ διὰ τοῦ δυνατοῦ.

Ὥσπερ δὲ τῶν ἐρωτώντων ἔνιοι, οὕτω τὴν ἑτέραν

---

[1] κινουμένοις Jaeger : γιγνομένοις libri.

---

[a] *i.e.* the objects of science ; *cf. An. Post.* 71 b 18 ff.

460

gression should be the last thing to happen in things that are moved, since the animal is moved and walks from desire or purpose, when some alteration has been caused as the result of sensation or imagination.

VII. But why is it that thought sometimes results in action and sometimes does not, sometimes in movement and sometimes not? Apparently the same kind of thing happens as when one thinks and forms an inference about immovable objects.[a] But in the latter case, the end is speculation (for when you have conceived the two premises, you immediately conceive and infer the conclusion) ; but in the former case the conclusion drawn from the two premises becomes the action. For example, when you conceive that every man ought to walk and you yourself are a man, you immediately walk ; or if you conceive that on a particular occasion no man ought to walk, and you yourself are a man, you immediately remain at rest. In both instances action follows unless there is some hindrance or compulsion. Again, I ought to create a good, and a house is a good, I immediately create a house. Again, I need a covering, and a cloak is a covering, I need a cloak. What I need I ought to make ; I need a cloak, I ought to make a cloak. And the conclusion " I ought to make a cloak " is an action. The action results from the beginning of the train of thought. If there is to be a cloak, such and such a thing is necessary, if this thing then something else ; and one immediately acts accordingly. That the action is the conclusion is quite clear ; but the premises which lead to the doing of something are of two kinds, through the good and through the possible.

And as those sometimes do who are eliciting con-

701 a

πρότασιν τὴν δήλην οὐδ' ἡ διάνοια ἐφιστᾶσα σκοπεῖ
οὐδέν· οἷον εἰ τὸ βαδίζειν ἀγαθὸν ἀνθρώπῳ, ὅτι
αὐτὸς ἄνθρωπος, οὐκ ἐνδιατρίβει. διὸ καὶ ὅσα μὴ
λογισάμενοι πράττομεν, ταχὺ πράττομεν. ὅταν γὰρ
ἐνεργήσῃ ἢ τῇ αἰσθήσει πρὸς τὸ οὗ ἕνεκα ἢ τῇ
30 φαντασίᾳ ἢ τῷ νῷ, οὗ ὀρέγεται, εὐθὺς ποιεῖ· ἀντ'
ἐρωτήσεως γὰρ ἢ νοήσεως ἡ τῆς ὀρέξεως γίνεται ἐν-
έργεια. ποτέον μοι, ἡ ἐπιθυμία λέγει· τοδὶ δὲ ποτόν,
ἡ αἴσθησις εἶπεν ἢ ἡ φαντασία ἢ ὁ νοῦς· εὐθὺς πίνει.
οὕτως μὲν οὖν ἐπὶ τὸ κινεῖσθαι καὶ πράττειν τὰ
ζῷα ὁρμῶσι, τῆς μὲν ἐσχάτης αἰτίας τοῦ κινεῖσθαι
35 ὀρέξεως οὔσης, ταύτης δὲ γινομένης ἢ δι' αἰσθήσεως
ἢ διὰ φαντασίας καὶ νοήσεως. τῶν δ' ὀρεγομένων
πράττειν τὰ μὲν δι' ἐπιθυμίαν ἢ θυμὸν τὰ δὲ δι'
701 b ὄρεξιν ἢ βούλησιν τὰ μὲν ποιοῦσι, τὰ δὲ πράττουσιν.

Ὥσπερ δὲ τὰ αὐτόματα κινεῖται μικρᾶς κινήσεως
γινομένης, λυομένων τῶν στρεβλῶν καὶ κρουουσῶν[1]
ἀλλήλας [τὰς στρέβλας],[2] καὶ τὸ ἁμάξιον, ὅπερ
5 ⟨τὸ⟩[3] ὀχούμενον αὐτὸ κινεῖ εἰς εὐθύ, καὶ πάλιν
κύκλῳ κινεῖται τῷ ἀνίσους ἔχειν τοὺς τροχούς
(ὁ γὰρ ἐλάττων ὥσπερ κέντρον γίνεται, καθάπερ
ἐν τοῖς κυλίνδροις), οὕτω καὶ τὰ ζῷα κινεῖται.
ἔχει γὰρ ὄργανα τοιαῦτα τήν τε τῶν νεύρων
φύσιν καὶ τὴν τῶν ὀστῶν, τὰ μὲν ὡς ἐκεῖ τὰ

---

[1] κρουουσῶν scripsi (Leon. renders *laxatis seque mutuo im-*
pellentibus vertebris): κρουόντων libri.
[2] τὰς στρέβλας seclusi.   [3] τὸ addidi.

---

[a] For this technical use of ἐρωτᾶν *cf. An. Prior.* 24 a 24.
[b] By the removal of the pegs (ξύλα), *cf.* below, 701 b 9, 10.
[c] The context seems to show that the toy-carriage was
on an axle which coupled two wheels of unequal diameter.
There is, however, no evidence for the existence of such toy-
carriages in antiquity.

clusions by questioning,[a] so here the mind does not stop and consider at all one of the two premisses, namely, the obvious one; for example, if walking is good for a man, one does not waste time over the premiss "I am myself a man." Hence such things as we do without calculation, we do quickly. For when a man acts for the object which he has in view from either perception or imagination or thought, he immediately does what he desires; the carrying out of his desire takes the place of inquiry or thought. My appetite says, I must drink; this is drink, says sensation or imagination or thought, and one immediately drinks. It is in this manner that animals are impelled to move and act, the final cause of their movement being desire; and this comes into being through either sensation or imagination and thought. And things which desire to act, at one time create something, and at another act, by reason either of appetite or of passion, or else through desire or wish.

The movement of animals resembles that of marionettes which move as the result of a small movement, when the strings are released [b] and strike one another; or a toy-carriage which the child that is riding upon it himself sets in motion in a straight direction, and which afterwards moves in a circle because its wheels are unequal, for the smaller wheel acts as a centre,[c] as happens also in the cylinders.[d] Animals have similar parts in their organs, namely, the growth of their sinews and bones, the latter corresponding to the pegs in the marionettes and the

[d] The marionettes seem to have been worked by means of cylinders round which weighted strings were wound, the cylinders being set in motion by the removal of pegs.

701 b

ξύλα καὶ ὁ σίδηρος, τὰ δὲ νεῦρα ὡς αἱ στρέβλαι·
10 ὧν λυομένων καὶ ἀνιεμένων κινοῦνται. ἐν μὲν
οὖν τοῖς αὐτομάτοις καὶ τοῖς ἁμαξίοις οὐκ ἔστιν
ἀλλοίωσις, ἐπεὶ εἰ ἐγίνοντο ἐλάττους οἱ ἐντὸς
τροχοὶ καὶ πάλιν μείζους, κἂν κύκλῳ τὸ αὐτὸ
ἐκινεῖτο· ἐν δὲ τῷ ζῴῳ δύναται τὸ αὐτὸ καὶ
μεῖζον καὶ ἔλαττον γίνεσθαι καὶ τὰ σχήματα μετα-
15 βάλλειν, αὐξανομένων τῶν μορίων διὰ θερμότητα
καὶ πάλιν συστελλομένων διὰ ψύξιν καὶ ἀλλοιουμένων. ἀλλοιοῦσι δ' αἱ φαντασίαι καὶ αἱ αἰσθήσεις
καὶ αἱ ἔννοιαι. αἱ μὲν γὰρ αἰσθήσεις εὐθὺς ὑπάρχουσιν ἀλλοιώσεις τινὲς οὖσαι, ἡ δὲ φαντασία καὶ
ἡ νόησις τὴν τῶν πραγμάτων ἔχουσι δύναμιν· τρό-
20 πον γάρ τινα τὸ εἶδος τὸ νοούμενον τὸ τοῦ θερμοῦ
ἢ ψυχροῦ ἢ ἡδέος ἢ φοβεροῦ τοιοῦτον τυγχάνει
ὂν οἷόν περ καὶ τῶν πραγμάτων ἕκαστον, διὸ καὶ
φρίττουσι καὶ φοβοῦνται νοήσαντες μόνον. ταῦτα
δὲ πάντα πάθη καὶ ἀλλοιώσεις εἰσίν. ἀλλοιουμένων δ' ἐν τῷ σώματι τὰ μὲν μείζω τὰ δ' ἐλάττω
25 γίνεται. ὅτι δὲ μικρὰ μεταβολὴ γινομένη ἐν ἀρχῇ
μεγάλας καὶ πολλὰς ποιεῖ διαφορὰς ἄποθεν, οὐκ
ἄδηλον· οἷον τοῦ οἴακος ἀκαριαῖόν τι μεθισταμένου
πολλὴ ἡ τῆς πρῴρας γίνεται μετάστασις. ἔτι δὲ
κατὰ θερμότητα ἢ ψύξιν ἢ κατ' ἄλλο τι τοιοῦτον
πάθος ὅταν γένηται ἀλλοίωσις περὶ τὴν καρδίαν,
30 καὶ ἐν ταύτῃ κατὰ μέγεθος ἐν ἀναισθήτῳ μορίῳ,
πολλὴν ποιεῖ τοῦ σώματος διαφορὰν ἐρυθήμασι
καὶ ὠχρότησι καὶ φρίκαις καὶ τρόμοις καὶ τοῖς
τούτων ἐναντίοις.

VIII. Ἀρχὴ μὲν οὖν, ὥσπερ εἴρηται, τῆς

---

^a The reference is probably to some part of the toy-carriage.

iron,[a] while the sinews correspond to the strings, the setting free and loosening of which causes the movement. In the marionettes and the toy-carriages no alteration takes place, though, if the inner wheels were to become smaller and then again larger, the same circular movement would take place. In the animal, however, the same part can become both greater and smaller and change its form, the members increasing through heat and contracting again through cold and thus altering. Alteration is caused by imagination and sensations and thoughts. For sensations are from the first a kind of alteration, and imagination and thought have the effect of the objects which they present ; for in a way the idea conceived—of hot or cold or pleasant or terrible—is really of the same kind as an object possessing one of these qualities, and so we shudder and feel fear simply by conceiving an idea ; and all these affections are alterations, and when an alteration takes place in the body some parts become larger, others smaller. Now it is clear that a small change taking place in an origin of movement [b] causes great and numerous changes at a distance ; just as, if the rudder of a boat is moved to an infinitesimal extent, the change resulting in the position of the bows is considerable. Furthermore, when, owing to heat or cold or a similar affection, an alteration is caused in the region of the heart—and even in an imperceptibly small part of it—it gives rise to a considerable change in the body, causing blushing or pallor or shuddering or trembling or the opposites of these.

VIII. The origin, then, of movement, as has already

[b] *i.e.* here, the heart, *cf.* below, 701 b 30 ; see also note on 698 b 1.

701 b

κινήσεως τὸ ἐν τῷ πρακτῷ διωκτὸν καὶ φευκτόν·
ἐξ ἀνάγκης δ' ἀκολουθεῖ τῇ νοήσει καὶ τῇ φαντασίᾳ
35 αὐτῶν θερμότης καὶ ψῦξις. τὸ μὲν γὰρ λυπηρὸν
φευκτόν, τὸ δ' ἡδὺ διωκτόν (ἀλλὰ λανθάνει περὶ
τὰ μικρὰ τοῦτο συμβαῖνον), ἔστι δὲ τὰ λυπηρὰ
702 a καὶ ἡδέα πάντα σχεδὸν μετὰ ψύξεώς τινος καὶ
θερμότητος. τοῦτο δὲ δῆλον ἐκ τῶν παθημάτων.
θάρρη γὰρ καὶ φόβοι καὶ ἀφροδισιασμοὶ καὶ τἆλλα
τὰ σωματικὰ λυπηρὰ καὶ ἡδέα τὰ μὲν κατὰ μόριον
μετὰ θερμότητος ἢ ψύξεώς ἐστι, τὰ δὲ καθ' ὅλον
5 τὸ σῶμα· μνῆμαι δὲ καὶ ἐλπίδες, οἷον εἰδώλοις
χρώμεναι τοῖς τοιούτοις, ὁτὲ μὲν ἧττον ὁτὲ δὲ
μᾶλλον αἴτιαι τῶν αὐτῶν εἰσίν. ὥστ' εὐλόγως
ἤδη δημιουργεῖται τὰ ἐντὸς καὶ τὰ περὶ τὰς ἀρχὰς
τῶν ὀργανικῶν μορίων μεταβάλλοντα ἐκ πεπηγότων
10 ὑγρὰ καὶ ἐξ ὑγρῶν πεπηγότα καὶ μαλακὰ καὶ
σκληρὰ ἐξ ἀλλήλων. τούτων δὲ συμβαινόντων
τὸν τρόπον τοῦτον, καὶ ἔτι τοῦ παθητικοῦ καὶ
ποιητικοῦ τοιαύτην ἐχόντων τὴν φύσιν οἵαν πολ-
λαχοῦ εἰρήκαμεν, ὁπόταν συμβῇ ὥστ' εἶναι τὸ
μὲν ποιητικὸν τὸ δὲ παθητικόν, καὶ μηδὲν ἀπολίπῃ
15 αὐτῶν ἑκάτερον τῶν ἐν τῷ λόγῳ, εὐθὺς τὸ μὲν
ποιεῖ τὸ δὲ πάσχει. διὰ τοῦτο δ' ἅμα ὡς εἰπεῖν
νοεῖ ὅτι πορευτέον καὶ πορεύεται, ἂν μή τι ἐμ-
ποδίζῃ ἕτερον. τὰ μὲν γὰρ ὀργανικὰ μέρη παρασκευάζει ἐπιτηδείως τὰ πάθη, ἡ δ' ὄρεξις τὰ
πάθη, τὴν δ' ὄρεξιν ἡ φαντασία· αὕτη δὲ γίνεται
ἢ διὰ νοήσεως ἢ δι' αἰσθήσεως. ἅμα δὲ καὶ ταχὺ
20 διὰ τὸ ⟨τὸ⟩[1] ποιητικὸν καὶ παθητικὸν τῶν πρὸς
ἄλληλα εἶναι τὴν φύσιν.

[1] τὸ add. Bonitz.

been said, is the object of pursuit or avoidance in the sphere of action, and heat and cold necessarily follow the thought and imagination of these objects. For what is painful is avoided, and what is pleasant is pursued. We do not, it is true, notice the effect of this in the minute parts of the body ; but practically anything painful or pleasant is accompanied by some degree of chilling or heating. This is clear from the effects produced. Reckless daring, terrors, sexual emotions and the other bodily affections, both painful and pleasant, are accompanied by heating or chilling, either local or throughout the body. Recollections too and anticipations, employing, as it were, the images of such feelings, are to a greater or less degree the cause of the same effects. So it is with good reason that the inner portions of the body and those which are situated near the origins of the motion of the organic parts are created as they are, changing as they do from solid to liquid and from liquid to solid and from soft to hard and *vice versa.* Since, then, these processes occur in this way, and since, moreover, the passive and the active principles have the nature which we have frequently ascribed to them, whenever it so happens that the one is active and the other passive and neither fails to fulfil its definition, immediately the one acts and the other is acted upon. So a man thinks he ought to go, and goes, practically at the same time, unless something else hinders him. For the affections fittingly prepare the organic parts, the desire prepares the affections, and the imagination prepares the desire, while the imagination is due to thought or sensation. The process is simultaneous and quick, because the active and the passive are by nature closely interrelated.

702 a

Τὸ δὲ κινοῦν πρῶτον τὸ ζῷον ἀνάγκη εἶναι ἔν τινι
ἀρχῇ. ἡ δὲ καμπὴ ὅτι μέν ἐστι τοῦ μὲν ἀρχὴ τοῦ
δὲ τελευτή, εἴρηται. διὸ καὶ ἔστι μὲν ὡς ἑνί, ἔστι
δ᾽ ὡς δυσὶ χρῆται ἡ φύσις αὐτῇ. ὅταν γὰρ κινῆται
25 ἐντεῦθεν, ἀνάγκη τὸ μὲν ἠρεμεῖν τῶν σημείων τῶν
ἐσχάτων, τὸ δὲ κινεῖσθαι· ὅτι γὰρ πρὸς ἠρεμοῦν δεῖ
ἀπερείδεσθαι τὸ κινοῦν, εἴρηται πρότερον. κινεῖται
μὲν οὖν καὶ οὐ κινεῖ τὸ ἔσχατον τοῦ βραχίονος, τῆς δ᾽
ἐν τῷ ὠλεκράνῳ κάμψεως τὸ μὲν κινεῖται τὸ ἐν αὐτῷ
τῷ ὅλῳ κινουμένῳ, ἀνάγκη δ᾽ εἶναί τι καὶ ἀκίνητον,
30 ὃ δή φαμεν δυνάμει μὲν ἓν εἶναι σημεῖον, ἐνεργείᾳ
δὲ γίνεσθαι δύο· ὥστ᾽ εἰ τὸ ζῷον ἦν βραχίων, ἐν-
ταῦθ᾽ ἄν που ἦν ἡ ἀρχὴ τῆς ψυχῆς ἡ κινοῦσα.
ἐπεὶ δ᾽ ἐνδέχεται καὶ πρὸς τὴν χεῖρα ἔχειν τι
οὕτως τῶν ἀψύχων, οἷον εἰ κινοίη τὴν βακτηρίαν
ἐν τῇ χειρί, φανερὸν ὅτι οὐκ ἂν εἴη ἐν οὐδετέρῳ
35 ἡ ψυχὴ τῶν ἐσχάτων, οὔτ᾽ ἐν τῷ ἐσχάτῳ τοῦ
κινουμένου οὔτ᾽ ἐν τῇ ἑτέρᾳ ἀρχῇ. καὶ γὰρ τὸ
702 b ξύλον ἔχει καὶ ἀρχὴν καὶ τέλος πρὸς τὴν χεῖρα.
ὥστε διά γε τοῦτο, εἰ μὴ ἐν τῇ βακτηρίᾳ ἡ κινοῦσα
ἀπὸ τῆς ψυχῆς ἀρχὴ ἔνεστιν, οὐδ᾽ ἐν τῇ χειρί·
ὁμοίως γὰρ ἔχει καὶ τὸ ἄκρον τῆς χειρὸς πρὸς τὸν
καρπόν, καὶ τοῦτο τὸ μέρος πρὸς τὸ ὠλέκρανον.
5 οὐδὲν γὰρ διαφέρει τὰ προσπεφυκότα τῶν μὴ

---

[a] *i.e.* the same relation as the forearm has to the elbow.
[b] *i.e.* the end of the stick where it meets the hand.
[c] *i.e.* the origin of the movement of the hand which is
situated in the wrist.
[d] It is impossible to find a word in English which covers
the double meaning given to ἀρχή here and in the previous line
(see note on 698 b 1). The sentence καὶ γὰρ τὸ ξύλον . . . χεῖρα
explains why the ἀρχὴ κινήσεως of the hand is called ἡ ἑτέρα
ἀρχή, viz. that there is another ἀρχή (in the sense of " beginning ") in the stick, namely, the point nearest the hand.

Now that which first causes movement in the animal must be situated in a definite beginning. Now it has already been stated that the joint is the beginning of one thing and the end of another ; wherefore nature employs it sometimes as one and sometimes as two. For when movement is being originated from it, one of its extreme points must be at rest, while the other must move ; for we have already said that what causes movement must be supported on something which is at rest. The extremity, therefore, of the forearm is moved and does not cause movement, but in the elbow-joint one part, namely that which is situated in the actual whole which is in motion, is moved, but there must also be something which is unmoved ; and this is what we mean when we say that a point is potentially one but becomes actually two. So if the forearm were a living creature, it is somewhere near this point that the origin of movement set in motion by the soul would be situated. Since, however, it is possible for an inanimate object to bear this same relation to the hand,[a] for instance if one moves a stick in one's hand, it is clear that the soul could not be situated in either of the extremities, neither in the extremity of that which is moved [b] nor in the other origin of movement (ἀρχή) [c] ; for the stick has an end and a beginning (ἀρχή) [d] in relation to the hand. So, for this reason, if the origin of movement set up by the soul is not situated in the stick, it is not situated in the hand either ; for the extremity of the hand [e] bears the same relation to the wrist as the latter does to the elbow. For there is no difference between what is attached by growth and what is not

[a] *i.e.* the point where the hand joins the stick.

702 b

γίνεται γὰρ ὥσπερ ἀφαιρετὸν μέρος ἡ βακτηρία.
ἀνάγκη ἄρα ἐν μηδεμιᾷ εἶναι ἀρχῇ, ἢ ἐστιν ἄλλου
τελευτή, μηδὲ εἴ τι ἐστὶν ἕτερον ἐκείνου ἐξωτέρω,
οἷον τοῦ μὲν τῆς βακτηρίας ἐσχάτου ἐν τῇ χειρὶ
ἡ ἀρχή, τούτου δ' ἐν καρπῷ. εἰ δὲ μηδ' ἐν τῇ
10 χειρί, ὅτι ἀνωτέρω ἔτι, ἡ ἀρχὴ οὐδ' ἐνταῦθα· ἔτι
γὰρ τοῦ ὠλεκράνου μένοντος κινεῖται ἅπαν τὸ
κάτω συνεχές.

IX. Ἐπεὶ δ' ὁμοίως ἔχει ἀπὸ τῶν ἀριστερῶν
καὶ ἀπὸ τῶν δεξιῶν, καὶ ἅμα τἀναντία κινεῖται,
ὥστε μὴ εἶναι τῷ ἠρεμεῖν τὸ δεξιὸν κινεῖσθαι τὸ
ἀριστερὸν μηδ' αὖ τῷ τοῦτο ἐκεῖνο, ἀεὶ δ' ἐν τῷ
15 ἀνωτέρω ἀμφοτέρων ἡ ἀρχή, ἀνάγκη ἐν τῷ μέσῳ
εἶναι τὴν ἀρχὴν τῆς ψυχῆς τῆς κινούσης· ἀμφοτέρων
γὰρ τῶν ἄκρων τὸ μέσον ἔσχατον. ὁμοίως δ' ἔχει
πρὸς τὰς κινήσεις τοῦτο καὶ τὰς ἀπὸ τοῦ ἄνω καὶ
κάτω, οἷον τὰς ἀπὸ τῆς κεφαλῆς καὶ[1] τὰς ἀπὸ τῆς
20 ῥάχεως τοῖς ἔχουσι ῥάχιν. καὶ εὐλόγως δὲ τοῦτο
συμβέβηκεν· καὶ γὰρ τὸ αἰσθητικὸν ἐνταῦθα εἶναί
φαμεν, ὥστ' ἀλλοιουμένου διὰ τὴν αἴσθησιν τοῦ
τόπου τοῦ περὶ τὴν ἀρχὴν καὶ μεταβάλλοντος τὰ
ἐχόμενα συμμεταβάλλει ἐκτεινόμενά τε καὶ συναγό-
μενα τὰ μόρια, ὥστ' ἐξ ἀνάγκης διὰ ταῦτα γίνεσθαι
25 τὴν κίνησιν τοῖς ζῴοις. τὸ δὲ μέσον τοῦ σώματος

---

[1] καὶ scripsi : πρὸς libri.

---

a This is simply a restatement of the doctrine of 702 b 1-4.
The true ἀρχή is not situated in the extremity of the stick
nearest to the hand (which is an ἀρχή as being the place
where the stick begins in relation to the hand), nor yet in any
other member, such as the wrist, which is still farther away
from the stick and is an ἀρχή as being the origin of motion
in the hand. The wrist, elbow, and shoulder are all of them

so attached ; for the stick becomes a kind of detached member. The origin of movement, therefore, cannot be situated in any origin which is the termination of something else, nor in any other part which is farther from it ; for example, the origin of movement of the extremity of the stick is in the hand, but the origin of the movement of the hand is in the wrist.[a] And so if the origin of movement is not in the hand, because it is still higher up,[b] neither is it in this higher position ; for, again, if the elbow is at rest, the continuous part below it can be set in motion as a whole.

IX. Now since there is similarity in the left and the right sides of the body, and the opposite parts can be moved simultaneously, so that it is impossible for the right side to move just because the left is at rest or *vice versa*, and the origin of movement must be in that which lies above both sides, it necessarily follows that the origin of movement in the moving soul must be between them ; for the middle is the limit of both extremes. And it stands in the same relation to the movements above as to those below, to those, for example, which proceed from the head and to those which proceed from the spine in animals which have a spine. And there is good reason for this ; for we say that the organ of sensation is also situated in the centre of the body ; and so if the region round about the origin of movement is altered by sense-perception and undergoes change, the parts which are attached to it change with it by extension or contraction, so that in this way movement necessarily takes place in animals. And the central part of the

ἀρχαί in relation to the parts below them, but the true ἀρχή is situated in the soul, which lies in the centre of the body.
[b] *i.e.* the wrist.

702 b

μέρος δυνάμει μὲν ἔν, ἐνεργείᾳ δ' ἀνάγκη γίνεσθαι
πλείω· καὶ γὰρ ἅμα κινεῖται τὰ κῶλα ἀπὸ τῆς
ἀρχῆς, καὶ θατέρου ἠρεμοῦντος θάτερον κινεῖται.
λέγω δ' οἷον ἐπὶ τῆς ΑΒΓ τὸ Β κινεῖται, κινεῖ
δὲ τὸ Α. ἀλλὰ μὴν δεῖ γέ τι ἠρεμεῖν, εἰ μέλλει
30 τὸ μὲν κινεῖσθαι τὸ δὲ κινεῖν. ἐν ἄρα δυνάμει ὂν
τὸ Α ἐνεργείᾳ δύο ἔσται, ὥστ' ἀνάγκη μὴ στιγμὴν
ἀλλὰ μέγεθός τι εἶναι. ἀλλὰ μὴν ἐνδέχεται τὸ Γ
ἅμα τῷ Β κινεῖσθαι, ὥστ' ἀνάγκη ἀμφοτέρας τὰς
ἀρχὰς τὰς ἐν τῷ Α κινουμένας κινεῖν. δεῖ τι ἄρα
35 εἶναι παρὰ ταύτας ἕτερον τὸ κινοῦν καὶ μὴ κινούμενον. ἀπερείδοιντο μὲν γὰρ ἂν τὰ ἄκρα καὶ αἱ
ἀρχαὶ αἱ ἐν τῷ Α πρὸς ἀλλήλας κινουμένων, ὥσπερ
703 a ἂν εἴ τινες τὰ νῶτα ἀντερείδοντες κινοῖεν τὰ σκέλη.
ἀλλὰ τὸ κινοῦν ἄμφω ἀναγκαῖον εἶναι. τοῦτο δ'
ἐστὶν ἡ ψυχή, ἕτερον μὲν οὖσα τοῦ μεγέθους τοῦ
τοιούτου, ἐν τούτῳ δ' οὖσα.

X. Κατὰ μὲν οὖν τὸν λόγον τὸν λέγοντα τὴν
5 αἰτίαν τῆς κινήσεως ἐστὶν ἡ ὄρεξις τὸ μέσον, ὃ
κινεῖ κινούμενον· ἐν δὲ τοῖς ἐμψύχοις σώμασι δεῖ
τι εἶναι σῶμα τοιοῦτον. τὸ μὲν οὖν κινούμενον
μὲν μὴ πεφυκὸς δὲ κινεῖν δύναται πάσχειν κατ'
ἀλλοτρίαν δύναμιν· τὸ δὲ κινοῦν ἀναγκαῖον ἔχειν
τινὰ δύναμιν καὶ ἰσχύν. πάντα δὲ φαίνεται τὰ
10 ζῷα καὶ ἔχοντα πνεῦμα σύμφυτον καὶ ἰσχύοντα
τούτῳ. (τίς μὲν οὖν ἡ σωτηρία τοῦ συμφύτου
πνεύματος, εἴρηται ἐν ἄλλοις.) τοῦτο δὲ πρὸς
τὴν ἀρχὴν τὴν ψυχικὴν ἔοικεν ὁμοίως ἔχειν ὥσπερ

―――――――――――――――――
ᵃ See Introd. p. 436.

472

body is potentially one, but actually must necessarily become more than one; for the limbs are set in motion simultaneously from the origin of movement, and when one is at rest the other is in motion. For example, in ABC, B is moved and A moves it; there must, however, be something at rest if one thing is to be moved and another is to move it. So A, though potentially one, will be actually two, so that it must be not a point but a magnitude. Again, C may be moved simultaneously with B, so that both the origins in A must cause movement by being moved; there must, therefore, be something other than these origins which causes movement without being itself moved. Otherwise, when movement took place, the extremities, or origins, in A would rest upon one another, like men standing back to back and moving their limbs. There must be something which moves them both, namely the soul, other than such a magnitude as we have described but situated in it.

X. In accordance with the definition which defines the cause of motion, desire is the central origin, which moves by being itself moved; but in animate bodies there must be some bodily substance which has these characteristics. That, then, which is moved but does not possess the natural quality of setting up movement may be affected by a power external to it, and that which causes movement must possess some power and strength. Now all animals clearly both possess an innate spirit and exercise their strength in virtue of it. (What it is that conserves the innate spirit has been explained elsewhere.[a]) This spirit seems to bear the same relation to the origin in the

τι ἐν ταῖς καμπαῖς σημεῖον, τὸ κινοῦν καὶ κινούμενον, πρὸς τὸ ἀκίνητον. ἐπεὶ δ' ἡ ἀρχὴ τοῖς μὲν ἐν τῇ καρδίᾳ τοῖς δ' ἐν τῷ ἀνάλογον, διὰ τοῦτο καὶ τὸ πνεῦμα τὸ σύμφυτον ἐνταῦθα φαίνεται ὄν.

15 πότερον μὲν οὖν ταὐτόν ἐστι τὸ πνεῦμα ἀεὶ ἢ γίνεται ἀεὶ ἕτερον, ἔστω ἄλλος λόγος (ὁ αὐτὸς γάρ ἐστι καὶ περὶ τῶν ἄλλων μορίων)· φαίνεται δ' εὐφυῶς ἔχον πρὸς τὸ κινητικὸν εἶναι καὶ παρέχειν ἰσχύν. τὰ δ' ἔργα τῆς κινήσεως ὦσις καὶ ἕλξις,

20 ὥστε δεῖ τὸ ὄργανον αὐξάνεσθαί τε δύνασθαι καὶ συστέλλεσθαι. τοιαύτη δ' ἐστὶν ἡ τοῦ πνεύματος φύσις· καὶ γὰρ ἀβίαστος συστελλομένη, καὶ βιαστικὴ καὶ ὠστικὴ διὰ τὴν αὐτὴν αἰτίαν, καὶ ἔχει καὶ βάρος πρὸς τὰ πυρώδη καὶ κουφότητα πρὸς τὰ ἐναντία. δεῖ δὲ τὸ μέλλον κινεῖν μὴ

25 ἀλλοιώσει τοιοῦτον εἶναι· κρατεῖ γὰρ κατὰ τὴν ὑπεροχὴν τὰ φυσικὰ σώματα ἀλλήλων, τὸ μὲν κοῦφον κάτω ὑπὸ τοῦ βαρυτέρου ἀπονικώμενον, τὸ δὲ βαρὺ ἄνω ὑπὸ τοῦ κουφοτέρου.

Ὧι μὲν οὖν κινεῖ κινουμένῳ μορίῳ ἡ ψυχή, εἴρηται, καὶ δι' ἣν αἰτίαν· ὑποληπτέον δὲ συνεστάναι τὸ

30 ζῷον ὥσπερ πόλιν εὐνομουμένην. ἔν τε γὰρ τῇ πόλει ὅταν ἅπαξ συστῇ¹ ἡ τάξις, οὐδὲν δεῖ κεχωρισμένου μονάρχου, ὃν δεῖ παρεῖναι παρ' ἕκαστον τῶν γινομένων, ἀλλ' αὐτὸς ἕκαστος ποιεῖ τὰ αὑτοῦ ὡς τέτακται, καὶ γίνεται τόδε μετὰ τόδε διὰ

¹ συστῇ P : στῇ ESY.

ᵃ For this meaning of ἀβίαστος cf. Plato, Tim. 61 A. The action of the πνεῦμα is represented as resembling that of the breath in the lungs ; when the breath contracts it lacks force and the lungs collapse, when it expands it thrusts outwards and exercises force.

ᵇ Namely, expansion.

soul as the point in the joints, which moves and is moved, bears to that which is unmoved. Now since the origin is in some animals situated in the heart, in others in what corresponds to the heart, it is therefore clear that the innate spirit also is situated there. Whether the spirit is always the same or is always changing must be discussed elsewhere (for the same question arises about the other parts of the body) ; at any rate it is clearly well adapted by nature to be a motive power and to exercise strength. Now the functions of movement are thrusting and pulling, so that the organ of movement must be able to increase and contract. And the nature of spirit has these qualities ; for when it contracts it is without force,[a] and one and the same cause [b] gives it force and enables it to thrust, and it possesses weight as compared with the fiery element, and lightness as compared with the contrary elements.[c] Now that which is to create movement without causing alteration must be of this kind ; for the natural bodies [d] overcome one another according as one of them prevails, the light being conquered and borne down by the heavier and the heavy borne up by the lighter.

We have now stated what is the part by the movement of which the soul creates movement and for what reason. The constitution of an animal must be regarded as resembling that of a well-governed city-state. For when order is once established in a city there is no need of a special ruler with arbitrary powers to be present at every activity, but each individual performs his own task as he is ordered, and one act succeeds another because of custom. And in the

[c] The contrary of fire is water, cf. De gen. et corrupt. 331 a 1.
[d] i.e. the elements.

ARISTOTLE

τὸ ἔθος· ἔν τε τοῖς ζῴοις τὸ αὐτὸ τοῦτο διὰ τὴν
φύσιν γίνεται καὶ τῷ πεφυκέναι ἕκαστον οὕτω
συστάντων ποιεῖν τὸ αὑτοῦ ἔργον, ὥστε μηδὲν
δεῖν ἐν ἑκάστῳ εἶναι ψυχήν, ἀλλ᾽ ἔν τινι ἀρχῇ τοῦ
σώματος οὔσης τἆλλα ζῆν μὲν τῷ προσπεφυκέναι,
ποιεῖν δὲ τὸ ἔργον τὸ αὑτῶν διὰ τὴν φύσιν.

XI. Πῶς μὲν οὖν κινεῖται τὰς ἑκουσίας κινήσεις
τὰ ζῷα, καὶ διὰ τίνας αἰτίας, εἴρηται· κινεῖται δέ
τινας καὶ ἀκουσίους ἔνια τῶν μερῶν, τὰς δὲ
πλείστας οὐχ ἑκουσίους. λέγω δ᾽ ἀκουσίους μὲν
οἷον τὴν τῆς καρδίας τε καὶ τὴν τοῦ αἰδοίου (πολλάκις
γὰρ φανέντος τινός, οὐ μέντοι κελεύσαντος τοῦ
νοῦ κινοῦνται), οὐχ ἑκουσίους δ᾽ οἷον ὕπνον καὶ
ἐγρήγορσιν καὶ ἀναπνοήν, καὶ ὅσαι ἄλλαι τοιαῦταί
εἰσιν.   οὐθενὸς γὰρ τούτων κυρία ἁπλῶς ἐστιν
οὔθ᾽ ἡ φαντασία οὔθ᾽ ἡ ὄρεξις, ἀλλ᾽ ἐπειδὴ ἀνάγκη
ἀλλοιοῦσθαι τὰ ζῷα φυσικὴν ἀλλοίωσιν, ἀλλοιουμένων δὲ τῶν μορίων τὰ μὲν αὔξεσθαι τὰ δὲ φθίνειν,
ὥστ᾽ ἤδη κινεῖσθαι καὶ μεταβάλλειν τὰς πεφυκυίας
ἔχεσθαι μεταβολὰς ἀλλήλων (αἰτίαι δὲ τῶν
κινήσεων θερμότητές τε καὶ ψύξεις, αἵ τε θύραθεν
καὶ αἱ ἐντὸς ὑπάρχουσαι φυσικαί), καὶ αἱ παρὰ
τὸν λόγον δὴ γινόμεναι κινήσεις τῶν ῥηθέντων
μορίων ἀλλοιώσεως συμπεσούσης γίνονται. ἡ γὰρ
νόησις καὶ ἡ φαντασία, ὥσπερ εἴρηται πρότερον, τὰ
ποιητικὰ τῶν παθημάτων προσφέρουσιν· τὰ γὰρ εἴδη
τῶν ποιητικῶν προσφέρουσιν.   μάλιστα δὲ τῶν
μορίων ταῦτα ποιεῖ ἐπιδήλως διὰ τὸ ὥσπερ ζῷον
κεχωρισμένον ἑκάτερον εἶναι τῶν μορίων [· τούτου

---

[a] See note on 698 b 1.
[b] Viz. the heart and the privy member.
[c] 701 b 18 ff.

animals the same process goes on because of nature, and because each part of them, since they are so constituted, is naturally suited to perform its own function ; so that there is no need of soul in each part, but since it is situated in a central origin of authority over the body,[a] the other parts live by their structural attachment to it and perform their own functions in the course of nature.

XI. We have now discussed the manner of the voluntary movements of animals, and the cause of them. Some of their parts, however, undergo certain involuntary movements, though most of these are really non-voluntary. By involuntary I mean such movements as those of the heart and of the privy member, which are often moved by the presentation of some image and not at the bidding of reason. By non-voluntary I mean sleeping and waking and respiration and the like. For neither imagination nor desire is strictly speaking responsible for any of these movements ; but, since animals must necessarily undergo physical alteration, and, when their parts undergo alteration, some increase and others decrease, and so their bodies immediately move and undergo the natural sequence of changes (the causes of their movements being the natural heatings and chillings, both external and internal), the movements too of the above-mentioned parts [b] which occur contrary to reason are due to the occurrence of a change. For thought and imagination, as has already been said,[c] induce the states which cause the affections ; for they present the images of the things which cause them. Now these parts act in this way much more conspicuously than any others, because each is as it were a separate vital organism[, the reason being that

703 b

δ' αἴτιον ὅτι ἔχουσιν ὑγρότητα ζωτικήν].[1] ἡ μὲν οὖν
καρδία φανερὸν δι' ἣν αἰτίαν· τὰς γὰρ[2] ἀρχὰς ἔχει
τῶν αἰσθήσεων· τὸ δὲ μόριον τὸ γεννητικὸν ὅτι
25 τοιοῦτόν ἐστι, σημεῖον· καὶ γὰρ ἐξέρχεται ἐξ
αὐτοῦ ὥσπερ ζῷόν τι ἡ τοῦ σπέρματος δύναμις.
αἱ δὲ κινήσεις τῇ τε ἀρχῇ ἀπὸ τῶν μορίων καὶ
τοῖς μορίοις ἀπὸ τῆς ἀρχῆς εὐλόγως συμβαίνουσι,
καὶ πρὸς ἀλλήλας οὕτως ἀφικνοῦνται. δεῖ γὰρ
νοῆσαι τὸ Α ἀρχήν. αἱ οὖν κινήσεις καθ' ἕκαστον
30 στοιχεῖον τῶν ἐπιγεγραμμένων ἐπὶ τὴν ἀρχὴν ἀφικνοῦνται, καὶ ἀπὸ τῆς ἀρχῆς κινουμένης καὶ μετα-
βαλλούσης, ἐπειδὴ πολλὰ δυνάμει ἐστίν, ἡ μὲν τοῦ
Β ἀρχὴ ἐπὶ τὸ Β, ἡ δὲ τοῦ Γ ἐπὶ τὸ Γ, ἡ δ' ἀμφοῖν
ἐπ' ἄμφω. ἀπὸ δὲ τοῦ Β ἐπὶ τὸ Γ τῷ[3] ἀπὸ μὲν τοῦ
35 Β ἐπὶ τὸ Α ἐλθεῖν ὡς ἐπ' ἀρχήν, ἀπὸ δὲ τοῦ Α ἐπὶ
τὸ Γ ὡς ἀπ' ἀρχῆς. ὅτι δὲ ὁτὲ μὲν ταὐτὰ[4] νοησάν-
των γίνεται ἡ κίνησις ἡ παρὰ τὸν λόγον ἐν τοῖς
704 a μορίοις, ὁτὲ δ' οὔ, αἴτιον τὸ ὁτὲ μὲν ὑπάρχειν τὴν
παθητικὴν ὕλην ὁτὲ δὲ μὴ τοσαύτην ἢ τοιαύτην.

Περὶ μὲν οὖν τῶν μορίων ἑκάστου τῶν ζῴων,
704 b καὶ περὶ ψυχῆς, ἔτι δὲ περὶ αἰσθήσεως καὶ ὕπνου
καὶ μνήμης καὶ τῆς κοινῆς κινήσεως, εἰρήκαμεν
τὰς αἰτίας· λοιπὸν δὲ περὶ γενέσεως εἰπεῖν.

---

[1] τούτου . . . ζωτικήν ut interpolamentum del. Jaeger.
[2] γὰρ om. EY.
[3] τῷ EP: τῷ δὲ Y: τὸ δὲ S.
[4] ταὐτὰ Jaeger: τὰ αὐτὰ P: ταῦτα ESY.

---

a These words are probably an interpolated gloss; they

478

each contains vital moisture].[a] The reason for this as regards the heart is plain, for it contains the origins of the senses. That the generative organ is of the same nature is shown by the fact that the seminal force comes forth from it, being as it were a living thing. Now it is only in accordance with reason that movements are set up both in the central origin by the parts and in the parts by the central origin, and thus reach one another. Let A be the central origin; the movements at each letter in the diagram drawn above [b] reach the central origin, and from the central origin, when it is moved or undergoes change (for it is potentially many), the origin of movement in B goes to B, and the origin of movement is C to C, and of both to both; but from B to C it travels by going from B to A as to a central origin, and from A to C as from a central origin. Movement, however, contrary to reason, sometimes takes place and sometimes does not take place in the organs as the result of the same thoughts, the reason being that the matter which is liable to be affected is sometimes present and sometimes not present in the proper quantity and quality.

We have now dealt with the reasons for the parts of each animal, the soul, and also sense-perception, sleep, memory, and general movement. It remains to deal with the generation of animals.

are unnecessary in view of the following sentences and contradictory in doctrine to them.

[b] See figure on p. 473.

PROGRESSION
OF ANIMALS

ANALYSIS OF CONTENTS

PROGRESSION OF ANIMALS

XI. Man, the only erect animal, compared with the birds. Winged human beings an impossible invention of the artists.

XII. Differences of flexion in the limbs of man and of the quadrupeds explained.

XIII. The different modes of flexion enumerated and illustrated by diagrams.

XIV. "Diagonal" movement of the legs of quadrupeds. Movement of crabs.

XV. Birds and quadrupeds compared. The structure of the legs of birds. Oblique attachment of wings and fins. The structure of oviparous quadrupeds.

XVI. Movement of bloodless animals. The peculiar movement of the crab.

XVII. Crabs, lobsters, flat-fish, and web-footed birds.

XVIII. Why birds have feet, while fishes have not. Fins and wings compared.

XIX. The movement of testaceans. Conclusion.

ABBREVIATIONS USED IN THE APPARATUS CRITICUS

Z = Codex Oxoniensis Collegii Corporis Christi W.A. 2. 7.
U = Codex Vaticanus 260.
S = Codex Laurentianus 81. 1.
P = Codex Vaticanus 1339.
Y = Codex Vaticanus 261.
Leon. = Latin translation of Nicholas Leonicus.
Mich. = Greek commentary of Michael Ephesius.

ΠΕΡΙ ΠΟΡΕΙΑΣ ΖΩΙΩΝ

I. Περὶ δὲ τῶν χρησίμων μορίων τοῖς ζῴοις
5 πρὸς τὴν κίνησιν τὴν κατὰ τόπον ἐπισκεπτέον διὰ
τίν' αἰτίαν τοιοῦτόν ἐστιν ἕκαστον αὐτῶν καὶ τίνος
ἕνεκεν ὑπάρχει αὐτοῖς, ἔτι δὲ περὶ τῶν διαφορῶν
τῶν τε πρὸς ἄλληλα τοῖς τοῦ αὐτοῦ καὶ ἑνὸς ζῴου
μορίοις, καὶ πρὸς τὰ τῶν ἄλλων τῶν τῷ γένει δια-
φόρων. πρῶτον δὲ λάβωμεν περὶ ὅσων ἐπισκεπτέον.

10 Ἔστι δὲ πρῶτον μὲν πόσοις ἐλαχίστοις τὰ ζῷα
κινεῖται σημείοις, ἔπειτα διὰ τί τὰ μὲν ἔναιμα
τέτταρσι τὰ δ' ἄναιμα πλείοσι, καὶ καθόλου δὲ διὰ
τίν' αἰτίαν τὰ μὲν ἄποδα τὰ δὲ δίποδα τὰ δὲ
τετράποδα τὰ δὲ πολύποδα τῶν ζῴων ἐστί, καὶ
διὰ τί πάντ' ἀρτίους ἔχει τοὺς πόδας, ὅσαπερ ἔχει
15 πόδας αὐτῶν· ὅλως δ' οἷς κινεῖται σημείοις, ἄρτια
ταῦτ' ἐστίν.

Ἔτι δὲ διὰ τίν' αἰτίαν ἄνθρωπος μὲν καὶ ὄρνις
δίπους, οἱ δ' ἰχθύες ἄποδές εἰσιν· καὶ τὰς κάμψεις
ὅ τε ἄνθρωπος καὶ ὁ ὄρνις δίποδες ὄντες ἐναντίας
ἔχουσι τῶν σκελῶν. ὁ μὲν γὰρ ἄνθρωπος ἐπὶ
20 τὴν περιφέρειαν κάμπτει τὸ σκέλος, ὁ δ' ὄρνις
ἐπὶ τὸ κοῖλον. καὶ ὁ ἄνθρωπος αὐτὸς αὑτῷ

PROGRESSION OF ANIMALS

I. We must next discuss the parts which are useful to animals for their movement from place to place, and consider why each part is of the nature which it is, and why they possess them, and further the differences in the various parts of one and the same animal and in those of animals of different species compared with one another. We must first decide what questions we have to discuss.

One question is, what is the smallest number of points at which animals move ; the next is, why red-blooded animals move at four points, while bloodless animals move at more than four ; and, in general, why some animals are without feet, others biped, others quadrupeds, and others polypods, and why all that have feet at all have an even number of feet ; and, in general, why the points at which movement is made are even in number.

We must further consider why a man and a bird are bipeds, while fishes are without feet ; and why a man and a bird, being both bipeds, have opposite bendings of the legs. For a man bends his legs in a convex direction, a bird in a concave direction ; and a man

485

704 a
ἐναντίως τὰ σκέλη καὶ τοὺς βραχίονας· τοὺς μὲν
γὰρ ἐπὶ τὸ κοῖλον, τὰ δὲ γόνατα ἐπὶ τὴν περιφέρειαν κάμπτει. καὶ τὰ τετράποδα τὰ ζῳοτόκα
τοῖς τ' ἀνθρώποις ἐναντίως κάμπτει καὶ αὐτὰ
αὑτοῖς· τὰ μὲν γὰρ πρόσθια σκέλη ἐπὶ τὸ κυρτὸν
704 b τῆς περιφερείας κάμπτει, τὰ δ' ὀπίσθια ἐπὶ τὸ
5 κοῖλον. ἔτι δὲ τῶν τετραπόδων ὅσα μὴ ζῳοτοκεῖ
ἀλλ' ᾠοτοκεῖ, ἰδίως καὶ εἰς τὸ πλάγιον κάμπτει.
πρὸς δὲ τούτοις διὰ τίν' αἰτίαν τὰ τετράποδα
κινεῖται κατὰ διάμετρον. περὶ δὴ πάντων τούτων,
καὶ ὅσα ἄλλα συγγενῆ τούτοις, τὰς αἰτίας θεωρη-
10 τέον. ὅτι μὲν γὰρ οὕτω ταῦτα συμβαίνει, δῆλον ἐκ
τῆς ἱστορίας τῆς φυσικῆς, διότι δέ, νῦν σκεπτέον.

II. Ἀρχὴ δὲ τῆς σκέψεως ὑποθεμένοις οἷς
εἰώθαμεν χρῆσθαι πολλάκις πρὸς τὴν μέθοδον
τὴν φυσικήν, λαβόντες τὰ τοῦτον ἔχοντα τὸν
τρόπον ἐν πᾶσι τοῖς τῆς φύσεως ἔργοις. τούτων
15 δ' ἐν μέν ἐστιν ὅτι ἡ φύσις οὐθὲν ποιεῖ μάτην,
ἀλλ' ἀεὶ ἐκ τῶν ἐνδεχομένων τῇ οὐσίᾳ περὶ ἕκαστον
γένος ζῴου τὸ ἄριστον· διόπερ εἰ βέλτιον ὡδί,
οὕτως καὶ ἔχει κατὰ φύσιν. ἔτι τὰς διαστάσεις
τοῦ μεγέθους, πόσαι καὶ ποῖαι ποίοις ὑπάρχουσι,
δεῖ λαβεῖν. εἰσὶ γὰρ διαστάσεις μὲν ἕξ, συζυγίαι
20 δὲ τρεῖς, μία μὲν τὸ ἄνω καὶ τὸ κάτω, δευτέρα δὲ
τὸ ἔμπροσθεν καὶ τὸ ὄπισθεν, τρίτη δὲ τὸ δεξιὸν
καὶ τὸ ἀριστερόν. πρὸς δὲ τούτοις ὅτι τῶν κινήσεων
τῶν κατὰ τόπον ἀρχαὶ ὦσις καὶ ἕλξις. καθ'
αὑτὰς μὲν οὖν αὗται, κατὰ συμβεβηκὸς δὲ κινεῖ-

a *i.e.* the front right foot with the left back foot, and the left
front with the right back.   b The *Historia Animalium.*
c Leon. renders *eodem . . . modo* which seems to imply that he was translating τὸν αὐτὸν ἔχοντα τρόπον.

himself bends his legs and his arms in opposite
directions, the arms concavely and the knees convexly. And viviparous quadrupeds bend their limbs
in the opposite way to a man's and in opposite
ways to one another ; for they bend their front legs
convexly and their back legs concavely. Further,
quadrupeds which are not viviparous but oviparous
have the peculiarity of bending their legs sideways.
A further question is why do quadrupeds move their
legs diagonally.[a]

We must examine the reasons of all these and
similar facts ; that they are facts is clear from our
Natural History,[b] and we have now to examine their
causes.

II. We must begin our inquiry by assuming the
principles which we are frequently accustomed to
employ in natural investigation, namely, by accept-
ing as true what occurs in accordance with these
principles [c] in all the works of nature. One of these
principles is that nature never creates anything
without a purpose, but always what is best in view
of the possibilities allowed by the essence of each
kind of animal ; therefore, if it is better to do a thing
in a particular manner, it is also in accordance with
nature. Further, we must accept the dimensions of
magnitude in the size and quality in which they are
present in various objects. For there are six dimensions grouped in three pairs, the first being the
superior and the inferior, the second the front and
the back, and the third the right and the left. We
must further postulate that the origins of movement
from place to place are thrusting and pulling. These
are movements *per se* ; that which is carried by

704 b

ται τὸ φερόμενον ὑπ᾽ ἄλλου· οὐ γὰρ αὐτὸ δοκεῖ
705 a κινεῖν αὐτὸ ἀλλ᾽ ὑπ᾽ ἄλλου κινεῖσθαι τὸ ὑπό τινος
φερόμενον.

III. Τούτων δὲ διωρισμένων λέγωμεν τὰ τούτων
ἐφεξῆς. τῶν δὴ ζῴων ὅσα μεταβάλλει κατὰ
τόπον, τὰ μὲν ἀθρόῳ παντὶ τῷ σώματι μεταβάλλει,
5 καθάπερ τὰ ἀλλόμενα, τὰ δὲ κατὰ μέρος,¹ καθάπερ
τῶν πορευομένων ἕκαστον. ἐν ἀμφοτέραις δὲ ταῖς
μεταβολαῖς ταύταις ἀεὶ μεταβάλλει τὸ κινούμενον
ἀποστηριζόμενον πρὸς τὸ ὑποκείμενον αὐτῷ.
διόπερ ἐάν τε ὑποφέρηται τοῦτο θᾶττον ἢ ὥστ᾽
10 ἔχειν ἀπερείσασθαι τὸ ποιούμενον ἐπ᾽ αὐτοῦ τὴν
κίνησιν, ἐάν θ᾽ ὅλως μηδεμίαν ἔχῃ τοῖς κινουμένοις
ἀντέρεισιν, οὐθὲν ἐπ᾽ αὐτοῦ δύναται κινεῖν ἑαυτό.
καὶ γὰρ τὸ ἀλλόμενον καὶ πρὸς αὐτὸ² ἀπερειδόμενον τὸ ἄνω καὶ πρὸς τὸ ὑπὸ τοὺς πόδας ποιεῖται
τὴν ἅλσιν· ἔχει γάρ τινα ἀντέρεισιν πρὸς ἄλληλα
15 τὰ μόρια ἐν ταῖς καμπαῖς, καὶ ὅλως τὸ πιέζον
πρὸς τὸ πιεζόμενον. διὸ καὶ οἱ πένταθλοι ἅλλονται
πλεῖον ἔχοντες τοὺς ἁλτῆρας ἢ μὴ ἔχοντες, καὶ
οἱ θέοντες θᾶττον θέουσι παρασείοντες τὰς χεῖρας·
γίνεται γάρ τις ἀπέρεισις ἐν τῇ διατάσει πρὸς τὰς
χεῖρας καὶ τοὺς καρπούς. ἀεὶ δὲ τὸ κινούμενον
20 δυσὶν ἐλαχίστοις χρώμενον ὀργανικοῖς μέρεσι
ποιεῖται τὴν μεταβολήν, τῷ μὲν ὡσπερανεὶ θλίβοντι,
τῷ δὲ θλιβομένῳ. τὸ μὲν γὰρ μένον θλίβεται διὰ

---

¹ κατὰ μέρος Z : μέρει S : τοῖς μορίοις cet.
² αὐτὸ PUY : αὑτὸ S : ἑαυτὸ Z.

ᵃ Special weights (ἁλτῆρες) or sometimes stones were held
in the hands and thrown backwards by jumpers while in the
air to add to their impetus ; *cf.* Norman Gardiner, *Greek*

something else is only moved accidentally, for what is carried by something else is regarded not as moving itself but as being moved by something else.

III. These points having been decided, let us proceed to the considerations which follow from them. Of the animals, then, which change their local position, some do so with their whole body at the same time, for instance those which jump; others move part by part, for example those that walk. In both these changes the animal that moves makes its change of position by pressing against that which is beneath it; and so, if the latter slips away too quickly to allow that which is setting itself in motion upon it to press against it, or if it offers no resistance at all to that which is moving, the animal cannot move itself at all upon it. For that which jumps performs that movement by pressing both on its own upper part and on that which is beneath its feet; for the parts in a way lean upon one another at their joints, and, in general, that which presses leans on that which is pressed. Hence athletes jump farther if they have the weights in their hands than if they have not,[a] and runners run faster if they swing their arms [b]; for in the extension of the arms there is a kind of leaning upon the hands and wrists. Now that which moves always makes its change of place by the employment of at least two organic parts, one as it were compressing and the other being compressed. For the part which remains still is compressed by

Athletic Sports and Festivals, pp. 298 ff., who proves by experiment the truth of the statement made in the present passage.

[b] On the importance attached by the Greeks to arm-action in running, especially in short races, *cf.* N. Gardiner, *op. cit.* p. 282.

705 a

τὸ φέρειν, τὸ δ' αἰρόμενον τείνεται τῷ φέροντι
τὸ φορτίον. διόπερ ἀμερὲς οὐδὲν οὕτω κινηθῆναι
δυνατόν· οὐ γὰρ ἔχει τὴν τοῦ πεισομένου καὶ τοῦ
25 ποιήσοντος ἐν αὑτῷ[1] διάληψιν.

IV. Ἐπεὶ δ' εἰσὶν αἱ διαστάσεις τὸν ἀριθμὸν
ἕξ, αἷς ὁρίζεσθαι πέφυκε τὰ ζῷα,[2] τό τε ἄνω καὶ
κάτω καὶ τὸ ἔμπροσθεν καὶ ὄπισθεν, ἔτι δὲ δεξιὸν
καὶ ἀριστερόν, τὸ μὲν ἄνω καὶ κάτω μόριον πάντ'
ἔχει τὰ ζῶντα. οὐ μόνον γὰρ ἐν τοῖς ζῴοις ἐστὶ
τὸ ἄνω καὶ κάτω, ἀλλὰ καὶ ἐν τοῖς φυτοῖς. δι-
30 είληπται δ' ἔργῳ, καὶ οὐ θέσει μόνον τῇ πρός τε
τὴν γῆν καὶ τὸν οὐρανόν. ὅθεν μὲν γὰρ ἡ τῆς τροφῆς
διάδοσις καὶ ἡ αὔξησις ἑκάστοις, ἄνω τοῦτ' ἐστίν·
705 b πρὸς ὃ δ' ἔσχατον αὕτη περαίνει, τοῦτο κάτω.
τὸ μὲν γὰρ ἀρχή τις, τὸ δὲ πέρας· ἀρχὴ δὲ τὸ ἄνω.
καίτοι δόξειεν ἂν τοῖς φυτοῖς οἰκεῖον εἶναι τὸ κάτω
μᾶλλον· οὐχ ὁμοίως γὰρ ἔχει τῇ θέσει τὸ ἄνω καὶ
κάτω τούτοις καὶ τοῖς ζῴοις. ἔχει δὲ πρὸς μὲν
5 τὸ ὅλον οὐχ ὁμοίως, κατὰ δὲ τὸ ἔργον ὁμοίως.
αἱ γὰρ ῥίζαι εἰσὶ τὸ ἄνω τοῖς φυτοῖς· ἐκεῖθεν γὰρ
ἡ τροφὴ διαδίδοται τοῖς φυομένοις, καὶ λαμβάνει
ταύταις αὐτήν, καθάπερ τὰ ζῷα τοῖς στόμασιν.

Ὅσα δὲ μὴ μόνον ζῇ ἀλλὰ καὶ ζῷά ἐστι, τοῖς
τοιούτοις ὑπάρχει τό τε ἔμπροσθεν καὶ τὸ ὄπισθεν.
10 αἴσθησιν γὰρ ἔχει ταῦτα πάντα, ὁρίζεται δὲ κατὰ
ταύτην τό τε ἔμπροσθεν καὶ τὸ ὄπισθεν· ἐφ' ὃ
μὲν γὰρ ἡ αἴσθησις πέφυκε καὶ ὅθεν ἐστὶν ἑκάστοις,

---

[1] αὑτῷ Jaeger : αὐτῷ libri.   [2] ζῷα Y : ζῶντα ceteri.

---

[a] Cf. above, 704 b 19 ff.   [b] Cf. De caelo, 294 b 17.
[c] More literally "personal."
[d] Cf. De vit. long. et brev. 467 b 2 ; Phys. 199 a 28.

having to carry the weight, and the part which is raised is extended by that which carries the weight. And so nothing that is without parts can move in this manner; for it does not contain in itself the distinction between what is to be passive and what is to be active.

IV. Now the dimensions by which animals are naturally bounded are six in number, namely, superior and inferior, front and back, and also right and left.[a] Now all living things have a superior and an inferior part; for the superior and the inferior is found not only in the animals but also in plants.[b] The distinction is one of function and not merely of position in relation to the earth and heavens. For the part from which is derived the distribution of nutriment and the growth in any particular thing is the superior; the part to which the growth extends and in which it finally ends is the inferior. The one is a kind of origin, the other a termination; and it is the superior which is an origin. It might, however, seem that in plants the inferior is the more essential [c] part; for the superior and the inferior are not in the same position in them as in the animals. Though in relation to the universe they have not the same position, they are similarly situated as regards function. For in plants the roots are the superior part [d]; for it is from them that the nutriment is distributed to the parts that grow, and it is from their roots that plants receive it, as do animals from their mouths.

Things which not only live but are also animals have both a front and a back. For all animals have sense-perception, and it is on account of sense-perception that the front and the back are distinguished; for the parts in which the sense-perception is implanted

491

705 b

ἔμπροσθεν ταῦτ' ἐστί, τὰ δ' ἀντικείμενα τούτοις ὄπισθεν.

Ὅσα δὲ τῶν ζώων μὴ μόνον αἰσθήσεως κοινωνεῖ,
15 ἀλλὰ δύναται ποιεῖσθαι τὴν κατὰ τόπον μεταβολὴν αὐτὰ δι' αὐτῶν, ἐν τούτοις δὴ¹ διώρισται
πρὸς τοῖς λεχθεῖσι τό τ' ἀριστερὸν καὶ τὸ δεξιόν,
ὁμοίως τοῖς πρότερον εἰρημένοις ἔργῳ τινὶ καὶ οὐ
θέσει διωρισμένον ἑκάτερον αὐτῶν· ὅθεν μὲν γάρ
20 ἐστι τοῦ σώματος ἡ τῆς κατὰ τόπον μεταβολῆς ἀρχὴ
φύσει, τοῦτο μὲν δεξιὸν ἑκάστῳ, τὸ δ' ἀντικείμενον
καὶ τούτῳ πεφυκὸς ἀκολουθεῖν ἀριστερόν. τοῦτο
δὲ διήρθρωται μᾶλλον ἑτέροις ἑτέρων. ὅσα μὲν
γὰρ ὀργανικοῖς μέρεσι χρώμενα (λέγω δ' οἷον
ποσὶν ἢ πτέρυξιν ἤ τινι ἄλλῳ τοιούτῳ) τὴν εἰρημένην μεταβολὴν ποιεῖται, περὶ μὲν τὰ τοιαῦτα
25 μᾶλλον διήρθρωται τὸ λεχθέν· ὅσα δὲ μὴ τοιούτοις
μορίοις, αὐτῷ δὲ τῷ σώματι διαλήψεις ποιούμενα
προέρχεται, καθάπερ ἔνια τῶν ἀπόδων, οἷον οἵ
τε ὄφεις καὶ τὸ τῶν καμπῶν γένος, καὶ πρὸς τούτοις
ἃ καλοῦσι γῆς ἔντερα, ὑπάρχει μὲν καὶ ἐν τούτοις
τὸ λεχθέν, οὐ μὴν διασεσάφηταί γ' ὁμοίως.

30 Ὅτι δ' ἐκ τῶν δεξιῶν ἡ ἀρχὴ τῆς κινήσεώς ἐστι,
σημεῖον καὶ τὸ φέρειν τὰ φορτία πάντας ἐπὶ τοῖς
ἀριστεροῖς· οὕτως γὰρ ἐνδέχεται κινεῖσθαι τὸ φέρον,
λελυμένου τοῦ κινήσοντος. (διὸ καὶ ἀσκωλιάζουσι
ῥᾷον ἐπὶ τοῖς ἀριστεροῖς· κινεῖν γὰρ πέφυκε τὸ
706 a δεξιόν, κινεῖσθαι δὲ τὸ ἀριστερόν.) ὥστε καὶ τὸ
φορτίον οὐκ ἐπὶ τῷ κινήσοντι ἀλλ' ἐπὶ τῷ κινησο-

¹ δὴ Jaeger: δὲ libri.

ᵃ Viz. superior and inferior.
ᵇ i.e. from place to place.

492

and whence every kind of creature derives it are at the front, and the opposite parts to these are at the back.

Those animals which not only partake of sense-perception but can also of themselves make the change from place to place, in addition to the distinctions already mentioned,[a] have a further distinction of left and right, these being each, like the above, distinctions of function and not of position. For the part of the body where the origin of change from place to place naturally arises is the right in each kind of animal, while the part which is opposed to this and naturally follows its lead is the left.

There is a greater differentiation between right and left in some animals than in others. All animals which make the above-mentioned change [b] by the use of instrumental parts—for example, feet or wings or the like—show a greater differentiation between right and left in such parts ; those, on the other hand, that progress not by means of such parts but by moving the body itself in sections—like some of the footless animals, such as snakes and the caterpillars, and also earthworms—possess, it is true, this differentiation, but it is not nearly so clearly defined.

That the origin of movement is from the right side is shown by the fact that men always carry burdens on the left shoulder ; for then it is possible for that which bears the weight to be set in motion, that which is to initiate the movement being free. (For this reason, too, it is easier to hop on the left leg ; for it is natural to the right leg to initiate movement, and to the left to be set in motion.) The burden, therefore, must rest not on the part which is to initiate movement, but on that which is to be set in

493

706 a

μένῳ δεῖ ἐπικεῖσθαι· ἐὰν δ' ἐπὶ τῷ κινοῦντι καὶ τῇ
ἀρχῇ τῆς κινήσεως ἐπιτεθῇ, ἤτοι ὅλως οὐ κινήσεται
5 ἢ χαλεπώτερον. σημεῖον δ' ὅτι ἀπὸ τῶν δεξιῶν
ἡ ἀρχὴ τῆς κινήσεως καὶ αἱ προβολαί· πάντες
γὰρ τὰ ἀριστερὰ προβάλλονται, καὶ ἑστῶτες προβεβλήκασι[1] τὰ ἀριστερὰ μᾶλλον, ἂν μὴ ἀπὸ τύχης
συμβῇ. οὐ γὰρ τῷ προβεβηκότι κινοῦνται, ἀλλὰ
τῷ ἀποβεβηκότι· καὶ ἀμύνονται τοῖς δεξιοῖς.
10 διὰ ταύτην δὲ τὴν αἰτίαν καὶ τὰ δεξιὰ ταυτά ἐστι
πάντων. ὅθεν μὲν γὰρ ἡ ἀρχὴ τῆς κινήσεως, τὸ
αὐτὸ πᾶσι καὶ ἐν τῷ αὐτῷ τὴν θέσιν ἔχει κατὰ φύσιν·
δεξιὸν δ' ἐστὶν ὅθεν ἡ ἀρχὴ τῆς κινήσεώς ἐστιν.
καὶ διὰ τοῦτο τὰ στρομβώδη τῶν ὀστρακοδέρμων
δεξιὰ πάντ' ἐστίν. οὐ γὰρ ἐπὶ τὴν ἕλίκην κινεῖται,
15 ἀλλ' ἐπὶ τὸ καταντικρὺ πάντα προέρχεται, οἷον
πορφύραι καὶ κήρυκες. κινουμένων οὖν πάντων
ἀπὸ τῶν δεξιῶν, κἀκείνων ἐπὶ ταὐτὰ κινουμένων
ἑαυτοῖς, ἀνάγκη πάντα δεξιὰ εἶναι ὁμοίως. ἀπολελυμένα δ' ἔχουσι τὰ ἀριστερὰ τῶν ζῴων μάλιστα ἄνθρωποι διὰ τὸ κατὰ φύσιν ἔχειν μάλιστα
20 τῶν ζῴων· φύσει δὲ βέλτιόν τε τὸ δεξιὸν τοῦ
ἀριστεροῦ καὶ κεχωρισμένον. διὸ καὶ τὰ δεξιὰ
ἐν τοῖς ἀνθρώποις μάλιστα δεξιά ἐστιν. διωρισμένων δὲ τῶν δεξιῶν εὐλόγως τὰ ἀριστερὰ ἀκινη-
τότερά ἐστι, καὶ ἀπολελυμένα μάλιστα ἐν τούτοις.
καὶ αἱ ἄλλαι δ' ἀρχαὶ μάλιστα κατὰ φύσιν δι-
25 ωρισμέναι ἐν τῷ ἀνθρώπῳ ὑπάρχουσι, τό τ' ἄνω
καὶ τὸ ἔμπροσθεν.

---

[1] προβεβλήκασι PSU : προβεβήκασι YZ.

---

[a] i.e. in the sense that man is right-handed.

motion; and if it be placed on that which causes
and is the origin of movement, it will either not be
moved at all or with greater difficulty. The manner
in which we step out also shows that the origin of
movement is in the right side; for all men put the
left foot foremost, and, when standing, preferably
place the left foot in front, unless they do otherwise
accidentally. For they are moved, not by the foot
which they put in front, but by that with which they
step off; also they defend themselves with their
right limbs. Therefore the right is the same in all;
for that from which the origin of movement is derived
is the same in all and has its position by nature in the
same place, and it is from the right that the origin
of movement is derived. For this reason, too, the
stromboid testaceans all have their shells on the
right; for they all move not in the direction of the
spiral but in the opposite direction, the purple-fish,
for example, and the trumpet-shell. Since, then,
movement in all animals starts from the right, and
the right moves in the same direction as the animal
itself, they must all alike be right-sided.[a] Now man
more than any other animal has his left limbs detached, because of all animals he is most in accordance with nature, and the right is naturally better
than the left and separated from it. Therefore the
right is most right-sided in man. And since the right
is differentiated, it is only reasonable that the left is
less easily set in motion and most detached in man.
Moreover the other principles,[b] the superior and the
front, are in man most in accord with nature and most
differentiated.

[b] The ἀρχαί here are the διαστάσεις of 704 b 19, 705 a 26,
from the point of view of function rather than position.

V. Οἷς μὲν οὖν τὸ ἄνω καὶ τὸ ἔμπροσθεν διώρισται, καθάπερ τοῖς ἀνθρώποις καὶ τοῖς ὄρνισι,
ταῦτα μὲν δίποδα (τῶν δὲ τεττάρων τὰ δύο σημεῖα
τοῖς μὲν πτέρυγες τοῖς δὲ χεῖρες καὶ βραχίονές
30 εἰσιν). ὅσα δ' ἐπὶ τὸ αὐτὸ τὸ πρόσθεν ἔχει καὶ
τὸ ἄνω, τετράποδα καὶ πολύποδα καὶ ἄποδα.
καλῶ γὰρ πόδα μέρος ἐπὶ σημείῳ πεζῷ κινητικῷ
κατὰ τόπον· καὶ γὰρ τὸ ὄνομα ἐοίκασιν εἰληφέναι
ἀπὸ τοῦ πέδου οἱ πόδες. ἔνια δ' ἐπὶ τὸ αὐτὸ
ἔχει τὸ πρόσθιον καὶ τὸ ὀπίσθιον, οἷον τά τε
706 b μαλάκια καὶ τὰ στρομβώδη τῶν ὀστρακοδέρμων·
εἴρηται δὲ περὶ αὐτῶν πρότερον ἐν ἑτέροις.

Τριῶν δ' ὄντων τόπων, τοῦ ἄνω καὶ μέσου
καὶ κάτω, τὰ μὲν δίποδα τὸ ἄνω πρὸς τὸ τοῦ
ὅλου ἄνω ἔχει, τὰ δὲ πολύποδα ἢ ἄποδα πρὸς
5 τὸ μέσον, τὰ δὲ φυτὰ πρὸς τὸ κάτω. αἴτιον δ'
ὅτι τὰ μὲν ἀκίνητα, πρὸς τὴν τροφὴν δὲ τὸ ἄνω,
ἡ δὲ τροφὴ ἐκ τῆς γῆς. τὰ δὲ τετράποδα ἐπὶ
τὸ μέσον, καὶ τὰ πολύποδα καὶ ἄποδα, διὰ τὸ
μὴ ὀρθὰ εἶναι. τὰ δὲ δίποδα πρὸς τὸ ἄνω διὰ
10 τὸ ὀρθὰ εἶναι, μάλιστα δ' ὁ ἄνθρωπος· μάλιστα
γὰρ κατὰ φύσιν ἐστὶ δίπους. εὐλόγως δὲ καὶ αἱ
ἀρχαί εἰσιν ἀπὸ τούτων τῶν μορίων· ἡ μὲν γὰρ
ἀρχὴ τίμιον, τὸ δ' ἄνω τοῦ κάτω καὶ τὸ πρόσθεν
τοῦ ὄπισθεν καὶ τὸ δεξιὸν τοῦ ἀριστεροῦ τιμιώτερον.
καλῶς δ' ἔχει καὶ τὸ ἀνάπαλιν λέγειν περὶ αὐτῶν,

---

[a] The whole of man is "front," and his "front" is divided
into superior and inferior; in a quadruped only that part
is "front" which is superior in man.

[b] P.A. 684 b 14 ff.; H.A. 523 b 21 ff.

[c] Ἀρχή has here the double meaning of "starting-point"
and "centre of authority"; see note on De mot. anim.
698 b 1.

V. Animals in which the superior and the front are differentiated, man, for example, and the birds, are bipeds (two of the four points being wings in birds, and hands and arms in man). But the animals in which the superior and the front are in the same position [a] are four-footed (quadrupeds), many-footed (polypods), and footless. By " foot " I mean the part that is at a point which has connexion with the ground and gives movement from place to place ; for the feet ($\pi\acute{o}\delta\epsilon s$) seem to have derived their name from the ground ($\pi\acute{\epsilon}\delta o\nu$). Some animals have their front and their back in the same position, for example the molluscs and the stromboid testaceans ; with these we have already dealt elsewhere.[b]

Now since there are three regions, the superior, the middle, and the inferior, bipeds have their superior part in a position corresponding to the superior region of the universe, polypods and footless animals in a position corresponding to the middle region, and plants in a position corresponding to the inferior region. The reason is that plants lack movement, and the superior part is situated with a view to nutriment, and their nutriment comes from the earth. Quadrupeds, polypods, and footless animals have their superior part in a position corresponding to the middle region because they are not erect ; bipeds have it in a position corresponding to the superior region because they are erect, especially man, the biped most in accordance with nature. And it is only reasonable that the origins [c] should come from these parts ; for the origin is honourable, and the superior is more honourable than the inferior, and the front than the back, and the right than the left. It is also true if we reverse the proposition and assert

706 b

15 ὡς διὰ τὸ τὰς ἀρχὰς ἐν τούτοις εἶναι ταῦτα τιμιώτερα τῶν ἀντικειμένων μορίων ἐστίν.

VI. Ὅτι μὲν οὖν ἐκ τῶν δεξιῶν ἡ τῆς κινήσεώς ἐστιν ἀρχή, φανερὸν ἐκ τῶν εἰρημένων. ἐπεὶ δ' ἀνάγκη παντὸς συνεχοῦς, οὗ τὸ μὲν κινεῖται τὸ δ' ἠρεμεῖ, ὅλου δυναμένου κινεῖσθαι ἑστῶτος 20 θατέρου, ᾗ ἄμφω κινεῖται ἐναντίας κινήσεις, εἶναί τι κοινὸν καθ' ὃ συνεχῆ ταῦτ' ἐστὶν ἀλλήλοις, κἀνταῦθ' ὑπάρχειν τὴν ἀρχὴν τῆς ἑκατέρου τῶν μερῶν κινήσεως (ὁμοίως δὲ καὶ τῆς στάσεως), δῆλον ὅτι,[1] καθ' ὅσας τῶν λεχθεισῶν ἀντιθέσεων ἰδία κίνησις ὑπάρχει τῶν ἀντικειμένων μερῶν 25 ἑκατέρῳ, πάντα ταῦτα κοινὴν ἀρχὴν ἔχει κατὰ[2] τὴν τῶν εἰρημένων μερῶν σύμφυσιν, λέγω δὲ τῶν τε δεξιῶν καὶ ἀριστερῶν καὶ τῶν ἄνω καὶ κάτω καὶ τῶν ἔμπροσθεν καὶ τῶν ὄπισθεν. κατὰ μὲν οὖν τὸ ἔμπροσθεν καὶ τὸ ὄπισθεν διάληψις οὐκ ἔστι τοιαύτη περὶ τὸ κινοῦν ἑαυτό, διὰ τὸ μηθενὶ 30 φυσικὴν ὑπάρχειν κίνησιν εἰς τὸ ὄπισθεν, μηδὲ διορισμὸν ἔχειν τὸ κινούμενον καθ' ὃν τὴν ἐφ' ἑκάτερα τούτων μεταβολὴν ποιεῖται· κατὰ δὲ τὸ δεξιόν γε καὶ ἀριστερὸν καὶ τὸ ἄνω καὶ τὸ κάτω ἐστίν. διὸ τῶν ζῴων ὅσα μέρεσιν ὀργανικοῖς 707 a χρώμενα προέρχεται, τῇ μὲν τοῦ ἔμπροσθεν καὶ ὄπισθεν διαφορᾷ οὐκ ἔχει διωρισμένα ταῦτα, ταῖς δὲ λοιπαῖς, ἀμφοτέραις μέν, προτέρᾳ δὲ τῇ κατὰ τὸ δεξιὸν καὶ ἀριστερὸν διοριζούσῃ, διὰ τὸ τὴν

---

[1] δῆλον ὅτι (Leon. *manifestum est quod*, etc.): δηλονότι libri.
[2] κατὰ P Leon.: om. ceteri.

---

[a] *i.e.* the three pairs of "dimensions" (704 b 19).

that, because the origins are situated in these parts, they are therefore more honourable than the opposite parts.

VI. It is clear, then, from what has been said that the origin of movement is on the right. Now in anything continuous of which part is in motion and part at rest (the whole being able to move while one part stands still), there must be, at the point where both parts move in opposite movements, something common to both which makes these parts continuous with one another (and at this point must be situated the origin of the movement of each of these parts, and likewise also of their immobility): it is evident, therefore, that in respect of whichever of the above-mentioned contraries [a] the individual movement of each of the opposite parts takes place, there is in all these cases a common origin of movement by reason of the interconnexion of the said parts, namely, of the right and the left, the superior and the inferior, the front and the back. The differentiation according to front and back is not one which applies to that which moves itself, because nothing possesses a natural movement backwards nor has the moving animal any distinction in accordance with which it can make a change from place to place in each of these two directions [b]; but there is a differentiation of right and left, superior and inferior. All animals, therefore, which progress by the employment of instrumental parts have these parts differentiated, not by the distinction between front and back, but by the other two pairs, first, by the distinction of right and left (for this must immediately exist where there are

[b] In other words an animal cannot divide itself into two parts, one of which goes forwards and the other backwards.

707 a

μὲν ἐν τοῖς δυσὶν εὐθέως ἀναγκαῖον εἶναι ὑπ-
5 άρχειν, τὴν δ' ἐν τοῖς τέτταρσι πρώτοις.

Ἐπεὶ οὖν τό τε ἄνω καὶ κάτω καὶ τὸ δεξιὸν καὶ
ἀριστερὸν τῇ αὐτῇ ἀρχῇ καὶ κοινῇ συνήρτηται πρὸς
αὐτά (λέγω δὲ ταύτην τὴν τῆς κινήσεως κυρίαν), δεῖ
δ' ἐν ἅπαντι τῷ μέλλοντι κατὰ τρόπον ποιεῖσθαι
τὴν ἀφ' ἑκάστου κίνησιν ὡρίσθαι πως καὶ τετά-
10 χθαι ταῖς ἀποστάσεσι ταῖς πρὸς τὰς ῥηθείσας
ἀρχάς, τάς τε ἀντιστοίχους καὶ τὰς συστοίχους
τῶν ἐν τοῖς μέρεσι τούτοις, τὸ τῶν λεχθεισῶν
κινήσεων ἀπασῶν αἴτιον (αὕτη δ' ἐστὶν ἀφ' ἧς
ἀρχῆς κοινῆς τῶν ἐν τῷ ζῴῳ ἥ τε τοῦ δεξιοῦ καὶ
ἀριστεροῦ κίνησίς ἐστιν, ὁμοίως δὲ καὶ ἡ τοῦ ἄνω
15 καὶ κάτω), ταύτην δ'¹ ἔχειν ἑκάστῳ ᾗ παραπλησίως
ἔχει² πρὸς ἑκάστην τῶν ἐν τοῖς ῥηθεῖσι μέρεσιν
ἀρχῶν, VII. δῆλον οὖν ὡς ἢ μόνοις ἢ μάλιστα
τούτοις ὑπάρχει τῶν ζῴων ἡ κατὰ τόπον κίνησις,
ἃ δυσὶν ἢ τέτταρσι ποιεῖται σημείοις τὴν κατὰ
τόπον μεταβολήν. ὥστ' ἐπεὶ σχεδὸν τοῖς ἐναίμοις
20 τοῦτο μάλιστα συμβέβηκε, φανερὸν ὅτι πλείοσί
τε σημείοις τεττάρων οὐθὲν οἷόν τε κινεῖσθαι τῶν
ἐναίμων ζῴων, καὶ εἴ τι τέτταρσι σημείοις κινεῖσθαι
πέφυκε μόνον, ἀναγκαῖον τοῦτ' εἶναι ἔναιμον.

Ὁμολογεῖ δὲ τοῖς λεχθεῖσι καὶ τὰ συμβαίνοντα
περὶ τὰ ζῷα. τῶν μὲν γὰρ ἐναίμων οὐδὲν εἰς
25 πλείω διαιρούμενον δύναται ζῆν οὐθένα χρόνον

---

¹ δ' PUZ: om. SY.   ² ἔχει Z: om. cet.

ᵃ *i.e.* the distinction of superior and inferior.
ᵇ Namely, the soul situated in the heart (Mich.).
ᶜ The legs move in pairs, either the front and back legs
on the same side together, or the front leg on one side with
the back leg on the other (*cf.* 704 b 7).

500

two things), and, secondly, by the distinction which must arise as soon as there are four things.[a]

Since, then, the superior and the inferior, and the right and the left are connected with one another by the same common origin (and by this I mean that which controls their movement [b]), and since in anything which is to carry out the movement of each part properly the cause of all the said movements must be somehow defined and arranged at the right distance in relation to the said origins, namely, those in the limbs, which are in pairs opposite or diagonal to one another,[c] (and the cause of their movement is the common origin from which the movement of left and right and likewise of superior and inferior in the animal's limbs is derived), and since this origin must in each animal be at a point where it is in more or less the same relation to each of the origins in the said parts,[d] (VII.) it is, therefore, clear that movement from place to place belongs either solely or chiefly to those animals which make their change of place by means of two or four points. And so, since this condition occurs almost exclusively in red-blooded animals, it is clear that no red-blooded animal can move by means of more than four points, and if an animal is so constituted by nature as to move by means of four points only, it must necessarily be red-blooded.

What actually occurs in animals is also in agreement with the above statement. For no red-blooded animal can live for any time worth mentioning if it be

---

[d] There are two kinds of ἀρχαί in, e.g., a quadruped, (a) those in each of the four legs and (b) the central ἀρχή in the heart; the former must each be approximately equidistant from the latter

707 a

ὡς εἰπεῖν, τῆς τε κατὰ τόπον κινήσεως, καθ' ἣν
ἐκινεῖτο συνεχὲς ὂν καὶ μὴ διῃρημένον, οὐ δύναται
κοινωνεῖν· τῶν δ' ἀναίμων τε καὶ πολυπόδων ἔνια
διαιρούμενα δύναται ζῆν πολὺν χρόνον ἑκάστῳ
τῶν μερῶν, καὶ κινεῖσθαι τὴν αὐτὴν ἥνπερ καὶ
30 πρὶν διαιρεθῆναι κίνησιν, οἷον αἵ τε καλούμεναι
σκολόπενδραι καὶ ἄλλα τῶν ἐντόμων καὶ προμήκων·
πάντων γὰρ τούτων καὶ τὸ ὄπισθεν μέρος ἐπὶ
707 b ταὐτὸ ποιεῖται τὴν πορείαν τῷ ἔμπροσθεν. αἴτιον
δὲ τοῦ διαιρούμενα ζῆν ὅτι, καθάπερ ἂν εἴ τι
συνεχὲς ἐκ πολλῶν εἴη ζῴων συγκείμενον, οὕτως
ἕκαστον αὐτῶν συνέστηκεν. φανερὸν δὲ τοῦτο ἐκ
τῶν πρότερον εἰρημένων, διότι τοῦτον ἔχει τὸν
5 τρόπον.

Δυσὶ γὰρ ἢ τέτταρσι σημείοις πέφυκε κινεῖσθαι
τὰ μάλιστα συνεστηκότα κατὰ φύσιν, ὁμοίως δὲ
καὶ ὅσα τῶν ἐναίμων ἄποδά ἐστιν. καὶ γὰρ ταῦτα
κινεῖται τέτταρσι σημείοις, δι' ὧν τὴν κίνησιν
ποιεῖται. δυσὶ γὰρ χρώμενα προέρχεται καμ-
10 παῖς· τὸ γὰρ δεξιὸν καὶ ἀριστερὸν καὶ τὸ πρόσθιον
καὶ ὄπισθιον ἐν τῷ πλάτει ἐστὶν ἐν ἑκατέρᾳ τῇ
καμπῇ αὐτοῖς, ἐν μὲν τῷ πρὸς τὴν κεφαλὴν
μέρει τὸ πρόσθιον σημεῖον δεξιόν τε καὶ ἀριστερόν, ἐν δὲ τῷ πρὸς τὴν οὐρὰν τὰ ὀπίσθια
σημεῖα. δοκεῖ δὲ δυοῖν σημείοιν κινεῖσθαι, τῇ τ'
ἔμπροσθεν ἁφῇ καὶ τῇ ὑστέρον. αἴτιον δ' ὅτι
15 στενὸν κατὰ πλάτος ἐστίν, ἐπεὶ καὶ ἐν τούτοις τὸ
δεξιὸν ἡγεῖται, καὶ ἀνταποδίδωσι κατὰ τὸ ὄπισθεν,
ὥσπερ ἐν τοῖς τετράποσιν. τῶν δὲ κάμψεων
αἴτιον τὸ μῆκος· ὥσπερ γὰρ οἱ μακροὶ τῶν ἀνθρώπων λορδοὶ βαδίζουσι, καὶ τοῦ δεξιοῦ ὤμου

_____

* Centipedes.

divided into several parts, and can no longer partake of the motion from place to place whereby it moved while it was still continuous and undivided. On the other hand, some of the bloodless animals and polypods can, when they are divided, live in each of these parts for a considerable time and move with the same motion as before they were divided, the so-called scolopendrae,[a] for example, and other elongated insects ; for the hinder part of all these continues to progress in the same direction as the fore-part. The reason why they live when they are divided is that each of them consists as it were of a continuous body made up of many animals. And the reason why they are of this kind is clear from what has been said above.

Animals which are constituted most in accordance with nature naturally move by means of two or four points, and likewise also those among the red-blooded animals which are footless ; for they too move at four points and so effect locomotion. For they progress by means of two bends ; for in each of their bends there is a right and a left, a front and a back in their breadth—a front point on the right and another on the left in the part towards the head, and the two hinder points in the part towards the tail. They appear to move at two points only, namely, the points of contact with the ground in front and behind. The reason for this is that they are narrow in breadth ; for in these animals too, as in the quadrupeds, the right leads the way and sets up a corresponding movement behind. The reason of their bendings is their length ; for just as tall men walk with their backs hollowed [b] and, while their right shoulder leads the

[b] λορδός is the opposite of κυφός, hunchbacked (Hippocr. *Fract.* 763).

503

707 b

εἰς τὸ πρόσθεν ἡγουμένου τὸ ἀριστερὸν ἰσχίον εἰς
20 τοὔπισθεν μᾶλλον ἀποκλίνει, καὶ τὸ μέσον κοῖλον
γίνεται καὶ λορδόν, οὕτω δεῖ νοεῖν καὶ τοὺς ὄφεις
κινουμένους ἐπὶ τῇ γῇ λορδούς. σημεῖον δ' ὅτι
ὁμοίως κινοῦνται τοῖς τετράποσιν · ἐν μέρει γὰρ
μεταβάλλουσι τὸ κοῖλον καὶ τὸ κυρτόν. ὅταν
25 γὰρ πάλιν τὸ ἀριστερὸν τῶν προσθίων ἡγήσηται,
ἐξ ἐναντίας πάλιν τὸ κοῖλον γίνεται· τὸ γὰρ δεξιὸν
ἐντὸς πάλιν γίνεται. σημεῖον δεξιὸν πρόσθιον
ἐφ' οὗ Α, ἀριστερὸν ἐφ' οὗ Β, ὀπίσθιον δεξιὸν ἐφ'
οὗ Γ, ἀριστερὸν ἐφ' οὗ Δ.

Οὕτω δὲ κινοῦνται τῶν μὲν χερσαίων οἱ ὄφεις,
τῶν δ' ἐνύδρων αἱ ἐγχέλεις καὶ οἱ γόγγροι καὶ αἱ
30 μύραιναι, καὶ τῶν ἄλλων ὅσα ἔχει τὴν μορφὴν
ὀφιωδεστέραν. πλὴν ἔνια μὲν τῶν ἐνύδρων τῶν
τοιούτων οὐδὲν ἔχει πτερύγιον, οἷον αἱ μύραιναι,
708 a ἀλλὰ χρῆται τῇ θαλάττῃ ὥσπερ οἱ ὄφεις τῇ γῇ
καὶ τῇ θαλάττῃ (νέουσι γὰρ οἱ ὄφεις ὁμοίως
καὶ ὅταν κινῶνται ἐπὶ τῆς γῆς)· τὰ δὲ δύ' ἔχει
πτερύγια μόνον, οἷον οἵ τε γόγγροι καὶ αἱ ἐγχέλεις καὶ γένος τι κεστρέων, οἳ γίνονται ἐν
5 τῇ λίμνῃ τῇ ἐν Σιφαῖς. καὶ διὰ τοῦτο ταῖς
καμπαῖς ἐλάττοσι κινοῦνται ἐν τῷ ὑγρῷ ἢ ἐν τῇ
γῇ τὰ ζῆν εἰωθότα ἐν τῇ γῇ, καθάπερ τὸ τῶν
ἐγχέλεων γένος. οἱ δὲ δύο πτερύγια ἔχοντες τῶν
κεστρέων τῇ καμπῇ ἀνισάζουσιν ἐν τῷ ὑγρῷ τὰ
10 τέτταρα σημεῖα. VIII. τοῖς δ' ὄφεσιν αἴτιον τῆς
ἀποδίας τό τε τὴν φύσιν μηθὲν ποιεῖν μάτην,

---

[a] On the Boeotian coast of the Corinthian Gulf, the Tipha
of Paus. ix. 32. 3.

[b] *i.e.* two of its "points" are fins and the other two are
made by bends.

way forward, their left hip inclines towards the rear and the middle of the body becomes concave and hollow, so we must suppose that snakes too move upon the ground with their backs hollowed. And that they move in the same manner as quadrupeds is shown by the fact that they change the concave into the convex and the convex into the concave. For when the left forward point is again leading the way, the concavity comes in turn on the other side, for the right again becomes the inner. Let the front point on the right be A, and that on the

left B, and the rear point on the right C, and that on the left D.

This is the way that snakes move as land-animals, and eels, conger-eels and lampreys and all the other snake-like creatures as water-animals. Some water-animals, however, of this class, lampreys for example, have no fin and use the sea as snakes use both the sea and the land ; for snakes swim in just the same manner as when they move on land. Others have two fins only, conger-eels for example, and ordinary eels and a species of mullet which occurs in the lake at Siphae.[a] For this reason too those which are accustomed to live on land, the eels for example, move with fewer bends in the water than on dry land. The kind of mullet which has only two fins makes up the number of four points in the water by its bends.[b] VIII. The reason why snakes are footless is, first, that nature creates nothing without

ἀλλὰ πάντα πρὸς τὸ ἄριστον ἀποβλέπουσαν ἑκάστῳ
τῶν ἐνδεχομένων, διασώζουσαν ἑκάστου τὴν ἰδίαν
οὐσίαν καὶ τὸ τί ἦν αὐτῷ εἶναι· ἔτι δὲ καὶ τὸ πρότερον ἡμῖν εἰρημένον, τὸ τῶν ἐναίμων μηθὲν οἷόν
τ' εἶναι πλείοσι κινεῖσθαι σημείοις ἢ τέτταρσιν.
ἐκ τούτων γὰρ φανερὸν ὅτι τῶν ἐναίμων ὅσα κατὰ
15 τὸ μῆκος ἀσύμμετρά ἐστι πρὸς τὴν ἄλλην τοῦ
σώματος φύσιν, καθάπερ οἱ ὄφεις, οὐθὲν αὐτῶν
οἷόν θ' ὑπόπουν εἶναι. πλείους μὲν γὰρ τεττάρων
οὐχ οἷόν τε αὐτὰ πόδας ἔχειν (ἄναιμα γὰρ ἂν ἦν),
ἔχοντα δὲ δύο πόδας ἢ τέτταρας σχεδὸν ἦν ἂν
ἀκίνητα πάμπαν· οὕτω βραδεῖαν ἀναγκαῖον εἶναι
20 καὶ ἀνωφελῆ τὴν κίνησιν.

"Απαν δὲ τὸ ὑπόπουν ἐξ ἀνάγκης ἀρτίους ἔχει
τοὺς πόδας· ὅσα μὲν γὰρ ἅλσει χρώμενα μόνον
ποιεῖται τὴν κατὰ τόπον μεταβολήν, οὐθὲν ποδῶν
πρός γε τὴν τοιαύτην δεῖται κίνησιν· ὅσα δὲ
χρῆται μὲν ἅλσει, μή ἐστι δ' αὐτοῖς αὐτάρκης
25 αὕτη ἡ κίνησις ἀλλὰ καὶ πορείας προσδέονται, δῆλον ὡς τοῖς μὲν βέλτιον τοῖς δ' ⟨ἄλλως⟩ ὅλως
ἀδύνατον[1] πορεύεσθαι. [διότι πᾶν ζῷον ἀναγκαῖον
ἀρτίους ἔχειν τοὺς πόδας.][2] οὔσης γὰρ τῆς
τοιαύτης μεταβολῆς κατὰ μέρος, ἀλλ' οὐκ ἀθρόῳ
παντὶ τῷ σώματι καθάπερ τῆς ἅλσεως, ἀναγκαῖον
30 ἐστι τοῖς μὲν μένειν μεταβαλλόντων τῶν ποδῶν
τοῖς δὲ κινεῖσθαι, καὶ τοῖς ἀντικειμένοις τούτων
ποιεῖν ἑκάτερον, μεταβάλλον ἀπὸ τῶν κινουμένων
ἐπὶ τὰ μένοντα τὸ βάρος. διόπερ οὔτε τρισὶ μὲν

---

[1] ⟨ἄλλως⟩ ὅλως ἀδύνατον] ὅλως ἀδύνατον ⟨ἄλλως⟩ Farquharson.
[2] διότι ... πόδας om. PSU: tanquam glossema del. Jaeger.

---

[a] Mich.'s explanation of this passage is that certain polypods, which can walk with an uneven number of legs (cf.

506

a purpose but always with a view to what is best for each thing within the bounds of possibility, preserving the particular essence and purpose of each ; and, secondly, as we have already said, because no red-blooded animal can move by means of more than four points. It is clear from this that all red-blooded animals whose length is out of proportion to the rest of their bodily constitution, like the snakes, can none of them have feet ; for they cannot have more than four feet (for if they had, they would be bloodless), whereas, if they had two or four feet, they would be practically incapable of any movement at all, so slow and useless would their movement necessarily be.

Every animal which has feet must necessarily have an even number of feet ; for those which move from place to place by jumping only do not require feet (at least not for this movement), while those which jump but do not find this mode of locomotion sufficient by itself and need to walk also, must clearly either progress better with an even number of legs or else cannot otherwise progress at all.[a] For since this kind of change from place to place is carried out by a part and not, like jumping, with the whole of the body at once, some of the feet during the change of position must remain at rest while others are in motion, and the animal must rest and move with opposite legs, transferring the weight from the legs in motion to those at rest. Hence no animal can

708 b 5 ff.), would walk better with an even number; quadrupeds and bipeds, on the other hand, cannot walk at all with an uneven number of legs. Farquharson's insertion of ἄλλως seems therefore a certain emendation : the omission of ἄλλως, however, in our mss. would be better accounted for if it is inserted before ὅλως rather than before πορεύεσθαι.

708 b οὐδὲν οὔθ' ἑνὶ¹ χρώμενον βαδίζειν οἷόν τε· τὸ μὲν
γὰρ οὐδὲν ὅλως ὑπόστημα ἔχει ἐφ' ᾧ τὸ τοῦ
σώματος ἕξει βάρος, τὸ δὲ κατὰ τὴν ἑτέραν
ἀντίθεσιν μόνην, ὥστ' ἀναγκαῖον αὐτὸ οὕτως
ἐπιχειροῦν κινεῖσθαι πίπτειν. ὅσα δὲ πολύποδά
5 ἐστιν, οἷον αἱ σκολόπενδραι, τούτοις δυνατὸν μὲν
καὶ ἀπὸ περιττῶν ποδῶν πορείαν γίνεσθαι, καθάπερ
φαίνεται ποιούμενα καὶ νῦν, ἄν τις αὐτῶν ἕνα
πηρώσῃ τῶν ποδῶν, διὰ τὸ τὴν τῶν ἀντιστοίχων
ποδῶν κολόβωσιν ἰᾶσθαι τῷ λοιπῷ πλήθει τῶν
ἐφ' ἑκάτερα ποδῶν· γίνεται γὰρ τούτοις οἷον
10 ἔφελξις τοῦ πεπηρωμένου μορίου τοῖς ἄλλοις,
ἀλλ' οὐ βάδισις. οὐ μὴν ἀλλὰ φανερὸν ὅτι βέλτιον
ἂν καὶ ταῦτα ποιοῖτο τὴν μεταβολὴν ἀρτίους
ἔχοντα τοὺς πόδας, καὶ μηθενὸς ἐλλείποντος, ἀλλ'
ἀντιστοίχους ἔχοντα τοὺς πόδας· οὕτω γὰρ ⟨ἂν⟩²
αὑτῶν ἀνισάζειν τε δύναιτο³ τὸ βάρος καὶ μὴ
15 ταλαντεύειν ἐπὶ θάτερα μᾶλλον, εἰ ἀντίστοιχα
ἐρείσματ' ἔχοι καὶ μὴ κενὴν τὴν ἑτέραν χώραν
τῶν⁴ ἀντικειμένων. προβαίνει δ' ἀφ' ἑκατέρου
τῶν μερῶν ἐναλλὰξ τὸ πορευόμενον· οὕτω γὰρ
εἰς ταὐτὸ τῷ ἐξ ἀρχῆς σχήματι γίνεται ἡ κατά-
στασις.
20 Ὅτι μὲν οὖν ἀρτίους ἔχει τοὺς πόδας πάντα,
καὶ διὰ τίν' αἰτίαν, εἴρηται· IX. ὅτι δ' εἰ μηθὲν
ἦν ἠρεμοῦν, οὐκ ἂν ἦν κάμψις οὐδ' εὔθυνσις, ἐκ
τῶνδε δῆλον. ἔστι γὰρ κάμψις μὲν ἡ ἐξ εὐθέος ἢ
εἰς περιφερὲς ἢ εἰς γωνίαν μεταβολή, εὔθυνσις
δ' ἡ ἐκ θατέρου τούτων εἰς εὐθύ. ἐν ἁπάσαις δὲ
25 ταῖς εἰρημέναις μεταβολαῖς ἀνάγκη πρὸς ἓν σημεῖον

---

¹ οὔτε τρισὶ μὲν οὐδὲν οὔθ' ἑνὶ Jaeger: οὐδὲ (οὐδὲ om. PYZ)
τρισὶ μὲν οὐδὲν οὐθενὶ libri.   ² ἂν add. Jaeger.

walk using either three legs or one leg; for if it uses
one leg it has absolutely no support on which it is to
rest the weight of the body, and if it uses three it will
rest it on a pair of opposite legs, so that, if it attempts
to move thus, it necessarily falls.   Polypods, however,
for instance the scolopendrae, can achieve progression
with an odd number of legs, as they can be immediately
seen to do if you mutilate one of their feet, because
the maiming of some of the feet in the opposing rows
is compensated by the greater number of feet still
remaining on either side; the result is that the
maimed leg is as it were dragged along by the others,
and the animal does not walk properly.   However,
it is clear that these maimed animals would achieve
the change of position better if they had an even
number of feet, that is, if none were lacking and they
had all the feet in the corresponding rows; for then
they would be able to distribute their weight evenly
and would not sway to one side, if they had corresponding supports on each side and had not one space
in the opposite rows devoid of a leg.   An animal,
then, when it walks progresses by means of each of
its limbs alternately; for thus its state is restored so
as to be identical with its original form.

It has now been established that all animals have
an even number of feet, and the reason for this has
been stated.   IX. That, if nothing were at rest, there
could be no bending or straightening is clear from the
following considerations.   Bending is the change from
what is straight to what is curved or angular; straightening is the change of either of these to what is straight.
In all the above changes the bending or straightening

    ᵃ δύναιτο scripsi: δύναται Z: δύναιντο ceteri.
    ⁴ τὴν ante τῶν add. Z.

τὴν κάμψιν ἢ τὴν εὔθυνσιν γίνεσθαι. ἀλλὰ μὴν
κάμψεώς γε μὴ οὔσης οὔτ' ἂν πορεία οὔτε νεῦσις
οὔτε πτῆσις ἦν. τὰ μὲν γὰρ ὑπόποδα ἐπειδὴ ἐν
ἑκατέρῳ τῶν ἀντικειμένων σκελῶν ἐν μέρει ἵσταται
καὶ τὸ βάρος ἴσχει, ἀναγκαῖον θατέρου προ-
30 βαίνοντος θατέρου ποιεῖσθαι κάμψιν. ἴσα τε γὰρ
πέφυκεν ἔχειν τῷ μήκει τὰ ἀντίστοιχα κῶλα, καὶ
ὀρθὸν δεῖ εἶναι τὸ ὑφεστὸς τῷ βάρει, οἷον κάθετον
πρὸς τὴν γῆν.   ὅταν δὲ προβαίνῃ, γίνεται ἡ
709 a ὑποτείνουσα καὶ δυναμένη τὸ μένον μέγεθος καὶ
τὴν μεταξύ. ἐπεὶ δ' ἴσα τὰ κῶλα, ἀνάγκη κάμψαι
τὸ μένον, ἢ ἐν τῷ γόνατι ἢ ἐν τῇ κάμψει, οἷον
εἴ τι ἀγόνατον εἴη τῶν βαδιζόντων. σημεῖον δ'
5 ὅτι οὕτως ἔχει· εἰ γάρ τις ἐν γῇ¹ βαδίζοι παρὰ
τοῖχον, ἡ γραφομένη ἔσται οὐκ εὐθεῖα ἀλλὰ σκολιά,
διὰ τὸ ἐλάττω μὲν κάμπτοντος γίνεσθαι τὴν
γραφομένην, μείζω δ' ἱσταμένου καὶ ἐξαίροντος.

Ἐνδέχεται μέντοι κινεῖσθαι καὶ μὴ ἔχοντος καμπὴν τοῦ σκέλους, ὥσπερ τὰ παιδία ἕρπουσιν. καὶ
10 περὶ τῶν ἐλεφάντων ὁ παλαιὸς ἦν λόγος τοιοῦτος,
οὐκ ἀληθὴς ὤν.   κινεῖται δὲ καὶ τὰ τοιαῦτα
κάμψεως γινομένης ἐν ταῖς ὠμοπλάταις ἢ τοῖς
ἰσχίοις. ἀλλ' ὀρθὸν οὐδὲν δύναιτ' ἂν πορευθῆναι
συνεχῶς καὶ ἀσφαλῶς, κινηθείη δ' ἂν οἷον ἐν
ταῖς παλαίστραις οἱ διὰ τῆς κόνεως προϊόντες ἐπὶ
τῶν γονάτων.   πολὺ γὰρ τὸ ἄνω μέρος, ὥστε

---

¹ ἐν γῇ libri: locus corruptus et lacuna mutilatus.

---

[a] It does not actually do so because it is not long enough
to reach the ground : and so, as is explained below, the other
leg must be bent to enable it to do so.

[b] Δύναμις in mathematics is used of a " power," generally
the second power, *i.e.* the square of a number : similarly in
geometry δύναμις and δύναμαι are used of the figure which

must necessarily be relative to a single point. Further, if there were no bending, there would be no walking or swimming or flying. For since animals with feet stand and rest their weight alternately on each of their two opposite legs, as one leg advances the other must necessarily be bent. For the corresponding legs on either side are naturally equal in length, and the leg which supports the weight must be straight, at right angles, as it were, to the ground. But when a leg advances, it is assuming the position of the side subtending a right angle,[a] the square upon which equals the squares [b] on the side which is at rest and the line between the two legs ; but since the legs are equal, the leg which is at rest must bend either at the knee or, in any kneeless animal that walks, at the joint. That this is so is shown by the fact that if a man were to walk on the ground alongside a wall [with a reed dipped in ink attached to his head],[c] the line traced [by the reed] would not be straight but zigzag, because it goes lower when he bends and higher when he stands upright and raises himself.

It is possible, however, to move even if the leg has no bend in it, as happens when children crawl. (The old account attributed such motion to elephants, but it is untrue.) Movement of this kind takes place through a bending in the shoulders or hips. But no creature could walk erect in this way continuously and safely, but could only move like those who drag themselves forward through the dust in the wrestling-school on their knees. For the upper portion of the

can be formed by constructing squares on the side of, *e.g.* a triangle.

[c] The text here is corrupt and something has fallen out in all our MSS.: the words here bracketed are supplied from the explanation given by Mich.

709 a

15 δεῖ μακρὸν εἶναι τὸ κῶλον· εἰ δὲ τοῦτο, κάμψιν
ἀναγκαῖον εἶναι. ἐπεὶ γὰρ ἔστηκε πρὸς ὀρθήν,
16 b εἰ ἄκαμπτον ἔσται τὸ κινούμενον εἰς τὸ πρόσθεν,[1]
ἢ καταπεσεῖται ἐλάττονος τῆς ὀρθῆς γινομένης, ἢ
οὐ προβήσεται. εἰ γὰρ ὀρθοῦ ὄντος θατέρου σκέλους
θάτερον ἔσται προβεβηκός, μεῖζον ἔσται, ἴσον ὄν·
δυνήσεται γὰρ τοῦτο τό τ᾿ ἠρεμοῦν καὶ τὴν ὑπο-
20 τείνουσαν. ἀνάγκη ἄρα κάμπτεσθαι τὸ προϊόν, καὶ
κάμψαν ἅμα ἐκτείνειν θάτερον, ἐκκλίνειν τε καὶ δια-
βεβηκέναι καὶ ἐπὶ τῆς καθέτου μένειν· ἰσοσκελὲς
γὰρ γίνεται τρίγωνον τὰ κῶλα, καὶ ἡ κεφαλὴ γίνεται κατώτερον, ὅταν κάθετος ᾖ ἐφ᾿ ἧς βέβηκεν.

25   Τὰ δ᾿ ἄποδα τὰ μὲν κυμαίνοντα προέρχεται
(τοῦτο δὲ διττῶς συμβαίνει· τὰ μὲν γὰρ ἐπὶ
τῆς γῆς, καθάπερ οἱ ὄφεις, τὰς καμπὰς ποιεῖται, τὰ δ᾿ εἰς τὸ ἄνω, ὥσπερ αἱ κάμπαι), ἡ δὲ
κύμανσις καμπή ἐστιν· τὰ δ᾿ ἰλυσπάσει χρώμενα,
30 καθάπερ τὰ καλούμενα γῆς ἔντερα καὶ βδέλλαι.
ταῦτα γὰρ τῷ μὲν ἡγουμένῳ προέρχεται, τὸ δὲ
λοιπὸν σῶμα πᾶν πρὸς τοῦτο συνάγουσι, καὶ τοῦτον τὸν τρόπον εἰς τόπον ἐκ τόπου μεταβάλλουσιν.
φανερὸν δ᾿ ὅτι εἰ μὴ αἱ δύο τῆς μιᾶς μείζους ἦσαν,

---

[1] εἰ ἄκαμπτον ἔσται τὸ κινούμενον εἰς τὸ πρόσθεν om. PSU
Bekker: εἰ et πρόσθεν om. Z.

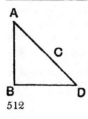

a Let AB be the stationary leg and
AC the advanced leg, which are by
hypothesis of equal length. If the right-
angled triangle ABD is constructed its
hypotenuse AD must be longer than
AC.

body is large, and therefore the leg must be long; and if this is so, there must necessarily be a bending. For since a standing position is perpendicular, the leg which is moved forward, if it is to be unbent, will either fall as the right angle becomes less, or else it will not advance at all ; for if, while one leg is at right angles, the other is advanced, the advanced leg will be greater and at the same time equal ; for it will be equal to the leg which is at rest and also to the side subtending the right angle.[a] The advancing leg must therefore be bent, and the animal, as it bends it, must at the same time stretch the other leg and lean forward and make a stride and remain in the perpendicular ; for the legs form an isosceles triangle and the head becomes lower when it is perpendicular to the base of the triangle.[b]

Of animals which are footless, some advance with an undulating motion—this can be of two kinds, for some animals, for example snakes, make their bends on the ground, while others, for instance caterpillars, make them upwards—and undulation is bending. Others move by crawling, like the earthworms and leeches ; for these advance with one part leading the way, and then draw up all the rest of their body to it, and in this manner make the change from place to place. It is plain that, if the two lines which they

[b] When the stride has been completed the result is an isosceles triangle formed by the two legs and the ground ; the head, which is necessarily lower than when the legs were together, is perpendicularly above the base.

709 b οὐκ ἂν ἐδύναντο κινεῖσθαι τὰ κυμαίνοντα τῶν
ζῴων. ἐκταθείσης γὰρ τῆς καμπῆς, εἰ ἴσην
κατεῖχεν, οὐθὲν ἂν προῄεσαν· νῦν δ' ὑπερβάλλει
ἐκταθεῖσα, καὶ ἠρεμήσαντος τούτου ἐπάγει τὸ
λοιπόν.

Ἐν πάσαις δὲ ταῖς λεχθείσαις μεταβολαῖς τὸ κινού-
5 μενον ὁτὲ μὲν ἐκτεινόμενον εἰς εὐθὺ προέρχεται,
ὁτὲ δὲ συγκαμπτόμενον, τοῖς μὲν ἡγουμένοις
μέρεσιν εὐθὺ γινόμενον, τοῖς δ' ἑπομένοις συγ-
καμπτόν. ποιεῖται δὲ καὶ τὰ ἀλλόμενα πάντα
κάμψιν ἐν τῷ ὑποκειμένῳ μέρει τοῦ σώματος,
καὶ τοῦτον τὸν τρόπον ἔχοντα ἅλλεται. καὶ τὰ
πετόμενα δὲ καὶ τὰ νέοντα, τὰ μὲν τὰς πτέρυγας
10 εὐθύνοντα καὶ κάμπτοντα πέταται, τὰ δὲ τοῖς
πτερυγίοις, καὶ τούτων τὰ μὲν τέτταρσι τὰ δὲ
δυσίν, ὅσα προμηκέστερα τὴν μορφήν, ὥσπερ τὸ
τῶν ἐγχελέων γένος· τὴν δὲ λοιπὴν κίνησιν ἀντὶ
τῶν δύο πτερυγίων τῷ λοιπῷ τοῦ σώματος καμπτό-
μενα νεῖ, καθάπερ εἴρηται πρότερον. οἱ δὲ πλατεῖς
15 τῶν ἰχθύων τῇ μὲν τῷ πλάτει χρῶνται τοῦ σώματος
ἀντὶ πτερυγίων, τῇ δὲ πτερυγίοις δυσίν. τὰ δὲ
πάμπαν πλατέα, καθάπερ ὁ βάτος, αὐτοῖς τοῖς
πτερυγίοις καὶ ταῖς ἐσχάταις τοῦ σώματος περι-
φερείαις εὐθύνοντα καὶ κάμπτοντα ποιεῖται τὴν
νεῦσιν.

20 X. Ἀπορήσειε δ' ἄν τις ἴσως πῶς κινοῦνται
τέτταρσι σημείοις οἱ ὄρνιθες, ἢ πετόμενοι ἢ πορευό-
μενοι, ὡς εἰρημένου ὅτι πάντα τὰ ἔναιμα κινεῖται
τέτταρσιν. οὐκ εἴρηται δέ, ἀλλ' ὅτι οὐ πλείοσιν.
οὐ μὴν ἀλλ' οὔτ' ἂν πέτεσθαι δύναιντο ἀφαιρε-

---

^a The bend is represented as two lines forming an angle:
514

form were not greater than the one,[a] movement would be impossible for animals which advance by undulations. For, when the bend is extended, they would not have made any advance, if it subtended an equal line ; whereas, in fact, it is longer when it is extended, and then, when this part has come to a standstill, the animal draws up the rest.

In all the above-mentioned changes that which moves advances by first extending itself straight out and then curving itself—straightening itself out with its leading parts and curving itself in the parts which follow. All animals, too, which jump make a bend in the lower part of their body and jump in this manner. Animals also which fly and those which swim, fly by straightening and bending their wings and swim with their fins, some fish having four fins and others, namely those which are of a more elongated form (eels for example), having two fins. The latter accomplish the rest of their movement by bending themselves in the rest of their body, as a substitute for the second pair of fins, as has already been said. Flat-fish use their two fins, and the flat part of their body instead of the second pair. Fish that are entirely flat, like the ray, manage to swim by using their actual fins and the outer periphery of their body, which they alternately straighten and bend.

X. A question might perhaps be asked as to how birds, whether flying or walking, can move at four points, in view of the statement that " all red-blooded animals move at four points." But this is not exactly what we stated ; what we said was " at not more than four points." However, they could not fly if their

these two lines together must be longer than the line which subtends their angle.

709 b

θέντων τῶν κώλων οὔτε πορεύεσθαι τῶν πτερύγων
25 ἀφαιρεθεισῶν, ἐπεὶ οὐδ' ἄνθρωπος βαδίζει μὴ
κινῶν τοὺς ὤμους. ἀλλὰ πάντα γε, καθάπερ
εἴρηται, κάμψει καὶ ἐκτάσει ποιεῖται τὴν μετα-
βολήν· ἅπαντα γὰρ εἰς τὸ ὑποκείμενον μέχρι τινὸς
οἱονεὶ συνυπεῖκον¹ προέρχεται, ὥστ' ἀναγκαῖον,
εἰ μὴ καὶ κατ' ἄλλο μόριον γίνεται ἡ κάμψις, ἀλλ'
30 ὅθεν γε ἡ ἀρχὴ τοῖς μὲν ὁλοπτέροις τοῦ πτεροῦ,
τοῖς δ' ὄρνισι τῆς πτέρυγος, τοῖς δ' ἄλλοις τοῦ
ἀνάλογον μορίου, καθάπερ τοῖς ἰχθύσιν. τοῖς δ',
ὥσπερ οἱ ὄφεις, ἐν ταῖς καμπαῖς τοῦ σώματός
710 a ἐστιν ἡ ἀρχὴ τῆς κάμψεως. τὸ δ' οὐροπύγιόν ἐστι
τοῖς πτηνοῖς πρὸς τὸ κατευθύνειν τὴν πτῆσιν,
καθάπερ τὰ πηδάλια τοῖς πλοίοις. ἀναγκαῖον δὲ
καὶ ταῦτα ἐν τῇ προσφύσει κάμπτειν. διόπερ τά
5 τε ὁλόπτερα καὶ τῶν σχιζοπτέρων οἷς τὸ οὐροπύγιον
ἀφυῶς ἔχει πρὸς τὴν εἰρημένην χρῆσιν, οἷον τοῖς τε
ταῶς καὶ τοῖς ἀλεκτρυόσι καὶ ὅλως τοῖς μὴ πτητικοῖς, οὐκ εὐθυποροῦσιν· τῶν μὲν γὰρ ὁλοπτέρων
ἁπλῶς οὐθὲν ἔχει οὐροπύγιον, ὥστε καθάπερ ἀ-
πήδαλον πλοῖον φέρεται, καὶ ὅπου ἂν τύχῃ ἕκαστον
10 αὐτῶν προσπίπτει, ὁμοίως τά τε κολεόπτερα,
οἷον κάνθαροι καὶ μηλολόνθαι, καὶ τὰ ἀνέλυτρα,
οἷον μέλιτται καὶ σφῆκες. καὶ τοῖς μὴ πτητικοῖς
ἀχρεῖον τὸ οὐροπύγιόν ἐστιν, οἷον τοῖς τε πορφυ-
ρίωσι καὶ ἐρωδιοῖς καὶ πᾶσι τοῖς πλωτοῖς· ἀλλ'
ἀντὶ τοῦ οὐροπυγίου πέτανται τοὺς πόδας ἀπο-

¹ οἱονεὶ συνυπεῖκον Z: οἷον εἰς ὑπεῖκον ceteri.

ᵃ Lit. "creatures with undivided wings." (The Greek
here has different words for the wings of insects and those
of birds.)

ᵇ Lit. creatures with cloven wings (i.e. made up of feathers)
as opposed to insects which have undivided wings.

legs were taken from them, or walk if their wings were taken from them, just as a man cannot walk without moving his shoulders to some extent. All things, as has been said, make their change of position by bending and stretching; for they all progress upon that which, being beneath them, also as it were gives way to them up to a certain point; so that, even if the bending does not take place in any other part, it must at any rate do so at the point where the wing begins in flying insects [a] and in birds, and where the analogous part begins in other animals, such as fishes. In other animals, snakes for example, the beginning of their bending is in the joints of the body.

In winged creatures the tail is used, like the rudder in a ship, to direct the flight; and this too must bend at the point where it joins the body. Flying insects also, therefore, and those birds [b] whose tails are ill-adapted for the purpose just mentioned, peacocks, for example, and domestic fowls and, generally, those birds which are not adapted for flight, cannot keep a straight course. Of the flying insects not a single one possesses a tail, so that they are carried along like rudderless ships and collide with anything that they happen to meet. The same is true of sheath-winged insects,[c] such as beetles and cockchafers, and the sheathless insects, such as bees and wasps. The tail is useless in such birds as are not adapted to flight, the porphyrio,[d] for example, and the heron and water-fowls in general; these fly stretching out

[c] Coleoptera.
[d] The identity of this bird is disputed. W. W. Merry (on Aristoph. *Aves*, 707) suggests some kind of coot; D'A. W. Thompson (on *H.A.* 509 a 11, 595 a 13) suggests the purple coot or the flamingo.

517

710 a

τείνοντα, καὶ χρῶνται ἀντ' οὐροπυγίου τοῖς
15 σκέλεσι πρὸς τὸ κατευθύνειν τὴν πτῆσιν. βρα-
δεῖα δ' ἡ πτῆσις τῶν ὁλοπτέρων ἐστὶ καὶ ἀσθενὴς
διὰ τὸ μὴ κατὰ λόγον ἔχειν τὴν τῶν πτερῶν φύσιν
πρὸς τὸ τοῦ σώματος βάρος, ἀλλὰ τὸ μὲν πολύ,
τὰ δὲ μικρὰ καὶ ἀσθενῆ. ὥσπερ ἂν οὖν εἰ ὁλ-
καδικὸν πλοῖον ἐπιχειροίη κώπαις ποιεῖσθαι τὸν
20 πλοῦν, οὕτω ταῦτα τῇ πτήσει χρῆται. καὶ ἡ
ἀσθένεια δὲ αὐτῶν τε τῶν πτερῶν καὶ ἡ τῆς
ἐκφύσεως συμβάλλεταί τι πρὸς τὸ λεχθέν. τῶν
δ' ὀρνίθων τῷ μὲν ταῷ τὸ οὐροπύγιον ὁτὲ μὲν
διὰ τὸ μέγεθος ἄχρηστον, ὁτὲ δὲ διὰ τὸ ἀπο-
βάλλειν οὐθὲν ὠφελεῖ. ὑπεναντίως δ' ἔχουσιν θι
25 ὄρνιθες τοῖς ὁλοπτέροις τὴν τῶν πτερῶν φύσιν,
μάλιστα δ' οἱ τάχιστα αὐτῶν πετόμενοι. τοιοῦτοι
δ' οἱ γαμψώνυχες· τούτοις γὰρ ἡ ταχυτὴς τῆς
πτήσεως χρήσιμος πρὸς τὸν βίον. ἀκόλουθα δ'
αὐτῶν ἔοικεν εἶναι καὶ τὰ λοιπὰ μόρια τοῦ σώματος πρὸς τὴν οἰκείαν κίνησιν, κεφαλὴ μὲν
30 ἁπάντων μικρὰ καὶ αὐχὴν οὐ παχύς, στῆθος δ'
ἰσχυρὸν καὶ ὀξύ, ὀξὺ μὲν πρὸς τὸ εὔτονον εἶναι,
καθάπερ ἂν εἰ πλοίου πρῷρα λεμβώδους, ἰσχυρὸν
δὲ τῇ περιφύσει τῆς σαρκός, ἵν' ἀπωθεῖν δύνηται
710 b τὸν προσπίπτοντα ἀέρα, καὶ τοῦτο ῥᾳδίως καὶ μὴ
μετὰ πόνου. τὰ δ' ὄπισθεν κοῦφα καὶ συνήκοντα
πάλιν εἰς στενόν, ἵν' ἐπακολουθῇ τοῖς ἔμπροσθεν,
μὴ σύροντα τὸν ἀέρα διὰ τὸ πλάτος.
5 XI. Καὶ περὶ μὲν τούτων διωρίσθω τὸν τρόπον
τοῦτον, τὸ δὲ μέλλον ζῷον ὀρθὸν βαδιεῖσθαι διότι
δίπουν τε ἀναγκαῖόν ἐστιν εἶναι, καὶ τὰ μὲν ἄνω
τοῦ σώματος μέρη κουφότερα ἔχειν τὰ δ' ὑφεστῶτα
τούτοις βαρύτερα, δῆλον· μόνως γὰρ ἂν οὕτως

their feet in place of a tail and use their legs instead of a tail to direct their flight. The flight of flying insects is slow and weak, because the growth of their wings is not in proportion to the weight of their body ; for their weight is considerable, while their wings are small and weak ; so they use their power of flight like a merchant-ship attempting to travel by means of oars. The weakness also of the wings themselves and of their manner of growth contributes to some extent to the result which we have described. Among birds, the peacock's tail is at one season of no service because of its size, at another useless because the bird moults. But birds are the exact opposite of winged insects in the nature of their wings, especially the swiftest flyers among them, namely, those with curved talons ; for their swiftness of flight is useful in enabling them to gain their livelihood. The other parts of their body, too, seem to be similarly adapted for their particular movement, the head being always small and the neck not thick and the breast strong and sharp—sharp so as to be compact like the prow of a light-built ship, and strong owing to the way the flesh grows—so as to thrust aside the air which meets it, and that easily and without effort ; but the hinder parts are light and contract again to a narrow point, in order that they may follow the forward parts without sweeping the air by their breadth.

XI. So much for the discussion of these topics. The reason why an animal which is to walk erect must both be a biped and also have the upper part of its body lighter and the parts situated beneath these heavier is obvious ; for only if it were so

710 b

ἔχον οἷόν τ' εἴη φέρειν ἑαυτὸ ῥᾳδίως. διόπερ
10 ἄνθρωπος μόνον ὀρθὸν τῶν ζῴων ὢν τὰ σκέλη
κατὰ λόγον ἔχει πρὸς τὰ ἄνω τοῦ σώματος μέγιστα
τῶν ὑποπόδων καὶ ἰσχυρότατα. δῆλον δὲ ποιεῖ
τοῦτο καὶ τὸ συμβαῖνον τοῖς παιδίοις· οὐ γὰρ
δύνανται βαδίζειν ὀρθὰ διὰ τὸ πάντα νανώδη εἶναι
καὶ μείζω καὶ ἰσχυρότερα ἔχειν ἢ κατὰ λόγον[1] τὰ
15 ἄνω μέρη τοῦ σώματος τῶν κάτωθεν. προϊούσης
δὲ τῆς ἡλικίας αὔξησιν λαμβάνει τὰ κάτω μᾶλλον,
μέχρι περ ἂν λάβωσι τὸ προσῆκον μέγεθος, καὶ
ποιοῦνται τότε τοῖς σώμασι τὴν βάδισιν ὀρθήν.
οἱ δ' ὄρνιθες κοῦφοι ὄντες δίποδές εἰσι διὰ τὸ
20 ὄπισθεν αὐτοῖς τὸ βάρος εἶναι, καθάπερ ἐργάζονται
τοὺς ἵππους τοὺς χαλκοῦς τοὺς τὰ πρόσθια ἠρκότας
τῶν σκελῶν. αἴτιον δὲ μάλιστα τοῦ δίποδας
ὄντας δύνασθαι ἑστάναι τὸ ἔχειν τὸ ἰσχίον ὅμοιον
μηρῷ καὶ τηλικοῦτον ὥστε δοκεῖν δύο μηροὺς
ἔχειν, τόν τ' ἐν τῷ σκέλει πρὸ τῆς καμπῆς καὶ τὸν
πρὸς τοῦτο τὸ μέρος ἀπὸ τῆς ἕδρας· ἔστι δ' οὐ
μηρὸς ἀλλ' ἰσχίον. εἰ γὰρ μὴ τηλικοῦτον ἦν,
25 οὐκ ἂν ἦν ὄρνις δίπους. ὥσπερ γὰρ τοῖς ἀνθρώποις καὶ τοῖς τετράποσι ζῴοις, εὐθὺς ἂν ἦν ἀπὸ
βραχέος ὄντος τοῦ ἰσχίου ὁ μηρὸς καὶ τὸ ἄλλο
σκέλος· λίαν οὖν ἦν ἂν τὸ σῶμα πᾶν προπετὲς
αὐτῶν. νῦν δὲ μακρὸν ὂν μέχρι ὑπὸ μέσην παρα-
τείνει τὴν γαστέρα, ὥστ' ἐντεῦθεν τὰ σκέλη ὑπ-
30 ερηρεισμένα φέρει τὸ σῶμα πᾶν. φανερὸν δ'
ἐκ τούτων καὶ ὅτι ὀρθὸν οὐκ ἐνδέχεται τὸν
ὄρνιθα εἶναι ὥσπερ τὸν ἄνθρωπον. ἡ γὰρ τῶν
πτερῶν φύσις ὡς ἔχουσι τὸ σῶμα νῦν οὕτως
711 a αὐτοῖς χρήσιμός ἐστιν, ὀρθοῖς δ' οὖσιν ἄχρηστος

---

[1] ἢ κατὰ λόγον om. PY.

520

constituted would it be able to carry itself easily.
Therefore man, the only erect animal, has legs larger
and stronger in proportion to the upper part of his
body than any of the other animals which have legs.
What happens with children illustrates this : they
cannot walk erect because they are always dwarfish
and have the upper parts of their body too big and
too strong in proportion to the lower parts. As they
grow older, the lower parts increase more quickly,
until they attain their proper size ; and it is only then
that they can walk with their bodies erect. Birds
are lightly built but can stand on two feet because
their weight is at the back, just like bronze horses
which are made by sculptors with their fore-legs
raised in the air. The chief reason why birds can
stand although they are bipeds is that their hip-joint
resembles a thigh and is of such a size that they seem
to have two thighs, one on the leg above the joint and
the other between this and the fundament ; but it is
not really a thigh but a hip. If it were not so large,
a bird could not be a biped ; for then, just as in man
and the quadrupeds, the thigh and the rest of the
leg would be directly attached to a short hip, and so
the whole body would tend to fall forward too much.
But, as it is, the hip, being long, extends up to the
middle of the belly, and so the legs form supports at
that point and carry the whole body. It is clear too
from this that it is impossible for a bird to stand erect
in the way that a man stands ; for the way that birds'
wings grow is useful to them in the position in which
they now hold themselves, but if they stood erect,

711 a

ἂν ἦν, ὥσπερ γράφουσι τοὺς ἔρωτας ἔχοντας πτέρυγας.

Ἅμα γὰρ τοῖς εἰρημένοις δῆλον ὅτι οὐδ' ἄνθρωπον οὐδ' εἰ ἄλλο τι τοιοῦτόν ἐστι τὴν μορφὴν δυνατὸν εἶναι πτερωτόν, οὐ μόνον ὅτι πλείοσι σημείοις κινή-
5 σεται ἢ τέτταρσιν ἔναιμον ὄν, ἀλλ' ὅτι ἄχρηστος αὐτοῖς ἡ τῶν πτερύγων ἕξις κατὰ φύσιν κινου-μένοις· ἡ δὲ φύσις οὐδὲν ποιεῖ παρὰ φύσιν.

XII. Ὅτι μὲν οὖν εἰ μὴ κάμψις ἦν ἐν τοῖς σκέλεσιν ἢ ἐν ταῖς ὠμοπλάταις καὶ ἰσχίοις, οὐθὲν οἷόν τ' ἦν ἂν τῶν ἐναίμων καὶ ὑποπόδων προ-
10 βαίνειν, εἴρηται πρότερον, καὶ ὅτι κάμψις οὐκ ἂν ἦν μηθενὸς ἠρεμοῦντος, ὅτι τε ἐναντίως οἵ τε ἄνθρωποι δίποδες ὄντες καὶ οἱ ὄρνιθες τὴν τῶν σκελῶν ποιοῦνται κάμψιν, ἔτι δὲ τὰ τετράποδα ὑπεναντίως καὶ αὐτοῖς καὶ τοῖς ἀνθρώποις. οἱ μὲν γὰρ ἄνθρωποι τοὺς μὲν βραχίονας κάμπτουσιν
15 ἐπὶ τὰ κοῖλα, τὰ δὲ σκέλη ἐπὶ τὸ κυρτόν, τὰ δὲ τετράποδα τὰ μὲν πρόσθια σκέλη ἐπὶ τὸ κυρτόν, τὰ δ' ὀπίσθια ἐπὶ τὸ κοῖλον· ὁμοίως δὲ καὶ οἱ ὄρνιθες. αἴτιον δ' ὅτι ἡ φύσις οὐδὲν δημιουργεῖ μάτην, ὥσπερ εἴρηται πρότερον, ἀλλὰ πάντα πρὸς τὸ βέλτιστον ἐκ τῶν ἐνδεχομένων. ὥστ' ἐπεὶ
20 πᾶσιν ὅσοις ὑπάρχει κατὰ φύσιν ἡ κατὰ τόπον μεταβολὴ τοῖν σκελοῖν, ἑστῶτος μὲν ἑκάστου τὸ βάρος ἐν τούτῳ ἐστί, κινουμένοις δ' εἰς τὸ πρόσθεν δεῖ τὸν πόδα τὸν ἡγούμενον τῇ θέσει κοῦφον εἶναι, συνεχοῦς δὲ τῆς πορείας γινομένης αὖθις ἐν τούτῳ τὸ βάρος ἀπολαμβάνειν, δῆλον ὡς ἀναγκαῖον ἐκ
25 τοῦ κεκάμφθαι τὸ σκέλος αὖθίς τε εὐθὺ γίνεσθαι, μένοντος τοῦ τε κατὰ τὸν προωσθέντα πόδα σημείου καὶ τῆς κνήμης. τοῦτο δὲ συμβαίνειν ἅμα

as winged cupids are represented in pictures, the wings would serve no purpose.

At the same time it is clear from what has been said that man, or any other creature of like form, cannot be winged, not only because, being red-blooded, he would then move at more points than four, but also because the possession of wings would be useless to him when moving in a natural manner. Now nature creates nothing unnatural.

XII. It has already been stated that, if there were no bending in the legs or shoulders and hips, none of the animals which are red-blooded and have feet could progress; and that bending would be impossible if something were not at rest; and that men and birds, being both bipeds, bend their legs in opposite directions; and, furthermore, that quadrupeds bend their pairs of legs in opposite directions to one another and in an opposite manner to men. For men bend their arms concavely and their legs convexly, but quadrupeds bend their front legs convexly and their back legs concavely; birds too do the latter. The reason is that nature never does anything without a purpose, as has been said before, but creates all things with a view to the best that circumstances allow. And so since in all creatures which possess by nature the power of locomotion by means of their two legs, when each leg is stationary the weight must be upon it, but when they move forward, the leading leg must have no weight upon it, and as progression continues it is necessary to transfer the weight on to this leg; it is clearly essential that the leg after being bent should become straight again, the point at which the leg is thrust forward and the shin remaining at rest. And it is possible

711 a
καὶ προϊέναι τὸ ζῷον εἰς τοὔμπροσθεν μὲν ἔχοντος
τὴν καμπὴν τοῦ ἡγουμένου σκέλους δυνατόν, εἰς
τοὔπισθεν δ' ἀδύνατον. οὕτω μὲν γὰρ προενεχθέν-
80 τος τοῦ σώματος ἡ ἔκτασις τοῦ σκέλους ἔσται,
ἐκείνως δ' ἀνενεχθέντος. ἔτι δ' εἰς τὸ ὄπισθεν
μὲν τῆς καμπῆς οὔσης διὰ δύο κινήσεων ἐγίγνετ'
ἂν ἡ τοῦ ποδὸς θέσις ὑπεναντίων τε αὐταῖς,[1] καὶ
711 b τῆς μὲν εἰς τὸ ὄπισθεν τῆς δὲ εἰς τὸ ἔμπροσθεν·
ἀναγκαῖον γὰρ ἐν τῇ συγκάμψει τοῦ σκέλους τοῦ
μὲν μηροῦ τὸ ἔσχατον εἰς τοὔπισθεν προάγειν,
τὴν δὲ κνήμην ἀπὸ τῆς καμπῆς εἰς τὸ ἔμπροσθεν
τὸν πόδα κινεῖν. εἰς τὸ ἔμπροσθεν δὲ τῆς καμπῆς
5 οὔσης, οὔθ' ὑπεναντίαις κινήσεσι μιᾷ τε τῇ εἰς
τὸ ἔμπροσθεν ἡ λεχθεῖσα πορεία συμβήσεται.

Ὁ μὲν οὖν ἄνθρωπος δίπους ὢν καὶ τὴν κατὰ
τόπον μεταβολὴν κατὰ φύσιν τοῖς σκέλεσι ποιού-
μενος διὰ τὴν εἰρημένην αἰτίαν κάμπτει εἰς τὸ ἔμπροσθεν τὰ σκέλη, τοὺς δὲ βραχίονας ἐπὶ τὸ κοῖλον
10 εὐλόγως· ἄχρηστοι γὰρ ἂν ἦσαν καμπτόμενοι τοὐναντίον πρός τε τὴν τῶν χειρῶν χρῆσιν καὶ πρὸς
τὴν τῆς τροφῆς λῆψιν. τὰ δὲ τετράποδα καὶ
ζῳοτόκα τὰ μὲν ἔμπροσθεν σκέλη, ἐπειδὴ ἡγεῖταί
τε τῆς πορείας αὐτῶν καὶ ἔστι ταῦτ' ἐν τῷ μέρει
τῷ ἔμπροσθεν τοῦ σώματος, ἀνάγκη κάμπτειν
15 ἐπὶ τὴν περιφέρειαν διὰ τὴν αὐτὴν αἰτίαν ἥνπερ
καὶ οἱ ἄνθρωποι· κατὰ γὰρ τοῦτο ὁμοίως ἔχουσιν.
διόπερ καὶ τὰ τετράποδα κάμπτουσιν εἰς τὸ
πρόσθεν τὸν εἰρημένον τρόπον. καὶ γὰρ οὕτως
μὲν τῆς κάμψεως αὐτῶν γινομένης ἐπὶ πολὺ
δυνήσονται τοὺς πόδας μετεωρίζειν· ἐναντίως δὲ

[1] ὑπεναντίων τε αὐταῖς Jaeger: ὑπεναντίως τε (δε UZ) αὐταῖ
libri.

for this to happen and for the animal at the same time to progress if the leading leg can bend forward, but impossible if it bends backwards. For in the first case the extension of the leg will take place with the forward movement of the body, in the second case with its backward movement. Further, if the bending were backwards, the planting of the foot would be carried out by two movements contrary to one another, one backwards and the other forwards. For in bending the leg it is necessary to draw the extremity of the thigh backwards, and the shin would move the foot forwards from the point of bending ; but if the bending be forward, the progression described above will take place not by two contrary movements but by a single forward movement.

Man then, being a biped and carrying out the change from place to place in a natural manner by means of his legs, bends his legs forwards for the reason already stated, but bends his arms concavely. This is only in accordance with reason ; for if they were bent in the opposite direction, they would be useless for the purpose of the hands and for taking food. But viviparous quadrupeds of necessity bend their front legs in an outward curve, because these legs lead the way when they walk, and are also situated in the front part of their bodies ; and the reason is the same as in man, for in this they resemble man. Thus the quadrupeds too bend their legs forward in the manner already described ; for indeed, since they bend their legs in this way, they will be able to raise their feet high in the air, whereas, if they bent them in the opposite direction, they would

711 a
20 κάμπτοντες μικρὸν ἀπὸ τῆς γῆς ἂν αὐτοὺς ἐμε-
τεώριζον διὰ τὸ τόν τε μηρὸν ὅλον καὶ τὴν
καμπήν, ἀφ' ἧς ἡ κνήμη πέφυκεν, ὑπὸ τῇ γαστρὶ
γίγνεσθαι προϊόντος αὐτοῦ. τῶν δ' ὄπισθεν σκελῶν
εἰ μὲν ἦν εἰς τὸ ἔμπροσθεν ἡ κάμψις, τῶν ποδῶν
ὁ μετεωρισμὸς ὁμοίως ἂν αὐτοῖς εἶχε τοῖς προ-
25 σθίοις (ἐπὶ βραχὺ γὰρ ἂν ἐγίγνετο καὶ τούτοις
κατὰ τὴν ἄρσιν τῶν σκελῶν, τοῦ τε μηροῦ καὶ
τῆς καμπῆς ἀμφοτέρων ὑπὸ τὸν τῆς γαστρὸς τόπον
ὑποπιπτόντων), εἰ δ' εἰς τὸ ὄπισθεν, καθάπερ καὶ
νῦν κάμπτουσιν, οὐθὲν ἐμπόδιον αὐτοῖς γίγνεται
πρὸς τὴν πορείαν ἐν τῇ τοιαύτῃ κινήσει τῶν ποδῶν.
ἔτι τοῖς γε θηλαζομένοις αὐτῶν καὶ πρὸς τὴν
30 τοιαύτην λειτουργίαν ἀναγκαῖον ἢ βέλτιόν γ' οὕτω
κεκάμφθαι τὰ σκέλη· οὐ γὰρ ῥᾴδιον τὴν κάμψιν
ποιουμένων ἐντὸς ὑφ' αὑτὰ ἔχειν τὰ τέκνα καὶ
σκεπάζειν.

712 a   XIII. Ὄντων δὲ τεττάρων τρόπων τῆς κάμψεως
κατὰ τοὺς συνδυασμοὺς[1] (ἀνάγκη γὰρ κάμπτειν
ἢ ἐπὶ τὸ κοῖλον καὶ τὰ πρόσθια καὶ τὰ ὀπίσθια,
καθάπερ ἐφ' οἷς Α, ἢ ἐπὶ τοὐναντίον ἐπὶ τὸ κυρτόν,
καθάπερ ἐφ' οἷς Β, ἢ ἀντεστραμμένως καὶ μὴ ἐπὶ
5 τὰ αὐτά, ἀλλὰ τὰ μὲν πρόσθια ἐπὶ τὸ κυρτόν, τὰ
δ' ὀπίσθια ἐπὶ τὸ κοῖλον, καθάπερ ἐφ' οἷς τὸ Γ,
ἢ τοὐναντίον τούτοις τὰ μὲν κυρτὰ πρὸς ἄλληλα,

---

[1] συνδυασμοὺς Z : συνδέσμους ceteri.

lift them only a little way from the ground, because the whole of the thigh and the joint from which the shin grows would come up against the belly as the animal advanced. On the other hand, if the bending of the back legs were forward, the raising of the feet would be similar to that of the front feet (for they could only be raised a short distance by lifting the legs, since the thigh and the joint of both legs would come up under the region of the belly), but the bending being, as it is, backwards, there is nothing to hinder their progression as they move the feet in this manner. Again, for those animals which are suckling their young, it is necessary, or at any rate better, that their legs should bend in this way with a view to this function; for if they bent their legs inwards, it would not be easy for them to keep their young underneath them and to protect them.

XIII. Now there are four ways of bending the legs taking them in pairs. Both the fore and the hind legs must bend either concavely, as in figure A; or in the opposite manner, that is convexly, as in B:

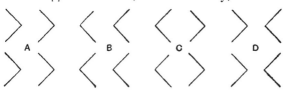

(Mich. supplies the figures which are lacking in the mss. In each group the front legs are the left pair, the hind legs the right.)

or inversely, that is to say, not in the same direction, but the forelegs bend convexly and the back legs concavely, as in C; or (the converse of C) with the convexities towards one another and the concavities

527

712 a

τὰ δὲ κοῖλα ἐκτός, καθάπερ ἔχει ἐφ' οἷς τὸ Δ),
ὡς μὲν ἔχει ἐφ' οἷς τὸ Α ἢ τὸ Β, οὐθὲν κάμπτεται
οὔτε τῶν διπόδων οὔτε τῶν τετραπόδων, ὡς δὲ
10 τὸ Γ, τὰ τετράποδα, ὡς δὲ τὸ Δ, τῶν μὲν τετραπόδων οὐθὲν πλὴν ἐλέφας, ὁ δ' ἄνθρωπος τοὺς βραχίονας καὶ τὰ σκέλη· τοὺς μὲν γὰρ ἐπὶ τὸ κοῖλον
κάμπτει, τὰ δὲ σκέλη ἐπὶ τὸ κυρτόν.

Ἀεὶ δ' ἐναλλὰξ ἐναντίως ἔχει τὰ κῶλα τὰς
κάμψεις τοῖς ἀνθρώποις, οἷον τὸ ὠλέκρανον ἐπὶ τὸ
15 κοῖλον, ὁ δὲ καρπὸς τῆς χειρὸς ἐπὶ τὸ κυρτόν, καὶ
πάλιν ὁ ὦμος ἐπὶ τὸ κυρτόν· ὡσαύτως δὲ καὶ ἐπὶ
τῶν σκελῶν ὁ μηρὸς ἐπὶ τὸ κοῖλον, τὸ δὲ γόνυ ἐπὶ
τὸ κυρτόν, ὁ δὲ ποὺς τοὐναντίον ἐπὶ τὸ κοῖλον. καὶ
τὰ κάτω δὴ πρὸς τὰ ἄνω φανερὸν ὅτι ἐναντίως·
ἡ γὰρ ἀρχὴ ὑπεναντίως, ὁ μὲν ὦμος ἐπὶ τὸ κυρτόν,
20 ὁ δὲ μηρὸς ἐπὶ τὸ κοῖλον· διὸ καὶ ὁ μὲν ποὺς
ἐπὶ τὸ κοῖλον, ὁ δὲ καρπὸς τῆς χειρὸς ἐπὶ τὸ
κυρτόν.

XIV. Αἱ μὲν οὖν κάμψεις τῶν σκελῶν τοῦτόν
τε τὸν τρόπον ἔχουσι καὶ διὰ τὰς αἰτίας τὰς
εἰρημένας, κινεῖται δὲ τὰ ὀπίσθια πρὸς τὰ ἔμ-
25 προσθεν κατὰ διάμετρον· μετὰ γὰρ τὸ δεξιὸν τῶν
ἔμπροσθεν τὸ ἀριστερὸν τῶν ὄπισθεν κινοῦσιν,
εἶτα τὸ ἀριστερὸν τῶν ἔμπροσθεν, μετὰ δὲ τοῦτο τὸ
δεξιὸν τῶν ὄπισθεν. αἴτιον δ' ὅτι εἰ μὲν τὰ
ἔμπροσθεν ἅμα καὶ πρῶτον, διεσπᾶτο ἂν ἢ καὶ
προπετὴς ἂν ἐγίνετο ἡ βάδισις οἷον ἐφελκομένοις
30 τοῖς ὄπισθεν. ἔτι δ' οὐ πορεία ἀλλὰ ἅλσις τὸ
τοιοῦτον· χαλεπὸν δὲ συνεχῆ ποιεῖσθαι τὴν μεταβολὴν ἁλλόμενα. σημεῖον δέ· ταχὺ γὰρ ἀπαγορεύ-
ουσι καὶ νῦν τῶν ἵππων ὅσοι τὸν τρόπον τοῦτον
ποιοῦνται τὴν κίνησιν, οἷον οἱ πομπεύοντες. χωρὶς

outwards, as in D. No biped or quadruped bends its limbs as in figure A or B, but quadrupeds bend them as in C. The bendings illustrated by figure D occur in none of the quadrupeds except the elephant, and in the movement of the arms and legs by man, for he bends his arms concavely and his legs convexly.

In man the bendings of the limbs always take place alternately in opposite directions ; for example, the elbow bends concavely but the wrist convexly, and the shoulder again convexly. Similarly in the legs, the thigh bends concavely, the knee convexly, and the foot, on the other hand, concavely. And obviously the lower limbs bend in opposite directions to the upper ; for the origin of movement bends in opposite directions, the shoulder convexly and the thigh concavely ; therefore also the foot bends concavely and the wrist convexly.

XIV. The bendings, then, of the legs take place in this manner and for the reasons stated. But the back legs move diagonally in relation to the front legs ; for after the right fore leg animals move the left hind leg, then the left fore leg, and after it the right hind leg. The reason is that, if they moved the fore legs at the same time and first, their progression would be interrupted or they would even stumble forward, with their hind legs as it were trailing behind. Further, such movement would not be walking but jumping ; and it is difficult to keep up a continuous movement from place to place by jumping. An illustration of this is that, in actual fact, horses that move in this manner,[a] for example in religious processions, soon become tired. For this reason, then, animals do

---

[a] *i.e.* prancing instead of walking.

712 a

μὲν οὖν τοῖς ἔμπροσθεν καὶ ὄπισθεν διὰ ταῦτα
712 b οὐ¹ ποιοῦνται τὴν κίνησιν· εἰ δὲ τοῖς δεξιοῖς ἀμφοτέροις πρώτοις, ἔξω ἂν ἐγίγνοντο τῶν ἐρεισμάτων καὶ ἔπιπτον ἄν. εἰ δὴ ἀνάγκη μὲν ἢ
τούτων τῶν τρόπων ὁποτερονοῦν ποιεῖσθαι τὴν
κίνησιν ἢ κατὰ διάμετρον, μὴ ἐνδέχεται δ' ἐκείνων
5 μηδέτερον, ἀνάγκη κινεῖσθαι κατὰ διάμετρον·
οὕτω γὰρ κινούμενα ὥσπερ εἴρηται οὐδέτερα τούτων
οἷόν τε πάσχειν. καὶ διὰ τοῦτο οἱ ἵπποι καὶ ὅσα
τοιαῦτα, ἵσταται προβεβηκότα κατὰ διάμετρον,
καὶ οὐ τοῖς δεξιοῖς ἢ τοῖς ἀριστεροῖς ἀμφοτέροις
ἅμα. τὸν αὐτὸν δὲ τρόπον καὶ ὅσα πλείους ἔχει
10 πόδας τεττάρων ποιεῖται τὴν κίνησιν· ἀεὶ γὰρ ἐν
τοῖς τέτταρσι τοῖς ἐφεξῆς τὰ ὀπίσθια πρὸς τὰ
ἔμπροσθεν κινεῖται κατὰ διάμετρον. δῆλον δ'
ἐπὶ τοῖς βραδέως κινουμένοις. καὶ οἱ καρκίνοι
γὰρ τὸν αὐτὸν τρόπον κινοῦνται· τῶν πολυπόδων
γάρ εἰσιν. ἀεὶ γὰρ καὶ οὗτοι κατὰ διάμετρον
15 κινοῦνται, ἐφ' ὅπερ ἂν ποιῶνται τὴν πορείαν.
ἰδίως γὰρ τοῦτο τὸ ζῷον ποιεῖται τὴν κίνησιν·
μόνον γὰρ οὐ κινεῖται ἐπὶ τὸ πρόσθεν τῶν ζῴων,
ἀλλ' ἐπὶ τὸ πλάγιον. ἀλλ' ἐπεὶ τοῖς ὄμμασι
διώρισται τὸ πρόσθιον, ἡ φύσις πεποίηκεν ἀκολουθεῖν δυναμένους τοὺς ὀφθαλμοὺς τοῖς κώλοις·
20 κινοῦνται γὰρ εἰς τὸ πλάγιον αὐτοῖς, ὥστε τρόπον
τινὰ καὶ τοὺς καρκίνους κινεῖσθαι διὰ τοῦτ' ἐπὶ
τὸ ἔμπροσθεν.

XV. Οἱ δ' ὄρνιθες τὰ σκέλη καθάπερ τὰ τετράποδα κάμπτουσιν. τρόπον γάρ τινα παραπλησίως

¹ οὐ P: om. SYUZ.

530

not move separately with their front and back legs[a];
and, if they moved with both their right legs first,
they would not be above their supporting limbs and
would fall. If, then, they must necessarily move in
one or other of these two ways or else diagonally,
and neither of the first two ways is possible, they
must necessarily move diagonally ; for if they move
thus they cannot, as has been explained, suffer
either of the above ill results. For this reason horses
and similar animals stand at rest with their legs
advanced diagonally and not with both right or both
left legs advanced at the same time. And those
animals which have more than four legs move in
the same manner ; for in any four adjoining legs the
back legs move diagonally with the fore legs, as can
be plainly seen in those which move slowly.

Crabs too move in the same fashion, for they are
among the polypods. They, too, always move on
the diagonal principle in whatever direction they are
proceeding. For this animal moves in a peculiar
manner, being the only animal to move obliquely
and not forward. But since " forward " is determined
in relation to the vision, nature has made the crab's
eyes able to conform with its limbs ; for its eyes
move obliquely, and so, for this reason, crabs too can,
in a sense, be said to move " forward."

XV. Birds bend their legs in the same manner as
quadrupeds ; for in a way their nature is closely

---

[a] *i.e.* do not move first the front legs together and then
their back legs together. The ᴍs. authority is strongly in
favour of the omission of the negative: but 712 b 4 " one or
other of these two ways " implies the alternative of movement
with the front legs together and then the back legs together,
or else with the right legs together and then the left legs
together.

ἡ φύσις αὐτῶν ἔχει· τοῖς γὰρ ὄρνισιν αἱ πτέρυγες
ἀντὶ τῶν προσθίων σκελῶν εἰσίν. διὸ καὶ κεκαμ-
25 μέναι τὸν αὐτὸν εἰσὶ τρόπον ὥσπερ ἐκείνοις τὰ
πρόσθια σκέλη, ἐπεὶ τῆς ἐν τῇ πορείᾳ κινήσεως
τούτοις ἀπὸ τῶν πτερύγων ἡ κατὰ φύσιν ἀρχὴ
τῆς μεταβολῆς ἐστίν· πτῆσις γάρ ἐστιν ἡ τούτων
οἰκεία κίνησις. διόπερ ἀφαιρεθεισῶν τούτων οὔθ'
30 ἑστάναι οὔτε προϊέναι δύναιτ' ἂν οὐθεὶς ὄρνις.

Ἔτι δίποδος ὄντος καὶ οὐκ ὀρθοῦ, καὶ τὰ ἔμπροσθεν μέρη τοῦ σώματος κουφότερα ἔχοντος, ἢ
ἀναγκαῖον ἢ βέλτιον πρὸς τὸ ἑστάναι δύνασθαι τὸν
μηρὸν οὕτως ὑποκείμενον ἔχειν ὡς νῦν ἔχει, λέγω
δ' ὅτι εἰς τὸ ὄπισθεν πεφυκότα. ἀλλὰ μὴν εἰ ἔδει
τοῦτον ἔχειν τὸν τρόπον, ἀνάγκη τὴν κάμψιν ἐπὶ
713 a τὸ κοῖλον γίνεσθαι τοῦ σκέλους, καθάπερ τοῖς
τετράποσιν ἐπὶ τῶν ὀπισθίων, διὰ τὴν αὐτὴν αἰτίαν
ἥνπερ εἴπομεν ἐπὶ τῶν τετραπόδων καὶ ζῳοτόκων.

Ὅλως δὲ οἵ τε ὄρνιθες καὶ τὰ ὀλόπτερα τῶν πε-
τομένων καὶ τὰ ἐν τῷ ὑγρῷ νευστικά, ὅσα αὐτῶν
5 δι' ὀργάνων τὴν ἐπὶ τοῦ ὑγροῦ ποιεῖται πορείαν, οὐ
χαλεπὸν ἰδεῖν ὅτι βέλτιον ἐκ πλαγίου τὴν τῶν εἰρημένων μερῶν πρόσφυσιν ἔχειν, καθάπερ καὶ
φαίνεται νῦν ὑπάρχειν αὐτοῖς ἐπί τε τῶν ὀρνίθων
καὶ τῶν ὀλοπτέρων. ταὐτὸ δὲ τοῦτο καὶ ἐπὶ τῶν
ἰχθύων· τοῖς μὲν γὰρ ὄρνισιν αἱ πτέρυγες, τοῖς δ'
10 ἐνύδροις τὰ πτερύγια, τὰ δὲ πτίλα τοῖς ὀλοπτέροις
ἐκ τοῦ πλαγίου προσπέφυκεν. οὕτω γὰρ ἂν τάχιστα καὶ ἰσχυρότατα διαστέλλοντα τὰ μὲν τὸν
ἀέρα τὰ δὲ τὸ ὑγρὸν ποιοῖτο τὴν κίνησιν· εἰς γὰρ
τὸ ἔμπροσθεν καὶ τὰ ὄπισθεν μόρια[1] τοῦ σώματος
ἐπακολουθοίη ἂν ὑπείκοντι φερόμενα τὰ μὲν ἐν
15 τῷ ὑγρῷ τὰ δ' ἐν τῷ ἀέρι. τὰ δὲ τρωγλοδυτικὰ

similar. For in birds the wings serve instead of front legs, and so they are bent in the same manner as the front legs of quadrupeds, since in the movement involved in progression the natural beginning of the change is from the wings, for their particular form of movement is flight. Hence, if the wings were taken away, no bird could stand or progress forward.

Further, since the bird is a biped and not erect, and the front parts of its body are lighter, it is either necessary (or at any rate more desirable), in order to enable it to stand, that the thigh should be placed, as it actually is, underneath, by which I mean growing towards the hinder part. But if the thigh is necessarily in this position, the bending of the leg must be in a concave direction, as in the back legs of quadrupeds, and for the same reason as we gave in dealing with viviparous quadrupeds.

Generally in birds and winged insects and creatures that swim in the water (all, that is to say, that progress in the water by means of their instrumental parts), it is not difficult to see that it is better that the attachment of such parts should be oblique, as in fact it seems actually to be in the birds and the flying insects. The same is also true of the fishes ; for the wings in birds, the fins in fishes, and the wings in flying insects all grow obliquely. This enables them to cleave the air or water with the greatest speed and force, and so effect their movement ; for the hinder parts, too, can thus follow in a forward direction, being carried along in the yielding water or air.

The oviparous quadrupeds which live in holes,

---

[1] καὶ τὰ ὄπισθεν μόρια Jaeger: καὶ τὸ ὄπισθεν τὰ (τὰ om. YZ) μόρια libri.

713 a

τῶν τετραπόδων καὶ ᾠοτόκων, οἷον οἵ τε κροκόδειλοι καὶ σαῦροι καὶ ἀσκαλαβῶται καὶ ἐμύδες
τε καὶ χελῶναι, πάντα ἐκ τοῦ πλαγίου προσπεφυκότα τὰ σκέλη ἔχει καὶ ἐπὶ τῇ γῇ κατατεταμένα,
20 καὶ κάμπτει εἰς τὸ πλάγιον, διὰ τὸ οὕτω χρήσιμα
εἶναι πρὸς τὴν τῆς ὑποδύσεως ῥᾳστώνην καὶ πρὸς
τὴν ἐπὶ τοῖς ᾠοῖς ἐφεδρείαν καὶ φυλακήν. ἔξω
δ᾽ ὄντων αὐτῶν, ἀναγκαῖον τοὺς μηροὺς προσ-
στέλλοντα[1] καὶ ὑποτιθέμενα ὑφ᾽ αὑτὰ τὸν μετεω-
ρισμὸν τοῦ ὅλου σώματος ποιεῖσθαι. τούτου δὲ
25 γινομένου κάμπτειν αὐτὰ οὐχ οἷόν τε ἄλλως ἢ
ἔξω.

XVI. Τὰ δ᾽ ἄναιμα τῶν ὑποπόδων ὅτι μὲν
πολύποδά ἐστι καὶ οὐθὲν αὐτῶν τετράπουν,
πρότερον ἡμῖν εἴρηται· διότι δ᾽ αὐτῶν ἀναγκαῖον
ἦν τὰ σκέλη πλὴν τῶν ἐσχάτων ἔκ τε τοῦ πλαγίου
προσπεφυκέναι καὶ εἰς τὸ ἄνω τὰς καμπὰς ἔχειν,
30 καὶ αὐτὰ ὑπόβλαισα εἶναι εἰς τὸ ὄπισθεν, φανερόν.
ἁπάντων γὰρ τῶν τοιούτων ἀναγκαῖόν ἐστι τὰ
μέσα τῶν σκελῶν καὶ ἡγούμενα εἶναι καὶ ἑπόμενα.
εἰ οὖν ὑπ᾽ αὐτοῖς ἦν, ἔδει αὐτὰ καὶ εἰς τὸ ἔμ-
713 b προσθεν καὶ εἰς τὸ ὄπισθεν τὴν καμπὴν ἔχειν, διὰ
μὲν τὸ ἡγεῖσθαι εἰς τὸ ἔμπροσθεν, διὰ δὲ τὸ
ἀκολουθεῖν εἰς τὸ ὄπισθεν. ἐπεὶ δ᾽ ἀμφότερα
συμβαίνειν ἀναγκαῖον αὐτοῖς, διὰ τοῦτο βεβλαίσω-
5 ταί τε καὶ εἰς τὸ πλάγιον ἔχει τὰς καμπάς, πλὴν
τῶν ἐσχάτων· ταῦτα δ᾽ ὥσπερ πέφυκε μᾶλλον,
τὰ μὲν ὡς ἑπόμενα τὰ δ᾽ ὡς ἡγούμενα. ἔτι δὲ
κέκαμπται τὸν τρόπον τοῦτον καὶ διὰ τὸ πλῆθος
τῶν σκελῶν· ἧττον γὰρ ἂν οὕτως ἐν τῇ πορείᾳ
ἐμπόδιά τε αὐτὰ αὑτοῖς εἴη καὶ προσκόπτοι. ἢ
10 τε βλαισότης αὐτῶν ἐστι διὰ τὸ τρωγλοδυτικὰ

such as the crocodile, the common and the spotted
lizard, and land and water tortoises, all have their
legs attached obliquely and stretched out upon the
ground ; and they bend them obliquely, since they
are thus useful in enabling them to crawl easily into
their holes and to sit upon and protect their eggs.
Since their legs project, they are obliged to raise
their whole body by drawing in their thighs and
placing them underneath them ; and in this process
they cannot bend them otherwise than outwards.

XVI. It has already been said that bloodless
animals which have legs are polypods, and none of
them quadrupeds. Their legs, except the two
extreme pairs, are necessarily attached obliquely
and bend upwards and are themselves bowed somewhat backwards ; and the reason for this is plain.
For in all such animals the middle legs must both
lead and follow. If, therefore, they were underneath them, they would have to bend both forwards and backwards—forwards because they lead,
and backwards because they follow. But since they
must do both these things, their legs are bowed
and make their bends obliquely, except the extreme
pairs, which are more in accordance with nature,
since the first pair leads and the last pair follows.
The number of legs is a further reason for their being
bent in this way ; for they would thus be less likely
to get in each other's way during movement and
collide with one another. The reason that these
animals are bow-legged is that they all, or most of

---

713 b

εἶναι πάντα ἢ τὰ πλεῖστα· οὐ γὰρ οἷόν τε ὑψηλὰ
εἶναι τὰ ζῶντα[1] τὸν τρόπον τοῦτον.

Οἱ δὲ καρκίνοι τῶν πολυπόδων περιττότατα πεφύ-
κασιν· οὔτε γὰρ εἰς τὸ πρόσθεν ποιοῦνται τὴν πορείαν
πλὴν ὥσπερ εἴρηται πρότερον, πολλούς τε τοὺς
ἡγουμένους ἔχουσι μόνοι τῶν ζῴων. τούτου δ᾽
15 αἴτιον ἡ σκληρότης τῶν ποδῶν, καὶ ὅτι χρῶνται
οὐ νεύσεως χάριν αὐτοῖς ἀλλὰ πορείας· πεζεύοντα
γὰρ διατελοῦσιν. πάντων μὲν οὖν τῶν πολυπόδων
εἰς τὸ πλάγιον αἱ καμπαί, ὥσπερ καὶ τῶν τετραπόδων ὅσα τρωγλοδυτικά· τοιαῦτα δ᾽ ἐστὶν οἷον
σαῦραι καὶ κροκόδειλοι καὶ τὰ πολλὰ τῶν ᾠο-
20 τοκούντων.   αἴτιον δ᾽ ὅτι τρωγλοδυτεῖ τὰ μὲν
τοῖς τόκοις, τὰ δὲ καὶ τῷ βίῳ παντί.

XVII. Ἀλλὰ τῶν μὲν ἄλλων βλαισοῦται τὰ κῶλα
διὰ τὸ μαλακὰ εἶναι, τῶν δὲ καράβων ὄντων σκλη-
ροδέρμων οἱ πόδες εἰσὶν ἐπὶ τῷ νεῖν καὶ οὐ τοῦ
βαδίζειν χάριν· τῶν δὲ καρκίνων ἡ κάμψις εἰς τὸ
25 πλάγιον, καὶ οὐ βεβλαίσωται ὥσπερ τοῖς ᾠοτόκοις
τῶν τετραπόδων καὶ τοῖς ἀναίμοις καὶ πολύποσι,
διὰ τὸ σκληρόδερμα εἶναι τὰ κῶλα καὶ ὀστρακώδη
ὄντι οὐ νευστικῷ καὶ τρωγλοδύτῃ· πρὸς τῇ γῇ γὰρ
ὁ βίος.   καὶ στρογγύλος δὲ τὴν μορφήν, καὶ οὐκ
ἔχων οὐροπύγιον ὥσπερ ὁ κάραβος· πρὸς τὴν
30 νεῦσιν γὰρ τοῖς καράβοις χρήσιμον, ὃ δ᾽ οὐ νευ-
στικός.   καὶ ὅμοιον δὲ τῷ ὄπισθεν τὸ πλάγιον
ἔχει μόνος, διὰ τὸ πολλοὺς ἔχειν τοὺς ἡγεμόνας

----

[1] τὰ ζῶντα om. SU.

----

[a] 712 b 20 f.
[b] Viz. two pairs of front legs.
[c] *i.e.* they walk both on dry land and in the sea.
[d] The whole of the section is obscure, and the text doubtful.

536

them, live in holes ; for creatures that live thus cannot be tall.

Crabs are the most strangely constituted of all the polypods ; for they do not progress forward (except in the sense already mentioned[a]), and they alone among animals have several leading legs.[b] The reason is the hardness of their feet and the fact that they use them not for swimming but for walking ; for they always go along the ground.[c] All the polypods bend their legs obliquely like the quadrupeds that live in holes ; lizards, for instance, and crocodiles and most oviparous quadrupeds are of this nature. The reason is that they live in holes, some only during the breeding season, others throughout their lives.

XVII. Now the other polypods[d] are bow-legged because they are soft-skinned, but the legs of the spiny lobster,[e] which is hard-skinned, are used for swimming and not for walking.[f] The bendings of crabs' legs are oblique but their legs are not bowed, as are those of viviparous quadrupeds and bloodless polypods, because their legs are hard-skinned and testaceous, the crab not being a swimming animal and living in holes, for it lives on the ground. Moreover, the crab is round in shape and does not possess a tail like the spiny lobster ; for the latter's tail is useful for swimming, but the crab does not swim. And it is the only animal in which the side is like a hinder part, because its leading feet are numerous.[g]

[e] There is no single word in English for this animal, the Latin *locusta* and the French *langouste*.

[f] And therefore are not bowed, as Mich. explains.

[g] Since the crab moves sidewise, one of its sides becomes as it were the back, but why it should be so for the reason given is obscure.

πόδας. τούτου δ' αἴτιον ὅτι οὐ κάμπτει εἰς τὸ
714 a πρόσθεν οὐδὲ βεβλαίωται. τοῦ δὲ μὴ βεβλαι-
ῶσθαι τὸ αἴτιον πρότερον εἴρηται, ἡ σκληρότης
καὶ τὸ ὀστρακῶδες τοῦ δέρματος. ἀνάγκη δὴ
διὰ ταῦτα πᾶσί τε προηγεῖσθαι καὶ εἰς τὸ πλάγιον,
εἰς μὲν τὸ πλάγιον ὅτι εἰς τὸ πλάγιον ἡ κάμψις,
5 πᾶσι δ' ὅτι ἐνεπόδιζον ἂν οἱ ἠρεμοῦντες πόδες
τοῖς κινουμένοις. οἱ δὲ ψηττοειδεῖς τῶν ἰχθύων,
ὥσπερ οἱ ἑτερόφθαλμοι βαδίζουσιν, οὕτω νέουσιν·
διέστραπται γὰρ αὐτῶν ἡ φύσις. οἱ δὲ στεγανόποδες τῶν ὀρνίθων νέουσι τοῖς ποσίν, καὶ διὰ μὲν
10 τὸ τὸν ἀέρα δέχεσθαι καὶ ἀναπνεῖν δίποδές εἰσι,
διὰ δὲ τὸ ἐν ὑγρῷ τὸν βίον ἔχειν στεγανόποδες·
ἀντὶ πτερυγίων γὰρ χρήσιμοι οἱ πόδες αὐτοῖς
τοιοῦτοι ὄντες. ἔχουσι δὲ τὰ σκέλη οὐχ ὥσπερ
οἱ ἄλλοι κατὰ μέσον, ἀλλ' ὄπισθεν μᾶλλον· βρα-
χυσκελῶν γὰρ αὐτῶν ὄντων ὄπισθεν ὄντα πρὸς
τὴν νεῦσιν χρήσιμα. βραχυσκελεῖς δ' εἰσὶν οἱ
15 τοιοῦτοι διὰ τὸ ἀπὸ τοῦ μήκους τῶν σκελῶν
ἀφελοῦσαν τὴν φύσιν προσθεῖναι εἰς τοὺς πόδας,
καὶ ἀντὶ τοῦ μήκους πάχος ἀποδοῦναι τοῖς σκέλεσι
καὶ πλάτος τοῖς ποσίν· χρήσιμοι γὰρ πλατεῖς[1]
ὄντες μᾶλλον ἢ μακροὶ πρὸς τὸ ἀποβιάζεσθαι τὸ
ὑγρόν, ὅταν νέωσιν.

20 XVIII. Εὐλόγως δὲ καὶ τὰ μὲν πτηνὰ πόδας
ἔχει, οἱ δ' ἰχθύες ἄποδες· τοῖς μὲν γὰρ ὁ βίος ἐν
τῷ ξηρῷ, μετέωρον δ' ἀεὶ μένειν ἀδύνατον, ὥστ'
ἀνάγκη πόδας ἔχειν· τοῖς δ' ἰχθύσιν ἐν τῷ ὑγρῷ
ὁ βίος, καὶ τὸ ὕδωρ δέχονται, οὐ τὸν ἀέρα. τὰ
714 b μὲν οὖν πτερύγια χρήσιμα πρὸς τὸ νεῖν, οἱ δὲ
πόδες ἄχρηστοι. εἰ δ' ἄμφω εἶχον, ἄναιμοι ἂν
ἦσαν. ὁμοίως δ' ἔχουσιν οἱ ὄρνιθες τρόπον τινὰ

The reason is that it does not bend its legs forwards and is not bow-legged. Why it is not bow-legged has been already explained before, namely, because its skin is hard and testaceous. For this reason it must lead off with all its legs and obliquely—obliquely because its bendings are oblique, and with all its legs, because otherwise those which were at rest would impede those which were moving.

Flat-fish swim as one-eyed men walk ; for their nature is distorted. Web-footed birds swim with their feet. They are bipeds, because they take in breath and respire ; they are web-footed, because they live in the water, for their feet being of this kind are of service to them in place of fins. They do not have their legs, as the other birds do, in the centre of the body, but placed rather towards the back ; for since they are short-legged, their legs being set back are useful for swimming. This class of bird is short-legged because nature has taken away from the length of their legs and added to their feet, and has given thickness instead of length to the legs and breadth to the feet ; for, being broad, they are more useful than if they were long, in order to force away the water when they are swimming.

XVIII. It is for a good reason, too, that winged animals have feet, while fishes have none. The former live on dry land and cannot always remain up in the air, and so necessarily have feet ; but fishes live in the water, and take in water and not air. Their fins, then, are useful for swimming, whereas feet would be useless. Also, if they had both feet and fins, they would be bloodless. Birds in a way

---

τοῖς ἰχθύσιν. τοῖς μὲν γὰρ ὄρνισιν ἄνω αἱ πτέρυγές
5 εἰσι, τοῖς δὲ πτερύγια δύο ἐν τῷ πρανεῖ· καὶ τοῖς
μὲν ἐν τοῖς ὑπτίοις οἱ πόδες, τοῖς δὲ ἔν τε τοῖς
ὑπτίοις καὶ ἐγγὺς τῶν πρανῶν πτερύγια τοῖς
πλείστοις· καὶ οἱ μὲν οὐροπύγιον ἔχουσιν, οἱ δ'
οὐραῖον.

XIX. Περὶ δὲ τῶν ὀστρακοδέρμων ἀπορήσειεν
ἄν τις τίς ἡ κίνησις, καὶ εἰ μὴ ἔχουσι δεξιὸν καὶ
ἀριστερόν, πόθεν κινοῦνται· φαίνονται δὲ κινού-
10 μενα. ἢ ὥσπερ ἀνάπηρον δεῖ τιθέναι πᾶν τὸ
τοιοῦτον γένος, καὶ κινεῖσθαι ὁμοίως οἷον εἴ τις
ἀποκόψειε τῶν ὑποπόδων τὰ σκέλη, ⟨ἢ⟩[1] ὥσπερ
ἡ φώκη καὶ ἡ νυκτερίς· καὶ γὰρ ταῦτα τετράποδα,
κακῶς δ' ἐστίν. τὰ δ' ὀστρακόδερμα κινεῖται μέν,
κινεῖται δὲ παρὰ φύσιν· οὐ γάρ ἐστι κινητικά, ἀλλ'
15 ὡς μὲν μόνιμα καὶ προσπεφυκότα κινητικά, ὡς δὲ
πορευτικὰ μόνιμα. ἔχουσι δὲ φαύλως καὶ οἱ καρκίνοι
τὰ δεξιά, ἐπεὶ ἔχουσί γε. δηλοῖ δ' ἡ χηλή· μείζων
γὰρ καὶ ἰσχυροτέρα ἡ δεξιά, ὡς βουλομένων δι-
ωρίσθαι τῶν δεξιῶν καὶ τῶν ἀριστερῶν.

20 Τὰ μὲν οὖν περὶ τῶν μορίων, τῶν τ' ἄλλων καὶ
τῶν περὶ τὴν πορείαν τῶν ζῴων καὶ περὶ πᾶσαν
τὴν κατὰ τόπον μεταβολήν, τοῦτον ἔχει τὸν τρόπον·
τούτων δὲ διωρισμένων ἐχόμενόν ἐστι θεωρῆσαι
περὶ ψυχῆς.

---

[1] ἢ addidi.

---

[a] *i.e.* a second pair of fins.
[b] See *H.A.* 527 b 35 ff., where land-snails, sea-snails, oysters
and sea-urchins are given as examples.
[c] See *H.A.* 498 a 31, *P.A.* 697 b 1 ff.
[d] These words can only refer to the *De anima*, which from
its citation in the *De generatione animalium*, *De partibus
animalium*, etc., must be regarded as an earlier work. This

resemble fishes. For birds have their wings in the upper part of their bodies, fishes have two fins in their fore-part; birds have feet on their under-part, most fishes have fins [a] in their under-part and near their front fins; also, birds have a tail, fishes a tail-fin.

XIX. A question may be raised as to what is the movement of testaceans,[b] and where their movement begins if they have no right and left; for they obviously do move. Must all this class be regarded as maimed and as moving in the same way as an animal with feet if one were to cut off its legs, or as analogous to the seal and bat, which are quadrupeds but malformed?[c] Now the testaceans move, but move in a way contrary to nature. They are not really mobile; but if you regard them as sedentary and attached by growth, you find that they are capable of movement; if you regard them as progressing, you find that they are sedentary.

Crabs show only a feeble differentiation of right and left, but they *do* show it. It can be seen in the claw; for the right claw is bigger and stronger, as though the left and right wished to be differentiated.

So much for our discussion of the parts of animals and particularly those which have to do with progression and all change from place to place. Now that these points have been settled, our next task is to consider soul.[d]

has led some critics (*e.g.* Brandis) to reject the whole of this paragraph as a later addition. Such a paragraph, however, is a characteristic conclusion in Aristotle, and should not be rejected as a whole. It is quite possible that the words περὶ ψυχῆς are corrupt, and indeed the word ψυχῆς has been supplied by a later hand in Z, whereas the first hand had left a blank and had written ζωησ (*sic*) in the margin, which would be a reference to the latter part of the group of treatises known as the *Parva Naturalia*.

INDEX TO PARTS OF ANIMALS

The Index is to be regarded as supplementary to the Summary on pages 12-18. Further references will sometimes be found in the notes on Terminology, pages 24-39.

The numbers 3 to 50 refer to the pages of the Introduction.

The numbers 39a to 97b (standing for 639a to 697b) refer to the pages and columns of the Berlin edition which are printed at the top of each page of the Greek text. The lines are referred to in units of five lines ; thus

$$40a1 = 640a1-640a4$$
$$40b5 = 640b5-640b9.$$

Such references include footnotes to the translation.

f, ff = following section or sections.

Under any heading, each entry is separated from the preceding by a dash (/), unless it has the same page number.

INDEX TO PARTS OF ANIMALS

INDEX TO PARTS OF ANIMALS

547

INDEX TO PARTS OF ANIMALS

INDEX TO PARTS OF ANIMALS

INDEX TO MOVEMENT AND PROGRESSION OF ANIMALS

MOVEMENT & PROGRESSION OF ANIMALS

opposite parts of the body move simultaneously 02b10

oviparous quadrupeds 04b1 / 13a15

passive)(active 02a10 / 05a20

peacock 10a5, 20

Physics, reference to (258b49) 98a10

plants, nutrition of 06b5, lack movement *ib.*, compared with animals 05b5, superior and inferior parts of 05a25, b1 ff

points at which movement takes place 04a10 / 07a15 ff, b5 ff / 09b20

poles, the 99a20 ff

polypods 04a10 / 06a30, b5 / 08b1 / 12b10 / 13a25, b15

porphyrio (a bird) 10a10

procession, religious 12a30

purple-fish 06a15

purpose 00b10 / 01a5, in nature 04b15 / 08a10

quadrupeds 04a10 ff / 06a30 / 07b15; bending of the legs of quadrupeds in walking 11b10; oviparous quadrupeds 04b1

ray (fish) 09b15

red-blooded animals 11a5, move at four points 04a10 / 07a15, b5 / 09b20, cannot live if divided into parts 07a25

resistance of earth, air or sea necessary to movement 98b15

rest)(motion 98b5

right)(left 05a25, b1 / 06b25 / 07a5; movement originates on the right side 05b30 / 06b15; right side superior to left 06a20, b10; right limbs used in defence 06a5

roots of plants 05b5

rudder, slight movement of, changes direction of boat 01b25

scolopendrae 07a30 / 08b5

seal 14b10

sensation 01a35, cause of alteration 01b15; origin of sensation situated in the centre of the body 02b20

sense-perception, in animals 05b10; objects of 98a10

sexual organs 03b5, 20

shin 98b1

shoulder 98 b1 / 09a10 / 11a5 / 12a10

sinews 01b5

Siphae 08a5

snakes 05b25 / 09a25, movement of 07b20 ff, why footless 08a5 ff

soul, movement of the 90b1, central position of 03a35, as origin of movement 02b1 ff / 03a1

spine 02b20

spirit, innate, in animals 03a10 ff

stromboid testaceans 06a10, b1

superior)(inferior parts 04b20 / 05a25 / 06b1 ff / 07a5

swimming 98a5, b15 / 09b5

syllogism, the practical 01a10 ff

555